Operator Theory: Advances and Applications
Vol. 171

The Extended Field of Operator Theory

Michael A. Dritschel
Editor

Birkhäuser
Basel · Boston · Berlin

Editor:

Michael A. Dritschel
School of Mathematics and Statistics
University of Newcastle
Newcastle upon Tyne NE1 7RU
UK
e-mail: m.a.dritschel@ncl.ac.uk

2000 Mathematics Subject Classification 30E, 30G, 35R, 41A, 45E, 45P, 46E, 47A, 47B, 47G, 65F

Library of Congress Control Number: 2006937468

Bibliographic information published by Die Deutsche Bibliothek
Die Deutsche Bibliothek lists this publication in the Deutsche Nationalbibliografie; detailed bibliographic data is available in the Internet at <http://dnb.ddb.de>.

ISBN 978-3-7643-7979-7 Birkhäuser Verlag AG, Basel – Boston – Berlin

Printed on acid-free paper produced from chlorine-free pulp. TCF ∞
Cover design: Heinz Hiltbrunner, Basel
Printed in Germany
ISBN-10: 3-7643-7979-0
ISBN-13: 978-3-7643-7979-7
e-ISBN-10: 3-7643-7980-4
e-ISBN-13: 978-3-7643-7980-3

9 8 7 6 5 4 3 2 1

www.birkhauser.ch

To my friend and colleague
Nicholas Young
on his retirement

Contents

Editorial Preface

As this volume demonstrates, at roughly 100 years of age operator theory remains a vibrant and exciting subject area with wide ranging applications. Many of the papers found here expand on lectures given at the 15th *International Workshop on Operator Theory and Its Applications*, held at the University of Newcastle upon Tyne from the 12th to the 16th of July 2004. The workshop was attended by close to 150 mathematicians from throughout the world, and is the first IWOTA to be held in the UK. Talks ranged over such subjects as operator spaces and their applications, invariant subspaces, Kreĭn space operator theory and its applications, multivariate operator theory and operator model theory, applications of operator theory to function theory, systems theory including inverse scattering, structured matrices, and spectral theory of non-selfadjoint operators, including pseudodifferential and singular integral operators. These interests are reflected in this volume. As with all of the IWOTA proceedings published by Birkhäuser Verlag, the papers presented here have been refereed to the same high standards as those of the journal *Integral Equations and Operator Theory.*

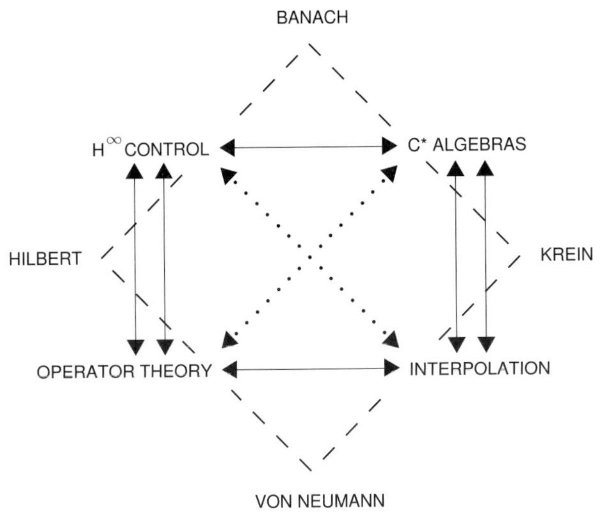

A few words about the above image which graced the workshop programme and bag. In commuting between home in Hexham and work in Newcastle, I often travel by train. The journeys have resulted in a number of friendships, including with Chris Dorsett, who is a member of the Fine Arts department at Northumbria University (also in Newcastle). A mathematics question led him to introduce me an article by the art critic and theorist, Rosalind Krauss titled "Sculpture in the Expanded Field", which was first published in art journal *October* in 1978, and is now recognized as a key work in contemporary art theory. To briefly summarize a portion of the thesis of her article, the term "sculpture" has been applied in the 20th century to such a broad collection of art objects as to become essentially meaningless. This leads her to propose a refined classification built from the idea of what sculpture is not (architecture, landscape) and the negation of these terms. The idea is encoded in a diagram much like the one given above, and is based on a model of the Klein Viergruppe, also known as the Piaget group due to its use by the Swiss developmental psychologist Jean Piaget in the 1940's to describe the development of logical reasoning in children. While the version pictured above makes a hash of the intended logic of the diagram (which would, for example, require that H^∞ control be the negation of operator theory), it is nevertheless an homage to Krauss and Chris Dorsett, indicates one of the unexpected ways that art and mathematics come together, and encapsulates for me some of the salient features of IWOTA.

Michael Dritschel

List of Participants

Alaa Abou-Hajar (*University of Newcastle*)
Tsuyoshi Ando (*Hokkaido University*) (*Emeritus*)
Jim Agler (*University of California, San Diego*)
Tuncay Aktosun (*Mississippi State University*)
Edin Alijagic (*Delft University of Technology*)
Calin-Grigore Ambrozie (*Institute of Mathematics of the Romanian Academy*)
Robert Archer (*University of Newcastle*)
Yury Arlinskii (*East Ukrainian National University*)
Damir Arov (*Weizmann Institute*)
William Arveson (*University of California, Berkeley*)
Tomas Azizov (*Voronezh State University*)
Catalin Badea (*University of Lille*)
Mihaly Bakonyi (*Georgia State University*)
Joseph Ball (*Virginia Tech*)
Oscar Bandtlow (*University of Nottingham*)
M. Amelia Bastos (*Instituto Superior Técnico*)
Ismat Beg (*Lahore University of Management Sciences*)
Jussi Behrndt (*TU Berlin*)
Sergio Bermudo (*Universidad Pablo de Olavide*)
Tirthankar Bhattacharyya (*Indian Institute of Science*)
Paul Binding (*University of Calgary*)
Debapriya Biswas (*University of Leeds*)
Danilo Blagojevic (*Edinburgh University*)
Albrecht Boettcher (*Technical University Chemnitz*)
Craig Borwick (*University of Newcastle upon Tyne*)
Lyonell Boulton (*University of Calgary*)
Maria Cristina Câmara (*Instituto Superior Técnico*)
Isabelle Chalendar (*University of Lyon*)

Yemon Choi (*University of Newcastle*)
Raul Curto (*University of Iowa*)
Stefan Czerwik (*Silesian University of Technology*)
Volodymyr Derkach (*Donetsk National University*)
Patrick Dewilde (*TU Delft*)
Aad Dijksma (*Rijksuniversiteit Groningen*)
Ronald Douglas (*Texas A & M University*)
Michael Dritschel (*University of Newcastle*)
Chen Dubi (*Ben Gurion University of the Negev*)
Harry Dym (*The Weizmann Institute of Science*)
Douglas Farenick (*University of Regina*)
Claudio Antonio Fernandes (*Faculdade de Ciencias e Tecnologia, Almada*)
Lawrence Fialkow (*SUNY New Paltz*)
Alastair Gillespie (*University of Edinburgh*)
Jim Gleason (*University of Tennessee*)
Israel Gohberg (*Tel Aviv University*)
Zen Harper (*University of Leeds*)
Mahir Hasanov (*Istanbul Technical University*)
Seppo Hassi (*University of Vaasa*)
Munmun Hazarika (*Tezpur University*)
Georg Heinig (*Kuwait University*)
William Helton (*University of California, San Diego*)
Sanne ter Horst (*Vrije Universiteit Amsterdam*)
Birgit Jacob (*University of Dortmund*)
Martin Jones (*University of Newcastle*)
Peter Junghanns (*Technical University Chemnitz*)
Marinus A. Kaashoek (*Vrije Universiteit Amsterdam*)
Michael Kaltenbaeck (*Technical University of Vienna*)
Dmitry Kalyuzhnyi-Verbovetzkii (*Ben Gurion University of the Negev*)
Alexei Karlovich (*Insituto Superior Tecnico*)
Yuri Karlovich (*Universidad Autonoma del Estado de Morelos*)
Iztok Kavkler (*Institute of Mathematics, Physics and Mechanics, Ljubljana*)
Laszlo Kerchy (*University of Szeged*)
Alexander Kheifets (*University of Massachusetts Lowell*)
Vladimir V. Kisil (*University of Leeds*)
Edward Kissin (*London Metropolitan University*)
Martin Klaus (*Virginia Tech*)
Artem Ivanovich Kozko (*Moscow State University*)
Ilya Krishtal (*Washington University*)
Matthias Langer (*Technical University of Vienna*)
Annemarie Luger (*Technical University of Vienna*)
Philip Maher (*Middlesex University*)
Carmen H. Mancera (*Universidad de Sevilla*)
Stefania Marcantognini (*Instituto Venezolano de Investigaciones Científicas*)
Laurent W. Marcoux (*University of Waterloo*)

John Maroulas (*National Technical University*)
Rui Carlos Marreiros (*Universidade do Algarve*)
Helena Mascarenhas (*Instituto Superior Técnico*)
Vladimir Matsaev (*Tel Aviv University*)
John E. McCarthy (*Washington University*)
Cornelis van der Mee (*University of Cagliari*)
Andrey Melnikov (*Ben Gurion University of the Negev*)
Christian Le Merdy (*Université de Besançon*)
Blaž Mojškerc (*University of Ljubljana*)
Manfred Möller (*University of the Witwatersrand*)
Joachim Moussounda Mouanda (*Newcastle University*)
Ana Moura Santos (*Technical University of Lisbon*)
Aissa Nasli Bakir (*University of Chlef*)
Zenonas Navickas (*Kaunas University of Technology*)
Ludmila Nikolskaia (*Université de Bordeaux 1*)
Nika Novak (*University of Ljubljana*)
Mark Opmeer (*Rijksuniversiteit Groningen*)
Peter Otte (*Ruhr-Universität Bochum*)
Jonathan R. Partington (*University of Leeds*)
Linda Patton (*Cal Poly San Luis Obispo*)
Jordi Pau (*Universitat Autonoma de Barcelona*)
Vern Paulsen (*University of Houston*)
Alexander Sergeevich Pechentsov (*Moscow State University*)
Vladimir Peller (*Michigan State University*)
Luis Pessoa (*Instituto Superior Técnico*)
Sandra Pott (*University of York*)
Geoffrey Price (*United States Naval Academy*)
Marek Ptak (*University of Agriculture, Kraków*)
Vladimir Rabinovich (*National Politechnic Institute of Mexico*)
André Ran (*Vrije Universiteit Amsterdam*)
Maria Christina Bernadette Reurings (*The College of William and Mary*)
Stefan Richter (*University of Tennessee*)
Guyan Robertson (*University of Newcastle*)
Helen Robinson (*University of York*)
Steffen Roch (*TU Darmstadt*)
Richard Rochberg (*Washington University*)
Leiba Rodman (*College of William and Mary*)
Karla Rost (*Technical University Chemnitz*)
James Rovnyak (*University of Virginia*)
Cora Sadosky (*Howard University*)
Lev Sakhnovich (*University of Connecticut*)
Natasha Samko (*University of Algarve*)
Stefan Samko (*University of Algarve*)
Kristian Seip (*Norwegian University of Science and Technology*)
Eugene Shargorodsky (*King's College London*)

Malcolm C. Smith (*University of Cambridge*)
Martin Paul Smith (*University of York*)
Rachael Caroline Smith (*University of Leeds*)
Henk de Snoo (*Rijksuniversiteit Groningen*)
Michail Solomyak (*The Weizmann Institute of Science*)
Olof Staffans (*Åbo Akademi*)
Jan Stochel (*Uniwersytet Jagielloński*)
Vladimir Strauss (*Simon Bolivar University*)
Elizabeth Strouse (*Université de Bordeaux 1*)
Franciszek Hugon Szafraniec (*Uniwersytet Jagielloński*)
Richard M. Timoney (*Trinity College Dublin*)
Dan Timotin (*Institute of Mathematics of the Romanian Academy*)
Vadim Tkachenko (*Ben Gurion University of the Negev*)
Christiane Tretter (*University of Bremen*)
Nikolai Vasilevski (*CINVESTAV del I.P.N., Mexico*)
Luis Verde-Star (*Universidad Autonoma Metropolitana*)
Sjoerd Verduyn Lunel (*Universiteit Leiden*)
Victor Vinnikov (*Ben Gurion University of the Negev*)
Jani Virtanen (*King's College London*)
Dan Volok (*Ben Gurion University of the Negev*)
George Weiss (*Imperial College London*)
Brett Wick (*Brown University*)
Henrik Winkler (*Rijksuniversiteit Groningen*)
Harald Woracek (*Technical University of Vienna*)
Dmitry V. Yakubovich (*Autonoma University of Madrid*)
Nicholas J. Young (*Newcastle University*)
Joachim Peter Zacharias (*University of Nottingham*)
László Zsidó (*University of Rome*)

IWOTA 2004

1
2
3
4
5
6
7
8
9
10
11
12
13
14
15
16
17
18
19
20
21
22
23
24
25
26
27
28
29
30
31
32
33
34
35
36
37
38
39
40
41
42
43
44
45
46
47
48
49
50
51
52
53
54
55
56
57
58
59
60
61
62
63
64
65
66
67
68
69
70
71
72
73
74
75
76
77
78
79
80
81
82
83
84
85
86
87
88
89
90
91
92
93
94
95
96
97
98
99
100
101
102
103
104

(1) László Kérchy
(2) Martin Smith
(3) André Ran
(4) Mahir Hasanov
(5) Zen Harper
(6) Yemon Choi
(7) Manfred Möller
(8) Sanne ter Horst
(9) Cornelis van der Mee
(10) Yuri Arlinskii
(11) Rachael Smith
(12) Vladimir Matsaev
(13) Seppo Hassi
(14) Henk de Snoo
(15) Stefan Richter
(16) Leiba Rodman
(17) Chen Dubi
(18) Ronald Douglas
(19) Joseph Ball
(20) Joachim Zacharias
(21) Jim Agler
(22) Aissa Nasli Bakir
(23) Eugene Shargorodsky
(24) Vladimir Peller
(25) Jani Virtanen
(26) Ilya Krishtal
(27) Harry Dym
(28) Martin Klaus
(29) Jan Stochel
(30) Peter Otte
(31) Richard Timoney
(32) Tomas Azizov
(33) Aad Dijksma
(34) Oscar Bandtlow
(35) James Rovnyak
(36) Cora Sadosky
(37) László Zsidó
(38) Tsuyoshi Ando
(39) Nika Novak
(40) Mihaly Bakonyi
(41) Zenonas Nacickas
(42) Blaž Mojškerc
(43) Volodymyr Derkach
(44) Iztok Kavkler
(45) Dimitry Yakubovich
(46) Jonathan Partington
(47) Alexander Pechentsov
(48) Mark Opmeer
(49) Maria Christina Reurings
(50) Joachim Mouanda
(51) Helen Robinson
(52) Alec Matheson
(53) Marek Ptak
(54) Maria Christina Câmara
(55) Artem Kozko
(56) William Arveson
(57) Vladimir Rabinovich
(58) Tirthankar Bhattacharyya
(59) Edin Alijagic
(60) Jordi Pau
(61) Albrecht Böttcher
(62) Dimitry Kalyuzhnyi-Verbovetzkii
(63) Calin-Grigore Ambrozie
(64) Helena Mascarenhas
(65) Georg Heinig
(66) Sandra Pott
(67) Peter Junghanns
(68) Brett Wick
(69) Franciszek Szafraniec
(70) Dan Timotin
(71) Isabelle Chalendar
(72) John Maroulas
(73) Sjoerd Verduyn Lunel
(74) Sergio Bermudo
(75) Jim Gleason
(76) Damir Arov
(77) Rui Marreiros
(78) Birgit Jacob
(79) Yuri Karlovich
(80) Olof Staffans
(81) Laurent Marcoux
(82) Vern Paulsen
(83) Douglas Farenick
(84) Matthias Kanger
(85) Jussi Behrndt
(86) Ana Moura Santos
(87) Ismat Beg
(88) Karla Rost
(89) Nicholas Young
(90) Dan Volok
(91) Tuncay Aktosun
(92) Debapriya Biswas
(93) Alexei Karlovich
(94) Marinus Kaashoek
(95) Israel Gohberg
(96) William Helton
(97) Munmun Hazarika
(98) Elizabeth Strouse
(99) Vladimir Strauss
(100) Patrick Dewilde
(101) Annemarie Luger
(102) Victor Vinnikov
(103) Michael Dritschel
(104) Jackie Roberts

Talk Titles

^{p} indicates plenary speakers, ^{s} semi-plenary speakers.

Jim Agler ^{p}
Function theory on analytic sets

Tuncay Aktosun
Inverse spectral-scattering problem with two sets of discrete spectral data

Edin Alijagic
Minimal factorizations of isometric operators

Calin-Grigore Ambrozie ^{p}
Invariant subspaces and reflexivity results in Banach spaces

Robert Archer
A commutant lifting theorem for n-tuples with regular dilations

Yury Arlinskii
Realizations of H_{-2}-perturbations of self-adjoint operators

Damir Arov ^{s}
Bi-inner dilations and bi-stable passive scattering realizations

William Arveson ^{p}
Free resolutions in multivariable operator theory

Tomas Azizov
A uniformly bounded C_0-semigroup of bi-contractions

Catalin Badea
Invertible extensions, growth conditions and the Cesaro operator

Mihaly Bakonyi
Factorization of operator-valued functions on ordered groups

Joseph Ball
Realization and interpolation for Schur-Agler class functions on domains in $\mathbb{C}^n$ with matrix polynomial defining function

Oscar Bandtlow
Estimates for norms of resolvents

M. Amelia Bastos
A symbol calculus for a C^*-algebra with piecewise oscillating coefficients

Ismat Beg
Convergence of iterative algorithms to a common random fixed point

Jussi Behrndt
Sturm-Liouville operators with an indefinite weight and a λ*-dependent boundary condition*

Sergio Bermudo
A parametrization for the symbols of an operator of Hankel type

Paul Binding
Variational principles for nondefinite eigenvalue problems

Albrecht Boettcher ^{p}
Asymptotical linear algebra illustrated with Toeplitz matrices

Lyonell Boulton
Effective approximation of eigenvalues in gaps of the essential spectrum

Maria Cristina Câmara
Polynomial almost periodic solutions for a class of oscillatory Riemann–Hilbert problems

Isabelle Chalendar ^{p}
Constrained approximation and invariant subspaces

Stefan Czerwik
Quadratic difference operators in L^p*-spaces*

Volodymyr Derkach
Boundary relations and Weyl families

Patrick Dewilde ^{s}
Matrix interpolation: the singular case revisited

Aad Dijksma
Orthogonal polynomials, operator realizations, and the Schur transform for generalized Nevanlinna functions

Ronald Douglas
Essentially reductive Hilbert modules

Chen Dubi
On commuting operators solving Gleason's problem

Harry Dym
A factorization theorem and its application to inverse problems

Douglas Farenick ^{s}
Some formulations of classical inequalities for singular values and traces of operators

Claudio Antonio Fernandes
Symbol calculi for nonlocal operator C^**-algebras with shifts*

Lawrence Fialkow
A survey of moment problems on algebraic and semialgebraic sets

Alastair Gillespie [p]
An operator-valued principle with applications

Jim Gleason
Quasinormality of Toeplitz tuples with analytic symbols

Israel Gohberg [p]
From infinite to finite dimensions

Zen Harper
The discrete Weiss conjecture and applications in operator theory

Mahir Hasanov
On perturbations of two parameter operator pencils of waveguide type

Seppo Hassi
On Kreĭn's extension theory of nonnegative operators

Munmun Hazarika
Recursively generated subnormal shifts

Georg Heinig
Fast algorithms for Toeplitz and Toeplitz-plus-Hankel matrices with arbitrary rank profile

William Helton [p]
Noncommutative convexity

Birgit Jacob [s]
What can semigroups do for nonlinear dissipative systems?

Peter Junghanns
$O(n \log n)$-algorithms for singular integral equations in the nonperiodic case

Marinus A. Kaashoek
Canonical systems with a pseudo-exponential potential

Michael Kaltenbaeck
Canonical differential equations of Hilbert-Schmidt type

Dmitry Kalyuzhnyi-Verbovetzkii
On the intersection of null spaces for matrix substitutions in a non-commutative rational formal power series

Yuri Karlovich
Essential spectra of pseudodifferential operators with symbols of limited smoothness

Alexei Karlovich
Banach algebras of singular integral operators in generalized Lebesgue spaces with variable exponent

Laszlo Kerchy
Supercyclic representations

Alexander Kheifets
Once again on the commutant

Vladimir V. Kisil
Spectra of non-selfadjoint operators and group representations

Edward Kissin
Lipschitz functions on Hermitian Banach $*$*-algebras*

Martin Klaus
Remarks on the eigenvalues of the Manakov system

Artem Ivanovich Kozko
Regularized traces of singular differential operators of higher orders

Ilya Krishtal
Methods of abstract harmonic analysis in the spectral theory of causal operators

Matthias Langer
Variational principles for eigenvalues of block operator matrices

Annemarie Luger
Realizations for a generalized Nevanlinna function with its only generalized pole not of positive type at ∞

Philip Maher
Operator approximation via spectral approximation and differentiation

Laurent W. Marcoux
Amenable operators on Hilbert space

Rui Carlos Marreiros
On the kernel of some one-dimensional singular integral operators with a non-Carleman shift

Helena Mascarenhas
Convolution type operators on cones and asymptotic properties

John E. McCarthy [p]
The Fantappie transform

Cornelis van der Mee
Approximation of solutions of Riccati equations

Andrey Melnikov
Overdetermined 2D systems which are time-invariant In one direction, and their transfer functions

Christian Le Merdy [p]
Semigroups and functional calculus on noncommutative L^p*-spaces*

Manfred Möller [s]
Problems in magnetohydrodynamics: The block matrix approach

Ana Moura Santos
Wave diffraction by a strip grating: the two-straight line approach

Aissa Nasli Bakir
Unitary nodes and contractions on Hilbert spaces

Zenonas Navickas
Representing solution of differential equations in terms of algebraic operators

Ludmila Nikolskaia
Toeplitz and Hankel matrices as Hadamard-Schur multipliers

Mark Opmeer
Ill-posed open dynamical systems

Peter Otte
Liouville-type formulae for section determinants and their relation to Szegő limit theorems

Jonathan R. Partington ^{p}
Semigroups, diagonal systems, controllability and interpolation

Vern Paulsen ^{p}
Frames, graphs and erasures

Alexander Sergeevich Pechentsov
Asymptotic behavior of spectral function in singular Sturm–Liouville problems

Vladimir Peller ^{p}
Extensions of the Koplienko-Neidhardt trace formulae

Luis Pessoa
Algebras generated by a finite number of poly and anti-poly Bergman projections and by operators of multiplication by piecewise continuous functions

Sandra Pott ^{s}
Dyadic BMO and Besov spaces on the bidisk

Marek Ptak
Reflexive finite dimensional subspaces are hyperreflexive

Vladimir Rabinovich
Wiener algebras of operator-valued band dominated operators and its applications

André Ran ^{s}
Inertia and hermitian block Toeplitz matrices

Stefan Richter
Analytic contractions, nontangential limits, and the index of invariant subspaces

Guyan Robertson
Singular masas of von Neumann algebras: examples from the geometry of spaces of nonpositive curvature

Steffen Roch
Finite sections of band-dominated operators

Richard Rochberg
Interpolation sequences for Besov spaces of the complex ball

Leiba Rodman
Polar decompositions in Krein spaces: An overview

Karla Rost
Inversion formulas for Toeplitz and Toeplitz-plus-Hankel matrices with symmetries

James Rovnyak
On indefinite cases of operator identities which arise in interpolation theory

Cora Sadosky
Bounded mean oscillation in product spaces

Lev Sakhnovich
Integrable operators and canonical systems

Natasha Samko
Singular integral operators in weighted spaces of continuous functions with a non-equilibrated continuity modulus

Stefan Samko
Maximal and potential operators in function spaces with non-standard growth

Kristian Seip
Extremal functions for kernels of Toeplitz operators

Eugene Shargorodsky [s]
Some open problems in the spectral theory of Toeplitz operators

Malcolm C. Smith [p]
Foundations of systems theory and robust control: An operator theory perspective

Martin Paul Smith
Schatten class paraproducts and commutators with the Hilbert transform

Henk de Snoo
Closed semibounded forms and semibounded selfadjoint relations

Michail Solomyak
On a spectral properties of a family of differential operators

Olof Staffans
State/signal representations of infinite-dimensional positive real relations

Jan Stochel
Selfadjointness of integral and matrix operators

Vladimir Strauss
On an extension of a J-orthogonal pair of definite subspaces that are invariant with respect to a projection family

Elizabeth Strouse
Products of Toeplitz operators

Franciszek Hugon Szafraniec
A look at indefinite inner product dilations of forms

Richard M. Timoney
Norms of elementary operators

Dan Timotin
A coupling argument for some lifting theorems

Vadim Tkachenko
On spectra of 1d periodic selfadjoint differential operators of order $2n$

Christiane Tretter
From selfadjoint to nonselfadjoint variational principles

Nikolai Vasilevski
Commutative C^**-algebras of Toeplitz operators on Bergman spaces*

Luis Verde-Star
Vandermonde matrices associated with polynomial sequences of interpolatory type sequences of interpolatory type

Sjoerd Verduyn Lunel [s]
New completeness results for classes of operators

Victor Vinnikov
Scattering systems with several evolutions, related to the polydisc, and multidimensional conservative input/state/output systems

Dan Volok
The Schur class of operators on a homogeneous tree: coisometric realizations

George Weiss
Strong stabilization of distributed parameter systems by colocated feedback

Brett Wick
Remarks on Product VMO

Henrik Winkler
Perturbations of selfadjoint relations with a spectral gap

Harald Woracek
Canonical systems in the indefinite setting

Dmitry V. Yakubovich [s]
Linearly similar functional models in a domain

Joachim Peter Zacharias
K*-theoretical duality for higher rank Cuntz-Krieger algebras*

László Zsidó
Weak mixing properties for vector sequences

Operator Theory:
Advances and Applications, Vol. 171, 1–16

Inverse Scattering to Determine the Shape of a Vocal Tract

Tuncay Aktosun

Abstract. The inverse scattering problem is reviewed for determining the cross sectional area of a human vocal tract. Various data sets are examined resulting from a unit-amplitude, monochromatic, sinusoidal volume velocity sent from the glottis towards the lips. In case of nonuniqueness from a given data set, additional information is indicated for the unique recovery.

Mathematics Subject Classification (2000). Primary 35R30; Secondary 34A55 76Q05.

Keywords. Inverse scattering, speech acoustics, shape of vocal tract.

1. Introduction

A fundamental inverse problem related to human speech is [6, 7, 18, 19, 22] to determine the cross sectional area of the human vocal tract from some data. The vocal tract can be visualized as a tube of 14–20 cm in length, with a pair of lips known as vocal cords at the glottal end and with another pair of lips at the mouth. In this review paper, we consider various types of frequency-domain scattering data resulting from a unit-amplitude, monochromatic, sinusoidal volume velocity input at the glottis (the opening between the vocal cords), and we examine whether each data set uniquely determines the vocal-tract area, or else, what additional information may be needed for the unique recovery.

Human speech consists of phonemes; for example, the word "book" consists of the three phonemes /b/, /u/, and /k/. The number of phonemes may vary from one language to another, and in fact the exact number itself of phonemes in a language is usually a subject of debate. In some sense, this is the analog of the number of colors in a rainbow. In American English the number of phonemes is 36, 39, 42, 45, or more or less, depending on the analyst and many other factors. The

The research leading to this article was supported in part by the National Science Foundation under grant DMS-0204437 and the Department of Energy under grant DE-FG02-01ER45951.

phonemes can be sorted into two main groups as vowels and consonants. The vowels can further be classified into monophthongs such as /e/ in "pet" and diphthong such as the middle sound in "boat." The consonants can further be classified into approximants (also known as semivowels) such as /y/ in "yes," fricatives such as /sh/ in "ship," nasals such as /ng/ in "sing," plosives such as /p/ in "put," and affricates such as /ch/ "church."

The production of human speech occurs usually by inhaling air into the lungs and then sending it back through the vocal tract and out of the mouth. The air flow is partially controlled by the vocal cords, by the muscles surrounding the vocal tract, and by various articulators such as the tongue and jaw. As the air is pushed out of the mouth the pressure wave representing the sound is created. The effect of articulators in vowel production is less visible than in consonant production, and one can use compensatory articulation, especially in vowel production, by ignoring the articulators and by producing a phoneme solely by controlling the shape of the vocal tract with the help of surrounding muscles. Such a technique is often used by ventriloquists.

From a mathematical point of view, we can assume [18, 19] that phonemes are basic units of speech, each phoneme lasts about 10–20 msec, the shape of the vocal tract does not change in time during the production of each phoneme, the vocal tract is a right cylinder whose cross sectional area A varies along the distance x from the glottis. We let $x = 0$ correspond to the glottis and $x = l$ to the lips. We can also assume that A is positive on $(0, l)$ and that both A and its derivative A' are continuous on $(0, l)$ and have finite limits at $x = 0^+$ and $x = l^-$. In fact, we will simply write $A(0)$ for the glottal area and $A(l)$ for the area of the opening at the lips because we will not use A or A' when $x \notin [0, l]$.

Besides $A(x)$, the primary quantities used in the vocal-tract acoustics are the time t, the pressure $p(x, t)$ representing the force per unit cross sectional area exerted by the moving air molecules, the volume velocity $v(x, t)$ representing the product of the cross sectional area and the average velocity of the air molecules crossing that area, the air density μ (about 1.2×10^{-3} gm/cm^3 at room temperature), and the speed of sound c (about 3.43×10^4 cm/sec in air at room temperature). In our analysis, we assume that the values of μ and c are already known and we start the time at $t = 0$. It is reasonable [18, 19] to assume that the propagation is lossless and planar and that the acoustics in the vocal tract is governed [6, 7, 18, 19, 22] by

$$\begin{cases} A(x)\, p_x(x,t) + \mu\, v_t(x,t) = 0, \\ A(x)\, p_t(x,t) + c^2 \mu\, v_x(x,t) = 0, \end{cases} \tag{1.1}$$

where the subscripts x and t denote the respective partial derivatives.

In order to recover A, we will consider various types of data for $k \in \mathbf{R}^+$ resulting from the glottal volume velocity

$$v(0,t) = e^{ickt}, \qquad t > 0. \tag{1.2}$$

Note that the quantity $\frac{ck}{2\pi}$ is the frequency measured in Hertz. Informally, we will refer to k as the frequency even though k is actually the angular wavenumber.

The recovery of A can be analyzed either as an inverse spectral problem or as an inverse scattering problem. When formulated as an inverse spectral problem, the boundary conditions are imposed both at the glottis and at the lips. The imposition of the boundary conditions at both ends of the vocal tract results in standing waves whose frequencies form an infinite sequence. It is known [9, 11, 14, 15, 17–19] that A can be recovered from a data set consisting of two infinite sequences. Such sequences can be chosen as the zeros and poles [15, 17] of the input impedance or the poles and residues [11] of the input impedance.

If the recovery of A is formulated as an inverse scattering problem, a boundary condition is imposed at only one end of the vocal tract – either at the glottis or at the lips. The data can be acquired either at the same end or at the opposite end. We have a reflection problem if the data acquisition and the imposed boundary condition occur at the same end of the vocal tract. If these occur at the opposite ends, we have a transmission problem. The inverse scattering problem may be solved either in the time domain or in the frequency domain, where the data set is a function of t in the former case and of k in the latter. The reader is referred to [4, 18–21, 23] for some approaches as time-domain reflection problems, to [16] for an approach as a time-domain transmission problem, and to [1] for an approach as a frequency-domain transmission problem.

The organization of our paper is as follows. In Section 2 we relate (1.1) to the Schrödinger equation (2.3), introduce the selfadjoint boundary condition (2.7) involving $\cot\alpha$ given in (2.8), and present the Jost solution f, the Jost function F_α, and the scattering coefficients T, L, and R. In Section 3 we introduce the normalized radius η of the vocal tract, relate it to the regular solution φ_α to the half-line Schrödinger equation, and also express η in terms of the Jost solution and the scattering coefficients. In Section 4 we present the expressions for the pressure and the volume velocity in the vocal tract in terms of the area, the Jost solution, and the Jost function; in that section we also introduce various data sets that will be used in Section 6. In Section 5 we review the recovery of the potential in the Schrödinger equation and the boundary parameter $\cot\alpha$ from the amplitude of the Jost function; for this purpose we outline the Gel'fand-Levitan method [3, 12, 13, 10] and also the method of [13]. In Sections 6 we examine the recovery of the potential, normalized radius of the tract, and the vocal-tract area from various data sets introduced in Section 4. Such data sets include the absolute values of the impedance at the lips and at the glottis, the absolute value of the pressure at the lips, the absolute value of a Green's function at the lips associated with (2.6), the reflectance at the glottis, and the absolute value of the transfer function from the glottis to the lips. Finally, in Section 7 we show that two data sets containing the absolute value of the same transfer function but with different logarithmic derivatives of the area at the lips correspond to two distinct potentials as well as two distinct normalized radii and two distinct areas for the vocal tract.

2. Schrödinger equation and Jost function

In this section we relate the acoustic system in (1.1) to the Schrödinger equation, present the selfadjoint boundary condition (2.7) identified by the logarithmic derivative of the area function at the glottis, and introduce the Jost solution, Jost function, and the scattering coefficients associated with the Schrödinger equation.

Letting

$$P(k,x) := p(x,t)\, e^{-ickt}, \quad V(k,x) := v(x,t)\, e^{-ickt}, \tag{2.1}$$

we can write (1.1) as

$$\begin{cases} A(x)\, P'(k,x) + ic\mu k\, V(k,x) = 0, \\ c^2\mu\, V'(k,x) + ick\, A(x)\, P(k,x) = 0, \end{cases} \tag{2.2}$$

where the prime denotes the x-derivative. Eliminating V in (2.2), we get

$$[A(x)\, P'(k,x)]' + k^2 A(x)\, P(k,x) = 0, \qquad x \in (0,l),$$

or equivalently

$$\psi''(k,x) + k^2\psi(k,x) = Q(x)\, \psi(k,x), \tag{2.3}$$

with

$$\psi(k,x) =: \sqrt{A(x)}\, P(k,x), \quad Q(x) := \frac{[\sqrt{A(x)}]''}{\sqrt{A(x)}}. \tag{2.4}$$

Alternatively, letting

$$\Phi(x,t) := \sqrt{A(x)}\, p(x,t), \tag{2.5}$$

we find that Φ satisfies the plasma-wave equation

$$\Phi_{xx}(x,t) - \frac{1}{c^2}\, \Phi_{tt}(x,t) = Q(x)\, \Phi(x,t), \qquad x \in (0,l), \quad t > 0. \tag{2.6}$$

We can analyze the Schrödinger equation in (2.3) on the full line $\mathbf{R}$ by using the extension $Q \equiv 0$ for $x < 0$ and $x > l$. We can also analyze it on the half line $\mathbf{R}^+$ by using the extension $Q \equiv 0$ for $x > l$ and by imposing the selfadjoint boundary condition

$$\sin\alpha \cdot \varphi'(k,0) + \cos\alpha \cdot \varphi(k,0) = 0, \tag{2.7}$$

where

$$\cot\alpha := -\frac{A'(0)}{2\, A(0)} = -\frac{[\sqrt{A(x)}]'\big|_{x=0}}{\sqrt{A(0)}}. \tag{2.8}$$

As solutions to the half-line Schrödinger equation, we can consider the regular solution φ_α satisfying the initial conditions

$$\varphi_\alpha(k,0) = 1, \quad \varphi_\alpha'(k,0) = -\cot\alpha, \tag{2.9}$$

and the Jost solution f satisfying the asymptotic conditions

$$f(k,x) = e^{ikx}[1 + o(1)], \quad f'(k,x) = ik\, e^{ikx}[1 + o(1)], \qquad x \to +\infty.$$

Since $Q \equiv 0$ for $x > l$, we have

$$f(k,l) = e^{ikl}, \quad f'(k,l) = ik\, e^{ikl}. \tag{2.10}$$

The Jost function F_α associated with the boundary condition (2.7) is defined [3, 10, 12, 13] as

$$F_\alpha(k) := -i[f'(k,0) + \cot\alpha \cdot f(k,0)], \tag{2.11}$$

and it satisfies [3, 10, 12, 13]

$$F_\alpha(k) = k + O(1), \qquad k \to \infty \text{ in } \overline{\mathbf{C}^+}, \tag{2.12}$$

$$F_\alpha(-k) = -F_\alpha(k)^*, \qquad k \in \mathbf{R}, \tag{2.13}$$

where $\mathbf{C}^+$ is the upper half complex plane, $\overline{\mathbf{C}^+} := \mathbf{C}^+ \cup \mathbf{R}$, and the asterisk denotes complex conjugation.

Associated with the full-line Schrödinger equation, we have the transmission coefficient T, the left reflection coefficient L, and the right reflection coefficient R that can be obtained from the Jost solution f via

$$f(k,0) = \frac{1+L(k)}{T(k)}, \quad f'(k,0) = ik\frac{1-L(k)}{T(k)}, \quad R(k) = -\frac{L(-k)\,T(k)}{T(-k)}. \tag{2.14}$$

The scattering coefficients satisfy [2, 5, 12, 13]

$$T(-k) = T(k)^*, \quad R(-k) = R(k)^*, \quad L(-k) = L(k)^*, \qquad k \in \mathbf{R}. \tag{2.15}$$

In the inverse scattering problem of recovery of A, the bound states for the Schrödinger equation do not arise. The absence of bound states for the full-line Schrödinger equation is equivalent for $1/T(k)$ to be nonzero on $\mathbf{I}^+$, where $\mathbf{I}^+ := i(0,+\infty)$ is the positive imaginary axis in $\mathbf{C}^+$. For the half-line Schrödinger equation with the boundary condition (2.7) the absence of bound states is equivalent [3, 10, 12, 13] for $F_\alpha(k)$ to be nonzero on $\mathbf{I}^+$. It is known [3, 10, 12, 13] that either $F_\alpha(0) \neq 0$ or F_α has a simple zero at $k = 0$; the former is known as the generic case and the latter as the exceptional case for the half-line Schrödinger equation. For the full-line Schrödinger equation we have $T(0) = 0$ generically and $T(0) \neq 0$ in the exceptional case. The exceptional case corresponds to the threshold at which the number of bound states may change by one under a small perturbation. In general, the full-line generic case and the half-line generic case do not occur simultaneously because the former is solely determined by Q whereas the latter jointly by Q and $\cot\alpha$.

3. Relative concavity, normalized radius, and area

In this section we introduce the normalized radius η of the vocal tract, relate it to the regular solution to the Schrödinger equation, and present various expressions for it involving the Jost solution and the scattering coefficients.

Let r denote the radius of the cross section of the vocal tract so that $A(x) = \pi\,[r(x)]^2$. Then, we can write the potential Q appearing in (2.4) as

$$Q(x) = \frac{r''(x)}{r(x)}, \qquad x \in (0,l),$$

and hence we can refer to Q also as the relative concavity of the vocal tract. Define

$$\eta(x) := \frac{\sqrt{A(x)}}{\sqrt{A(0)}}, \tag{3.1}$$

or equivalently

$$A(x) = A(0)\,[\eta(x)]^2, \qquad x \in (0,l). \tag{3.2}$$

We can refer to η as the normalized radius of the vocal tract and η^2 as the normalized area of the tract. From (2.4) we see that η satisfies

$$y'' = Q(x)\,y, \qquad x \in (0,l),$$

with the initial conditions

$$\eta(0) = 1, \quad \eta'(0) = -\cot\alpha.$$

A comparison with (2.9) shows that η is nothing but the zero-energy regular solution, i.e.,

$$\eta(x) = \varphi_\alpha(0,x), \qquad x \in (0,l). \tag{3.3}$$

With the help of the expression [3, 10, 12, 13]

$$\varphi_\alpha(k,x) = \frac{1}{2k}\left[F_\alpha(k)\,f(-k,x) - F_\alpha(-k)\,f(k,x)\right],$$

we can write (3.3) as

$$\eta(x) = \dot F_\alpha(0)\,f(0,x) - F_\alpha(0)\,\dot f(0,x), \qquad x \in (0,l),$$

where an overdot indicates the k-derivative. We can also express η with the help of the scattering coefficients. In the full-line generic case we get [1]

$$\eta(x) = \begin{vmatrix} 0 & -\frac{i}{2}\,\dot T(0)\,f(0,x) & i\,\dot f(0,x) - \frac{i}{2}\,\dot R(0)\,f(0,x) \\ 1 & f(0,0) & 1 \\ -\cot\alpha & f'(0,0) & 0 \end{vmatrix}, \qquad x \in (0,l),$$

and in the full-line exceptional case we have [1]

$$\eta(x) = f(0,x)\begin{vmatrix} 0 & 1 & \displaystyle\int_0^x \frac{dz}{[f(0,z)]^2} \\ -1 & f(0,0) & 0 \\ \cot\alpha & 0 & \dfrac{1}{f(0,0)} \end{vmatrix}, \qquad x \in (0,l). \tag{3.4}$$

It is also possible to express η in other useful forms. Let $g(k,x)$ denote the corresponding Jost solution when we replace the zero fragment of Q for $x \in (l,+\infty)$ by another piece which is integrable, has a finite first moment, and does not yield any bound states. Let $\tau(k)$, $\ell(k)$, and $\rho(k)$ be the corresponding transmission coefficient, the left reflection coefficient, and the right reflection coefficient, respectively. In the generic case, i.e., when $\tau(0) = 0$, we have [1]

$$\eta(x) = \begin{vmatrix} 0 & -\frac{i}{2}\,\dot\tau(0)\,g(0,x) & i\,\dot g(0,x) - \frac{i}{2}\,\dot\rho(0)\,g(0,x) \\ 1 & g(0,0) & 1 \\ -\cot\alpha & g'(0,0) & 0 \end{vmatrix}, \qquad x \in (0,l),$$

and in the exceptional case, i.e., when $\tau(0) \neq 0$, we have [1]

$$\eta(x) = g(0,x) \begin{vmatrix} 0 & 1 & \displaystyle\int_0^x \frac{dz}{[g(0,z)]^2} \\ -1 & g(0,0) & 0 \\ \cot\alpha & 0 & \dfrac{1}{g(0,0)} \end{vmatrix}, \qquad x \in (0,l).$$

4. Pressure and volume velocity in the vocal tract

In this section we present the expressions for the pressure and the volume velocity in the vocal tract in terms of the Jost function and the Jost solution. We also introduce various data sets associated with the values of the pressure and the volume velocity at the glottal end of the vocal tract or at the lips.

From (1.2) and (2.1), we see that $V(k,0) = 1$. Further, under the reasonable assumption that there is no reflected pressure wave at the mouth and all the pressure wave is transmitted out of the mouth, we get [1]

$$P(k,x) = -\frac{c\mu k\, f(-k,x)}{\sqrt{A(0)}\,\sqrt{A(x)}\,F_\alpha(-k)}, \qquad x \in (0,l), \tag{4.1}$$

$$V(k,x) = -\frac{i\,\sqrt{A(x)}}{\sqrt{A(0)}\,F_\alpha(-k)} \left[f'(-k,x) - \frac{A'(x)}{2\,A(x)}\, f(-k,x) \right], \qquad x \in (0,l), \tag{4.2}$$

where we recall that f is the Jost solution to the Schrödinger equation and F_α is the Jost function appearing in (2.11). Then, the pressure $p(x,t)$ and the volume velocity $v(x,t)$ in the vocal tract are obtained by using (4.1) and (4.2) in (2.1).

The transfer function from the glottis to the point x is defined as the ratio $v(x,t)/v(0,t)$, and in our case, as seen from (1.2) and (2.1), that transfer function is nothing but $V(k,x)$. In particular, the transfer function $\mathbf{T}(k,l)$ at the lips is obtained by using (2.10) in (4.2), and we have

$$\mathbf{T}(k,l) = \frac{\sqrt{A(l)}\,e^{-ikl}}{\sqrt{A(0)}\,F_\alpha(-k)} \left[-k + \frac{i}{2}\,\frac{A'(l)}{A(l)} \right],$$

$$|\mathbf{T}(k,l)|^2 = \frac{A(l)}{A(0)\,|F_\alpha(k)|^2} \left[k^2 + \frac{[A'(l)]^2}{4[A(l)]^2} \right], \qquad k \in \mathbf{R}. \tag{4.3}$$

Using (2.12) in (4.3), we get

$$A(0)=\frac{A(l)}{\lim_{k\to+\infty}|\mathbf{T}(k,l)|},\tag{4.4}$$

and hence we can write (4.3) as

$$|F_\alpha(k)|^2=\frac{\lim_{k\to+\infty}|\mathbf{T}(k,l)|^2}{|\mathbf{T}(k,l)|^2}\left[k^2+\frac{1}{4}\frac{[A'(l)]^2}{[A(l)]^2}\right],\qquad k\in\mathbf{R}.\tag{4.5}$$

The impedance at the point x is defined as $p(x,t)/v(x,t)$, which is seen to be equal to $P(k,x)/V(k,x)$ because of (2.1). In particular, with the help of (1.2) we see that the glottal impedance $Z(k,0)$ is equal to $P(k,0)$, and hence we have

$$Z(k,0)=-\frac{c\mu k\,f(-k,0)}{A(0)\,F_\alpha(-k)},$$

$$|Z(k,0)|=\frac{c\mu\,|k|\,|f(k,0)|}{A(0)\,|F_\alpha(k)|},\qquad k\in\mathbf{R}.\tag{4.6}$$

Using (2.12) and the fact [2, 5, 10, 12, 13] that $f(k,0)=1+O(1/k)$ as $k\to\infty$ in $\overline{\mathbf{C}^+}$, from (4.6) we get

$$A(0)=\frac{c\mu}{\lim_{k\to+\infty}|Z(k,0)|},\tag{4.7}$$

and thus we can write (4.6) in the equivalent form

$$\left|\frac{k\,f(k,0)}{F_\alpha(k)}\right|=\frac{|Z(k,0)|}{\lim_{k\to+\infty}|Z(k,0)|},\qquad k\in\mathbf{R}.\tag{4.8}$$

In a similar manner, we can evaluate $Z(k,l)$, the impedance at the lips, by using (2.10) in (4.1) and (4.2). We get

$$Z(k,l)=\frac{2ic\mu k}{2ik\,A(l)+A'(l)},$$

$$|Z(k,l)|^2=\frac{4c^2k^2\mu^2}{4k^2[A(l)]^2+[A'(l)]^2},\qquad k\in\mathbf{R}.\tag{4.9}$$

We have already seen that the pressure at the glottis is the same as the impedance at the glottis because $V(k,0)=1$. So, let us only analyze the pressure at the lips. Using (2.10) in (4.1) we get

$$P(k,l)=-\frac{c\mu k\,e^{-ikl}}{\sqrt{A(0)}\,\sqrt{A(x)}\,F_\alpha(-k)},$$

and hence

$$|P(k,l)|=\frac{c\mu\,|k|}{\sqrt{A(0)\,A(l)}\,|F_\alpha(k)|},\qquad k\in\mathbf{R}.\tag{4.10}$$

Using (2.12) in (4.10), we obtain

$$\sqrt{A(0)\,A(l)} = \frac{c\mu}{\lim\limits_{k\to+\infty} |P(k,l)|}, \tag{4.11}$$

and hence we can write (4.10) in the equivalent form

$$|F_\alpha(k)| = \frac{|k|}{|P(k,l)|}\left(\lim_{k\to+\infty} |P(k,l)|\right), \qquad k \in \mathbf{R}. \tag{4.12}$$

The reflectance at x is defined as the ratio of the left-moving pressure wave to the right-moving pressure wave. As seen from (2.10) and (4.1), there is no left-moving pressure wave at the lips and hence the reflectance at the lips is zero. Since $Q \equiv 0$ for $x < 0$, we have [2, 5, 12, 13]

$$f(-k,x) = \frac{e^{-ikx}}{T(-k)} + \frac{L(-k)\,e^{ikx}}{T(-k)}, \qquad x \le 0,$$

and hence the reflectance at the glottis is equal to $L(-k)$.

The Green function $\mathbf{G}(k,l;t)$ associated with the vocal-tract acoustics can be defined [8] as the solution to the plasma-wave equation (2.6) with the input (1.2). From (2.1), (2.5), and (4.1) we obtain

$$\mathbf{G}(k,l;t) = \frac{-c\mu k\, e^{ik(ct-l)}}{\sqrt{A(0)}\,F_\alpha(-k)},$$

$$|\mathbf{G}(k,l;t)| = \frac{c\mu\,|k|}{\sqrt{A(0)}\,|F_\alpha(k)|}, \qquad k \in \mathbf{R}. \tag{4.13}$$

Note that the expression $|\mathbf{G}(k,l;t)|$ is independent of t. Using (2.12) in (4.13) we get

$$A(0) = \frac{c^2\mu^2}{\lim\limits_{k\to+\infty} |\mathbf{G}(k,l;t)|^2}, \tag{4.14}$$

and hence we can write (4.13) in the equivalent form

$$|F_\alpha(k)| = \frac{|k| \lim\limits_{k\to+\infty} |\mathbf{G}(k,l;t)|}{|\mathbf{G}(k,l;t)|}, \qquad k \in \mathbf{R}. \tag{4.15}$$

5. Recovery from the Jost function

In this section we review the recovery of the potential Q and the boundary parameter $\cot\alpha$ from the absolute value of the Jost function given for $k \in \mathbf{R}^+$. The methods presented include the Gel'fand-Levitan method [3, 10, 12, 13] and the method of [3].

In the absence of bound states, one can use the data $\{|F_\alpha(k)| : k \in \mathbf{R}\}$, or equivalently $\{|F_\alpha(k)| : k \in \mathbf{R}^+\}$ as a consequence of (2.13), to recover the potential Q and the boundary parameter $\cot\alpha$ in the half-line Schrödinger equation.

For example, in the Gel'fand-Levitan method [3, 10, 12, 13], the potential Q is obtained as

$$Q(x) = 2\,\frac{dh_\alpha(x,x^-)}{dx}, \qquad x>0, \tag{5.1}$$

the boundary parameter $\cot\alpha$ appearing in (2.7) is recovered as

$$\cot\alpha = -h_\alpha(0,0), \tag{5.2}$$

and the regular solution $\varphi_\alpha(k,x)$ appearing in (2.9) is constructed as

$$\varphi_\alpha(k,x) = \cos kx + \int_0^x dy\, h_\alpha(x,y)\,\cos ky, \qquad x\ge 0, \tag{5.3}$$

where $h_\alpha(x,y)$ is obtained by solving the Gel'fand-Levitan integral equation

$$h_\alpha(x,y) + G_\alpha(x,y) + \int_0^x dz\, G_\alpha(y,z)\,h_\alpha(x,z) = 0, \qquad 0\le y<x,$$

with the kernel $G_\alpha(x,y)$ given by

$$G_\alpha(x,y) := \frac{2}{\pi}\int_0^\infty dk\,\left[\frac{k^2}{|F_\alpha(k)|^2}-1\right]\cos(kx)\,\cos(ky).$$

Comparing (5.1)–(5.3) with (2.4), (2.8), (3.2), and (3.3) we see that

$$\frac{[\sqrt{A(x)}]''}{\sqrt{A(x)}} = 2\,\frac{dh_\alpha(x,x^-)}{dx}, \qquad x\in(0,l),$$

$$\frac{A'(0)}{A(0)} = 2\,h_\alpha(0,0),$$

$$\eta(x) = 1 + \int_0^x dy\, h_\alpha(x,y), \qquad x\in(0,l),$$

$$A(x) = A(0)\left[1+\int_0^x dy\, h_\alpha(x,y)\right]^2, \qquad x\in(0,l).$$

Alternatively, we can proceed [1, 3] as follows. Using the data $\{|F_\alpha(k)| : k\in\mathbf{R}\}$, we evaluate the integral on the right-hand side of

$$\Lambda_\alpha(k) = \frac{1}{\pi i}\int_{-\infty}^\infty \frac{ds}{s-k-i0^+}\left[\frac{s^2}{|F_\alpha(s)|^2}-1\right], \qquad k\in\overline{\mathbf{C}^+},$$

where the quantity $i0^+$ indicates that the values for real k should be obtained as limits from $\mathbf{C}^+$. The function $\Lambda_\alpha(k)$ is equivalent to

$$\Lambda_\alpha(k) = \frac{k\,f(k,0)}{F_\alpha(k)} - 1, \qquad k\in\overline{\mathbf{C}^+}. \tag{5.4}$$

Next, $F_\alpha(k)$ is obtained from $|F_\alpha(k)|$ by using

$$F_\alpha(k) = k\,\exp\left(\frac{-1}{\pi i}\int_{-\infty}^\infty ds\,\frac{\log|s/F_\alpha(s)|}{s-k-i0^+}\right), \qquad k\in\overline{\mathbf{C}^+}.$$

Then, we have

$$f(k,0) = \frac{1}{k}\, F_\alpha(k)\,[1+\Lambda_\alpha(k)], \qquad k \in \overline{\mathbf{C}^+},$$

$$f'(k,0) = i\,F_\alpha(k)\left[1+\frac{1+\Lambda_\alpha(k)}{k}\lim_{k\to\infty}[k\,\Lambda_\alpha(k)]\right], \qquad k \in \overline{\mathbf{C}^+}.$$

$$\cot\alpha = -i\lim_{k\to\infty}[k\,\Lambda_\alpha(k)], \tag{5.5}$$

where the limit in (5.5) can be evaluated in any manner in $\overline{\mathbf{C}^+}$. Having both $f(k,0)$ and $f'(k,0)$ at hand, all the quantities relevant to the scattering for the Schrödinger equation can be constructed. For example, as seen from (2.14), we have

$$T(k) = \frac{2ik}{ik\,f(k,0)+f'(k,0)}, \qquad L(k) = \frac{ik\,f(k,0)-f'(k,0)}{ik\,f(k,0)+f'(k,0)},$$

$$R(k) = \frac{-ik\,f(-k,0)-f'(-k,0)}{ik\,f(k,0)+f'(k,0)}.$$

Having obtained such quantities, the potential can be constructed via any one of the methods [2, 5, 12, 13] to solve the inverse scattering problem.

6. Recovery from other data sets

In this section we consider the recovery of Q, η, and A from various data sets introduced in Section 4 related to the values of the pressure and the volume velocity at the ends of the vocal tract. For this purpose we use the result of Section 5 that the data set $\{|F_\alpha(k)| : k \in \mathbf{R}^+\}$ uniquely determines the corresponding Q and η, and that the same set along with the value of $A(0)$ uniquely determines A.

(i) Impedance at the lips

If we use the data set $\{|Z(k,l)| : k \in \mathbf{R}^+\}$ coming from the impedance at the lips, as seen from (4.9), we can only recover $A(l)$ and $|A'(l)|$. Using (4.9) at two distinct real k-values, say k_1 and k_2, we obtain $A(l)$ and $|A'(l)|$ algebraically as

$$A(l) = \sqrt{\frac{c^2\mu^2}{k_1^2-k_2^2}\left[\frac{k_1^2}{|Z(k_1,l)|^2}-\frac{k_2^2}{|Z(k_2,l)|^2}\right]}, \tag{6.1}$$

$$|A'(l)| = \sqrt{\frac{4c^2\mu^2k_1^2k_2^2}{k_1^2-k_2^2}\left[\frac{1}{|Z(k_2,l)|^2}-\frac{1}{|Z(k_1,l)|^2}\right]}. \tag{6.2}$$

No other information can be extracted from $\{|Z(k,l)| : k \in \mathbf{R}^+\}$ related to Q, η, or A.

(ii) Impedance at the glottis

The information contained in $\{|Z(k,0)|: \ k \in \mathbf{R}^+\}$ enables us to uniquely construct Q, η, and A, which is seen as follows. From (4.8) and the evenness of $|f(k,0)/F_\alpha(k)|$ in $k \in \mathbf{R}$, we see that the recovery from $\{|Z(k,0)|: \ k \in \mathbf{R}^+\}$ is equivalent to the recovery from $\{|k\,f(k,0)/F_\alpha(k)|: \ k \in \mathbf{R}\}$. Since we assume that the half-line Schrödinger equation with the boundary condition (2.7) does not have any bound states, it is known [3] that $kf(k,0)/F_\alpha(k)$ is analytic in $\mathbf{C}^+$, continuous in $\overline{\mathbf{C}^+}$, nonzero in $\overline{\mathbf{C}^+} \setminus \{0\}$, either nonzero at $k=0$ or has a simple zero there, and

$$\frac{k\,f(k,0)}{F_\alpha(k)} = 1 + O(1/k), \qquad k \to \infty \text{ in } \overline{\mathbf{C}^+}.$$

As a result, we can recover $kf(k,0)/F_\alpha(k)$ for $k \in \overline{\mathbf{C}^+}$ from its amplitude known for $k \in \mathbf{R}$ via

$$\frac{k\,f(k,0)}{F_\alpha(k)} = \exp\left(\frac{1}{2\pi}\int_{-\infty}^{\infty} dt\,\frac{\log|t\,f(t,0)/F_\alpha(t)|}{t-k-i0^+}\right), \qquad k \in \overline{\mathbf{C}^+}.$$

Having constructed $kf(k,0)/F_\alpha(k)$ for $k \in \mathbf{R}$, we can use (5.4) and obtain $\Lambda_\alpha(k)$ for $k \in \mathbf{R}$. Next, by taking the real part of $\Lambda_\alpha(k)$ and using

$$\operatorname{Re}[\Lambda_\alpha(k)] = \frac{k^2}{|F_\alpha(k)|^2} - 1, \qquad k \in \mathbf{R},$$

we construct $|F_\alpha(k)|$ for $k \in \mathbf{R}$. Then, as described in Section 5, we can construct Q and η. Finally, since $A(0)$ is available via (4.7), we can also construct A.

(iii) Pressure at the lips

If we use the data set $\{|P(k,l)|: \ k \in \mathbf{R}^+\}$ coming from the pressure at the lips, as seen from (4.12) we have $\{|F_\alpha(k)|: \ k \in \mathbf{R}^+\}$ at hand, and hence Q and η are uniquely determined. Then, having the value $\eta(l)$ and using (3.1) and (4.11), we obtain $A(0)$ as

$$A(0) = \frac{1}{\eta(l)}\,\frac{c\mu}{\lim_{k\to+\infty}|P(k,l)|}.$$

Thus, A is also uniquely determined.

(iv) Green's function at the lips

If we use the data set $\{|\mathbf{G}(k,l;t)|: \ k \in \mathbf{R}^+\}$ coming from Green's function at the lips, as seen from (4.15) we have $\{|F_\alpha(k)|: \ k \in \mathbf{R}^+\}$ at hand, and hence Q and η are uniquely determined. From (4.14) we also have $A(0)$, and hence A is also uniquely determined.

(v) Reflectance at the glottis

We know from Section 4 that the reflectance at the glottis is given by $L(-k)$. Because of (2.15), the data sets $\{L(-k) : k \in \mathbf{R}^+\}$ and $\{L(k) : k \in \mathbf{R}\}$ are equivalent. The latter set uniquely determines Q, for example, via the Faddeev-Marchenko method [2, 5, 12, 13]. Either of the real and imaginary parts of the reflectance at the glottis known for $k \in \mathbf{R}^+$ also enables us to uniquely construct Q. This is because $\mathrm{Re}[L(k)]$ is an even function of k on $\mathbf{R}$ and $\mathrm{Im}[L(k)]$ is an odd function, and L can be recovered from either its real or imaginary part via the Schwarz integral formula as

$$L(k) = \frac{1}{\pi i}\int_{-\infty}^{\infty} \frac{ds\,\mathrm{Re}[L(s)]}{s-k-i0^+} = \frac{1}{\pi}\int_{-\infty}^{\infty} \frac{ds\,\mathrm{Im}[L(s)]}{s-k-i0^+}, \qquad k \in \overline{\mathbf{C}^+},$$

due to the fact [2, 5] that $L(k)$ is analytic in $\mathbf{C}^+$, continuous in $\overline{\mathbf{C}^+}$, and $o(1/k)$ as $k \to \infty$ in $\overline{\mathbf{C}^+}$. The reflectance contains no information related to $A(0)$ or $\cot\alpha$ given in (2.8). As a result, η is not uniquely determined and we have the corresponding one-parameter family for η with $\cot\alpha$ being the parameter. We also have the corresponding two-parameter family for A, where the parameters can be chosen, for example, as $A(0)$ and $A'(0)$, or as $A(l)$ and $A'(l)$.

(vi) Transfer function at the lips

From (4.5) we see that the data set $\{|\mathbf{T}(k,l)| : k \in \mathbf{R}^+, |A'(l)|/A(l)\}$ uniquely determines $\{|F_\alpha(k)| : k \in \mathbf{R}^+\}$ and hence also Q and η. On the other hand, there is the corresponding one-parameter set for A where $A(0)$ can be viewed as a parameter. In view of (4.4) and (4.5), we see that the data set $\{|\mathbf{T}(k,l)| : k \in \mathbf{R}^+, A(l), |A'(l)|\}$ uniquely determines each of Q, η, and A. Corresponding to the data set $\{|\mathbf{T}(k,l)| : k \in \mathbf{R}^+, A(l)\}$, in general there exists a one-parameter family for each of Q, η, and A, where $|A'(l)|$ can be chosen as the parameter. Corresponding to the data set $\{|\mathbf{T}(k,l)| : k \in \mathbf{R}^+\}$, in general there exists a two-parameter family for each of Q, η, and A, where $A(l)$ and $|A'(l)|$ can be chosen as the parameters. Note that we assume that $A(l)$ and $|A'(l)|$ do not change with k, and hence they are constants. As indicated in (i), they can be obtained via (6.1) and (6.2) by measuring the absolute value of the impedance at the lips at two different frequencies.

7. Transfer function and nonuniqueness

Let the data set $\{|\mathbf{T}(k,l)| : k \in \mathbf{R}^+, \theta\}$ with $\theta := |A'(l)|/[2A(l)]$ correspond to the potential Q, the boundary parameter $\cot\alpha$, the Jost function F_α, the normalized radius η, and the area A. From Section 6, we know that all these quantities are uniquely determined with the exception of A, which is determined up to the multiplicative constant $A(0)$. Let the data set $\{|\mathbf{T}(k,l)| : k \in \mathbf{R}^+, \tilde{\theta}\}$ with $\tilde{\theta} := |\tilde{A}'(l)|/[2\tilde{A}(l)]$ correspond to the potential $\tilde{Q}$, the boundary parameter $\cot\tilde{\alpha}$, the Jost function $F_{\tilde\alpha}$, the normalized radius $\tilde\eta$, and the area $\tilde A$. We assume that $\theta \neq \tilde\theta$. In this section we show that we must have $Q \not\equiv \tilde Q$.

If we had $Q \equiv \tilde Q$, then from (4.5) we would get

$$\frac{|F_\alpha(k)|^2}{|F_{\tilde\alpha}(k)|^2} = \frac{k^2+\theta^2}{k^2+\tilde\theta^2}, \qquad k \in \mathbf{R}, \tag{7.1}$$

or equivalently, with the help of (2.13) we would have

$$\frac{(k+i\tilde\theta)\,F_\alpha(k)}{(k+i\theta)\,F_{\tilde\alpha}(k)} = \frac{(k-i\theta)\,F_{\tilde\alpha}(-k)}{(k-i\tilde\theta)\,F_\alpha(-k)}, \qquad k \in \mathbf{R}.$$

It is known [3, 5, 10, 12, 13] that F_α and $F_{\tilde\alpha}$ are analytic in $\mathbf{C}^+$, continuous in $\overline{\mathbf{C}^+}$, and satisfy the asymptotics given in (2.12). Furthermore, since there are no bound states, F_α and $F_{\tilde\alpha}$ are nonzero in $\overline{\mathbf{C}^+}\setminus\{0\}$, and each of them has either a simple zero at $k=0$ or is nonzero there. In fact, as seen from (7.1) we would have $F_\alpha(0)=0$ if and only if $\theta=0$, and similarly $F_{\tilde\alpha}(0)=0$ if and only if $\tilde\theta=0$. From Liouville's theorem it would follow that

$$\frac{(k+i\tilde\theta)\,F_\alpha(k)}{(k+i\theta)\,F_{\tilde\alpha}(k)} = 1, \qquad k \in \mathbf{C}. \tag{7.2}$$

From (3.18) of [3] we have

$$\operatorname{Re}\left[\frac{i\,F_{\tilde\alpha}(k)}{F_\alpha(k)}\right] = \frac{k\,(\cot\tilde\alpha-\cot\alpha)}{|F_\alpha(k)|^2}, \qquad k \in \mathbf{R}. \tag{7.3}$$

Using (7.2) in (7.3), with the help of (2.12) we would get

$$\theta-\tilde\theta = \cot\tilde\alpha - \cot\alpha,$$

$$|F_\alpha(k)|^2 = k^2+\theta^2, \qquad k \in \mathbf{R},$$

implying $F_\alpha(k) = k+i\theta$ for $k \in \overline{\mathbf{C}^+}$. The unique potential corresponding to this particular $F_\alpha(k)$ is given by $Q \equiv 0$ and the boundary parameter $\cot\alpha$ is given by $\cot\alpha = -\theta$. Similarly, we would also have $\cot\tilde\alpha = -\tilde\theta$, $F_{\tilde\alpha}(k) = k+i\tilde\theta$, $\tilde Q \equiv 0$. Then, using (3.4) we would get

$$\eta(x) = 1+\theta x, \quad \tilde\eta(x) = 1+\tilde\theta x, \qquad x \in (0,l),$$

$$A(x) = A(0)\,[1+\theta x]^2, \quad \tilde A(x) = \tilde A(0)\,[1+\tilde\theta x]^2, \qquad x \in (0,l), \tag{7.4}$$

where $A(0)$ and $\tilde A(0)$ are arbitrary positive constants. Putting $|A'(l)|/[2A(l)]$ for θ in (7.4), we would obtain $\theta = 0$, and similarly we would get $\tilde\theta = 0$. However, this contradicts our assumption $\theta \neq \tilde\theta$, and hence we must have $Q \not\equiv \tilde Q$.

References

[1] T. Aktosun, *Inverse scattering for vowel articulation with frequency-domain data,* Inverse Problems **21** (2005), 899–914.

[2] T. Aktosun and M. Klaus, *Chapter* 2.2.4, *Inverse theory: problem on the line,* in: E.R. Pike and P.C. Sabatier (eds.), *Scattering,* Academic Press, London, 2001, pp. 770–785.

[3] T. Aktosun, and R. Weder, *Inverse spectral-scattering problem with two sets of discrete spectra for the radial Schrödinger equation,* Inverse Problems **22** (2006), 89–114.

[4] R. Burridge, *The Gelfand-Levitan, the Marchenko, and the Gopinath-Sondhi integral equations of inverse scattering theory, regarded in the context of inverse impulse-response problems,* Wave Motion **2** (1980), 305–323.

[5] K. Chadan and P.C. Sabatier, *Inverse problems in quantum scattering theory,* 2nd ed., Springer, New York, 1989.

[6] G. Fant, *Acoustic theory of speech production,* Mouton, The Hague, 1970.

[7] J.L. Flanagan, *Speech analysis synthesis and perception,* 2nd ed., Springer, New York, 1972.

[8] B.J. Forbes, E.R. Pike, and D.B. Sharp, *The acoustical Klein-Gordon equation: The wave-mechanical step and barrier potential functions,* J. Acoust. Soc. Am. **114** (2003), 1291–1302.

[9] L. Gårding, *The inverse of vowel articulation,* Ark. Mat. **15** (1977), 63–86.

[10] I.M. Gel'fand and B.M. Levitan, *On the determination of a differential equation from its spectral function,* Am. Math. Soc. Transl. (ser. 2) **1** (1955), 253–304.

[11] B. Gopinath and M.M. Sondhi, *Determination of the shape of the human vocal tract shape from acoustical measurements,* Bell Sys. Tech. J. **49** (1970), 1195–1214.

[12] B.M. Levitan, *Inverse Sturm-Liouville problems,* VNU Science Press, Utrecht, 1987.

[13] V.A. Marchenko, *Sturm-Liouville operators and applications,* Birkhäuser, Basel, 1986.

[14] J.R. McLaughlin, *Analytical methods for recovering coefficients in differential equations from spectral data,* SIAM Rev. **28** (1986), 53–72.

[15] P. Mermelstein, *Determination of the vocal-tract shape from measured formant frequencies,* J. Acoust. Soc. Am. **41** (1967), 1283–1294.

[16] Rakesh, *Characterization of transmission data for Webster's horn equation,* Inverse Problems **16** (2000), L9–L24.

[17] M.R. Schroeder, *Determination of the geometry of the human vocal tract by acoustic measurements,* J. Acoust. Soc. Am. **41** (1967), 1002–1010.

[18] J. Schroeter and M.M. Sondhi, *Techniques for estimating vocal-tract shapes from the speech signal,* IEEE Trans. Speech Audio Process. **2** (1994), 133–149.

[19] M.M. Sondhi, *A survey of the vocal tract inverse problem: theory, computations and experiments,* in: F. Santosa, Y.H. Pao, W.W. Symes, and C. Holland (eds.), *Inverse problems of acoustic and elastic waves,* SIAM, Philadelphia, 1984, pp. 1–19.

[20] M.M. Sondhi and B. Gopinath, *Determination of vocal-tract shape from impulse response at the lips,* J. Acoust. Soc. Am. **49** (1971), 1867–1873.

[21] M.M. Sondhi and J.R. Resnick, *The inverse problem for the vocal tract: numerical methods, acoustical experiments, and speech synthesis,* J. Acoust. Soc. Am. **73** (1983), 985–1002.

[22] K.N. Stevens, *Acoustic phonetics,* MIT Press, Cambridge, MA, 1998.

[23] W.W. Symes, *On the relation between coefficient and boundary values for solutions of Webster's Horn equation,* SIAM J. Math. Anal. **17** (1986), 1400–1420.

Tuncay Aktosun
Department of Mathematics
University of Texas at Arlington
Arlington, TX 76019-0408, USA
e-mail: `aktosun@uta.edu`

Operator Theory:
Advances and Applications, Vol. 171, 17–35

Positivity and the Existence of Unitary Dilations of Commuting Contractions

J. Robert Archer

Abstract. The central result of this paper is a method of characterizing those commuting tuples of operators that have a unitary dilation, in terms of the existence of a positive map with certain properties. Although this positivity condition is not necessarily easy to check given a concrete example, it can be used to find practical tests in some circumstances. As an application, we extend a dilation theorem of Sz.-Nagy and Foiaş concerning regular dilations to a more general setting.

1. Introduction

The Sz.-Nagy dilation theorem is a seminal result in the theory of contractions on Hilbert space. It states that every contraction has a unitary dilation. An elegant generalization was given by Andô who proved that every pair of commuting contractions has a unitary dilation. It is somewhat surprising that this phenomenon does not generalize further: Parrott gave an example of three commuting contractions that do not have a unitary dilation. This raises the question, when does a tuple of commuting operators have a unitary dilation?

After giving some background to the theory of unitary dilations of commuting operators and related topics, we present and analyse one answer to this question that employs a positive kernel condition. We use this to give a proof of Andô's theorem. Then we turn to the notion of regular unitary dilations with the ultimate aim of presenting an extension to a theorem of Sz.-Nagy and Foiaş.

Before getting to the theory proper, we start by introducing some mainly standard notation. Let $\mathbb{Z}^d$ and $\mathbb{Z}_+^d$, for d a natural number, denote the Cartesian product of d copies of the integers $\mathbb{Z}$ and the nonnegative integers $\mathbb{Z}_+$ respectively. Take $\mathbb{Z}^d$ to be an additive group in the obvious way, and define a partial order on $\mathbb{Z}^d$ by letting $\mathbb{Z}_+^d$ be the cone of positive elements, i.e., $\alpha \leq \beta \iff \beta - \alpha \in \mathbb{Z}_+^d$. Where there is no ambiguity in doing so, we refer to the tuple of consisting of all zeroes as 0. If α is in $\mathbb{Z}^d$, define α_j to be the jth component of α, so that

$\alpha = (\alpha_1, \ldots, \alpha_d)$. For $\alpha, \beta \in \mathbb{Z}^d$, define

$$\alpha \wedge \beta = (\min\{\alpha_1, \beta_1\}, \ldots, \min\{\alpha_d, \beta_d\}),$$
$$\alpha^- = -(\alpha \wedge 0) \quad \text{and} \quad \alpha^+ = (-\alpha)^-.$$

Note that α^-, α^+ are in $\mathbb{Z}_+^d$ and $\alpha = \alpha^+ - \alpha^-$.

Uppercase script letters, such as $\mathcal{H}$ or $\mathcal{K}$, are Hilbert spaces (over the complex field $\mathbb{C}$) unless stated otherwise. We use the word *subspace* to mean a closed linear manifold. We denote by $B(\mathcal{H}, \mathcal{K})$ the space of operators, i.e., bounded linear transformations, from $\mathcal{H}$ to $\mathcal{K}$, and write $B(\mathcal{H})$ for $B(\mathcal{H}, \mathcal{H})$.

The set of (pairwise) commuting d-tuples of operators from $B(\mathcal{H})$ we denote by $B^d(\mathcal{H})$. Let $\mathbf{T}$ be in $B^d(\mathcal{H})$, then $\mathbf{T}$ is said to be *unitary* if it consists of unitary operators. Likewise for *isometric*, *coisometric*, and *contractive*. We let $T_1, \ldots, T_d \in B(\mathcal{H})$ stand for the individual components of $\mathbf{T}$, so that then $\mathbf{T} = (T_1, \ldots, T_d)$. Note here that we implicitly associate the bold and not-bold versions of a symbol, with the former being used for the d-tuple itself, and the latter used in denoting associated operators from $B(\mathcal{H})$. Continuing with this convention, we let $T^\alpha \in B(\mathcal{H})$, $\alpha \in \mathbb{Z}_+^d$, be defined by

$$T^\alpha = T_1^{\alpha_1} \cdots T_d^{\alpha_d}.$$

If the T_j are all invertible then $T_j T_k^{-1} = T_k^{-1} T_k T_j T_k^{-1} = T_k^{-1} T_j T_k T_k^{-1} = T_k^{-1} T_j$, for all $j, k = 1, \ldots, d$, and hence it is natural to extend this definition to all $\alpha \in \mathbb{Z}^d$. Similarly, we let z^α stand for the function $(z_1, \ldots, z_d) \mapsto z_1^{\alpha_1} \cdots z_d^{\alpha_d}$ in $C(\mathbb{T}^d)$, the continuous $\mathbb{C}$-valued functions on the d-torus $\mathbb{T}^d$. This definition also applies for all $\alpha \in \mathbb{Z}^d$.

Definition 1.1. Let $\mathbf{T}$ be in $B^d(\mathcal{H})$ and $\mathbf{W}$ be in $B^d(\mathcal{K})$ where $\mathcal{K}$ contains $\mathcal{H}$ as a subspace. Denote by $P_{\mathcal{H}}$ the orthogonal projection $\mathcal{K} \to \mathcal{H}$. We say that $\mathbf{W}$ is a *weak dilation* of $\mathbf{T}$ if

$$T_j = P_{\mathcal{H}} W_j|_{\mathcal{H}}, \quad \text{for all } j = 1, \ldots, d.$$

We say $\mathbf{W}$ is a *dilation of* $\mathbf{T}$ if

$$T^\alpha = P_{\mathcal{H}} W^\alpha|_{\mathcal{H}}, \quad \text{for all } \alpha \in \mathbb{Z}_+^d.$$

By making the obvious identification of single operators with 1-tuples, this definition extends to the classical case of single operators. Historically, what we call a weak dilation was sometimes called simply a 'dilation', and the stronger notion of dilation this defines was termed 'power dilation'. Our definition agrees with the standard meaning in contemporary usage.

A necessary and sufficient condition for a weak dilation $\mathbf{W}$ to be a dilation is that $\mathcal{H}$ be 'semi-invariant' for $\mathbf{W}$, i.e., the difference of two invariant subspaces. The concept of a semi-invariant subspace, and the single variable version of this very useful result, are due to Sarason [9]. We give a proof of a multivariable version below.

Definition 1.2. Let $\mathbf{W}$ be in $B^d(\mathcal{K})$, and suppose we have subspaces $\mathcal{H} \subseteq \mathcal{K}$ and $\mathcal{M} \subseteq \mathcal{K} \ominus \mathcal{H}$. We say $\mathcal{H}$ is *semi-invariant for* $\mathbf{W}$ *with respect to* $\mathcal{M}$ if both $\mathcal{M}$ and $\mathcal{M} \oplus \mathcal{H}$ are invariant for $\mathbf{W}$.

Lemma 1.3. *The following are equivalent.*

(a) $\mathbf{W}$ *is a dilation of* $\mathbf{T}$.

(b) $\mathbf{W}$ *is a weak dilation of* $\mathbf{T}$ *and* $\mathcal{H}$ *is semi-invariant for* $\mathbf{W}$ *with respect to some subspace* $\mathcal{M} \subseteq \mathcal{K} \ominus \mathcal{H}$.

(c) *There exists a subspace* $\mathcal{M} \subseteq \mathcal{K} \ominus \mathcal{H}$ *such that each* W_j *has the form*

$$\begin{pmatrix} * & * & * \\ 0 & T_j & * \\ 0 & 0 & * \end{pmatrix}$$

with respect to the decomposition $\mathcal{M} \oplus \mathcal{H} \oplus (\mathcal{K} \ominus (\mathcal{M} \oplus \mathcal{H}))$.

In this case, we can always choose $\mathcal{M}$ *to be*

$$\overline{\bigvee}\{W^\alpha \mathcal{H} : \alpha \in \mathbb{Z}_+^d\} \ominus \mathcal{H}. \tag{1.4}$$

Proof. (b)$\Rightarrow$(c): This is straightforward.

(c)$\Rightarrow$(a): It is easy to verify that W^α, $\alpha \in \mathbb{Z}_+^d$, has the form

$$\begin{pmatrix} * & * & * \\ 0 & T^\alpha & * \\ 0 & 0 & * \end{pmatrix}$$

and hence $P_\mathcal{H} W^\alpha|_\mathcal{H} = T^\alpha$.

(a)$\Rightarrow$(b): Let $\mathcal{M}$ be as in (1.4) and put $\mathcal{N} = \mathcal{M} \oplus \mathcal{H}$. To complete the proof we need to show that $\mathcal{M}$ and $\mathcal{N}$ are invariant for $\mathbf{W}$. For $\mathcal{N}$ this is obvious. For $\mathcal{M}$, consider $f \in \mathcal{M} \subseteq \mathcal{N}$. For each $j = 1, \dots, d$, we already have $W_j f \in \mathcal{N}$ so it only remains to show $W_j f \in \mathcal{K} \ominus \mathcal{H}$.

Given $\varepsilon > 0$, choose $h_\alpha \in \mathcal{H}$, $\alpha \in \mathbb{Z}_+^d$, such that $\|f - \sum W^\alpha h_\alpha\| < \varepsilon$. Then

$$\begin{aligned}
\left\|P_\mathcal{H} W_j \sum W^\alpha h_\alpha\right\| &= \left\|T_j \sum T^\alpha h_\alpha\right\| \\
&= \left\|P_\mathcal{H} W_j P_\mathcal{H} \sum W^\alpha h_\alpha\right\| \\
&\le \left\|W_j\right\| \left\|P_\mathcal{H} \sum W^\alpha h_\alpha\right\| \\
&= \left\|W_j\right\| \left\|P_\mathcal{H} f - P_\mathcal{H} \sum W_\alpha h_\alpha\right\| \\
&\le \left\|W_j\right\| \varepsilon
\end{aligned}$$

and therefore $\|P_\mathcal{H} W_j f\| \le \|P_\mathcal{H} W_j f - P_\mathcal{H} W_j \sum W^\alpha h_\alpha\| + \|P_\mathcal{H} W_j \sum W^\alpha h_\alpha\| \le 2\|W_j\|\varepsilon$. As $\varepsilon > 0$ was arbitrary, it follows that $P_\mathcal{H} W_j f = 0$, or equivalently that $W_j f$ is in $\mathcal{K} \ominus \mathcal{H}$, as required. □

If $\mathbf{T}$ has a unitary dilation then each T_j must be a contraction operator, i.e., $\|T_j\| \le 1$.

Theorem 1.5 (Sz.-Nagy Dilation Theorem). *Every contraction on Hilbert space has a unitary dilation.*

This was first proved in [12]. Sz.-Nagy later gave another proof [13] using of the theory of positive functions on $\mathbb{Z}$. In the next section, we recover the idea behind this method as the $d = 1$ case of the characterization of commuting d-tuples that have unitary dilations.

There is another proof of this theorem, due to Schäffer [10], which is based on a 'geometric' approach, i.e., the unitary dilation is explicitly constructed as an operator matrix. Another achievement of geometric techniques is the following delightful result [1].

Theorem 1.6 (Andô's Dilation Theorem). *Every pair of commuting contractions has a unitary dilation.*

We give a proof of this theorem later. Somewhat surprisingly, the analogue for three or more commuting contractions is not true in general. Examples of three commuting contractions that do not have unitary dilations were first found by Parrott [7].

Note also that the notions of unitary dilation and coisometric extension are closely related. It is well known that a commuting tuple of operators has a unitary dilation if and only if it can be extended to a commuting tuple of coisometries.

2. Characterizing tuples with unitary dilations

Definition 2.1. Suppose $\mathbf{T}$ is in $B^d(\mathcal{H})$ and let K be a map $\mathbb{Z}^d \to B(\mathcal{H})$. We say K is a *kernel for* $\mathbf{T}$ if

$$K(\alpha) = T^\alpha, \quad \text{for all } \alpha \in \mathbb{Z}^d_+.$$

We say K is *positive* if

$$\sum_{\alpha,\beta\in\mathbb{Z}^d_+} \langle K(\alpha-\beta)x(\beta), x(\alpha)\rangle \geq 0,$$

for all $x : \mathbb{Z}^d_+ \to \mathcal{H}$ that are finitely nonzero. This is equivalent to saying the operator matrix $\big(K(\alpha-\beta)\big)_{\alpha,\beta\in F}$ is positive for every finite subset F of $\mathbb{Z}^d_+$. (Some authors use the phrase 'positive definite' to describe this property, and some use 'positive semi-definite'.)

Theorem 2.2. *If K is a map $\mathbb{Z}^d \to B(\mathcal{H})$, then K is positive if and only if there exists $\mathcal{K} \supseteq \mathcal{H}$ and $\mathbf{U} \in B^d(\mathcal{K})$ consisting of unitary operators such that*

$$P_\mathcal{H} U^\alpha|_\mathcal{H} = K(\alpha), \quad \textit{for all } \alpha \in \mathbb{Z}^d. \tag{2.3}$$

Later we shall give two proofs of this result. The following is an immediate corollary.

Corollary 2.4. *Let $\mathbf{T}$ be in $B^d(\mathcal{H})$, then there exists a positive kernel for $\mathbf{T}$ if and only if there exists a unitary dilation of $\mathbf{T}$.*

If K is positive then $K(-\alpha) = K(\alpha)^*$, $\alpha \in \mathbb{Z}^d$. Thus for $d = 1$, if a kernel K for T is to be a positive then it is completely determined by T, and its positivity is equivalent to the positivity of the infinite operator valued Toeplitz matrix

$$\begin{pmatrix} 1 & T^* & T^{*2} & T^{*3} & \cdots \\ T & 1 & T^* & T^{*2} & \\ T^2 & T & 1 & T^* & \ddots \\ T^3 & T^2 & T & 1 & \ddots \\ \vdots & & \ddots & \ddots & \ddots \end{pmatrix}.$$

This in turn is positive if and only if the finite matrices

$$\begin{pmatrix} 1 & T^* & \cdots & T^{n*} \\ T & 1 & \ddots & \vdots \\ \vdots & \ddots & \ddots & T^* \\ T^n & \cdots & T & 1 \end{pmatrix}, \quad n = 0, 1, 2, \dots,$$

are all positive.

It is straightforward to verify that these matrices can be factored as

$$\begin{pmatrix} 1 & 0 & \cdots & 0 \\ T & 1 & \ddots & \vdots \\ \vdots & \ddots & \ddots & 0 \\ T^n & \cdots & T & 1 \end{pmatrix} \begin{pmatrix} 1 & 0 & \cdots & 0 \\ 0 & 1-TT^* & \ddots & \vdots \\ \vdots & \ddots & \ddots & 0 \\ 0 & \cdots & 0 & 1-TT^* \end{pmatrix} \begin{pmatrix} 1 & T^* & \cdots & T^{n*} \\ 0 & 1 & \ddots & \vdots \\ \vdots & \ddots & \ddots & T^* \\ 0 & \cdots & 0 & 1 \end{pmatrix}$$

and therefore K is positive if and only if T is a contraction. Taken together with Corollary 2.4, this proves the Sz.-Nagy dilation theorem.

In the case $d \geq 2$ we have an operator valued Toeplitz matrix where each operator is itself an operator valued Toeplitz matrix, and so on, some of the entries of which are not determined by $\mathbf{T}$. We will say something about the nature of these entries in the next section.

We now turn our attention to proving Theorem 2.2. Our first approach is based on the theory of completely positive maps.

Definition 2.5. Let $\mathcal{A}$ be C^*-algebra with unit $1_{\mathcal{A}}$. A linear manifold $\mathcal{S}$ of $\mathcal{A}$ is called an *operator system* if it is unital ($1_{\mathcal{A}} \in \mathcal{S}$) and self-adjoint ($a \in \mathcal{S} \Rightarrow a^* \in \mathcal{S}$). Note that if $\mathcal{S}$ is an operator system then so is $M_n(\mathcal{S})$, the n-by-n matrices with entries in $\mathcal{S}$, regarded as a subset of the C^*-algebra $M_n(\mathcal{A})$.

Let $\mathcal{S}$ be an operator system, $\mathcal{B}$ a C^*-algebra, and $\varphi : \mathcal{S} \to \mathcal{B}$ a linear map. We say φ is *unital* if $\mathcal{B}$ has unit $1_{\mathcal{B}}$ and $\varphi(1_{\mathcal{A}}) = 1_{\mathcal{B}}$. We say φ is *positive* if it maps positive elements to positive elements.

Associate with φ the collection of maps

$$\varphi_n : M_n(\mathcal{S}) \to M_n(\mathcal{B}), \quad n = 1, 2, \ldots,$$

defined by

$$\varphi_n\big((a_{j,k})\big) = \big(\varphi(a_{j,k})\big).$$

We say φ is *completely positive* if the φ_n are all positive.

It is an elementary fact that positive maps are necessarily bounded.

Proposition 2.6. *Every positive map $\varphi : \mathcal{S} \to \mathcal{B}$ is bounded, with $\|\varphi\| \leq 2\|\varphi(1)\|$.*

The notion of completely positive maps is due to Stinespring [11]. A key result in the theory is his dilation theorem.

Theorem 2.7 (Stinespring's Dilation Theorem). *Let $\mathcal{A}$ be C^*-algebra with unit. If $\varphi : \mathcal{A} \to B(\mathcal{H})$ is unital and completely positive, then there exists $\mathcal{K} \supseteq \mathcal{H}$ and a unital $*$-homomorphism $\pi : \mathcal{A} \to B(\mathcal{K})$ such that*

$$\varphi(a) = P_{\mathcal{H}}\pi(a)|_{\mathcal{H}} \quad \text{for all } a \in \mathcal{A}.$$

The following is another fact established by Stinespring.

Theorem 2.8. *Every positive map on $C(\Omega)$, the continuous $\mathbb{C}$-valued functions on a compact Hausdorff space Ω, is completely positive.*

An introduction to the subject of complete positivity can be found in [8]. These last three results are given there as Proposition 2.1, and Theorems 4.1 and 3.11. Note also the commonality between the approach to proving the Sz.-Nagy dilation theorem demonstrated there in Theorems 2.6 and 4.3, and our first proof of Theorem 2.2 below, in the case $d = 1$.

Let $\mathcal{P}_d$ denote the space of the trigonometric polynomials in d-variables with coefficients in $\mathbb{C}$. We think of $\mathcal{P}_d$ as a subset of $C(\mathbb{T}^d)$; in this way $\mathcal{P}_d$ is an operator system. The following result, which generalizes one aspect of the Fejér-Riesz theorem on the factorization of nonnegative trigonometric polynomials in one variable, is an easy consequence of Corollary 5.2 in [4].

Lemma 2.9. *If $p \in \mathcal{P}_d$ is strictly positive (i.e., $p(z) > 0$ for all $z \in \mathbb{T}^d$) then there exists $n \geq 1$ and polynomials q_j, $j = 1, \ldots, n$, such that $p = \sum_{j=1}^{n} q_j q_j^*$.*

We can now turn to the first proof of Theorem 2.2.

Proof of Theorem 2.2. If $\mathbf{U}$ is unitary then it is straightforward to verify that the map K defined by (2.3) is positive.

Conversely, suppose that $K : \mathbb{Z}^d \to B(\mathcal{H})$ is positive. Consider the map $\varphi_K : \mathcal{P}_d \to B(\mathcal{H})$ defined by setting

$$\varphi_K(z^\alpha) = K(\alpha), \quad \alpha \in \mathbb{Z}^d,$$

and extending linearly. Note that φ_K is unital.

We shall show that φ_K is positive. Let $p \in \mathcal{P}_d$ be strictly positive and use Lemma 2.9 to write $p = \sum_{j=1}^{n} q_j q_j^*$, where

$$q_j(z) = \sum_{\alpha \in \mathbb{Z}_+^d} \lambda_{j,\alpha} z^\alpha$$

for some $\lambda_{j,\alpha} \in \mathbb{C}$. Then

$$p(z) = \sum_{j=1}^{n} \sum_{\alpha,\beta \in \mathbb{Z}_+^d} \lambda_{j,\alpha} \lambda_{j,\beta}^* z^{\alpha-\beta},$$

and so

$$\varphi_K(p) = \sum_j \sum_{\alpha,\beta} \lambda_{j,\alpha} \lambda_{j,\beta}^* K(\alpha - \beta).$$

Let f be in $\mathcal{H}$ and define $x_j : \mathbb{Z}_+^d \to \mathcal{H}$, $j = 1, \ldots, n$, by $x_j(\alpha) = \lambda_{j,\alpha}^* f$, then

$$\begin{aligned} \langle \varphi_K(p) f, f \rangle &= \sum_j \sum_{\alpha,\beta} \langle K(\alpha - \beta) \lambda_{j,\beta}^* f, \lambda_{j,\alpha}^* f \rangle \\ &= \sum_j \Big[\sum_{\alpha,\beta} \langle K(\alpha - \beta) x_j(\beta), x_j(\alpha) \rangle \Big] \\ &\geq 0, \end{aligned}$$

since the term in the square brackets is positive for each j. Hence $\varphi_K(p)$ is positive. Since any positive polynomial in $\mathcal{P}_d$ is the limit of strictly positive polynomials that all lie in some finite dimensional subspace, and φ_K restricted to such subspaces is obviously continuous, it follows that φ_K is positive on the whole of $\mathcal{P}_d$.

By Proposition 2.6, the positivity of φ_K implies boundedness, and moreover $\mathcal{P}_d$ is dense in $C(\mathbb{T}^d)$, so φ_K extends uniquely to a continuous positive function on the whole of $C(\mathbb{T}^d)$, which we also denote by φ_K. By Theorem 2.8, φ_K is completely positive, so we can apply Stinespring's dilation theorem to get a unital $*$-homomorphism $\pi_K : C(\mathbb{T}^d) \to B(\mathcal{K})$, such that

$$\varphi_K(f) = P_{\mathcal{H}} \pi_K(f)|_{\mathcal{H}}, \quad f \in C(\mathbb{T}^d).$$

Put $U_j = \pi_K(z_j)$, $j = 1, \ldots, d$, then $\mathbf{U} = (U_1, \ldots, U_d)$ is a commuting d-tuple of unitary operators, as can easily be verified. Furthermore, for all $\alpha \in \mathbb{Z}_+^d$,

$$K(\alpha) = \varphi_K(z^\alpha) = P_{\mathcal{H}} \pi_K(z^\alpha)|_{\mathcal{H}} = P_{\mathcal{H}} U^\alpha|_{\mathcal{H}}. \qquad \square$$

An alternative proof is possible, based on the dilation theorem of Naimark [6] that is stated below as Theorem 2.11.

Let G be an additive abelian group.

Definition 2.10. A map $U : G \to B(\mathcal{K})$ is called a *unitary representation* of G on $\mathcal{K}$ if $U(g)$ is unitary for all $g \in G$, $U(0) = 1$, and $U(g+h) = U(g)U(h)$ for all $g, h \in G$.

We say that a map $K : G \to B(\mathcal{H})$ is *positive* if, for every finite subset F of G, the operator matrix $\big(K(g-h)\big)_{g,h\in F}$ is positive. This agrees with our previous definition in the case $G = \mathbb{Z}^d$.

Theorem 2.11. *Suppose $K : G \to B(\mathcal{H})$ has $K(0) = 1$, then K is positive if and only if there exists $\mathcal{K} \supseteq \mathcal{H}$ and a unitary representation U of G on $\mathcal{K}$ such that $K(g) = P_{\mathcal{H}}U(g)|_{\mathcal{H}}$ for all $g \in G$.*

We can use this to prove Theorem 2.2: put $G = \mathbb{Z}^d$ and note the correspondence between the unitary representations U of G on $B(\mathcal{K})$ and the $\mathbf{U} \in B^d(\mathcal{K})$, given by $U_j = U(e_j)$ and $U(\alpha) = U^\alpha$, where $e_j \in \mathbb{Z}^d$ is the tuple with jth-entry equal to 1 and other entries equal to 0.

3. Defect operators

There is a relationship between those entries of a positive kernel for $\mathbf{T}$ that are not determined by $\mathbf{T}$, and defect operators for powers of $\mathbf{T}$.

Definition 3.1. Let T be in $B(\mathcal{H})$. We say that $D \in B(\mathcal{D}, \mathcal{H})$, for some auxiliary Hilbert space $\mathcal{D}$, is a *defect operator for T* if

$$DD^* = 1 - TT^*.$$

It is easy to see that there exist defect operators for T if and only if T is a contraction. Note that our definition of the term 'defect operator' is nonstandard. The usual requirement is that $D^*D = 1 - T^*T$, i.e., that D^* is a defect operator for T^* in the above sense. The usual meaning is less convenient to work with for our purposes.

Let us recall a well-known result due to Douglas [3], which we state as follows.

Lemma 3.2 (Douglas' Lemma). *Let A be in $B(\mathcal{D}, \mathcal{H})$ and B be in $B(\mathcal{E}, \mathcal{H})$. Then*

$$AA^* \le BB^* \tag{3.3}$$

if and only if there exists a contraction $C \in B(\mathcal{D}, \mathcal{E})$ such that

$$A = BC.$$

The case where equality holds in (3.3) can be generalized to collections of operators.

Lemma 3.4. *Take Λ to be some index set. For each $\lambda \in \Lambda$, let A_λ be in $B(\mathcal{D}, \mathcal{H}_\lambda)$ and B_λ be in $B(\mathcal{E}, \mathcal{H}_\lambda)$. Define*

$$\mathcal{D}_0 = \overline{\bigvee_\lambda} \operatorname{ran} A_\lambda^* \quad \textit{and} \quad \mathcal{E}_0 = \overline{\bigvee_\lambda} \operatorname{ran} B_\lambda^*.$$

Then

$$A_\lambda A_\mu^* = B_\lambda B_\mu^*, \quad \textit{for all } \lambda, \mu \in \Lambda, \tag{3.5}$$

and

$$\dim \mathcal{D} \ominus \mathcal{D}_0 \ge \dim \mathcal{E} \ominus \mathcal{E}_0 \tag{3.6}$$

if and only if there exists a coisometry $C \in B(\mathcal{D}, \mathcal{E})$ such that

$$A_\lambda = B_\lambda C \quad \textit{for all } \lambda \in \Lambda. \tag{3.7}$$

Moreover, equality holds in (3.6) *if and only if C can be chosen to be unitary.*

Proof. Note that if F is a finite subset of Λ and f_λ is in $\mathcal{H}_\lambda$ for each $\lambda \in F$, then from (3.5) it follows that

$$\begin{aligned}\Big\| \sum_{\lambda \in F} B_\lambda^* f_\lambda \Big\|^2 &= \sum_{\lambda, \mu \in F} \langle B_\lambda^* f_\lambda, B_\mu^* f_\mu \rangle \\ &= \sum \langle A_\lambda^* f_\lambda, A_\mu^* f_\mu \rangle \\ &= \Big\| \sum A_\lambda^* f_\lambda \Big\|^2.\end{aligned}$$

Hence we may define an isometric linear transformation from $\bigvee_{\lambda \in \Lambda} \operatorname{ran} B_\lambda^*$ onto $\bigvee_{\lambda \in \Lambda} \operatorname{ran} A_\lambda^*$ by

$$\sum B_\lambda^* f_\lambda \mapsto \sum A_\lambda^* f_\lambda,$$

which can be extended to a unitary $U \in B(\mathcal{E}_0, \mathcal{D}_0)$. From (3.6) it follows that we can choose an isometry $V \in B(\mathcal{E}, \mathcal{D})$ that extends U by means of a direct sum. If equality holds we can choose V to be unitary. Take $C = V^*$ then we have $C^* B_\lambda^* = U B_\lambda^* = A_\lambda^*$, $\lambda \in \Lambda$, and so C satisfies (3.7).

In proving the converse, showing (3.5) is trivial, and (3.6) follows because $C^*(\mathcal{E} \ominus \mathcal{E}_0) \subseteq \mathcal{D} \ominus \mathcal{D}_0$ and $C(\mathcal{D} \ominus \mathcal{D}_0) \subseteq \mathcal{E} \ominus \mathcal{E}_0$, as can easily be verified. □

The following classical result about positive 2-by-2 operator matrices can be proved as a consequence of Douglas' lemma. The simple proof given here is modelled on that of Lemma 1.2 in [4].

Lemma 3.8. *Consider an operator T on $\mathcal{H} = \mathcal{H}_1 \oplus \mathcal{H}_2$ with the form*

$$\begin{pmatrix} AA^* & B^* \\ B & CC^* \end{pmatrix}$$

for some $A \in B(\mathcal{D}_1, \mathcal{H}_1)$, $B \in B(\mathcal{H}_1, \mathcal{H}_2)$, and $C \in B(\mathcal{D}_2, \mathcal{H}_2)$. Then T is positive if and only if there exists a contraction $G \in B(\mathcal{D}_1, \mathcal{D}_2)$ such that

$$B^* = AG^*C^*.$$

Proof. If there exists such a G then

$$T = \begin{pmatrix} A & 0 \\ CG & C \end{pmatrix} \begin{pmatrix} 1 & 0 \\ 0 & 1 - GG^* \end{pmatrix} \begin{pmatrix} A^* & G^*C^* \\ 0 & C^* \end{pmatrix},$$

which is clearly positive.

Assume, conversely, that T is positive. Then we may write $T = RR^*$ with $R \in B(\mathcal{H})$ of the form

$$\begin{pmatrix} R_1 \\ R_2 \end{pmatrix}$$

for some $R_1 \in B(\mathcal{H},\mathcal{H}_1)$ and $R_2 \in B(\mathcal{H},\mathcal{H}_2)$. Since $R_1R_1^* = AA^*$ and $R_2R_2^* = CC^*$, there exist contractions $G_1 \in B(\mathcal{H},\mathcal{D}_1)$ and $G_2 \in B(\mathcal{H},\mathcal{D}_2)$ such that $R_1 = AG_1$ and $R_2 = CG_2$. Since $B^* = R_1R_2^* = AG_1G_2^*C^*$, we may take $G = G_2G_1^* \in B(\mathcal{D}_1,\mathcal{D}_2)$ to be the required contraction. □

Our next result highlights the close connection between positive kernels and defect operators.

Theorem 3.9. *Let* $\mathbf{T}$ *be in* $B^d(\mathcal{H})$. *Suppose* $\mathcal{D}$ *is a Hilbert space and that* $\{D_\alpha : \alpha \in \mathbb{Z}_+^d\}$ *is a collection of operators* $\mathcal{D} \to \mathcal{H}$. *Then the following statements are equivalent.*

(a) *There exists a Hilbert space* $\mathcal{M}$ *and a unitary dilation* $\mathbf{U}$ *of* $\mathbf{T}$ *on* $\mathcal{M}\oplus\mathcal{H}\oplus\mathcal{D}$ *such that each* U^α *has the form*

$$\begin{pmatrix} * & * & * \\ 0 & T^\alpha & D_\alpha \\ 0 & 0 & * \end{pmatrix}, \qquad \alpha \in \mathbb{Z}_+^d.$$

(b) *There exists a commuting coisometric extension* $\mathbf{W}$ *of* $\mathbf{T}$ *on* $\mathcal{H}\oplus\mathcal{D}$ *such that each* W^α *has the form*

$$\begin{pmatrix} T^\alpha & D_\alpha \\ 0 & * \end{pmatrix}, \qquad \alpha \in \mathbb{Z}_+^d.$$

(c) *It is the case that*

$$D_0 = 0,$$

and

$$T^\alpha T^{\beta *} + D_\alpha D_\beta^* = T^{(\alpha-\beta)^+} T^{(\alpha-\beta)^- *} + D_{(\alpha-\beta)^+} D^*_{(\alpha-\beta)^-}, \tag{3.10}$$

for all $\alpha,\beta \in \mathbb{Z}_+^d$.

In this case, each D_α *is a defect operator for* T^α, $\alpha \in \mathbb{Z}_+^d$, *and the map* $K : \mathbb{Z}^d \to B(\mathcal{H})$ *given by*

$$K(\alpha) = T^{\alpha^+} T^{\alpha^- *} + D_{\alpha^+} D^*_{\alpha^-} \tag{3.11}$$

is a positive kernel for $\mathbf{T}$.

Proof. (a)⇒(b): Define $\mathbf{W}$ by setting $W_j = P_{\mathcal{H}\oplus\mathcal{D}} U_j|_{\mathcal{H}\oplus\mathcal{D}}$, $j = 1,\dots,n$. It is straightforward to verify that $\mathbf{W}$ has the required properties.

(b)⇒(c): Clearly $D_0 = P_{\mathcal{H}} 1|_{\mathcal{D}} = 0$. We also see that, for all $\alpha,\beta \in \mathbb{Z}_+^d$,

$$\begin{aligned} T^\alpha T^{\beta *} + D_\alpha D_\beta^* &= P_{\mathcal{H}} W^\alpha W^{\beta *}|_{\mathcal{H}} \\ &= P_{\mathcal{H}} W^{\alpha-(\alpha\wedge\beta)} W^{\beta-(\alpha\wedge\beta)*}|_{\mathcal{H}} \quad \text{(since } \mathbf{W} \text{ is coisometric)} \\ &= P_{\mathcal{H}} W^{(\alpha-\beta)^+} W^{(\alpha-\beta)^- *}|_{\mathcal{H}} \\ &= T^{(\alpha-\beta)^+} T^{(\alpha-\beta)^- *} + D_{(\alpha-\beta)^+} D^*_{(\alpha-\beta)^-}. \end{aligned}$$

(c)⇒(a): Consider K defined as in (3.11). Since $D_0 = 0$, we have K is a kernel for $\mathbf{T}$. Using (3.10) we get that

$$K(\alpha-\beta) = T^\alpha T^{\beta*} + D_\alpha D_\beta^*, \quad \text{for all } \alpha,\beta \in \mathbb{Z}_+^d.$$

It is straightforward to use this to show that K is positive. (By setting $\alpha = \beta$ in (c), it is obvious that D_α is a defect operator for T^α, $\alpha \in \mathbb{Z}_+^d$.)

By Theorem 2.2, there exists $\mathbf{U} \in B^d(\mathcal{K})$, consisting of unitaries, such that

$$P_\mathcal{H} U^\alpha|_\mathcal{H} = K(\alpha), \quad \alpha \in \mathbb{Z}^d. \tag{3.12}$$

Since K is a kernel for $\mathbf{T}$, clearly $\mathbf{U}$ is a dilation of $\mathbf{T}$.

Let $\mathcal{N}$ be the space given in (1.4) with $\mathbf{W}$ replaced by $\mathbf{U}^*$, then $\mathcal{N}$ and $\mathcal{N}\oplus\mathcal{H}$ are invariant for $\mathbf{U}^*$. Put $\mathcal{M} = \mathcal{K} \ominus (\mathcal{N}\oplus\mathcal{H})$ so that $\mathcal{K} = \mathcal{M}\oplus\mathcal{H}\oplus\mathcal{N}$. It is not hard to see that U^α, $\alpha \in \mathbb{Z}_+^d$, has the form

$$\begin{pmatrix} * & * & * \\ 0 & T^\alpha & X_\alpha \\ 0 & 0 & * \end{pmatrix}$$

with respect to this decomposition, for some $X_\alpha \in B(\mathcal{N},\mathcal{H})$. Note that $\mathcal{N} = \bigvee \operatorname{ran} X_\alpha^*$. By adjoining a Hilbert space to $\mathcal{N}$ and replacing each U_j by its direct sum with the identity on this space, we can assume without loss that

$$\dim\Big(\mathcal{N} \ominus \overline{\bigvee} \operatorname{ran} X_\alpha^*\Big) = \dim\Big(\mathcal{D} \ominus \overline{\bigvee} \operatorname{ran} D_\alpha^*\Big).$$

In particular, (3.12) still holds after making this substitution. Therefore $T^\alpha T^{\beta*} + X_\alpha X_\beta^* = P_\mathcal{H} U^\alpha U^{\beta*}|_\mathcal{H} = P_\mathcal{H} U^{\alpha-\beta}|_\mathcal{H} = K(\alpha-\beta) = T^\alpha T^{\beta*} + D_\alpha D_\beta^*$, and hence

$$X_\alpha X_\beta^* = D_\alpha D_\beta^*, \quad \text{for all } \alpha,\beta \in \mathbb{Z}_+^d.$$

Apply Lemma 3.4 to get a unitary $C \in B(\mathcal{N},\mathcal{D})$ such that

$$X_\alpha = D_\alpha C, \quad \alpha \in \mathbb{Z}_+^d.$$

Let $\widetilde{C} \in B(\mathcal{M}\oplus\mathcal{H}\oplus\mathcal{N}, \mathcal{M}\oplus\mathcal{H}\oplus\mathcal{D})$ be the unitary with matrix

$$\begin{pmatrix} 1 & 0 & 0 \\ 0 & 1 & 0 \\ 0 & 0 & C \end{pmatrix}.$$

It is straightforward to show that $(\widetilde{C}U_1\widetilde{C}^*, \ldots, \widetilde{C}U_d\widetilde{C}^*)$ is a unitary dilation of $\mathbf{T}$ whose powers have the required form. □

As the following proposition shows, for $d = 2$ it is enough to consider kernels truncated to the finite index set $\{\alpha \in \mathbb{Z}_+^2 : \alpha \le (1,1)\}$. Again, there is a relationship between these truncated kernels and defect operators.

Proposition 3.13. *Let T_1, T_2 be commuting contractions on $\mathcal{H}$ and suppose that $D_1, D_2 \in B(\mathcal{D}, \mathcal{H})$ are defect operators for T_1, T_2 respectively. The matrix*

$$\begin{pmatrix} 1 & T_1^* & T_2^* & T_2^*T_1^* \\ T_1 & 1 & T_1T_2^* + D_1D_2^* & T_2^* \\ T_2 & T_2T_1^* + D_2D_1^* & 1 & T_1^* \\ T_1T_2 & T_2 & T_1 & 1 \end{pmatrix} \tag{3.14}$$

is positive if and only if T_1, T_2 have commuting contractive extensions to $\mathcal{H} \oplus \mathcal{D}$ with matrices

$$\begin{pmatrix} T_1 & D_1 \\ 0 & C_1 \end{pmatrix} \quad \text{and} \quad \begin{pmatrix} T_2 & D_2 \\ 0 & C_2 \end{pmatrix}, \tag{3.15}$$

for some operators $C_1, C_2 \in B(\mathcal{D})$.

In this case we may choose the extensions such that $C_1C_2 = C_2C_1 = 0$. There also exist commuting coisometric extensions of the form

$$\begin{pmatrix} T_1 & D_1 & 0 \\ 0 & C_1 & * \\ 0 & 0 & * \end{pmatrix} \quad \text{and} \quad \begin{pmatrix} T_2 & D_2 & 0 \\ 0 & C_2 & * \\ 0 & 0 & * \end{pmatrix}. \tag{3.16}$$

Proof. We apply Lemma 3.8 to (3.14), taking $\mathcal{H}_1$ to be the first three copies of $\mathcal{H}$, and $\mathcal{H}_2$ to be the last copy. This tells us that the positivity of (3.14) is equivalent to the existence of an operator $E \in B(\mathcal{D}, \mathcal{H})$ such that

$$\begin{pmatrix} T_2^*T_1^* \\ T_2^* \\ T_1^* \end{pmatrix} = \begin{pmatrix} 1 & 0 \\ T_1 & D_1 \\ T_2 & D_2 \end{pmatrix} \begin{pmatrix} T_2^*T_1^* \\ E^* \end{pmatrix} (1)$$

and $EE^* \leq 1 - T_1T_2T_2^*T_1^*$. Firstly let us assume that (3.14) is positive and that we have fixed such an E.

We have $T_2^* = T_1T_2^*T_1^* + D_1E^*$ and hence

$$ED_1^* = T_2D_1D_1^*. \tag{3.17}$$

Let $A_1 = E - T_2D_1$ and $B_1 = D_2$ then

$$\begin{aligned} A_1A_1^* &= EE^* - ED_1^*T_2^* - T_2D_1E^* + T_2D_1D_1^*T_2^* \\ &= EE^* - T_2D_1D_1^*T_2^* \quad \text{(using (3.17))} \\ &\leq 1 - T_1T_2T_2^*T_1^* - T_2D_1D_1^*T_2^* \\ &= 1 - T_2T_2^* = B_1B_1^*. \end{aligned}$$

Thus there exists a contraction $C_1 \in B(\mathcal{D})$ such that $A_1 = B_1C_1$. We can, without loss, assume that $\ker C_1 \supseteq \ker A_1$ and $\overline{\operatorname{ran}} C_1 \subseteq \mathcal{D} \ominus \ker B_1$. Since (3.17) implies that $A_1D_1^* = 0$, we have that $\overline{\operatorname{ran}} D_1^* \subseteq \ker A_1 \subseteq \ker C_1$. In summary, we have shown that C_1 satisfies

$$\begin{aligned} T_2D_1 + D_2C_1 &= E, \\ C_1D_1^* &= 0, \\ \text{and} \quad \overline{\operatorname{ran}} C_1 &\subseteq \overline{\operatorname{ran}} D_2^*. \end{aligned}$$

Now, by swapping the roles of 1 and 2 in the previous paragraph, we get a contraction $C_2 \in B(\mathcal{D})$ satisfying

$$\begin{aligned} T_1 D_2 + D_1 C_2 &= E, \\ C_2 D_2^* &= 0, \\ \text{and} \quad \overline{\operatorname{ran}} C_2 &\subseteq \overline{\operatorname{ran}} D_1^*. \end{aligned}$$

Note that $\overline{\operatorname{ran}} C_2 \subseteq \overline{\operatorname{ran}} D_1^* \subseteq \ker C_1$ and so $C_1 C_2 = 0$. Similarly, $C_2 C_1 = 0$. Therefore, $C_1 C_2 = C_2 C_1$ and moreover $T_1 D_2 + D_1 C_2 = T_2 D_1 + D_2 C_1$, so the operators in (3.15) commute. To see that they are contractions, calculate

$$\begin{pmatrix} T_j & D_j \\ 0 & C_j \end{pmatrix} \begin{pmatrix} T_j^* & 0 \\ D_j^* & C_j^* \end{pmatrix} = \begin{pmatrix} 1 & 0 \\ 0 & C_j C_j^* \end{pmatrix} \leq \begin{pmatrix} 1 & 0 \\ 0 & 1 \end{pmatrix}, \qquad j = 1, 2.$$

Next, for the converse, use Andô's theorem to get commuting coisometries of the form in (3.16). This places us in the situation of Theorem 3.9(b). The positivity of (3.14) follows from the positivity of the kernel K in Theorem 3.9. □

We do not know if there is a 'finite' positivity condition along these lines that would suffice in general to characterize the existence of unitary dilations. Is the positivity of a kernel perhaps equivalent to the positivity of some finite truncated part in general, as it is when $d = 1, 2$?

We conclude this section by giving a proof of Andô's theorem based on explicitly constructing a collection of operators that satisfy Theorem 3.9(c) for $d = 2$. Despite the indirection, this is essentially similar to the geometric approach of Andô's original proof.

Lemma 3.18. *If T_1, T_2 are commuting contractions on $\mathcal{H}$, and $E_1, E_2 : \mathcal{H} \to \mathcal{H}$ are respective defect operators, then there exists a Hilbert space $\mathcal{J}$ containing $\mathcal{H}$ and unitary operators U_1, U_2 on $\mathcal{J} \oplus \mathcal{J}$ such that*

$$(T_2 E_1 P \quad E_2 P) U_1 = (T_1 E_2 P \quad E_1 P) U_2 \quad : \quad \mathcal{J} \oplus \mathcal{J} \to \mathcal{H}, \tag{3.19}$$

where P is the projection of $\mathcal{J}$ onto $\mathcal{H}$.

Proof. Take a Hilbert space $\mathcal{I}$ such that $\dim \mathcal{I} = \dim(\mathcal{H} \oplus \mathcal{I})$. Put $\mathcal{J} = \mathcal{H} \oplus \mathcal{I}$. Let $D, D' : \mathcal{J} \oplus \mathcal{J} \to \mathcal{H}$ be defined by $D = (T_2 E_1 P \quad E_2 P)$ and $D' = (T_1 E_2 P \quad E_1 P)$. Then $DD^* = 1 - T_1 T_2 T_2^* T_1^* = D' D'^*$ and, since $\mathcal{I} \subseteq \ker D, \ker D'$, we have $\dim \ker D = \dim \mathcal{J} = \dim \ker D'$. Hence, by Lemma 3.4, there exists U such that $D = D'U$. Put $U_1 = 1$ and $U_2 = U$. □

The dilation theorems of Sz.-Nagy and Andô are an immediate consequence of the following proposition and Theorem 3.9.

Proposition 3.20. *If $d = 1$ or 2 and $\mathbf{T}$ is in $B^d(\mathcal{H})$, then there exists a Hilbert space $\mathcal{D}$ and a collection of operators $D_\alpha : \mathcal{D} \to \mathcal{H}$, $\alpha \in \mathbb{Z}_+^d$, such that $D_0 = 0$ and*

$$T^\alpha T^{\beta *} + D_\alpha D_\beta^* = T^{\alpha - (\alpha \wedge \beta)} T^{\beta - (\alpha \wedge \beta) *} + D_{\alpha - (\alpha \wedge \beta)} D^*_{\beta - (\alpha \wedge \beta)}. \tag{3.21}$$

Proof. We shall only prove the more difficult case, namely when $d = 2$. Choose defect operators $E_1, E_2 : \mathcal{H} \to \mathcal{H}$ for our pair of contractions, and let $\mathcal{J}$, P, and U_1, U_2 be as in Lemma 3.18. Set $\mathcal{D} = \bigoplus_1^\infty (\mathcal{J} \oplus \mathcal{J})$. Extend U_1 to a unitary $U_{(1,0)}$ on $\mathcal{D}$ by setting $U_{(1,0)} = \bigoplus_1^\infty U_1$, and likewise extend U_2 to $U_{(0,1)}$. In the sequel we freely identify $\mathcal{D}$ with $\mathcal{J} \oplus \mathcal{D}$ with $(\mathcal{J} \oplus \mathcal{J}) \oplus \mathcal{D}$ via the isomorphisms that identify $\big((f_1 \oplus f_2) \oplus (f_3 \oplus f_4) \oplus \cdots\big)$ with $f_1 \oplus \big((f_2 \oplus f_3) \oplus (f_4 \oplus f_5) \oplus \cdots\big)$ with $(f_1 \oplus f_2) \oplus \big((f_3 \oplus f_4) \oplus \cdots\big)$, for $f_j \in \mathcal{J}$ satisfying $\sum \|f_j\|^2 < \infty$.

Set $D_{(0,0)} = 0$. We wish to define D_α for $\alpha > (0,0)$ so that

$$D_\alpha = \begin{cases} \big(T^{\alpha-(1,0)}E_1P \quad D_{\alpha-(1,0)}U^*_{(0,1)}\big)U_{(1,0)}, & \text{if } \alpha \geq (1,0); \\ \big(T^{\alpha-(0,1)}E_2P \quad D_{\alpha-(0,1)}U^*_{(1,0)}\big)U_{(0,1)}, & \text{if } \alpha \geq (0,1). \end{cases} \tag{3.22}$$

If we suppose, inductively, that we have chosen D_α with $|\alpha| \leq n$ so as to satisfy (3.22), then to extend to $|\alpha| = n+1$, we just define the new D_α according to (3.22). For this to make sense we need only check that if $\alpha \geq (1,1)$ the two possible definitions of D_α agree, which indeed they do as

$$\begin{aligned} &\big(T^{\alpha-(1,0)}E_1P \quad D_{\alpha-(1,0)}U^*_{(0,1)}\big)U_{(1,0)} \\ &\qquad = \big(T^{\alpha-(1,0)}E_1P \quad (T^{\alpha-(1,1)}E_2P \quad D_{\alpha-(1,1)}U^*_{(1,0)})U_{(0,1)}U^*_{(0,1)}\big)U_{(1,0)} \\ &\qquad = \big((T^{\alpha-(1,0)}E_1P \quad T^{\alpha-(1,1)}E_2P) \quad D_{\alpha-(1,1)}U^*_{(1,0)}\big)U_{(1,0)} \\ &\qquad = \big(T^{\alpha-(1,1)}(T_2E_1P \quad E_2P)U_1 \quad D_{\alpha-(1,1)}\big), \end{aligned}$$

and similarly

$$\begin{aligned} &\big(T^{\alpha-(0,1)}E_2P \quad D_{\alpha-(0,1)}U^*_{(1,0)}\big)U_{(0,1)} \\ &\qquad = \big(T^{\alpha-(1,1)}(T_1E_2P \quad E_1P)U_2 \quad D_{\alpha-(1,1)}\big), \end{aligned}$$

which agree because (3.19) holds.

It remains to show that these D_α satisfy (3.21). We prove this by induction on $|\alpha \wedge \beta|$. Clearly (3.21) holds when $|\alpha \wedge \beta| = 0$. Suppose it holds when $|\alpha \wedge \beta| = n$, and consider α, β with $|\alpha \wedge \beta| = n+1$. Choose $\delta \in \{(1,0),(0,1)\}$ with $\delta < \alpha \wedge \beta$. Noting that $(\alpha - \delta) \wedge (\beta - \delta) = (\alpha \wedge \beta) - \delta$, we then see

$$\begin{aligned} D_\alpha D^*_\beta &= (D_\alpha U^*_\delta)(D_\beta U^*_\delta)^* \\ &= \big(T^{\alpha-\delta}E_\delta P \quad D_{\alpha-\delta}U^*_{(1,1)-\delta}\big)\big(T^{\beta-\delta}E_\delta P \quad D_{\beta-\delta}U^*_{(1,1)-\delta}\big)^* \\ &= T^{\alpha-\delta}E_\delta E^*_\delta T^{\beta-\delta*} + D_{\alpha-\delta}D^*_{\beta-\delta} \\ &= T^{\alpha-\delta}(1 - T^\delta T^{\delta^*})T^{\beta-\delta*} \\ &\qquad - T^{\alpha-\delta}T^{\beta-\delta*} + T^{\alpha-(\alpha\wedge\beta)}T^{\beta-(\alpha\wedge\beta)*} + D_{\alpha-(\alpha\wedge\beta)}D^*_{\beta-(\alpha\wedge\beta)} \\ &= \ -T^\alpha T^{\beta*} + T^{\alpha-(\alpha\wedge\beta)}T^{\beta-(\alpha\wedge\beta)*} + D_{\alpha-(\alpha\wedge\beta)}D^*_{\beta-(\alpha\wedge\beta)}. \end{aligned}$$

This completes the proof. □

4. A theorem of Sz.-Nagy and Foiaş

Definition 4.1. A unitary dilation $\mathbf{U} \in B^d(\mathcal{K})$ of $\mathbf{T} \in B^d(\mathcal{H})$ is called *$*$-regular* (respectively *regular*) if

$$P_{\mathcal{H}} U^{\alpha}|_{\mathcal{H}} = T^{\alpha^+} T^{\alpha^- *} \quad \text{(respectively} = T^{\alpha^- *} T^{\alpha^+}) \quad \text{for all } \alpha \in \mathbb{Z}^d.$$

For $\mathbf{T} \in B^d(\mathcal{H})$, let us denote by $\mathbf{T}^*$ the tuple $(T_1^*, \ldots, T_d^*)$. Note that $\mathbf{X}$ is a $*$-regular unitary dilation of $\mathbf{T}$ if and only if $\mathbf{X}^*$ is a regular unitary dilation of $\mathbf{T}^*$. Although we are going to work mainly with $*$-regular dilations, most statements have equivalents for regular dilations that are easily obtained by taking adjoints. In view of Theorem 2.2, it is easy to see that a necessary and sufficient condition for $\mathbf{T}$ to have a $*$-regular unitary dilation is that the kernel K for $\mathbf{T}$ defined by $K(\alpha) = T^{\alpha^+} T^{\alpha^- *}$ is positive.

The concept of regular dilations was introduced by Brehmer [2]. He gave a simple characterization of those commuting tuples that have regular unitary dilations. We only have need of a special case: a commuting pair of contractions (T_1, T_2) has a $*$-regular unitary dilation if and only if

$$1 - T_1 T_1^* - T_2 T_2^* + T_1 T_2 T_2^* T_1^* \geq 0. \tag{4.2}$$

In passing let us mention that $*$-regular unitary dilations need not exist, even for pairs of commuting contractions. For an easy example of this, consider (T, T), $T = \left(\begin{smallmatrix} 0 & 0 \\ 1 & 0 \end{smallmatrix}\right)$, for which $1 - 2TT^* + T^2 T^{2*} = \left(\begin{smallmatrix} 1 & 0 \\ 0 & -1 \end{smallmatrix}\right) \not\geq 0$. For a simple case when such dilations do exist, note that if $\mathbf{T} = (T_1, T_2)$ is a row contraction, i.e., $T_1 T_1^* + T_2 T_2^* \leq 1$, then $\mathbf{T}$ has a $*$-regular unitary dilation.

More information on regular dilations can be found in [14]. The following theorem appears there in chapter 1 as Proposition 9.2.

Theorem 4.3. *Suppose that $\mathbf{T}' = (T_1, \ldots, T_{d-1}, T_d)$ is in $B^d(\mathcal{H})$ and that $\mathbf{T} = (T_1, \ldots, T_{d-1})$ has a $*$-regular (regular) unitary dilation.*

(a) *If T_d is a coisometry (isometry) then $\mathbf{T}'$ has a $*$-regular (regular) unitary dilation.*
(b) *If T_d is a contraction that doubly commutes with $\mathbf{T}$, i.e., T_d commutes with both $\mathbf{T}$ and $\mathbf{T}^*$, then $\mathbf{T}'$ has a $*$-regular (regular) unitary dilation.*

In this section we present an analogous result for arbitrary (i.e., not necessarily $*$-regular) unitary dilations. Our proof makes use of the positive kernel condition of Corollary 2.4. A key role is played by Banach limits, a well-known tool that we define as follows.

Definition 4.4. A *Banach limit* is a linear functional b on $\ell_\infty(\mathbb{C})$ (the space of all bounded sequences of complex numbers) satisfying these two properties.

(a) $b\big((\lambda_j)_{j=0}^\infty\big) = b\big((\lambda_{j+1})_{j=0}^\infty\big)$.
(b) $b\big((\lambda_j)\big)$ is contained in the closed convex hull of $\{\lambda_j : j = 0, 1, \ldots\}$.

It is a classical result that Banach limits exist, often proved as an application of the Hahn-Banach theorem.

Lemma 4.5. *There exists a Banach limit.*

For the remainder of this section, fix a Banach limit b. This generalizes the notion of a limit to arbitrary bounded sequences. If (λ_j) converges to λ then $b(\lambda_j) = \lambda$.

Suppose $(A_j)_{j=0}^{\infty}$ is a uniformly bounded sequence in $B(\mathcal{H})$. As

$$(f, g) \mapsto b\big(\langle A_j f, g\rangle\big)$$

is a bounded sesquilinear form on $\mathcal{H}$, there exists a unique operator $b(A_j)$ on $\mathcal{H}$ such that

$$\big\langle b(A_j)f, g\big\rangle = b\big(\langle A_j f, g\rangle\big) \quad \text{for all } f, g \in \mathcal{H}.$$

This last equation implies that $b(B_1 A_j B_2) = B_1 b(A_j) B_2$ for all $B_1, B_2 \in B(\mathcal{H})$. If A_j converges weakly to A then $b(A_j) = A$.

We now come to our main result of this section.

Theorem 4.6. *Suppose* $\mathbf{T}' = (T_1, \ldots, T_{d-1}, T_d)$ *is in* $B^d(\mathcal{H})$ *and* $\mathbf{T} = (T_1, \ldots, T_{d-1})$ *has a unitary dilation. If there exists* $D \in B(\mathcal{H})$ *such that* $DD^* = 1 - T_d T_d^*$ *and* D *commutes with* $\mathbf{T}$ *then* $\mathbf{T}'$ *has a unitary dilation.*

Proof. Firstly consider the case when T_d is a coisometry. There exists a positive kernel K for $\mathbf{T}$. Fix α in $\mathbb{Z}^{d-1}$. Define $K_j(\alpha) \in B(\mathcal{H})$, $j = 0, 1, \ldots,$ by

$$K_j(\alpha) = T_d^j K(\alpha) T_d^{j*}.$$

Note that $\|K_j(\alpha)\| \leq \|K(\alpha)\|$ for all j, and hence $b\big(K_j(\alpha)\big)$ exists. Call this $\widetilde{K}(\alpha)$.

In this way we define $\widetilde{K} : \mathbb{Z}^{d-1} \to B(\mathcal{H})$ that we claim is also a positive kernel for $\mathbf{T}$. To see this, note that

(a) if α is in $\mathbb{Z}_+^{d-1}$ then

$$\widetilde{K}(\alpha) = b\big(T_d^j K(\alpha) T_d^{j*}\big) = b\big(T^\alpha T_d^j T_d^{j*}\big) = b(T^\alpha) = T^\alpha;$$

and

(b)

$$\sum_{\alpha,\beta \in \mathbb{Z}_+^{d-1}} \langle \widetilde{K}(\alpha - \beta) x(\beta), x(\alpha)\rangle$$

$$= b\Big[\sum \langle K(\alpha - \beta) T_d^{j*} x(\beta), T_d^{j*} x(\alpha)\rangle\Big] \geq 0$$

(since the term in the square brackets is positive for each j).

Furthermore, we have, for all $\alpha \in \mathbb{Z}^{d-1}$,

$$T_d \widetilde{K}(\alpha) T_d^* = T_d b\big(K_j(\alpha)\big) T_d^* = b\big(T_d K_j(\alpha) T_d^*\big) = b\big(K_{j+1}(\alpha)\big) = \widetilde{K}(\alpha),$$

and hence, for all $k = 0, 1, \ldots,$

$$T_d^k \widetilde{K}(\alpha) T_d^{k*} = \widetilde{K}(\alpha).$$

As a notational convenience, given $\alpha' = (\alpha_1, \ldots, \alpha_{d-1}, \alpha_d) \in \mathbb{Z}^d$, we let α stand for $(\alpha_1, \ldots, \alpha_{d-1}) \in \mathbb{Z}^{d-1}$. Let $K' : \mathbb{Z}^d \to B(\mathcal{H})$ be defined by

$$K'(\alpha') = T_d^{\alpha_d^+} \widetilde{K}(\alpha) T_d^{\alpha_d^-*},$$

then

(a) if α' is in $\mathbb{Z}_+^d$ then

$$K'(\alpha') = T_d^{\alpha_d^+} \widetilde{K}(\alpha) T_d^{\alpha_d^-*} = T_d^{\alpha_d} T^{\alpha} = T'^{\alpha'};$$

and

(b)

$$\begin{aligned} K'(\alpha' - \beta') &= T_d^{(\alpha_d-\beta_d)^+} \widetilde{K}(\alpha - \beta) T_d^{(\alpha_d-\beta_d)^- *} \\ &= T_d^{(\alpha_d-\beta_d)^+} T_d^{\alpha_d \wedge \beta_d} \widetilde{K}(\alpha - \beta) T_d^{\alpha_d \wedge \beta_d *} T_d^{(\alpha_d-\beta_d)^- *} \\ &= T_d^{\alpha_d} \widetilde{K}(\alpha - \beta) T_d^{\beta_d *} \end{aligned}$$

so

$$\begin{aligned} &\sum_{\alpha', \beta' \in \mathbb{Z}_+^d} \langle K'(\alpha' - \beta') x(\beta'), x(\alpha') \rangle \\ &\qquad = \sum_{\alpha_d, \beta_d} \Big[\sum_{\alpha, \beta} \langle \widetilde{K}(\alpha - \beta) T_d^{\beta_d *} x(\beta'), T_d^{\alpha_d *} x(\alpha') \rangle \Big] \geq 0. \end{aligned}$$

Hence K' is a positive kernel for $\mathbf{T}'$, and so $\mathbf{T}'$ has a unitary dilation.

Now, for the general case, consider the d operators on $\bigoplus_1^\infty \mathcal{H}$ given by

$$\begin{pmatrix} T_j & & & 0 \\ & T_j & & \\ & & T_j & \\ 0 & & & \ddots \end{pmatrix}, \quad j = 1, \ldots, d-1,$$

and

$$\begin{pmatrix} T_d & D & & & 0 \\ & 0 & 1 & & \\ & & 0 & 1 & \\ 0 & & & \ddots & \ddots \end{pmatrix}.$$

By hypothesis, these are commuting contractions, the first $d-1$ of which have a unitary dilation, and the last is a coisometry. Hence, by the first part of the proof they have a unitary dilation, which is then also a unitary dilation of $\mathbf{T}'$. □

As our final topic, we prove a result that further generalizes the $d = 3$ case of the last theorem.

Lemma 4.7. *Let (T, X) be a pair of contractions in $B^2(\mathcal{H})$, and let $D : \mathcal{D} \to \mathcal{H}$ be a defect operator for X. Then (T, X) has a $*$-regular unitary dilation if and only if there exists a contraction $R \in B(\mathcal{D})$ such that*

$$TD = DR.$$

Proof. Let $D : \mathcal{D} \to \mathcal{H}$ be any defect operator for X, then

$$\begin{aligned}
&(T, X) \text{ has a } *\text{-regular dilation,}\\
\iff\ & 1 - TT^* - XX^* + TXX^*T^* \geq 0, \quad \text{(by (4.2))}\\
\iff\ & DD^* \geq TDD^*T^*,\\
\iff\ & \text{there exists a contraction } R \text{ such that } TD = DR;
\end{aligned}$$

where the last statement follows from Douglas' lemma. □

The next theorem now follows easily from Theorem 4.6. It was originally shown in [5] by a different method.

Theorem 4.8. *Suppose $\mathbf{T} = (T_1, T_2, T_3)$ is a commuting triple of operators on $\mathcal{H}$. If (T_1, T_3) and (T_2, T_3) each have a $*$-regular unitary dilation then $\mathbf{T}$ has a unitary dilation.*

Proof. Let $D_3 = (1 - T_3T_3^*)^{1/2}$. There exist contractions $R_j \in B(\mathcal{H})$ such that $T_jD_3 = D_3R_j$, $j = 1, 2$. Apply Theorem 4.6 to the operators on $\bigoplus_1^\infty \mathcal{H}$ given by

$$\begin{pmatrix} T_j & & & 0 \\ & R_j & & \\ & & R_j & \\ 0 & & & \ddots \end{pmatrix}, \quad j = 1, 2,$$

and

$$\begin{pmatrix} T_3 & D_3 & & & 0 \\ & 0 & 1 & & \\ & & 0 & 1 & \\ 0 & & & \ddots & \ddots \end{pmatrix}.$$

Note that we know here by Andô's theorem that the first two of these operators have a unitary dilation. □

The condition that T_3 has pairwise $*$-regular dilations with each of T_1, T_2 is strictly weaker than insisting that there exists a D that satisfies $DD^* = 1 - T_3T_3^*$ and commutes with T_1, T_2. Strictness can be seen by considering, for example, $T_1 = T_2 = \begin{pmatrix} 0 & 0 \\ 1/\sqrt{3} & 0 \end{pmatrix}$, $T_3 = \begin{pmatrix} 0 & 0 \\ \sqrt{2/3} & 0 \end{pmatrix}$, then (T_1, T_3) and (T_2, T_3) have $*$-regular unitary dilations since they are row contractions, however it is straightforward to show that there is no D that satisfies the latter conditions. Thus the last theorem is a genuine generalization of Theorem 4.6 in the case of three variables.

References

[1] T. Andô. On a pair of commutative contractions. *Acta Sci. Math. (Szeged)*, 24:88–90, 1963.

[2] S. Brehmer. Über vetauschbare Kontraktionen des Hilbertschen Raumes. *Acta Sci. Math. Szeged*, 22:106–111, 1961.

[3] R.G. Douglas. On majorization, factorization, and range inclusion of operators on Hilbert space. *Proc. Amer. Math. Soc.*, 17:413–415, 1966.

[4] Michael A. Dritschel. On factorization of trigonometric polynomials. *Integral Equations Operator Theory*, 49(1):11–42, 2004.

[5] Dumitru Gaşpar and Nicolae Suciu. On the intertwinings of regular dilations. *Ann. Polon. Math.*, 66:105–121, 1997. Volume dedicated to the memory of Włodzimierz Mlak.

[6] M. Naimark. Positive definite operator functions on a commutative group. *Bull. Acad. Sci. URSS Sér. Math.* [*Izvestia Akad. Nauk SSSR*], 7:237–244, 1943.

[7] Stephen Parrott. Unitary dilations for commuting contractions. *Pacific J. Math.*, 34:481–490, 1970.

[8] Vern Paulsen. *Completely bounded maps and operator algebras*, volume 78 of *Cambridge Studies in Advanced Mathematics*. Cambridge University Press, Cambridge, 2002.

[9] Donald Sarason. On spectral sets having connected complement. *Acta Sci. Math. (Szeged)*, 26:289–299, 1965.

[10] J. J. Schäffer. On unitary dilations of contractions. *Proc. Amer. Math. Soc.*, 6:322, 1955.

[11] W. Forrest Stinespring. Positive functions on C^*-algebras. *Proc. Amer. Math. Soc.*, 6:211–216, 1955.

[12] Béla Sz.-Nagy. Sur les contractions de l'espace de Hilbert. *Acta Sci. Math. Szeged*, 15:87–92, 1953.

[13] Béla Sz.-Nagy. Transformations de l'espace de Hilbert, fonctions de type positif sur un groupe. *Acta Sci. Math. Szeged*, 15:104–114, 1954.

[14] Béla Sz.-Nagy and Ciprian Foiaş. *Harmonic analysis of operators on Hilbert space*. Translated from the French and revised. North-Holland Publishing Co., Amsterdam, 1970.

J. Robert Archer
Department of Mathematics
University of Glasgow
University Gardens
Glasgow G12 8QW, UK
e-mail: r.archer@maths.gla.ac.uk

Operator Theory:
Advances and Applications, Vol. 171, 37–72

The Infinite-dimensional Continuous Time Kalman–Yakubovich–Popov Inequality

Damir Z. Arov and Olof J. Staffans

Abstract. We study the set M_Σ of all generalized positive self-adjoint solutions (that may be unbounded and have an unbounded inverse) of the KYP (Kalman–Yakubovich–Popov) inequality for a infinite-dimensional linear time-invariant system Σ in continuous time with scattering supply rate. It is shown that if M_Σ is nonempty, then the transfer function of Σ coincides with a Schur class function in some right half-plane. For a minimal system Σ the converse is also true. In this case the set of all $H \in M_\Sigma$ with the property that the system is still minimal when the original norm in the state space is replaced by the norm induced by H is shown to have a minimal and a maximal solution, which correspond to the available storage and the required supply, respectively. The notions of strong H-stability, H-$*$-stability and H-bistability are introduced and discussed. We show by an example that the various versions of H-stability depend crucially on the particular choice of $H \in M_\Sigma$. In this example, depending on the choice of the original realization, some or all $H \in M_\Sigma$ will be unbounded and/or have an unbounded inverse.

Keywords. Kalman–Yakubovich–Popov inequality, passive, available storage, required supply, bounded real lemma, pseudo-similarity, Cayley transform.

1. Introduction

Linear finite-dimensional time-invariant systems in continuous time are typically modelled by the equations

$$\begin{aligned} \dot{x}(t) = Ax(t) + Bu(t), \quad y(t) = Cx(t) + Du(t), \quad t \geq s, \\ x(s) = x_s, \end{aligned} \tag{1}$$

Damir Z. Arov thanks Åbo Akademi for its hospitality and the Academy of Finland for its financial support during his visits to Åbo in 2003–2005. He also gratefully acknowledges the partial financial support by the joint grant UM1-2567-OD-03 from the U.S. Civilian Research and Development Foundation (CRDF) and the Ukrainian Government. Olof J. Staffans gratefully acknowledges the financial support from the Academy of Finland, grant 203991.

on a triple of finite-dimensional vector spaces, namely, the *input* space $\mathcal{U}$, the *state* space $\mathcal{X}$, and the *output* space $\mathcal{Y}$. We have $u(t) \in \mathcal{U}$, $x(t) \in \mathcal{X}$ and $y(t) \in \mathcal{Y}$. We are interested in the case where, in addition to the dynamics described by (1), the components of the system satisfy an energy inequality. In this paper we shall use the *scattering supply rate*

$$j(u,y) = \|u\|^2 - \|y\|^2 = \left\langle \left[\begin{smallmatrix} u \\ y \end{smallmatrix}\right], \left[\begin{smallmatrix} 1_{\mathcal{U}} & 0 \\ 0 & -1_{\mathcal{Y}} \end{smallmatrix}\right] \left[\begin{smallmatrix} u \\ y \end{smallmatrix}\right] \right\rangle \tag{2}$$

and the *storage (or Lyapunov) function*

$$E_H(x) = \langle x, Hx \rangle, \tag{3}$$

where $H > 0$ (i.e., $E_H(x) > 0$ for $x \neq 0$). A system is *scattering H-passive* (or simply scattering passive if $H = 1_{\mathcal{X}}$) if for any admissible data $(x_0, u(\cdot))$ the solution of the system (1) satisfies the condition

$$\frac{\mathrm{d}}{\mathrm{d}t} E_H(x(t)) \leq j(u(t), y(t)) \text{ a.e. on } (s, \infty). \tag{4}$$

This inequality is often written in integrated form

$$E_H(x(t)) - E_H(x(s)) \leq \int_s^t j(u(v), y(v))\, dv, \quad s \leq t. \tag{5}$$

It is not difficult to see that the inequality (4) with supply rate (2) is equivalent to the inequality

$$2\Re\langle Ax + Bu, Hx \rangle + \|Cx + Du\|^2 \leq \|u\|^2, \; x \in \mathcal{X}, u \in \mathcal{U}, \tag{6}$$

which is usually rewritten in the form

$$\begin{bmatrix} HA + A^*H + C^*C & HB + C^*D \\ B^*H + D^*C & D^*D - 1_{\mathcal{U}} \end{bmatrix} \leq 0. \tag{7}$$

This is the standard KYP (Kalman–Yakubovich–Popov) inequality for continuous time and scattering supply rate. If $R := 1_{\mathcal{U}} - D^*D > 0$, then (7) is equivalent to the Riccati inequality

$$HA + A^*H + C^*C + (B^*H + D^*C)^* R^{-1} (B^*H + D^*C) \leq 0. \tag{8}$$

This inequality is often called the *bounded real* Riccati inequality when all the matrices are real. There is a rich literature on this finite-dimensional version of this inequality and the corresponding equality; see, e.g., [PAJ91], [IW93], and [LR95], and the references mentioned there. This inequality is named after Kalman [Kal63], Popov [Pop73], and Yakubovich [Yak62].

In the development of the theory of absolute stability (or hyperstability) of systems which involve nonlinear feedback those linear systems which are H-passive with respect to scattering supply rate are of special interest, especially in H^∞-control. One of the main problems is to find conditions on the coefficients A, B, C, and D under which the KYP inequality has at least one solution $H > 0$.

To formulate a classical result about the solution of this problem we introduce the main frequency characteristic of the system (1), namely its *transfer function* defined by

$$\mathfrak{D}(\lambda) = D + C(\lambda - A)^{-1}B, \qquad \lambda \in \rho(A). \tag{9}$$

We also introduce the *Schur class* $\mathcal{S}(\mathcal{U}, \mathcal{Y}; \mathbb{C}^+)$ of *holomorphic contractive* functions $\mathfrak{D}$ defined on $\mathbb{C}^+ := \{\lambda \in \mathbb{C} \mid \Re\lambda > 0\}$ with values in $\mathcal{B}(\mathcal{U}, \mathcal{Y})$. If $\mathcal{X}, \mathcal{U}$, and $\mathcal{Y}$ are finite-dimensional, then the transfer function is rational and $\dim \mathcal{X} \geq \deg \mathfrak{D}$, where $\deg \mathfrak{D}$ is the MacMillan degree of $\mathfrak{D}$. A finite-dimensional system is *minimal* if $\dim X = \deg \mathfrak{D}$. The state space of a minimal system has the smallest dimension among all systems with the same transfer function $\mathfrak{D}$.

The (finite-dimensional) system (1) is *controllable* if, given any $z_0 \in \mathcal{X}$ and $T > 0$, there exists some continuous function u on $[0, T]$ such that the solution of (1) with $x(0) = 0$ satisfies $x(T) = z_0$. It is *observable* if it has the following property: if both the input function u and the output function y vanish on some interval $[0, T]$ with $T > 0$, then necessarily the initial state x_0 is zero.

Theorem 1.1 (Kalman). *A finite-dimensional system is minimal if and only if it is controllable and observable.*

Theorem 1.2 (Kalman–Yakubovich–Popov). *Let* $\Sigma = (\left[\begin{smallmatrix} A & B \\ C & D \end{smallmatrix}\right]; \mathcal{X}, \mathcal{U}, \mathcal{Y})$ *be a finite-dimensional system with transfer function* $\mathfrak{D}$.

(i) *If the KYP inequality* (7) *has a solution* $H > 0$, *i.e., if* Σ *is scattering* H*-passive for some* $H > 0$, *then* $\mathbb{C}^+ \subset \rho(A)$ *and* $\mathfrak{D}|_{\mathbb{C}^+} \in \mathcal{S}(\mathcal{U}, \mathcal{Y}; \mathbb{C}^+)$.

(ii) *If* Σ *is minimal and* $\mathfrak{D}|_{\mathbb{C}^+} \in \mathcal{S}(\mathcal{U}, \mathcal{Y}; \mathbb{C}^+)$, *then the KYP inequality* (7) *has a solution* H, *i.e.,* Σ *is scattering* H*-passive for some* $H > 0$.

Here $\mathfrak{D}|_\Omega$ is the restriction of $\mathfrak{D}$ to $\Omega \subset \rho(A)$. In the engineering literate this theorem is known under the name *bounded real lemma* (in the case where all the matrices are real).

It is can be shown that $H > 0$ is a solution of (7) if and only if $\tilde{H} = H^{-1}$ is a solution of the the dual KYP inequality

$$\begin{bmatrix} \tilde{H}A^* + A\tilde{H} + BB^* & \tilde{H}C^* + BD^* \\ C\tilde{H} + DB^* & DD^* - 1_{\mathcal{Y}} \end{bmatrix} \leq 0. \tag{10}$$

The *discrete time* scattering KYP inequality is given by

$$\begin{bmatrix} A^*HA + C^*C - H & A^*HB + C^*D \\ B^*HA + D^*C & D^*D + B^*HB - 1_{\mathcal{U}} \end{bmatrix} \leq 0. \tag{11}$$

The corresponding Kalman–Yakubovich–Popov theorem is still valid with $\mathbb{C}^+$ replaced by $\mathbb{D}^+ = \{z \in \mathbb{C} \mid |z| > 1\}$ and with the transfer function defined by the same formula (9).[1]

[1] This is the standard "engineering" version of the transfer function. In the mathematical literature one usually replace λ by $1/z$ and $\mathbb{D}^+$ by the unit disk $\mathbb{D} = \{z \in \mathbb{C} \mid |z| < 1\}$.

In the seventies the classical results on the KYP inequalities were extended to systems with $\dim \mathcal{X} = \infty$ by V. A. Yakubovich and his students and collaborators (see [Yak74, Yak75, LY76] and the references listed there). There is now also a rich literature on this subject; see, e.g., the discussion in [Pan99] and the references cited there. However, as far as we know, in these and all later generalizations it was assumed (until [AKP05]) that *either H itself is bounded or H^{-1} is bounded.*[2] This is not always a realistic assumption. The operator H is very sensitive to the choice of the state space $\mathcal{X}$ and its norm, and the boundedness of H and H^{-1} depend entirely on this choice. By allowing both H and H^{-1} to be unbounded we can use an analogue of the standard finite-dimensional procedure to determine whether a given transfer function θ is a Schur function or not, namely to *choose an arbitrary minimal realization of θ, and then check whether the KYP inequality* (7) *has a positive* (*generalized*) *solution.* This procedure would not work if we require H or H^{-1} to be bounded, because Theorem 5.4 below is not true in that setting. We shall discuss this further in Section 7 by means of an example.

A generalized solution of the discrete time KYP inequality (11) that permits both H and H^{-1} to be unbounded was developed by Arov, Kaashoek and Pik in [AKP05]. There it was required that

$$A\mathcal{D}(\sqrt{H}) \subset \mathcal{D}(\sqrt{H}) \text{ and } \mathcal{R}(B) \subset \mathcal{D}(\sqrt{H}), \tag{12}$$

and (11) was rewritten using the corresponding quadratic form defined on $\mathcal{D}(\sqrt{H}) \oplus \mathcal{U}$. *Here we extend this approach to continuous time.*

In this paper we only study the *scattering* case. Similar results are true in the *impedance* and *transmission* settings, as can be shown by using the technique developed in [AS05c, AS05d]. We shall return to this question elsewhere. We shall also return elsewhere with a discussion of the connection between the generalized KYP inequality and solutions of the algebraic Riccati inequality and equality, and a with a infinite-dimensional version of the *strict* bounded real lemma.

A summary of our results have been presented in [AS05b].

Notation. The space of bounded linear operators from the Hilbert space $\mathcal{X}$ to the Hilbert space $\mathcal{Y}$ is denoted by $\mathcal{B}(\mathcal{X};\mathcal{Y})$, and we abbreviate $\mathcal{B}(\mathcal{X};\mathcal{X})$ to $\mathcal{B}(\mathcal{X})$. The domain of a linear operator A is denoted by $\mathcal{D}(A)$, the range by $\mathcal{R}(A)$, the kernel by $\mathcal{N}(A)$, and the resolvent set by $\rho(A)$. The restriction of a linear operator A to some subspace $\mathcal{Z} \subset \mathcal{D}(A)$ is denoted by $A|_{\mathcal{Z}}$. Analogously, we denote the restriction of a function ϕ to a subset Ω of its original domain by $\phi|_{\Omega}$. The identity operator on $\mathcal{X}$ is denoted by $1_{\mathcal{X}}$. We denote the orthogonal projection onto a closed subspace $\mathcal{Y}$ of a space $\mathcal{X}$ by $P_{\mathcal{Y}}$.

The orthogonal cross product of the two Hilbert spaces $\mathcal{X}$ and $\mathcal{Y}$ is denoted by $\left[\begin{smallmatrix}\mathcal{X}\\ \mathcal{Y}\end{smallmatrix}\right]$, and we identify a vector $\left[\begin{smallmatrix}x\\ 0\end{smallmatrix}\right] \in \left[\begin{smallmatrix}\mathcal{X}\\ 0\end{smallmatrix}\right]$ with $x \in \mathcal{X}$ and a vector $\left[\begin{smallmatrix}0\\ y\end{smallmatrix}\right] \in \left[\begin{smallmatrix}0\\ \mathcal{Y}\end{smallmatrix}\right]$ with $y \in \mathcal{Y}$. The closed linear span or linear span of a sequence of subsets $\mathfrak{R}_n \subset \mathcal{X}$ where n runs over some index set Λ is denoted by $\vee_{n\in\Lambda}\mathfrak{R}_n$ and $\operatorname{span}_{n\in\Lambda}\mathfrak{R}_n$, respectively.

[2] Results where H^{-1} is bounded are typically proved by replacing the primal KYP inequality by the dual KYP inequality (10).

By a *component* of an open set $\Omega \subset \mathbb{C}$ we mean a *connected component* of Ω.

We denote $\mathbb{R} = (-\infty, \infty)$, $\mathbb{R}^+ = [0, \infty)$, and $\mathbb{R}^- = (-\infty, 0]$. The complex plane is denoted by $\mathbb{C}$, and $\mathbb{C}^+ = \{\lambda \in \mathbb{C} \mid \Re\lambda > 0\}$.

2. Continuous time system nodes

In discrete time one always assumes that A, B, C, and D are bounded operators. In continuous time this assumption is not reasonable. Below we will use a natural continuous time setting, earlier used in, e.g., [AN96], [MSW05], [Sal89], [Šmu86], and [Sta05] (in slightly different forms).

In the sequel, we think about the block matrix $S = \left[\begin{smallmatrix} A & B \\ C & D \end{smallmatrix}\right]$ as *one single closed (possibly unbounded) linear operator* from $\left[\begin{smallmatrix} \mathcal{X} \\ \mathcal{U} \end{smallmatrix}\right]$ (the cross product of $\mathcal{X}$ and $\mathcal{U}$) to $\left[\begin{smallmatrix} \mathcal{X} \\ \mathcal{Y} \end{smallmatrix}\right]$ with dense domain $\mathcal{D}(S) \subset \left[\begin{smallmatrix} \mathcal{X} \\ \mathcal{U} \end{smallmatrix}\right]$, and write (1) in the form

$$\begin{bmatrix} \dot{x}(t) \\ y(t) \end{bmatrix} = S \begin{bmatrix} x(t) \\ u(t) \end{bmatrix}, \quad t \geq s, \quad x(s) = x_s. \tag{13}$$

In the infinite-dimensional case such an operator S need not have a four block decomposition corresponding to the decompositions $\left[\begin{smallmatrix} \mathcal{X} \\ \mathcal{U} \end{smallmatrix}\right]$ and $\left[\begin{smallmatrix} \mathcal{X} \\ \mathcal{Y} \end{smallmatrix}\right]$ of the domain and range spaces. However, we shall throughout assume that the operator

$$\begin{aligned} Ax &:= P_{\mathcal{X}} S \left[\begin{smallmatrix} x \\ 0 \end{smallmatrix}\right], \\ x \in \mathcal{D}(A) &:= \{x \in \mathcal{X} \mid \left[\begin{smallmatrix} x \\ 0 \end{smallmatrix}\right] \in \mathcal{D}(S)\}, \end{aligned} \tag{14}$$

is closed and densely defined in $\mathcal{X}$ (here $P_{\mathcal{X}}$ is the orthogonal projection onto $\mathcal{X}$). We define $\mathcal{X}^1 := \mathcal{D}(A)$ with the graph norm of A, $\mathcal{X}^1_* := \mathcal{D}(A^*)$ with the graph norm of A^*, and let $\mathcal{X}^{-1}$ to be the dual of $\mathcal{X}^1_*$ when we identify the dual of $\mathcal{X}$ with itself. Then $\mathcal{X}^1 \subset \mathcal{X} \subset \mathcal{X}^{-1}$ with continuous and dense embeddings, and the operator A has a unique extension to an operator $\widehat{A} = (A^*)^* \in \mathcal{B}(\mathcal{X}; \mathcal{X}^{-1})$ (with the same spectrum as A), where we interpret A^* as an operator in $\mathcal{B}(\mathcal{X}^1_*; \mathcal{X})$.[3] Additional assumptions on A will be added in Definition 2.1 below.

The remaining blocks of S will be only partially defined. The 'block' B will be an operator in $\mathcal{B}(\mathcal{U}; \mathcal{X}^{-1})$. In particular, it may happen that $\mathcal{R}(B) \cap \mathcal{X} = \{0\}$. The 'block' C will be an operator in $\mathcal{B}(\mathcal{X}^1; \mathcal{Y})$. We shall make no attempt to define the 'block' D in general since this can be done only under additional assumptions (see, e.g., [Sta05, Chapter 5] or [Wei94a, Wei94b]). Nevertheless, we still use a modified block notation $S = \left[\begin{smallmatrix} A\&B \\ C\&D \end{smallmatrix}\right]$, where $A\&B = P_{\mathcal{X}} S$ and $C\&D = P_{\mathcal{Y}} S$.

Definition 2.1. By a *system node* we mean a colligation $\Sigma := (S; \mathcal{X}, \mathcal{U}, \mathcal{Y})$, where $\mathcal{X}$, $\mathcal{U}$ and $\mathcal{Y}$ are Hilbert spaces and the *system operator* $S = \left[\begin{smallmatrix} A\&B \\ C\&D \end{smallmatrix}\right]$ is a (possibly unbounded) linear operator from $\left[\begin{smallmatrix} \mathcal{X} \\ \mathcal{U} \end{smallmatrix}\right]$ to $\left[\begin{smallmatrix} \mathcal{X} \\ \mathcal{Y} \end{smallmatrix}\right]$ with the following properties:

(i) S is closed.

(ii) The operator A defined in (14) is the generator of a C_0 semigroup $t \mapsto \mathfrak{A}^t$, $t \geq 0$, on $\mathcal{X}$.

[3] This construction is found in most of the papers listed in the bibliography (in slightly different but equivalent forms), including [AN96], [MSW05], and [Sal87]–[WT03].

(iii) $A\&B$ has an extension $\begin{bmatrix}\widehat{A} & B\end{bmatrix} \in \mathcal{B}(\left[\begin{smallmatrix}\mathcal{X}\\ \mathcal{U}\end{smallmatrix}\right]; \mathcal{X}^{-1})$ (where $B \in \mathcal{B}(\mathcal{U}; \mathcal{X}^{-1})$).
(iv) $\mathcal{D}(S) = \left\{\left[\begin{smallmatrix}x\\ u\end{smallmatrix}\right] \in \left[\begin{smallmatrix}\mathcal{X}\\ \mathcal{U}\end{smallmatrix}\right] \mid \widehat{A}x + Bu \in \mathcal{X}\right\}$, and $A\&B = \begin{bmatrix}\widehat{A} & B\end{bmatrix}|_{\mathcal{D}(S)}$;

As we will show below, (ii)–(iv) imply that the domain of S is dense in $\left[\begin{smallmatrix}\mathcal{X}\\ \mathcal{U}\end{smallmatrix}\right]$. It is also true that if (ii)–(iv) holds, then (i) is equivalent to the following condition:
(v) $C\&D \in \mathcal{B}(\mathcal{D}(S); \mathcal{Y})$, where we use the graph norm

$$\left\|\left[\begin{smallmatrix}x\\ u\end{smallmatrix}\right]\right\|^2_{\mathcal{D}(A\&B)} = \left\|A\&B\left[\begin{smallmatrix}x\\ u\end{smallmatrix}\right]\right\|^2_{\mathcal{X}} + \|x\|^2_{\mathcal{X}} + \|u\|^2_{\mathcal{U}} \tag{15}$$

of $A\&B$ on $\mathcal{D}(S)$.

It is not difficult to see that the graph norm of $A\&B$ on $\mathcal{D}(S)$ is equivalent to the full graph norm

$$\left\|\left[\begin{smallmatrix}x\\ u\end{smallmatrix}\right]\right\|^2_{\mathcal{D}(S)} = \left\|A\&B\left[\begin{smallmatrix}x\\ u\end{smallmatrix}\right]\right\|^2_{\mathcal{X}} + \left\|C\&D\left[\begin{smallmatrix}x\\ u\end{smallmatrix}\right]\right\|^2_{\mathcal{Y}} + \|x\|^2_{\mathcal{X}} + \|u\|^2_{\mathcal{U}} \tag{16}$$

of S.

We call $A \in \mathcal{B}(\mathcal{X}^1; \mathcal{X})$ the *main operator* of Σ, $t \mapsto \mathfrak{A}^t$, $t \geq 0$, is the *evolution semigroup*, $B \in \mathcal{B}(\mathcal{U}; \mathcal{X}^{-1})$ is the *control operator*, and $C\&D \in \mathcal{B}(V; \mathcal{Y})$ is the *combined observation/feedthrough operator*. From the last operator we can extract $C \in \mathcal{B}(\mathcal{X}^1; \mathcal{Y})$, the *observation operator* of Σ, defined by

$$Cx := C\&D\begin{bmatrix}x\\ 0\end{bmatrix}, \qquad x \in \mathcal{X}^1. \tag{17}$$

A short computation shows that for each $\alpha \in \rho(A)$, the operator

$$E_\alpha := \begin{bmatrix}1_{\mathcal{X}} & (\alpha - \widehat{A})^{-1}B\\ 0 & 1_{\mathcal{U}}\end{bmatrix} \tag{18}$$

is a bounded bijection from $\left[\begin{smallmatrix}\mathcal{X}\\ \mathcal{U}\end{smallmatrix}\right]$ onto itself and also from $\left[\begin{smallmatrix}\mathcal{X}^1\\ \mathcal{U}\end{smallmatrix}\right]$ onto $\mathcal{D}(S)$. In particular, for each $u \in \mathcal{U}$ there is some $x \in \mathcal{X}$ such that $\left[\begin{smallmatrix}x\\ u\end{smallmatrix}\right] \in \mathcal{D}(S)$. Since $\left[\begin{smallmatrix}\mathcal{X}^1\\ \mathcal{U}\end{smallmatrix}\right]$ is dense in $\left[\begin{smallmatrix}\mathcal{X}\\ \mathcal{U}\end{smallmatrix}\right]$, this implies that also $\mathcal{D}(S)$ is dense in $\left[\begin{smallmatrix}\mathcal{X}\\ \mathcal{U}\end{smallmatrix}\right]$. Since the second column of E_α maps $\mathcal{U}$ into $\mathcal{D}(S)$, we can define the *transfer function* of S by

$$\widehat{\mathfrak{D}}(\lambda) := C\&D\begin{bmatrix}(\lambda - \widehat{A})^{-1}B\\ 1_{\mathcal{U}}\end{bmatrix}, \qquad \lambda \in \rho(A), \tag{19}$$

which is an $\mathcal{B}(\mathcal{U}; \mathcal{Y})$-valued analytic function. If $B \in \mathcal{B}(\mathcal{U}; \mathcal{X})$, then $\mathcal{D}(S) = \left[\begin{smallmatrix}\mathcal{X}_1\\ \mathcal{U}\end{smallmatrix}\right]$, and we can define the operator $D \in \mathcal{B}(\mathcal{U}; \mathcal{Y})$ by $D = P_{\mathcal{Y}}S|_{\left[\begin{smallmatrix}0\\ \mathcal{U}\end{smallmatrix}\right]}$, after which formula (19) can be rewritten in the form (9). By the resolvent identity, for any two α, $\beta \in \rho(A)$,

$$\begin{aligned}\widehat{\mathfrak{D}}(\alpha) - \widehat{\mathfrak{D}}(\beta) &= C[(\alpha - \widehat{A})^{-1} - (\beta - \widehat{A})^{-1}]B\\ &= (\beta - \alpha)C(\alpha - A)^{-1}(\beta - \widehat{A})^{-1}B.\end{aligned} \tag{20}$$

Let

$$\begin{aligned}F_\alpha :&= \left(\begin{bmatrix}\alpha & 0\\ 0 & 1_{\mathcal{U}}\end{bmatrix} - \begin{bmatrix}A\&B\\ 0 \quad 0\end{bmatrix}\right)^{-1}\\ &= \begin{bmatrix}(\alpha - A)^{-1} & (\alpha - \widehat{A})^{-1}B\\ 0 & 1_{\mathcal{U}}\end{bmatrix}, \qquad \alpha \in \rho(A).\end{aligned} \tag{21}$$

Then, for all $\alpha \in \rho(A)$, F_α is a bounded bijection from $\left[\begin{smallmatrix}\mathcal{X}\\ \mathcal{U}\end{smallmatrix}\right]$ onto $\mathcal{D}(S)$, and

$$\begin{bmatrix} A\&B \\ C\&D \end{bmatrix} F_\alpha = \begin{bmatrix} A(\alpha - A)^{-1} & \alpha(\alpha - \widehat{A})^{-1}B \\ C(\alpha - A)^{-1} & \widehat{\mathfrak{D}}(\alpha) \end{bmatrix}, \qquad \alpha \in \rho(A). \tag{22}$$

One way to construct a system operator $S = \left[\begin{smallmatrix} A\&B \\ C\&D \end{smallmatrix}\right]$ is to give a generator A of a C_0 semigroup on $\mathcal{X}$, a control operator $B \in \mathcal{B}(\mathcal{U};\mathcal{X}^{-1})$, and an observation operator $C \in \mathcal{B}(\mathcal{X}_1;\mathcal{Y})$, to fix some $\alpha \in \rho(A)$ and an operator $D_\alpha \in \mathcal{B}(\mathcal{U};\mathcal{Y})$, to define $\mathcal{D}(S)$ and $A\&B$ by (iv), and to finally define $C\&D\left[\begin{smallmatrix} x \\ u \end{smallmatrix}\right]$ for all $\left[\begin{smallmatrix} x \\ u \end{smallmatrix}\right] \in \mathcal{D}(S)$ by

$$C\&D \begin{bmatrix} x \\ u \end{bmatrix} := C(x - (\alpha - \hat{A})^{-1}Bu) + D_\alpha u. \tag{23}$$

The transfer function $\mathfrak{D}$ of this system node satisfies $\mathfrak{D}(\alpha) = D_\alpha$ (see [Sta05, Lemma 4.7.6]).

Lemma 2.2. *Let $\Sigma := (S;\mathcal{X},\mathcal{U},\mathcal{Y})$ be a system node with main operator A, control operator B, observation operator C, transfer function $\mathfrak{D}$, and evolution semigroup $t \mapsto \mathfrak{A}^t$, $t \geq 0$. Then $\Sigma^* := (S^*;\mathcal{X},\mathcal{U},\mathcal{Y})$ is another system node, which we call the* adjoint *of Σ. The main operator of Σ^* is A^*, the control operator of Σ^* is C^*, the observation operator of Σ^* is B^*, the transfer function of Σ^* is $\widehat{\mathfrak{D}}(\overline{\alpha})^*$, $\alpha \in \rho(A^*)$, and the evolution semigroup of Σ^* is $t \mapsto (\mathfrak{A}^t)^*$, $t \geq 0$.*

For a proof (and for more details), see, e.g., [AN96, Section 3], [MSW05, Proposition 2.3], or [Sta05, Lemma 6.2.14].

If $\Sigma := (S;\mathcal{X},\mathcal{U},\mathcal{Y})$ is a system node, then (13) has (smooth) trajectories of the following type. Note that we can use the operators $A\&B$ and $C\&D$ to split (13) into

$$\begin{aligned} \dot{x}(t) &= A\&B \begin{bmatrix} x(t) \\ u(t) \end{bmatrix}, \quad t \geq s, \quad x(s) = x_s, \\ y(t) &= C\&D \begin{bmatrix} x(t) \\ u(t) \end{bmatrix}, \quad t \geq s. \end{aligned} \tag{24}$$

Below we use the following notation: $W^{1,2}_{\mathrm{loc}}([s,\infty);\mathcal{U})$ is the set of $\mathcal{U}$-valued functions on $[s,\infty)$ which are locally absolutely continuous and have a derivative in $L^2_{\mathrm{loc}}([s,\infty);\mathcal{U})$. An equivalent formulation is to say that $u \in W^{1,2}_{\mathrm{loc}}([s,\infty);\mathcal{U})$ if $u \in L^2_{\mathrm{loc}}([s,\infty);\mathcal{U})$ and the distribution derivative of the function u consists of a point mass of size $u(s)$ at s plus a function in $L^2_{\mathrm{loc}}([s,\infty);\mathcal{U})$ (first extend u by zero to $(-\infty,s)$ before taking the distribution derivative). The space $W^{2,2}_{\mathrm{loc}}([s,\infty);\mathcal{U})$ consists of those $w \in W^{1,2}_{\mathrm{loc}}([s,\infty);\mathcal{U})$ which are locally absolutely continuous and have $u' \in W^{1,2}_{\mathrm{loc}}([s,\infty);\mathcal{U})$, too.

Lemma 2.3. *Let $\Sigma := (S;\mathcal{X},\mathcal{U},\mathcal{Y})$ be a system node. Then for each $s \in \mathbb{R}$, $x_s \in \mathcal{X}$ and $u \in W^{2,2}_{\mathrm{loc}}([s,\infty);\mathcal{U})$ such that $\left[\begin{smallmatrix} x_s \\ u(s) \end{smallmatrix}\right] \in \mathcal{D}(S)$, there is a unique function $x \in C^1([s,\infty);\mathcal{X})$ (called a* state trajectory*) satisfying $x(s) = x_s$, $\left[\begin{smallmatrix} x(t) \\ u(t) \end{smallmatrix}\right] \in \mathcal{D}(S)$, $t \geq s$, and $\dot{x}(t) = A\&B\left[\begin{smallmatrix} x(t) \\ u(t) \end{smallmatrix}\right]$, $t \geq s$. If we define the output by $y(t) = C\&D\left[\begin{smallmatrix} x(t) \\ u(t) \end{smallmatrix}\right]$, $t \geq s$, then $y \in C([s,\infty);\mathcal{Y})$, and the three functions u, x, and y satisfy* (13).

This lemma is contained in [Sta05, Lemmas 4.7.7–4.7.8], which are actually slightly stronger: it suffices to have $u \in W^{2,1}_{\mathrm{loc}}([s,\infty);\mathcal{U})$ (the second derivative is locally in L^1 instead of locally in L^2). (Equivalently, both u and u' are locally absolutely continuous.)

In addition to the classical solutions of (13) presented in Lemma 2.3 we shall also need generalized solutions. A generalized solution of (13) exists for all initial times $s \in \mathbb{R}$, all initial states $x_s \in \mathcal{X}$ and all input functions $u \in W^{1,2}_{\mathrm{loc}}([s,\infty);\mathcal{U})$. The state trajectory $x(t)$ is continuous in $\mathcal{X}$, and the output y belongs $W^{-1,2}_{\mathrm{loc}}([s,\infty);\mathcal{Y})$. This is the space of all distribution derivatives of functions in $L^2_{\mathrm{loc}}([s,\infty);\mathcal{Y})$ (first extended the functions to all of $\mathbb{R}$ by zero on $(-\infty,s)$). This space can also be interpreted as the space of all distributions in $W^{-1,2}_{\mathrm{loc}}(\mathbb{R};\mathcal{Y})$ which are supported on $[s,\infty)$. It is the dual of the space $W^{1,2}_c([s,\infty);\mathcal{Y})$, where the subindex c means that the functions in this space have compact support.[4]

The construction of generalized solutions of (13) is carried out as follows. It suffices to consider two separate cases where either x_s or u is zero, since we get the general case by adding the two special solutions. We begin with the case where $u = 0$. For each $x_s \in \mathcal{X}$ we define the corresponding state trajectory x by $x(t) = \mathfrak{A}^{t-s}x_s$, where $\mathfrak{A}^t$, $t \ge 0$, is the semigroup generated by the main operator A. The corresponding output $y \in W^{-1,2}_{\mathrm{loc}}([s,\infty);\mathcal{Y})$ is defined as follows. First we observe that the function $\int_s^t x(v)\,dv = \int_s^t \mathfrak{A}^{v-s}x_s\,dv$ is a continuous function on $[s,\infty)$ with values in $\mathcal{X}^1$ vanishing at s, hence $C\int_s^t \mathfrak{A}^{v-s}x_s\,dv$ is continuous with values in $\mathcal{Y}$. We can therefore define the output y to be given by the following distribution derivative:

$$y = \frac{\mathrm{d}}{\mathrm{d}t}\Big(t \mapsto C\int_s^t \mathfrak{A}^{v-s}x_s\,dv\Big);$$

here $\frac{\mathrm{d}}{dt}$ stands for a distribution derivative. In particular, $y \in W^{-1,2}_{\mathrm{loc}}([s,\infty);\mathcal{Y})$ and the map from x_s to y is continuous from $\mathcal{X}$ to $W^{-1,2}_{\mathrm{loc}}([s,\infty);\mathcal{Y})$. Of course, if $x_s \in \mathcal{X}^1$, then $y(t) = C\mathfrak{A}^{t-s}x_s$ for all $t \ge s$. For more details, see [Sta05, Lemma 4.7.9].)

Next suppose that $x_s = 0$ and that $u \in W^{1,2}_{\mathrm{loc}}([s,\infty);\mathcal{U})$. We then define the state trajectory x and the output distribution y as follows. We first replace u by $u_1(t) = \int_s^t u(v)\,dv$, let x_1 and y_1 be the state and output given by Lemma 2.3 with $x_s = 0$ and u replaced by u_1 (note that $u_1(s) = 0$), and then define

$$x = x_1', \qquad y = \frac{\mathrm{d}}{\mathrm{d}t}y_1,$$

where the differentiation is interpreted in the distribution sense. Again we find that $x \in C([s,\infty);\mathcal{X})$ and that $y \in W^{-1,2}_{\mathrm{loc}}([s,\infty);\mathcal{Y})$.

[4]Note that $W^{-1,2}_{\mathrm{loc}}([s,\infty);\mathcal{Y})$ is not the same space as $W^{-1,2}_{\mathrm{loc}}((s,\infty);\mathcal{Y})$, which is the dual of the space of all functions in $W^{1,2}_c([s,\infty);\mathcal{Y})$ which vanish at s. The space $W^{-1,2}_{\mathrm{loc}}((s,\infty);\mathcal{Y})$ is the quotient of $W^{-1,2}_{\mathrm{loc}}([s,\infty);\mathcal{Y})$ over all point evaluation functionals at s.

Given $x_0 \in \mathcal{X}$ and $u \in W^{1,2}_{\mathrm{loc}}([s,\infty);\mathcal{U})$ we shall refer to the functions $x \in C([s,\infty);\mathcal{X})$ and $y \in W^{-1,2}_{\mathrm{loc}}([s,\infty);\mathcal{Y})$ constructed above as the *generalized solution and output* of (24), respectively. A *generalized trajectory* of (24) consists of the triple (x,u,y) described above. A trajectory is *smooth* if it is of the type described in Lemma 2.3.

By the *system* induced by a system node $\Sigma := (S;\mathcal{X},\mathcal{U},\mathcal{Y})$ we mean the node itself together with all its generalized trajectories. We use the same notation Σ for the system as for the node.

Above we already introduced the notation $\mathfrak{A}^t$, $t \geq 0$, for the *semigroup* generated by the main operator A. The *output map* $\mathfrak{C}$ maps $\mathcal{X}$ into $W^{-1,2}_{\mathrm{loc}}(\mathbb{R}^+;\mathcal{Y})$, and it is the mapping from x_0 to y (i.e., take both the initial time $s = 0$ and the input function $u = 0$). Thus,

$$\mathfrak{C}x_0 = \frac{\mathrm{d}}{\mathrm{d}t}\Big(t \mapsto C\int_0^t \mathfrak{A}^v x_0\,dv\Big),$$

and if $x_0 \in \mathcal{X}^1$, then $\mathfrak{C}x_0 = t \mapsto C\mathfrak{A}^t x_0$, $t \geq 0$. This map is continuous from $\mathcal{X}$ into $W^{-1,2}_{\mathrm{loc}}(\mathbb{R}^+;\mathcal{Y})$ and from $\mathcal{X}^1$ into $C[\mathbb{R}^+;\mathcal{Y})$.

The *input map* $\mathfrak{B}$ is defined for all $u \in W^{1,2}_c(\mathbb{R}^-;\mathcal{U})$, i.e., functions $u \in W^{1,2}(\mathbb{R}^-;\mathcal{U})$ whose support is bounded to the left. It is the map from u to $x(0)$ (take the initial time to be $s < 0$ and the initial state to be zero). To get an explicit formula for this map we argue as follows. By Definition 2.1, we can rewrite the first equation in (24) in the form

$$\dot{x}(t) = \widehat{A}x(t) + Bu(t), \quad t \geq s, \quad x(s) = x_s, \tag{25}$$

where we now allow the equation to take its values in $\mathcal{X}^{-1}$. The operator $\widehat{A}$ generates a C_0 semigroup in $\mathcal{X}^{-1}$, which we denote by $\widehat{\mathfrak{A}}^t$, $t \geq 0$, and $B \in \mathcal{B}(\mathcal{U};\mathcal{X}^{-1})$. We can therefore use the variation of constants formula to solve for $\mathfrak{B}u = x(0)$ (take $x_s = 0$ and define $u(v)$ to be zero for $v < s$)

$$\mathfrak{B}u = \int_{-\infty}^0 \widehat{\mathfrak{A}}^{-v}Bu(v)\,dv. \tag{26}$$

Here the integral is computed in $\mathcal{X}^{-1}$, but the final result belongs to $\mathcal{X}$, and $\mathfrak{B}$ is continuous from $W^{1,2}_c(\mathbb{R}^-;\mathcal{U})$ to $\mathcal{X}$. (It is also possible to use (26) to extend $\mathfrak{B}$ to a continuous map from $L^2_c(\mathbb{R}^-;\mathcal{U})$ to $\mathcal{X}^{-1}$ as is done in [Sta05].)

Finally, the *input/output map* $\mathfrak{D}$ is defined for all $u \in W^{1,2}_{\mathrm{loc}}(\mathbb{R};\mathcal{U})$ whose support is bounded to the left, and it is the map from u to y (take the initial time to the left of the support of u, and the initial state to be zero). It maps this set of functions continuously into the set of distributions in $W^{-1,2}_{\mathrm{loc}}(\mathbb{R};\mathcal{Y})$ whose support is bounded to the left.

Our following lemma describes the connection between the input/output map $\mathfrak{D}$ and the transfer function $\widehat{\mathfrak{D}}$.

Lemma 2.4. *Let $\Sigma_i := (S_i;\mathcal{X}_i,\mathcal{U},\mathcal{Y})$, $i = 1,2$, be two system nodes with main operators A_i, input/output maps $\mathfrak{D}_i$, and transfer functions $\widehat{\mathfrak{D}}_i$. Let Ω_∞ be the component of $\rho(A_1) \cap \rho(A_2)$ which contains some right half-plane.*

(i) *If* $\mathfrak{D}_1 = \mathfrak{D}_2$, *then* $\widehat{\mathfrak{D}}_1(\lambda) = \widehat{\mathfrak{D}}_2(\lambda)$ *for all* $\lambda \in \Omega_\infty$.
(ii) *Conversely, if the set* $\{\lambda \in \Omega_\infty \mid \widehat{\mathfrak{D}}_1(\lambda) = \widehat{\mathfrak{D}}_2(\lambda)\}$ *has an interior cluster point, then* $\mathfrak{D}_1 = \mathfrak{D}_2$.

Proof. Fix some real $\alpha > \beta$, where β is the maximum of the growth bounds of the two semigroups $\mathfrak{A}_i^t$, $t \geq 0$, $i = 1, 2$, and suppose that $(t \mapsto e^{-\alpha t}u(t)) \in W_0^{2,1}(\mathbb{R}^+;\mathcal{U}) := \{u \in W^{2,1}(\mathbb{R}^+;\mathcal{U}) \mid u(0) = u'(0) = 0\}$. Define $y_1 = \mathfrak{D}_1 y$ and $y_2 = \mathfrak{D}_2 u$. Then, by [Sta05, Lemma 4.7.12], the functions $t \mapsto e^{-\alpha t}y_i(t)$, with $i = 1, 2$, are bounded, and the Laplace transforms of these functions satisfy $\hat{y}_i(\lambda) = \widehat{\mathfrak{D}}_i(\lambda)\hat{u}(\lambda)$ in the half plane $\Re\lambda > \alpha$.

If $\mathfrak{D}_1 = \mathfrak{D}_2$, then $y_1 = y_2$, and hence we conclude that $\widehat{\mathfrak{D}}_1(\lambda)\hat{u}(\lambda) = \widehat{\mathfrak{D}}_2(\lambda)\hat{u}(\lambda)$ for all u of the type described above and for all $\Re\lambda \geq \alpha$. This implies that $\widehat{\mathfrak{D}}_1(\lambda) = \widehat{\mathfrak{D}}_2(\lambda)$ for all $\Re\lambda > \alpha$, and by analytic continuation, for all $\lambda \in \Omega_\infty$.

Conversely, if set $\{\lambda \in \Omega_\infty \mid \widehat{\mathfrak{D}}_1(\lambda) = \widehat{\mathfrak{D}}_2(\lambda)\}$ has an interior cluster point, then by analytic extension theory, $\widehat{\mathfrak{D}}_1(\lambda) = \widehat{\mathfrak{D}}_2(\lambda)$ for all $\lambda \in \Omega_\infty$. Thus, $\hat{y}_1(\lambda) = \hat{y}_2(\lambda)$ for all $\Re\lambda > \alpha$. Since the Laplace transform is injective, this implies that $y_1 = y_2$. Hence, $\mathfrak{D}_1 u = \mathfrak{D}_2 u$ for all u of the type described above. By using the bilateral shift-invariance of $\mathfrak{D}_1$ and $\mathfrak{D}_2$ we find that the same identity is true for all $u \in W_{\mathrm{loc}}^{2,1}(\mathbb{R};\mathcal{U})$ whose support it bounded to the left. This set is dense in the common domain of $\mathfrak{D}_1$ and $\mathfrak{D}_2$, and so we must have $\mathfrak{D}_1 = \mathfrak{D}_2$. □

Remark 2.5. The system operator S is determined uniquely by the semigroup $\mathfrak{A}^t$, $t \geq 0$, the input map $\mathfrak{B}$, the output map $\mathfrak{C}$, and the input/output map $\mathfrak{D}$ of the system Σ, or alternatively, by $\mathfrak{A}^t$, $t \geq 0$, $\mathfrak{B}$, $\mathfrak{C}$ and the transfer function $\widehat{\mathfrak{D}}$. The corresponding operators for the adjoint system node Σ^* are closely related to those of Σ. The semigroup of Σ^* is $(\mathfrak{A}^t)^*$, $t \geq 0$, the input map of Σ^* is $\boldsymbol{\mathfrak{R}}\mathfrak{C}^*$, the output map of Σ^* is $\mathfrak{B}^*\boldsymbol{\mathfrak{R}}$, and the input/output map of Σ^* is $\boldsymbol{\mathfrak{R}}\mathfrak{D}^*\boldsymbol{\mathfrak{R}}$, where $\boldsymbol{\mathfrak{R}}$ is the time reflection operator: $(\boldsymbol{\mathfrak{R}}u)(t) = u(-t)$, $t \in \mathbb{R}$. As we already remarked earlier, the transfer function of Σ^* is $\widehat{\mathfrak{D}}(\overline{\alpha})^*$, $\alpha \in \rho(A^*)$.

We call Σ (*approximately*) *controllable* if the range of its input map $\mathfrak{B}$ is dense in $\mathcal{X}$ and (*approximately*) *observable* if its output map $\mathfrak{C}$ is injective. Finally, Σ is *minimal* if it is both controllable and observable.[5]

Lemma 2.6. *The system node* Σ *is controllable or observable if and only if* Σ^* *is observable or controllable, respectively. In particular,* Σ *is minimal if and only if* Σ^* *is minimal.*

Proof. This is true because the duality between the input and output maps of Σ and Σ^* (see Remark 2.5). □

Lemma 2.7. *Let* $\Sigma := (S; \mathcal{X}, \mathcal{U}, \mathcal{Y})$ *be a system node with main operator* A, *control operator* B, *and observation operator* C. *Let* $\rho_\infty(A)$ *be the component of* $\rho(A)$ *which contains some right half-plane.*

---

[5]There is another equivalent and more natural definition of minimality of a system: it should not be a nontrivial dilation of some other system (see [AN96, Section 7]).

(i) Σ *is observable if and only if*

$$\cap_{\lambda\in\rho_\infty(A)}\mathcal{N}\left(C(\lambda-A)^{-1}\right)=\{0\}.$$

(ii) Σ *is controllable if and only if*

$$\vee_{\lambda\in\rho_\infty(A)}\mathcal{R}\left((\lambda-\widehat{A})^{-1}B\right)=\mathcal{X},$$

where $\vee$ *stands for the closed linear span.*

Proof. Proof of (i): We have $x_0\in\mathcal{N}(\mathfrak{C})$ if and only if $\frac{\mathrm{d}}{\mathrm{d}t}C\int_0^t\mathfrak{A}^v x_0\,dv$ vanishes identically, or equivalently, if and only if $C\int_0^t\mathfrak{A}^v x_0\,dv$ vanishes identically, or equivalently, the Laplace transform of this function vanishes identically to the right of the growth-bound of this function. This Laplace transform is given by $\lambda^{-1}C(\lambda-A)^{-1}x_0$, and it vanishes to the right of the growth bound of $\mathfrak{A}^t$, $t\ge 0$, if and only if it vanishes on $\rho_\infty(A)$, or equivalently, $C(\lambda-A)^{-1}x_0$ vanishes identically on $\rho_\infty(A)$.

Proof of (ii): That (ii) holds follows from (i) by duality (see Lemma 2.6). □

3. The Cayley transform

The proofs of some of the results of this paper are based on a reduction by means of the Cayley transform of the continuous time case to the corresponding discrete time case studied in [AKP05]. In a linear time-independent discrete time system the input $u=\{u_n\}_{n=0}^\infty$, the state $x=\{x_n\}_{n=0}^\infty$, and the output $y=\{y_n\}_{n=0}^\infty$ are sequences with values in the Hilbert spaces $\mathcal{U}$, $\mathcal{X}$, and $\mathcal{Y}$, respectively. The discrete time system $\mathbf{\Sigma}$ is a colligation $\mathbf{\Sigma}:=\left(\left[\begin{smallmatrix}\mathbf{A}&\mathbf{B}\\\mathbf{C}&\mathbf{D}\end{smallmatrix}\right],\mathcal{X},\mathcal{U},\mathcal{Y})\right)$, where the *system operator* $\left[\begin{smallmatrix}\mathbf{A}&\mathbf{B}\\\mathbf{C}&\mathbf{D}\end{smallmatrix}\right]\in\mathcal{B}(\left[\begin{smallmatrix}\mathcal{X}\\\mathcal{U}\end{smallmatrix}\right];\left[\begin{smallmatrix}\mathcal{X}\\\mathcal{U}\end{smallmatrix}\right])$. The dynamics of this system is described by

$$\begin{aligned}x_{n+1}&=\mathbf{A}x_n+\mathbf{B}u_n,\\y_n&=\mathbf{C}x_n+\mathbf{D}u_n,\qquad n=0,1,2,\ldots,\\x_0&=\text{given.}\end{aligned}\tag{27}$$

We still call $\mathbf{A}$ the main operator, $\mathbf{B}$ the control operator, $\mathbf{C}$ the observation operator, and $\mathbf{D}$ the feedthrough operator. We define the *transfer function* $\widehat{\mathbf{D}}$ of $\mathbf{\Sigma}$ in the same way as in (9), namely by[6]

$$\widehat{\mathbf{D}}(z)=\mathbf{C}(z-\mathbf{A})^{-1}\mathbf{B}+\mathbf{D},\qquad z\in\rho(\mathbf{A}).$$

Observability, controllability, and minimality of a discrete time system is defined in exactly the same way as in continuous time, with continuous time trajectories replaced by discrete time trajectories. Thus, $\mathbf{\Sigma}$ is (approximately) controllable if the subspace of all states x_n reachable from the zero state in finite time (by a suitable choice of input sequence) is dense in $\mathcal{X}$, and it is (approximately) observable if it has the following property: if both the input sequence and the output

[6]This is the standard "engineering" version of the transfer function. In the mathematical literature one usually replace z by $1/z$.

sequence are zero, then necessarily $x_0 = 0$. Finally, it is minimal if it is both controllable and observable. The following discrete time version of Lemma 2.7 is well known: if we denote the unbounded component of the resolvent set of $\mathbf{A}$ by $\rho_\infty(\mathbf{A})$, then Σ is observable if and only if

$$\cap_{z\in\rho_\infty(\mathbf{A})}\mathcal{N}\left(\mathbf{C}(z-\mathbf{A})^{-1}\right)=\{0\},$$

and that $\boldsymbol{\Sigma}$ is controllable if and only if

$$\vee_{z\in\rho_\infty(\mathbf{A})}\mathcal{R}\left((z-\mathbf{A})^{-1}\mathbf{B}\right)=\mathcal{X}.$$

Given a system node $\Sigma := (S;\mathcal{X},\mathcal{U},\mathcal{Y})$ with main operator A, for each $\alpha\in\rho(A)\cap\mathbb{C}^+$ it is possible to define the (internal) *Cayley transform of* Σ *with parameter* α. This is the discrete time system $\boldsymbol{\Sigma}(\alpha) := \left(\left[\begin{smallmatrix}\mathbf{A}(\alpha) & \mathbf{B}(\alpha)\\ \mathbf{C}(\alpha) & \mathbf{D}(\alpha)\end{smallmatrix}\right];\mathcal{X},\mathcal{U},\mathcal{Y}\right)$ whose coefficients are given by

$$\begin{aligned}\mathbf{A}(\alpha)&=(\overline{\alpha}+A)(\alpha-A)^{-1}, & \mathbf{B}(\alpha)&=\sqrt{2\Re\alpha}\,(\alpha-\widehat{A})^{-1}B,\\ \mathbf{C}(\alpha)&=\sqrt{2\Re\alpha}\,C(\alpha-A)^{-1}, & \mathbf{D}(\alpha)&=\widehat{\mathfrak{D}}(\alpha).\end{aligned}\tag{28}$$

Note that $\mathbf{A}(\alpha)+1=2\Re\alpha(\alpha-A)^{-1}$, so that $\mathbf{A}(\alpha)+1$ is injective and has dense range. The transfer function $\widehat{\mathbf{D}}$ of $\boldsymbol{\Sigma}(\alpha)$ satisfies

$$\widehat{\mathbf{D}}(z)=\widehat{\mathfrak{D}}(\lambda),\quad z=\frac{\overline{\alpha}+\lambda}{\alpha-\lambda},\quad \lambda=\frac{\alpha z-\overline{\alpha}}{z+1},\quad \lambda\in\rho(A),\quad z\in\rho(\mathbf{A}(\alpha)).\tag{29}$$

An equivalent way to write the Cayley transform is

$$\begin{bmatrix}\mathbf{A}(\alpha)+1 & \mathbf{B}(\alpha)\\ \mathbf{C}(\alpha) & \mathbf{D}(\alpha)\end{bmatrix}=\begin{bmatrix}\sqrt{2\Re\alpha} & 0\\ 0 & 1\end{bmatrix}\begin{bmatrix}1 & 0\\ C\&D\end{bmatrix}F_\alpha\begin{bmatrix}\sqrt{2\Re\alpha} & 0\\ 0 & 1\end{bmatrix},\tag{30}$$

where F_α is the operator defined in (21).

The (internal) *inverse Cayley transform* with parameter $\alpha\in\mathbb{C}^+$ of a discrete time system $(\left[\begin{smallmatrix}\mathbf{A} & \mathbf{B}\\ \mathbf{C} & \mathbf{D}\end{smallmatrix}\right];\mathcal{X},\mathcal{U},\mathcal{Y})$ is defined whenever $\mathbf{A}+1$ is injective and has dense range. It is designed to reproduce the original system node Σ when applied to its Cayley transform $\left(\left[\begin{smallmatrix}\mathbf{A}(\alpha) & \mathbf{B}(\alpha)\\ \mathbf{C}(\alpha) & \mathbf{D}(\alpha)\end{smallmatrix}\right];\mathcal{X},\mathcal{U},\mathcal{Y}\right)$. The system operator $S=\left[\begin{smallmatrix}A\&B\\ C\&D\end{smallmatrix}\right]$ of this node is given by

$$\begin{bmatrix}A\&B\\ C\&D\end{bmatrix}-\begin{bmatrix}\alpha & 0\\ 0 & 0\end{bmatrix}=\begin{bmatrix}\sqrt{2\Re\alpha} & 0\\ 0 & 1\end{bmatrix}\begin{bmatrix}-1 & 0\\ \mathbf{C} & \mathbf{D}\end{bmatrix}\begin{bmatrix}\mathbf{A}+1 & \mathbf{B}\\ 0 & 1\end{bmatrix}^{-1}\begin{bmatrix}\sqrt{2\Re\alpha} & 0\\ 0 & 1\end{bmatrix}.\tag{31}$$

More specifically, the different operators which are part the node Σ are given by

$$\begin{aligned}A&=(\alpha\mathbf{A}-\overline{\alpha})(\mathbf{A}+1)^{-1}, & B&=\frac{1}{\sqrt{2\Re\alpha}}(\alpha-\widehat{A})\mathbf{B},\\ C&=\frac{1}{\sqrt{2\Re\alpha}}\mathbf{C}(\alpha-A), & \widehat{\mathfrak{D}}(\alpha)&=\mathbf{D}.\end{aligned}\tag{32}$$

If A is the generator of a C_0-semigroup (and only in this case) the operator S defined in this way is the system operator of a system node $\Sigma=(S;\mathcal{X},\mathcal{U},\mathcal{Y})$.[7]

[7] Otherwise it will be an operator node in the sense of [Sta05, Definition 4.7.2].

Lemma 3.1. *Let $\Sigma := (S;\mathcal{X},\mathcal{U},\mathcal{Y})$ be a system node with main operator A, and let $\alpha \in \rho_\infty(A)\cap\mathbb{C}^+$, where $\rho_\infty(A)$ is the component of $\rho(A)$ which contains some right half-plane. Let $\mathbf{\Sigma}(\alpha) := \left(\left[\begin{smallmatrix}\mathbf{A}(\alpha) & \mathbf{B}(\alpha)\\ \mathbf{C}(\alpha) & \mathbf{D}(\alpha)\end{smallmatrix}\right];\mathcal{X},\mathcal{U},\mathcal{Y}\right)$ be the Cayley transform of Σ with parameter α. Then $\mathbf{\Sigma}(\alpha)$ is controllable if and only if Σ is controllable, $\mathbf{\Sigma}(\alpha)$ is observable if and only if Σ is observable, and $\mathbf{\Sigma}(\alpha)$ is minimal if and only if Σ is minimal.*

This follows from Lemma 2.7. (The linear fractional transformation from the continuous time frequency variable λ to the discrete time frequency variable z in (29) maps $\rho_\infty(A)$ one-to-one onto $\rho_\infty(\mathbf{A}(\alpha))$.)

For more details on Cayley transforms we refer the reader to [AN96, Section 5], [Sta02, Section 7], or [Sta05, Section 12.3].

4. Pseudo-similar systems and system nodes

A linear operator Q acting from the Hilbert space $\mathcal{X}$ to the Hilbert space $\mathcal{Y}$ is called a *pseudo-similarity* if it is closed and injective, its domain $\mathcal{D}(Q)$ is dense in $\mathcal{X}$, and its range $\mathcal{R}(Q)$ is dense in $\mathcal{Y}$.

Definition 4.1. We say that two systems Σ_i, $i=1,2$, with state spaces $\mathcal{X}_i$, semigroups $\mathfrak{A}_i^t$, $t\ge 0$, input maps $\mathfrak{B}_i$, output maps $\mathfrak{C}_i$, and input/output maps $\mathfrak{D}_i$, are *pseudo-similar* if there is a pseudo-similarity $Q\colon \mathcal{X}_1\supset\mathcal{D}(Q)\to\mathcal{R}(Q)\subset\mathcal{X}_2$ with the following properties:

(i) $\mathcal{D}(Q)$ is invariant under $\mathfrak{A}_1^t$, $t\ge0$, and $\mathcal{R}(Q)$ is invariant under $\mathfrak{A}_2^t$, $t\ge0$;
(ii) $\mathcal{R}(\mathfrak{B}_1)\subset\mathcal{D}(Q)$ and $\mathcal{R}(\mathfrak{B}_2)\subset\mathcal{R}(Q)$;
(iii) The following intertwining conditions hold:

$$\begin{aligned}\mathfrak{A}_2^tQ &= Q\mathfrak{A}_1^t|_{\mathcal{D}(Q)}, \qquad t\ge 0,\\ \mathfrak{C}_2Q &= \mathfrak{C}_1|_{\mathcal{D}(Q)},\quad \mathfrak{B}_2 = Q\mathfrak{B}_1,\quad \mathfrak{D}_2=\mathfrak{D}_1.\end{aligned} \tag{33}$$

Theorem 4.2. *Let $\Sigma_i := (S_i;\mathcal{X}_i,\mathcal{U},\mathcal{Y})$, $i=1,2$, be two systems with main operators A_i, control operators B_i, observation operators C_i, semigroups $\mathfrak{A}_i^t$, $t\ge0$, and transfer functions $\widehat{\mathfrak{D}}_i$. Let $Q\colon \mathcal{X}_1\supset\mathcal{D}(Q)\to\mathcal{R}(Q)\subset\mathcal{X}_2$ be pseudo-similarity, with the graph*

$$G(Q) := \left\{\left[\begin{smallmatrix}Qx\\ x\end{smallmatrix}\right] \,\middle|\, x\in\mathcal{D}(Q)\right\}.$$

Let Ω_∞ be the component of $\rho(A_1)\cap\rho(A_2)$ which contains some right half-plane. Then the following conditions are equivalent:

(i) *The systems Σ_1 and Σ_2 are pseudo-similar with pseudo-similarity operator Q.*
(ii) *The following inclusion holds for some $\lambda\in\Omega_\infty$:*

$$\begin{bmatrix}(\lambda-A_2)^{-1} & 0 & (\lambda-\widehat{A}_2)^{-1}B_2\\ 0 & (\lambda-A_1)^{-1} & (\lambda-\widehat{A}_1)^{-1}B_1\\ C_2(\lambda-A_2)^{-1} & -C_1(\lambda-A_1)^{-1} & \widehat{\mathfrak{D}}_2(\lambda)-\widehat{\mathfrak{D}}_1(\lambda)\end{bmatrix}\begin{bmatrix}G(Q)\\ \mathcal{U}\end{bmatrix}\subset\begin{bmatrix}G(Q)\\ 0\end{bmatrix}. \tag{34}$$

(iii) *The inclusion (34) holds for all $\lambda\in\Omega_\infty$.*

Remark 4.3. It is easy to see that condition (34) is equivalent to the following set of conditions:

$$(\lambda - A_1)^{-1}\mathcal{D}(Q) \subset \mathcal{D}(Q), \quad (\lambda - \widehat{A}_1)^{-1}B_1\mathcal{U} \subset \mathcal{D}(Q), \tag{35}$$

and

$$\begin{aligned} (\lambda - A_2)^{-1}Q &= Q(\lambda - A_1)^{-1}|_{\mathcal{D}(Q)}, \\ C_2(\lambda - A_2)^{-1}Q &= C_1(\lambda - A_1)^{-1}|_{\mathcal{D}(Q)}, \\ (\lambda - \widehat{A}_2)^{-1}B_2 &= Q(\lambda - \widehat{A}_1)^{-1}B_1, \\ \widehat{\mathfrak{D}}_2(\lambda) &= \widehat{\mathfrak{D}}_1(\lambda). \end{aligned} \tag{36}$$

Proof of Theorem 4.2. Proof of (i) $\Rightarrow$ (ii): Fix an arbitrary $\lambda \in \mathbb{C}$ with $\Re\lambda > \beta$, where β is the maximum of the growth bounds of the two semigroups $\mathfrak{A}_i^t$, $t \geq 0$, $i = 1, 2$.

We begin by showing that $\left[\begin{smallmatrix}(\lambda - A_2)^{-1}x_2 \\ (\lambda - A_1)^{-1}x_1\end{smallmatrix}\right] \in G(Q)$ whenever $\left[\begin{smallmatrix}x_2 \\ x_1\end{smallmatrix}\right] \in G(Q)$. Take $\left[\begin{smallmatrix}x_2 \\ x_1\end{smallmatrix}\right] \in G(Q)$, i.e., $x_1 \in \mathcal{D}(Q)$ and $x_2 = Qx_1$. The first intertwining condition in (33) gives $\mathrm{e}^{-\lambda t}\mathfrak{A}_2^t x_2 = Q\mathrm{e}^{-\lambda t}\mathfrak{A}_1^t x_1$. Integrating this identity over $\mathbb{R}^+$ and use the fact that Q is closed we get $(\lambda - A_2)^{-1}x_2 = Q(\lambda - A_1)^{-1}x_1$. Thus, $\left[\begin{smallmatrix}(\lambda - A_2)^{-1}x_2 \\ (\lambda - A_1)^{-1}x_1\end{smallmatrix}\right] \in G(Q)$.

We next show that $C_2(\lambda - A_2)^{-1}x_2 = C_1(\lambda - A_1)^{-1}x_1$ whenever $\left[\begin{smallmatrix}x_2 \\ x_1\end{smallmatrix}\right] \in G(Q)$. We first fix some real $\alpha > \beta$, and considering the case where x_1 is replaced by $x_{1,\alpha} = \alpha(\alpha - A_1)^{-1}x_1$ for some $x_1 \in \mathcal{D}(Q)$ and x_2 is replace by $x_{2,\alpha} = Qx_{1,\alpha}$. Then, by what we have proved so far, $x_{1,\alpha} \in \mathcal{X}_1^1 \cap \mathcal{D}(Q)$ and $x_{2,\alpha} \in \mathcal{X}_2^1$. This implies that for all $t \geq 0$,

$$C_2\mathfrak{A}_2^t x_{2,\alpha} = (\mathfrak{C}_2 x_{2,\alpha})(t) = (\mathfrak{C}_1 x_{1,\alpha})(t) = C_1\mathfrak{A}_1^t x_{1,\alpha}.$$

Multiply this by $\mathrm{e}^{-\lambda t}$ and integrate over $\mathbb{R}^+$ to get

$$C_2(\lambda - A_2)^{-1}x_{2,\alpha} = C_1(\lambda - A_1)^{-1}x_{1,\alpha}.$$

Let $\alpha \to +\infty$ along the real axis. Then $x_{1,\alpha} = \alpha(\alpha - A_1)^{-1}x_1 \to x_1$ in $\mathcal{X}_1$ and $x_{2,\alpha} = Q\alpha(\alpha - A_2)^{-1}x_1 = \alpha(\alpha - A_2)^{-1}Qx_1 \to Qx_1$ in $\mathcal{X}_2$. This implies that $C_2(\lambda - A_2)^{-1}Qx_1 = C_1(\lambda - A_1)^{-1}x_1$ for all $x_1 \in \mathcal{D}(Q)$.

Next we show that $\left[\begin{smallmatrix}(\lambda - \widehat{A}_2)^{-1}B_2u_0 \\ (\lambda - \widehat{A}_1)^{-1}B_1u_0\end{smallmatrix}\right] \in G(Q)$ for all $u_0 \in \mathcal{U}$. By the third intertwining condition in (33), for all $\alpha > \beta$, all $t \in \mathbb{R}^+$, and all $u_0 \in \mathcal{U}$,

$$\begin{bmatrix}\mathrm{e}^{-\lambda t}\int_{-t}^0 \widehat{\mathfrak{A}}_2^{-v}(\mathrm{e}^{\lambda(t+v)} - \mathrm{e}^{\alpha(t+v)})B_2u_0\,dv \\ \mathrm{e}^{-\lambda t}\int_{-t}^0 \widehat{\mathfrak{A}}_2^{-v}(\mathrm{e}^{\lambda(t+v)} - \mathrm{e}^{\alpha(t+v)})B_1u_0\,dv\end{bmatrix} \in G(Q).$$

Here, with $i = 1, 2$,

$$\begin{aligned} \mathrm{e}^{-\lambda t}&\int_{-t}^0 \widehat{\mathfrak{A}}_i^{-v}(\mathrm{e}^{\lambda(t+v)} - \mathrm{e}^{\alpha(t+v)})B_iu_0\,dv \\ &= (1_{\mathcal{X}_i} - \mathrm{e}^{-\lambda t}\mathfrak{A}_i^t)(\lambda - \widehat{A}_i)^{-1}B_iu_0 \\ &\quad - \mathrm{e}^{(\alpha-\lambda)t}(1_{\mathcal{X}_i} - \mathrm{e}^{-\alpha t}\mathfrak{A}_i^t)(\alpha - \widehat{A}_i)^{-1}B_iu_0. \end{aligned}$$

Choose α and λ so that $\beta < \alpha < \Re\lambda$, and let $t \to \infty$. Then the above expression tends to $(\lambda - \widehat{A}_i)^{-1}B_i u_0$ in $\mathcal{X}$, and the closedness of $G(Q)$ implies that $\left[\begin{smallmatrix} (\lambda-\widehat{A}_2)^{-1}B_2u_0 \\ (\lambda-\widehat{A}_1)^{-1}B_1u_0 \end{smallmatrix}\right] \in G(Q)$.

Finally, since $\mathfrak{D}_1 = \mathfrak{D}_2$, by Lemma 2.4 we also have $\widehat{\mathfrak{D}}_2(\lambda) = \widehat{\mathfrak{D}}_1(\lambda)$.

Proof of (ii) $\Rightarrow$ (iii): Fix some $\lambda_0 \in \Omega_\infty$ for which (34) holds. Equivalently, $(\lambda_0 - A_1)^{-1}\mathcal{D}(Q) \subset \mathcal{D}(Q)$, and

$$(\lambda_0 - A_2)^{-1}Q = Q(\lambda_0 - A_1)^{-1}|_{\mathcal{D}(Q)}.$$

By iterating this equation, using the fact that $(\lambda_0 - A_1)^{-1}\mathcal{D}(Q) \subset \mathcal{D}(Q)$, we find that,

$$(\lambda_0 - A_2)^{-k}Q = Q(\lambda_0 - A_1)^{-k}|_{\mathcal{D}(Q)}, \quad k = 1, 2, \ldots. \tag{37}$$

Fix $\left[\begin{smallmatrix} x_2 \\ x_1 \end{smallmatrix}\right] G(Q)$. The function $\lambda \mapsto \left[\begin{smallmatrix} (\lambda-A_2)^{-1}x_2 \\ (\lambda-A_1)^{-1}x_1 \end{smallmatrix}\right]$ is a holomorphic $\left[\begin{smallmatrix} \mathcal{X}_1 \\ \mathcal{X}_2 \end{smallmatrix}\right]$-valued function on Ω_∞, and it follows from (37) that this function itself together with all its derivatives belong to $G(Q)$ at λ_0. Therefore this function must belong to $G(Q)$ for all $\lambda \in \Omega_\infty$: the inner product of this function with any vector in $G(Q)^\perp$ is an analytic function which vanishes together with all its derivatives at λ_0; hence it must vanish everywhere on Ω_∞. This means that the first inclusion in (35) and the first identity in (36) hold for all $\lambda \in \Omega_\infty$.

The proofs of the facts that also the second inclusion in (35) and the second and third identities in (36) hold for all $\lambda \in \Omega_\infty$ are similar to the one above, and we leave them to the reader.

It remains to show that $\widehat{\mathfrak{D}}_2(\lambda) = \widehat{\mathfrak{D}}_1(\lambda)$ for all $\lambda \in \Omega_\infty$. But this follows from (20) and the other intertwining conditions in (36), which give

$$\begin{aligned}
\widehat{\mathfrak{D}}_2(\lambda) &= \widehat{\mathfrak{D}}_2(\lambda_0) + (\lambda_0 - \lambda)C_2(\lambda_0 - A_2)^{-1}(\lambda - \widehat{A}_2)^{-1}B_2 \\
&= \widehat{\mathfrak{D}}_2(\lambda_0) + (\lambda_0 - \lambda)C_2(\lambda_0 - A_2)^{-1}Q(\lambda - \widehat{A}_1)^{-1}B_1 \\
&= \widehat{\mathfrak{D}}_2(\lambda_0) + (\lambda_0 - \lambda)C_2Q(\lambda_0 - A_1)^{-1}(\lambda - \widehat{A}_1)^{-1}B_1 \\
&= \widehat{\mathfrak{D}}_1(\lambda_0) + (\lambda_0 - \lambda)C_1(\lambda_0 - A_1)^{-1}(\lambda - \widehat{A}_1)^{-1}B_1 \\
&= \widehat{\mathfrak{D}}_1(\lambda).
\end{aligned}$$

Proof of (iii) $\Rightarrow$ (i): Fix some real $\Lambda > 0$ so that $[\Lambda, \infty) \in \Omega_\infty$.

We begin by showing that Q intertwines the two semigroups. Take $x_1 \in \mathcal{D}(Q)$. Then, for $\lambda \geq \Lambda$, $(\lambda - A_2)^{-1}Qx_1 = Q(\lambda - A_1)^{-1}x_1$. Iterating this identity we get $(\lambda - A_2)^{-n}Qx_1 = Q(\lambda - A_1)^{-n}x_1$ for all $n \in \mathbb{Z}^+$. In particular, for all $t > 0$ and all sufficiently large n,

$$\left(1 - \frac{t}{n}A_2\right)^{-n}Qx_1 = Q\left(1 - \frac{t}{n}A_1\right)^{-n}x_1.$$

Let $n \to \infty$ to find that $\mathfrak{A}_1^t x_1 \in \mathcal{D}(Q)$, $\mathfrak{A}_1^t Qx_1 \in \mathcal{R}(Q)$, and that $\mathfrak{A}_2^t Qx_1 = Q\mathfrak{A}_1^t x_1$ for all $t \geq 0$.

Next we look at the second intertwining condition in (33). We know that, for all $x_1 \in \mathcal{D}(Q)$,

$$C_2(\lambda - A_2)^{-1}Qx_1 = C_1(\lambda - A_1)^{-1}x_1$$

for $\lambda \geq \Lambda$. Let $\alpha \in \Omega_\infty$, and replace x_1 by $x_{1,\alpha} = \alpha(\alpha - A_1)^{-1}x_1$ where $x_1 \in \mathcal{D}(Q)$. Then (as we saw in the corresponding part of the proof of the implication (i) $\Rightarrow$ (ii)), the above identity is the Laplace transformed version of the identity $\mathfrak{C}_2 Q x_{1,\alpha} = \mathfrak{C}_1 x_{1,\alpha}$, which must then also hold. Let $\alpha \to \infty$. Then $x_{1,\alpha} \to x_1$ in $\mathcal{X}_1$ and $Qx_{1,\alpha} \to Qx_1$ in $\mathcal{X}_2$ (see the proof of the implication (i) $\Rightarrow$ (ii)). By the continuity of $\mathfrak{C}_1$ and $\mathfrak{C}_2$, $\mathfrak{C}_2 Q x_1 = \mathfrak{C}_1 x_1$, $x_1 \in \mathcal{D}(Q)$.

The third intertwining condition in (33) requires us to show that $\mathcal{R}(\mathfrak{B}_1) \subset \mathcal{D}(Q)$ and that $\mathfrak{B}_2 = Q\mathfrak{B}_1$. Actually, it suffices to show this for functions u which vanish on some interval $(-\infty, -t)$ and are given by $u(v) = (\mathrm{e}^{\lambda(t+v)} - \mathrm{e}^{\alpha(t+v)})u_0$ on $[-t, 0]$ for some real $\lambda \geq \alpha \geq \Lambda$, because the span of functions of this type is dense in $W^{1,2}_c(\mathbb{R}^-;\mathcal{U})$, $\mathfrak{B}_1$ and $\mathfrak{B}_2$ are continuous from $W^{1,2}_c(\mathbb{R}^-;\mathcal{U})$ to $\mathcal{X}_1$ and $\mathcal{X}_2$, respectively, and Q is closed. However, for $i = 1, 2$, applying $\mathfrak{B}_i$ to the above function we get

$$\begin{aligned}\mathfrak{B}_i u = \mathrm{e}^{\lambda t}(1_{\mathcal{X}_i} - \mathrm{e}^{-\lambda t}\mathfrak{A}_i^t)(\lambda - \widehat{A}_i)^{-1}B_i u_0 \\ - \mathrm{e}^{\alpha t}(1_{\mathcal{X}_i} - \mathrm{e}^{-\alpha t}\mathfrak{A}_i^t)(\alpha - \widehat{A}_i)^{-1}B_i u_0.\end{aligned}$$

This, together with the first condition in (33), condition (35), and the third condition in (36) implies that $\mathfrak{B}_1 u \in \mathcal{D}(Q)$ and that $\mathfrak{B}_2 = Q\mathfrak{B}_1$.

Finally, that $\mathfrak{D}_1 = \mathfrak{D}_2$ follows from Lemma 2.4. □

In the sequel it shall be important how the operator F_α defined in (21) interacts with the pseudo-similarity operator Q, and, in particular, with its domain. Our following two lemmas address this issue.

Lemma 4.4. *Let $\Sigma := (S; \mathcal{X}, \mathcal{U}, \mathcal{Y})$ be an system node with system operator $S = \left[\begin{smallmatrix} A\&B \\ C\&D \end{smallmatrix}\right]$, main operator A, and control operator B, and let $\mathcal{Z}$ be a subspace of $\mathcal{X}$. Let $\alpha \in \rho(A)$ and define F_α as in (21).*

(i) *$\left[\begin{smallmatrix} \mathcal{Z} \\ \mathcal{U} \end{smallmatrix}\right]$ is invariant under F_α if and only if*

$$(\alpha - A)^{-1}\mathcal{Z} \subset \mathcal{Z}, \quad (\alpha - \widehat{A})^{-1}B\mathcal{U} \subset \mathcal{Z}. \tag{38}$$

(ii) *If (38) holds, then $\left[\begin{smallmatrix} x \\ u \end{smallmatrix}\right]$ belongs to the range of $F_\alpha|_{\left[\begin{smallmatrix} \mathcal{Z} \\ \mathcal{U} \end{smallmatrix}\right]}$ if and only if*

$$\left[\begin{smallmatrix} x \\ u \end{smallmatrix}\right] \in \mathcal{D}(S), \quad x \in \mathcal{Z}, \quad A\&B\left[\begin{smallmatrix} x \\ u \end{smallmatrix}\right] \in \mathcal{Z}. \tag{39}$$

In particular, the range of $F_\alpha|_{\left[\begin{smallmatrix} \mathcal{Z} \\ \mathcal{U} \end{smallmatrix}\right]}$ does not depend on the particular $\alpha \in \rho(A)$, as long as $\left[\begin{smallmatrix} \mathcal{Z} \\ \mathcal{U} \end{smallmatrix}\right]$ is invariant under F_α.

Proof. That (i) holds follows directly from (21), so it suffices to prove (ii).

Suppose first that $\left[\begin{smallmatrix} x \\ u \end{smallmatrix}\right] = F_\alpha\left[\begin{smallmatrix} z \\ u \end{smallmatrix}\right]$ for some $z \in \mathcal{Z} \subset \mathcal{X}$ and $u \in \mathcal{U}$. Then $\left[\begin{smallmatrix} x \\ u \end{smallmatrix}\right] \in \mathcal{D}(S)$ (since F_α maps $\left[\begin{smallmatrix} \mathcal{X} \\ \mathcal{U} \end{smallmatrix}\right]$ into $\mathcal{D}(S)$) and $x \in \mathcal{Z}$ (by the assumed invariance condition). Furthermore, by (21), $\left(\left[\begin{smallmatrix} \alpha & 0 \\ 0 & 1_{\mathcal{U}} \end{smallmatrix}\right] - \left[\begin{smallmatrix} A\&B \\ 0 \quad 0 \end{smallmatrix}\right]\right)\left[\begin{smallmatrix} x \\ u \end{smallmatrix}\right] = \left[\begin{smallmatrix} z \\ u \end{smallmatrix}\right]$. In particular, $A\&B\left[\begin{smallmatrix} x \\ u \end{smallmatrix}\right] = \alpha x - z \in \mathcal{Z}$. Thus, $\left[\begin{smallmatrix} x \\ u \end{smallmatrix}\right] \in \mathcal{D}(S)$, $x \in \mathcal{Z}$, and $A\&B\left[\begin{smallmatrix} x \\ u \end{smallmatrix}\right] \in \mathcal{Z}$ whenever $\left[\begin{smallmatrix} x \\ u \end{smallmatrix}\right]$ belongs to the range of $F_\alpha|_{\left[\begin{smallmatrix} \mathcal{Z} \\ \mathcal{U} \end{smallmatrix}\right]}$.

Conversely, suppose that $\left[\begin{smallmatrix} x \\ u \end{smallmatrix}\right] \in \mathcal{D}(S)$, $x \in \mathcal{Z}$, and $A\&B \left[\begin{smallmatrix} x \\ u \end{smallmatrix}\right] \in \mathcal{Z}$. Define z by $z = \alpha x - A\&B \left[\begin{smallmatrix} x \\ u \end{smallmatrix}\right]$. Then $z \in \mathcal{Z}$ and $\left[\begin{smallmatrix} x \\ u \end{smallmatrix}\right] = F_\alpha \left[\begin{smallmatrix} z \\ u \end{smallmatrix}\right]$, so $\left[\begin{smallmatrix} x \\ u \end{smallmatrix}\right]$ belongs to the range of $F_\alpha|_{\left[\begin{smallmatrix} \mathcal{Z} \\ \mathcal{U} \end{smallmatrix}\right]}$. □

Lemma 4.5. *Let $\Sigma_i := (S_i; \mathcal{X}_i, \mathcal{U}, \mathcal{Y})$, $i = 1, 2$, be two pseudo-similar system nodes with main operators A_i, control operators B_i, and pseudo-similarity operator Q. Let Ω_∞ be the component of $\rho(A_1) \cap \rho(A_2)$ which contains some right half-plane. For each $\lambda \in \Omega_\infty$, define $F_{i,\lambda}$, $i = 1, 2$, by*

$$F_{i,\lambda} = \begin{bmatrix} (\lambda - A_i)^{-1} & (\lambda - \widehat{A}_i)^{-1} B_i \\ 0 & 1_{\mathcal{U}} \end{bmatrix}. \tag{40}$$

Then, for each $\lambda \in \Omega_\infty$, $F_{1,\lambda}$ maps $\left[\begin{smallmatrix} \mathcal{D}(Q) \\ \mathcal{U} \end{smallmatrix}\right]$ into itself, $F_{2,\lambda}$ maps $\left[\begin{smallmatrix} \mathcal{R}(Q) \\ \mathcal{U} \end{smallmatrix}\right]$ into itself, and

$$F_{2,\lambda} \begin{bmatrix} Q & 0 \\ 0 & 1_{\mathcal{U}} \end{bmatrix} = \begin{bmatrix} Q & 0 \\ 0 & 1_{\mathcal{U}} \end{bmatrix} F_{1,\lambda}|_{\left[\begin{smallmatrix} \mathcal{D}(Q) \\ \mathcal{U} \end{smallmatrix}\right]}. \tag{41}$$

In particular, $\left[\begin{smallmatrix} Q & 0 \\ 0 & 1_{\mathcal{U}} \end{smallmatrix}\right]$ maps the range of $F_{1,\lambda}|_{\left[\begin{smallmatrix} \mathcal{D}(Q) \\ \mathcal{U} \end{smallmatrix}\right]}$ one-to-one onto the range of $F_{2,\lambda}|_{\left[\begin{smallmatrix} \mathcal{R}(Q) \\ \mathcal{U} \end{smallmatrix}\right]}$.

Proof. That $F_{1,\lambda}$ maps $\left[\begin{smallmatrix} \mathcal{D}(Q) \\ \mathcal{U} \end{smallmatrix}\right]$ into itself follows from the two inclusions $(\lambda - A_1)^{-1}\mathcal{D}(Q) \subset \mathcal{D}(Q)$ and $(\lambda - \widehat{A}_1)^{-1} B_1 \mathcal{U} \subset \mathcal{D}(Q)$ (see Remark 4.3). Analogously, that $F_{2,\lambda}$ maps $\left[\begin{smallmatrix} \mathcal{R}(Q) \\ \mathcal{U} \end{smallmatrix}\right]$ into itself follows from the two inclusions $(\lambda - A_2)^{-1}\mathcal{R}(Q) \subset \mathcal{R}(Q)$, $(\lambda - \widehat{A}_2)^{-1} B_2 \mathcal{U} \subset \mathcal{R}(Q)$. Finally, (41) follows from (40) and the first and third identities in (36). □

Our next theorem gives a characterization of pseudo-similarity which is given directly in terms of the system operators involved.

Theorem 4.6. *Let $\Sigma_i := (S_i; \mathcal{X}_i, \mathcal{U}, \mathcal{Y})$, $i = 1, 2$, be two systems with system operators $S_i = \left[\begin{smallmatrix} [A\&B]_i \\ [C\&D]_i \end{smallmatrix}\right]$, main operators A_i, and control operators B_i. Let $Q \colon \mathcal{X}_1 \supset \mathcal{D}(Q) \to \mathcal{R}(Q) \subset \mathcal{X}_2$ be a pseudo-similarity, and let Ω_∞ be the component of $\rho(A_1) \cap \rho(A_2)$ which contains some right half-plane. Then the following conditions are equivalent:*

(i) *Σ_1 and Σ_2 are pseudo-similar with pseudo-similarity operator Q.*
(ii) *The following two conditions hold:*
  (a) *(35) holds for some $\lambda \in \Omega_\infty$.*
  (b) *For all $\left[\begin{smallmatrix} x_1 \\ u \end{smallmatrix}\right] \in \mathcal{D}(S_1)$ such that $x_1 \in \mathcal{D}(Q)$ and $[A\&B]_1 \left[\begin{smallmatrix} x_1 \\ u \end{smallmatrix}\right] \in \mathcal{D}(Q)$ we have*

$$S_2 \begin{bmatrix} Q & 0 \\ 0 & 1_{\mathcal{U}} \end{bmatrix} \begin{bmatrix} x_1 \\ u \end{bmatrix} = \begin{bmatrix} Q & 0 \\ 0 & 1_{\mathcal{Y}} \end{bmatrix} S_1 \begin{bmatrix} x_1 \\ u \end{bmatrix}. \tag{42}$$

Proof. Proof of (i) ⇒ (ii). Assume (i). By Theorem 4.2 and Remark 4.3, (35) holds for all $\lambda \in \Omega_\infty$. By Lemma 4.4, $F_{1,\lambda} \left[\begin{smallmatrix} \mathcal{D}(Q) \\ \mathcal{U} \end{smallmatrix}\right] \subset \left[\begin{smallmatrix} \mathcal{D}(Q) \\ \mathcal{U} \end{smallmatrix}\right]$, and the condition imposed on $\left[\begin{smallmatrix} x_1 \\ u \end{smallmatrix}\right]$ in (b) is equivalent to the requirement that $\left[\begin{smallmatrix} x_1 \\ u \end{smallmatrix}\right]$ belongs to the range of $F_{1,\lambda}|_{\left[\begin{smallmatrix}\mathcal{D}(Q)\\ \mathcal{U}\end{smallmatrix}\right]}$. If we replace $\left[\begin{smallmatrix}x_1\\ u\end{smallmatrix}\right]$ in (42) by $F_{1,\lambda}\left[\begin{smallmatrix}x_1\\ u\end{smallmatrix}\right]$ with $x_1 \in \mathcal{D}(Q)$, then a straightforward computation based on (22) shows that the right-hand side becomes

$$\begin{bmatrix} Q & 0 \\ 0 & 1_{\mathcal{Y}} \end{bmatrix} S_1 F_{1,\lambda} \begin{bmatrix} x_1 \\ u \end{bmatrix} = \begin{bmatrix} QA_1(\alpha - A_1)^{-1} & Q\alpha(\alpha - \widehat{A}_1)^{-1}B_1 \\ C_1(\alpha - A_1)^{-1} & \widehat{\mathfrak{D}}_1(\alpha) \end{bmatrix} \begin{bmatrix} x_1 \\ u \end{bmatrix}. \tag{43}$$

A similar computation which also uses (41) shows that

$$S_2 \begin{bmatrix} Q & 0 \\ 0 & 1_{\mathcal{U}} \end{bmatrix} F_{1,\lambda} \begin{bmatrix} x_1 \\ u \end{bmatrix} = \begin{bmatrix} A_2(\alpha - A_2)^{-1}Q & \alpha(\alpha - \widehat{A}_2)^{-1}B_2 \\ C_2(\alpha - A_2)^{-1}Q & \widehat{\mathfrak{D}}_2(\alpha) \end{bmatrix} \begin{bmatrix} x_1 \\ u \end{bmatrix}. \tag{44}$$

By (36), the right-hand sides of (43) and (44) are equal, and this implies (42).

Proof of (ii) ⇒ (i): Assume (ii). Then it follows from (42) with $\left[\begin{smallmatrix}x_1\\ u\end{smallmatrix}\right]$ replaced by $F_{1,\lambda}\left[\begin{smallmatrix}x_1\\ u\end{smallmatrix}\right]$ that for all $x_1 \in \mathcal{D}(Q)$ and all $u \in \mathcal{U}$ (recall (21))

$$F_{2,\lambda}^{-1} \begin{bmatrix} Q & 0 \\ 0 & 1_{\mathcal{U}} \end{bmatrix} F_{1,\lambda} \begin{bmatrix} x_1 \\ u \end{bmatrix} = \begin{bmatrix} Q & 0 \\ 0 & 1_{\mathcal{U}} \end{bmatrix} F_{1,\lambda}^{-1} F_{1,\lambda} \begin{bmatrix} x_1 \\ u \end{bmatrix}.$$

Multiplying this by $F_{2,\lambda}$ to the left we get (41). It follows from (42) that the left-hand sides of (43) and (44) are equal, and by using (41) we conclude that also the right-hand sides of (43) and (44) are equal. This implies (36). By Theorem 4.2, Σ_1 and Σ_2 are pseudo-similar with pseudo-similarity operator Q. □

Definition 4.7. Two *system nodes* $\Sigma_i := (S_i; \mathcal{X}_i, \mathcal{U}, \mathcal{Y})$ with system operators $S_i = \left[\begin{smallmatrix}[A\&B]_i\\ [C\&D]_i\end{smallmatrix}\right]$, $i = 1, 2$, are called *pseudo-similar with pseudo-similarity operator* Q if conditions (ii)(a) and (ii)(b) in Theorem 4.6 hold.

Thus, with this terminology, Theorem 4.6 says that two *systems* Σ_i, $i = 1, 2$, *are pseudo-similar if and only if the corresponding system nodes are pseudo-similar*, with the same pseudo-similarity operator. Two other equivalent characterization of the pseudo-similarity of two system nodes are given by conditions (ii) and (iii) in Theorem 4.2.

Theorem 4.6 can be used to recover S_2 from S_1 or S_1 from S_2 if we know the pseudo-similarity operator Q.

Corollary 4.8. *Let* $\Sigma_i := (S_i; \mathcal{X}_i, \mathcal{U}, \mathcal{Y})$, $i = 1, 2$ *be two pseudo-similar system nodes with system operators* $S_i = \left[\begin{smallmatrix}[A\&B]_i\\ [C\&D]_i\end{smallmatrix}\right]$ *and pseudo-similarity operator* Q. *Then* S_1 *and* S_2 *can be reconstructed from each other in the following way:*

(i) S_1 *is the closure of the restriction of* $\left[\begin{smallmatrix}Q^{-1} & 0\\ 0 & 1_{\mathcal{Y}}\end{smallmatrix}\right] S_2 \left[\begin{smallmatrix}Q & 0\\ 0 & 1_{\mathcal{U}}\end{smallmatrix}\right]$ *to the set of all* $\left[\begin{smallmatrix}x_1\\ u\end{smallmatrix}\right] \in \left[\begin{smallmatrix}\mathcal{D}(Q)\\ \mathcal{U}\end{smallmatrix}\right]$ *such that* $\left[\begin{smallmatrix}Qx_1\\ u\end{smallmatrix}\right] \in \mathcal{D}(S_2)$ *and* $[A\&B]_2\left[\begin{smallmatrix}Qx_1\\ u\end{smallmatrix}\right] \in \mathcal{R}(Q)$.

(ii) S_2 *is the closure of the restriction of* $\left[\begin{smallmatrix}Q & 0\\ 0 & 1_{\mathcal{Y}}\end{smallmatrix}\right] S_1 \left[\begin{smallmatrix}Q^{-1} & 0\\ 0 & 1_{\mathcal{U}}\end{smallmatrix}\right]$ *to the set of all* $\left[\begin{smallmatrix}x_2\\ u\end{smallmatrix}\right] \in \left[\begin{smallmatrix}\mathcal{R}(Q)\\ \mathcal{U}\end{smallmatrix}\right]$ *such that* $\left[\begin{smallmatrix}Q^{-1}x_2\\ u\end{smallmatrix}\right] \in \mathcal{D}(S_1)$ *and* $[A\&B]_1\left[\begin{smallmatrix}Q^{-1}x_2\\ u\end{smallmatrix}\right] \in \mathcal{D}(Q)$.

Proof. Because of the symmetry of the two statements it suffices to prove, for example, (i). As we observed in the proof of Theorem 4.6, the set of conditions imposed on $\left[\begin{smallmatrix}x_1\\ u\end{smallmatrix}\right]$ in condition (ii) in that theorem is equivalent to the requirement that $\left[\begin{smallmatrix} x_1 \\ u \end{smallmatrix}\right]$ belongs to the range of $F_{1,\lambda}|_{\left[\begin{smallmatrix} \mathcal{D}(Q) \\ \mathcal{U} \end{smallmatrix}\right]}$. By Lemma 4.5, this is equivalent to the requirement that $\left[\begin{smallmatrix} Qx_1 \\ u \end{smallmatrix}\right]$ belongs to the range of $F_{2,\lambda}|_{\left[\begin{smallmatrix} \mathcal{R}(Q) \\ \mathcal{U} \end{smallmatrix}\right]}$, and by Lemma 4.4, this is equivalent to the set of conditions on $\left[\begin{smallmatrix} x_1 \\ u \end{smallmatrix}\right]$ listed in (i). By Theorem 4.6, and since S_1 is closed, S_1 is a closed extension of the restriction of $\left[\begin{smallmatrix} Q^{-1} & 0 \\ 0 & 1_{\mathcal{U}} \end{smallmatrix}\right] S_2 \left[\begin{smallmatrix} Q & 0 \\ 0 & 1_{\mathcal{U}} \end{smallmatrix}\right]$ to the range of $F_{1,\lambda}|_{\left[\begin{smallmatrix} \mathcal{D}(Q) \\ \mathcal{U} \end{smallmatrix}\right]}$. That this is the minimal closed extension follows from the fact that the range of $F_{1,\lambda}|_{\left[\begin{smallmatrix} \mathcal{D}(Q) \\ \mathcal{U} \end{smallmatrix}\right]}$ is dense in $\mathcal{D}(S_1)$ with respect to the graph norm (because $\mathcal{D}(Q)$ is dense in $\mathcal{X}_1$, and $F_{1,\lambda}$ is a bounded bijection of $\left[\begin{smallmatrix} \mathcal{X}_1 \\ \mathcal{U} \end{smallmatrix}\right]$ onto $\mathcal{D}(S_1)$). □

Theorem 4.9. *Let $\Sigma_i := (S_i; \mathcal{X}_i, \mathcal{U}, \mathcal{Y})$, $i = 1, 2$ be two pseudo-similar system nodes with system operators $S_i = \left[\begin{smallmatrix} [A\&B]_i \\ [C\&D]_i \end{smallmatrix}\right]$ and pseudo-similarity operator Q. Let $s \in \mathbb{R}$ and $u \in W^{2,2}_{\mathrm{loc}}([s,\infty); \mathcal{U})$, and let $\left[\begin{smallmatrix} x_{1,s} \\ u(s) \end{smallmatrix}\right] \in \mathcal{D}(S_1)$ with $x_{1,s} \in \mathcal{D}(Q)$ and $[A\&B]_1 \left[\begin{smallmatrix} x_{1,s} \\ u(s) \end{smallmatrix}\right] \in \mathcal{D}(Q)$. Define $x_{2,s} := Qx_{1,s}$. Then the following conclusions hold.*

(i) *$\left[\begin{smallmatrix} x_{2,s} \\ u(s) \end{smallmatrix}\right] \in \mathcal{D}(S_2)$, so that we can let x_i and y_i, $i = 1, 2$, be the state trajectory and the output of S_i of described in Lemma 2.3 with initial state $x_{i,s}$ and input function u.*

(ii) *For all $t \geq s$, the solutions defined in (i) satisfy $\left[\begin{smallmatrix} x_1(t) \\ u(t) \end{smallmatrix}\right] \in \mathcal{D}(S_1)$, $\left[\begin{smallmatrix} x_2(t) \\ u(t) \end{smallmatrix}\right] \in \mathcal{D}(S_2)$, $x_1(t)$, $\dot{x}_1(t) \in \mathcal{D}(Q)$, $x_2(t)$, $\dot{x}_2(t) \in \mathcal{R}(Q)$, and*

$$x_2(t) = Qx_1(t), \quad \dot{x}_2(t) = Q\dot{x}_1(t), \quad y_2(t) = y_1(t), \qquad t \geq s.$$

Thus, in particular, $[A\&B]_1 \left[\begin{smallmatrix} x_1(t) \\ u(t) \end{smallmatrix}\right] \subset \mathcal{D}(Q)$ and $[A\&B]_2 \left[\begin{smallmatrix} x_2(t) \\ u(t) \end{smallmatrix}\right] \subset \mathcal{R}(Q)$ for all $t \geq s$.

Proof. That (i) holds follows from Lemmas 4.4 and 4.5. Thus, we can define the solution as explained in (i). By Lemma 2.3, $\left[\begin{smallmatrix} x_i(t) \\ u(t) \end{smallmatrix}\right] \in \mathcal{D}(S_i)$ and x_i is continuously differentiable in $\mathcal{X}_i$ for $i = 1, 2$.

We claim that $x_1(t) \in \mathcal{D}(Q)$ and $x_2(t) = Qx_1(t)$ for all $t \geq 0$. To prove this we split each of the two solutions into three parts: one where $x_{i,s} \neq 0$ and $u = 0$, one where $x_{i,s} = 0$ and the input function is $\mathrm{e}^{\lambda(t-s)}u(s)$, and one where $x_{i,s} = 0$ and the input function is $u(t) - \mathrm{e}^{\lambda(t-s)}u(s)$; here $\lambda \in \Omega_\infty$ and $i = 1, 2$. In the first case we have $x_i(t) = \mathfrak{A}_i^{t-s}x_{i,s}$, and the first intertwining condition in (33) implies that $x_1(t) \in \mathcal{D}(Q)$ and $x_2(t) = Qx_1(t)$ for $t \geq s$. In the second case we have

$$x_i(t) = \mathrm{e}^{\lambda(t-s)}(1_{\mathcal{X}_i} - \mathrm{e}^{-\lambda(t-s)}\mathfrak{A}_i^{t-s})(\lambda - \widehat{A}_i)^{-1}B_i u(s),$$

and again we have $x_1(t) \in \mathcal{D}(Q)$ and $x_2(t) = Qx_1(t)$ for $t \geq s$ because of the first condition in (33) and the third condition in (36). In the third case we have

$$x_i(t) = \int_{s-t}^{0} \widehat{\mathfrak{A}}_i^{-v} B_i[u(t+v) - \mathrm{e}^{\lambda(t-s+v)}u(s)]\,dv.$$

This is $\mathfrak{B}_i$ applied to a function in $W_c^{-1,2}(\mathbb{R}^-;\mathcal{U})$, and by the third condition in (33), again $x_1(t) \in \mathcal{D}(Q)$ and $x_2(t) = Qx_1(t)$ for $t \geq s$. Adding these three special solutions we find that the original solutions x_1 and x_2 satisfy $x_1(t) \in \mathcal{D}(Q)$ and $x_2(t) = Qx_1(t)$ for $t \geq s$.

Since both x_1 and $x_2 = Qx_1$ are continuously differentiable and Q is closed, we must have $\dot{x}_1(t) \in \mathcal{D}(Q)$ and $\dot{x}_1(t) = Q\dot{x}_1(t)$ for all $t \geq s$. In particular, $\dot{x}_1(t) = [A\&B]_1 \left[\begin{smallmatrix} x_1(t) \\ u(t) \end{smallmatrix}\right] \subset \mathcal{D}(Q)$ and $\dot{x}_1(t) = [A\&B]_2 \left[\begin{smallmatrix} x_2(t) \\ u(t) \end{smallmatrix}\right] \subset \mathcal{R}(Q)$ for all $t \geq s$. Finally, by (42), $y_2(t) = y_1(t)$ for all $t \geq s$. □

Let us end this section with a short discussion of the pseudo-similarity of two discrete-times systems, based on [AKP05]. We say that two discrete-time systems $\left(\left[\begin{smallmatrix} \mathbf{A}_1 & \mathbf{B}_1 \\ \mathbf{C}_1 & \mathbf{D}_1 \end{smallmatrix}\right];\mathcal{X}_1,\mathcal{U},\mathcal{Y}\right)$ and $\left(\left[\begin{smallmatrix} \mathbf{A}_2 & \mathbf{B}_2 \\ \mathbf{C}_2 & \mathbf{D}_2 \end{smallmatrix}\right];\mathcal{X}_2,\mathcal{U},\mathcal{Y}\right)$ are pseudo-similar if there is a pseudo-similarity $Q\colon \mathcal{X}_1 \supset \mathcal{D}(Q) \to \mathcal{R}(Q) \subset \mathcal{X}_2$ such that $\mathbf{A}_1\mathcal{D}(Q) \subset \mathcal{D}(Q)$, $\mathcal{R}(\mathbf{B}_1) \subset \mathcal{D}(Q)$, and

$$\begin{aligned} \mathbf{A}_2 Q &= Q\mathbf{A}_1|_{\mathcal{D}(Q)}, \\ \mathbf{C}_2 Q &= \mathbf{C}_1|_{\mathcal{D}(Q)}, \\ \mathbf{B}_2 &= Q\mathbf{B}_1, \\ \mathbf{D}_2 &= \mathbf{D}_1. \end{aligned} \tag{45}$$

Theorem 4.10. *Let $\Sigma_i := (S_i;\mathcal{X}_i,\mathcal{U},\mathcal{Y})$, $i = 1,2$, be two system nodes with main operators A_i. Let $Q\colon \mathcal{X}_1 \supset \mathcal{D}(Q) \to \mathcal{R}(Q) \subset \mathcal{X}_2$ be a pseudo-similarity. Let Ω_∞ be the component of $\rho(A_1) \cap \rho(A_2)$ which contains some right half-plane. Then the following conditions are equivalent:*[8]

(i) *Σ_1 and Σ_2 are pseudo-similar with pseudo-similarity operator Q.*
(ii) *For some $\alpha \in \mathbb{C}^+ \cap \Omega_\infty$, the Cayley transforms of Σ_1 and Σ_2 with parameter α defined by (28) are pseudo-similar with pseudo-similarity operator Q.*
(iii) *For all $\alpha \in \mathbb{C}^+ \cap \Omega_\infty$, the Cayley transforms of Σ_1 and Σ_2 with parameter α are pseudo-similar with pseudo-similarity operator Q.*

Proof. This follows directly from Theorem 4.2. □

Theorem 4.11. *Let $\Sigma_i := (S_i;\mathcal{X}_i,\mathcal{U},\mathcal{Y})$, $i = 1,2$, be two minimal systems with main operators A_i, input/output maps $\mathfrak{D}_i$, and transfer functions $\widehat{\mathfrak{D}}_i$. Let Ω_∞ be the component of $\rho(A_1) \cap \rho(A_2)$ which contains some right half-plane. Then the following conditions are equivalent:*

(i) *Σ_1 and Σ_2 are pseudo-similar.*
(ii) *The set $\{\lambda \in \Omega_\infty \mid \widehat{\mathfrak{D}}_1(\lambda) = \widehat{\mathfrak{D}}_2(\lambda)\}$ has an interior cluster point.*
(iii) *$\widehat{\mathfrak{D}}_1(\lambda) = \widehat{\mathfrak{D}}_2(\lambda)$ for all $\lambda \in \Omega_\infty$.*
(iv) *$\mathfrak{D}_1 = \mathfrak{D}_2$.*

Proof. If Σ_1 and Σ_2 are pseudo-similar, then it follows directly from Definition 4.1 that (iv) holds. By Lemma 2.4, (ii), (iii) and (iv) are equivalent. Thus, it only remains to show that (iii) ⇒ (i).

---

[8] See also [AN96, Proposition 7.9].

Assume (iii). By Lemma 3.1, the Cayley transforms of Σ_1 and Σ_2 with parameter $\lambda \in \mathbb{C}^+ \cap \Omega_\infty$ are two minimal discrete-time systems, whose transfer functions coincide in a neighborhood of ∞. According to [Aro79, Proposition 6], these two discrete-time systems are pseudo-similar with some pseudo-similarity operator Q. By Theorem 4.10, Σ_1 and Σ_2 are pseudo-similar with the same pseudo-similarity operator Q. □

5. H-passive systems

The following definition is a closely related to the corresponding definition in the two classical papers [Wil72a, Wil72b] (Willems allows the system to be nonlinear and his storage functions are locally bounded).

By a *nonnegative operator* in a Hilbert space $\mathcal{X}$ we mean a (possibly unbounded) self-adjoint operator H satisfying $\langle x, Hx\rangle_{\mathcal{X}} \geq 0$ for all $x \in \mathcal{D}(H)$. If, in addition, $\langle x, Hx\rangle_{\mathcal{X}} > 0$ for all nonzero $x \in \mathcal{D}(H)$, then we call H *positive*. The (unique) nonnegative self-adjoint square root of such a nonnegative operator H is denoted by $\sqrt{H}$.

Definition 5.1. A system node (or system) $\Sigma := (S; \mathcal{X}, \mathcal{U}, \mathcal{Y})$ with system operator $S = \left[\begin{smallmatrix} A\&B \\ C\&D \end{smallmatrix}\right]$ is (scattering) H*-passive* (or simply *passive* if $H = 1_{\mathcal{X}}$) if the following conditions hold:

(i) H is a positive operator on $\mathcal{X}$. Let $Q = \sqrt{H}$.
(ii) If $u \in W^{2,2}_{\mathrm{loc}}([s,\infty); \mathcal{U})$ and $\left[\begin{smallmatrix} x_s \\ u(s) \end{smallmatrix}\right] \in \mathcal{D}(S)$ with $x_s \in \mathcal{D}(Q)$ and $A\&B \left[\begin{smallmatrix} x_s \\ u(s) \end{smallmatrix}\right] \in \mathcal{D}(Q)$, then the solution x in Lemma 2.3 satisfies $x(t)$, $\dot{x}(t) \in \mathcal{D}(Q)$ for all $t \geq s$, and both Qx and its derivative are continuous in $\mathcal{X}$ on $[s, \infty)$.
(iii) Each solution of the type described in (ii) satisfies for all $s \leq t$,

$$\langle Qx(t), Qx(t)\rangle_{\mathcal{X}} + \int_s^t \|y(v)\|^2_{\mathcal{Y}}\, dv \leq \langle Qx(s), Qx(s)\rangle_{\mathcal{X}} + \int_s^t \|u(v)\|^2_{\mathcal{U}}\, dv. \tag{46}$$

If (46) holds in the form of an equality for all $s \leq t$,

$$\langle Qx(t), Qx(t)\rangle_{\mathcal{X}} + \int_s^t \|y(v)\|^2_{\mathcal{Y}}\, dv = \langle Qx(s), Qx(s)\rangle_{\mathcal{X}} + \int_s^t \|u(v)\|^2_{\mathcal{U}}\, dv. \tag{47}$$

then Σ is (scattering) *forward H-conservative,*

We denote the set of all positive operators H for which Σ is H passive by M_Σ.

As our following theorem shows, a system is H-passive (i.e., $H \in M_\Sigma$) if and only if it is pseudo similar to a passive system.

Theorem 5.2. *Let $\Sigma := (S; \mathcal{X}, \mathcal{U}, \mathcal{Y})$ be a system node.*

(i) *If Σ is pseudo-similar to a passive system node $\Sigma_1 := (S_1; \mathcal{X}_1, \mathcal{U}, \mathcal{Y})$ with pseudo-similarity operator Q, then Σ is H-passive with $H := Q^*Q$.*
(ii) *Conversely, if Σ is H-passive, and if $Q\colon \mathcal{X} \to \mathcal{X}_Q$ is an arbitrary pseudo-similarity satisfying $Q^*Q = H$ (for example, we can take $\mathcal{X}_Q = \mathcal{X}$ and $Q = \sqrt{H}$), then Σ is pseudo-similar to a unique passive system node $\Sigma_Q =$*

$(S_Q; \mathcal{X}_Q, \mathcal{U}, \mathcal{Y})$, with pseudo-similarity operator Q. The system operator S_Q is the closure of the restriction of $\left[\begin{smallmatrix} Q & 0 \\ 0 & 1_{\mathcal{Y}} \end{smallmatrix}\right] S \left[\begin{smallmatrix} Q^{-1} & 0 \\ 0 & 1_{\mathcal{U}} \end{smallmatrix}\right]$ to the set of all $\left[\begin{smallmatrix} x \\ u \end{smallmatrix}\right] \in \left[\begin{smallmatrix} \mathcal{R}(Q) \\ \mathcal{U} \end{smallmatrix}\right]$ such that $\left[\begin{smallmatrix} Q^{-1}x \\ u \end{smallmatrix}\right] \in \mathcal{D}(S)$ and $A\&B \left[\begin{smallmatrix} Q^{-1}x \\ u \end{smallmatrix}\right] \in \mathcal{D}(Q)$.

Proof. Proof of (i): Under the assumption of (i) it follows from Theorem 4.9 that conditions (ii) and (iii) in Definition 5.1 hold for the given operator Q. Define $H := Q^*Q$. Then H is a positive operator on $\mathcal{X}$, and Q has the polar decomposition $Q = U\sqrt{H}$, where U is a unitary operator $\mathcal{X} \to \mathcal{X}_1$ and $\mathcal{D}(Q) = \mathcal{D}(\sqrt{H})$ (see, e.g., [Kat80, pp. 334–336] or [Sta05, Lemma A.2.5]). This implies conditions (i)–(iii) in Definition 5.1.

Proof of (ii): Suppose that Σ is H-passive, and that $Q \colon \mathcal{X} \to \mathcal{X}_Q$ is an arbitrary pseudo-similarity satisfying $Q^*Q = H$. Denote the main operator of Σ by A. By condition (ii) in Definition 5.1, for each $x_0 \in \mathcal{X}^1 \cap \mathcal{D}(Q)$ with $Ax_0 \in \mathcal{D}(Q)$ and $t \in \mathbb{R}^+$ we can define $\mathfrak{A}_Q^t x_0 := Qx(t)$ and $(\mathfrak{C}_Q x_0)(t) := y(t)$, where $x(\cdot)$ is the state trajectory and $\mathfrak{C}_Q x_0$ is the output function of Σ with initial state $Q^{-1}x_0$ and zero input function u. In other words,

$$\mathfrak{A}_Q^t x_0 = Q\mathfrak{A}^t Q^{-1} x_0, \qquad (\mathfrak{C}_Q x_0)(t) = C\mathfrak{A}^t Q^{-1} x_0, \qquad t \in \mathbb{R}^+.$$

By (47), for all $t \in \mathbb{R}^+$, $\mathfrak{A}_Q^t$ is a contraction on its domain (with the norm of $\mathcal{X}$) into $\mathcal{X}$, and $\mathfrak{C}_Q$ is a contraction from its domain (with the norm of $\mathcal{X}$) into $L^2(\mathbb{R}^+; \mathcal{Y})$. Moreover, it is easy to see that $\mathfrak{A}_Q^t$, $t \geq 0$, is a C_0 semigroup on its domain. Therefore, this semigroup can be extended (being densely defined and uniformly bounded) to a C_0 semigroup on $\mathcal{X}$, and likewise, $\mathfrak{C}_Q$ can be extended to a contraction mapping from all of $\mathcal{X}$ into $L^2(\mathbb{R}^+; \mathcal{Y})$.

We next let $u \in W^{2,2}(\mathbb{R}; \mathcal{U})$ have a support which is bounded to the left. We take some initial time $s < 0$ to the left of the support of u, and let x be the state trajectory and y the output of Σ with initial state $x_s = 0$ and input function u. It follows from Definition 5.1 that $x(0) \in \mathcal{D}(Q)$. This permits us to define $\mathfrak{B}_Q u_- := Qx(0)$ where $u_- = u|_{\mathbb{R}^-}$ and $\mathfrak{D}_Q u = y$. Thus,

$$\mathfrak{B}_Q u = Q\mathfrak{B} u, \qquad \mathfrak{D}_Q u = \mathfrak{D} u.$$

By condition (iii) in Definition 5.1, these two operators are contractions on their domains (with the norm of $L^2(\mathbb{R}; \mathcal{U})$) into their range spaces, so by density and continuity we can extend them to contraction operators defined on all of $L^2(\mathbb{R}^-, \mathcal{U})$ and $L^2(\mathbb{R}, \mathcal{U})$, respectively.

It is easy to see that the quadruple $\left[\begin{smallmatrix} \mathfrak{A}_Q & \mathfrak{B}_Q \\ \mathfrak{C}_Q & \mathfrak{D} \end{smallmatrix}\right]$ is an L^2-well-posed linear system in the sense of [Sta05, Definition 2.2.1], i.e., that $t \mapsto \mathfrak{A}_Q^t$ is a C_0 semigroup, that $\mathfrak{A}^t$, $\mathfrak{B}_Q$ and $\mathfrak{C}_Q$ satisfy the intertwining conditions

$$\mathfrak{A}_Q^t \mathfrak{B}_Q = \mathfrak{B}_Q \tau_-^t, \quad \mathfrak{C}_Q \mathfrak{A}_Q^t = \tau_+^t \mathfrak{C}_Q, \qquad t \geq 0,$$

where τ_-^t is the left-shift on $L^2(\mathbb{R}^-; \mathcal{U})$ and τ_+^t is the left-shift on $L^2(\mathbb{R}^+; \mathcal{Y})$, and that $\mathfrak{C}_Q \mathfrak{B}_Q = \pi_+ \mathfrak{D} \pi_-$ where π_- is the orthogonal projection of $L^2(\mathbb{R}; \mathcal{U})$ onto $L^2(\mathbb{R}^-; \mathcal{U})$ and π_+ is the orthogonal projection of $L^2(\mathbb{R}; \mathcal{Y})$ onto $L^2(\mathbb{R}^+; \mathcal{U})$ (thus, the Hankel operator induced by $\mathfrak{D}$ is $\mathfrak{C}_Q\mathfrak{B}_Q$). This well-posed linear system is induced by some system node $\Sigma_Q := (S_Q; \mathcal{X}, \mathcal{U}, \mathcal{Y})$ (see, e.g., [Sta05, Theorem 4.6.5]). The main operator A_Q of this system node is the generator of $t \mapsto \mathfrak{A}_Q^t$, the observation operator C_Q is given by $C_Q x = (\mathfrak{C}_Q x)(0)$ for $x \in \mathcal{D}(A_Q)$, the control operator B_Q is determined by the fact that $(B_Q^* x_*) = (\mathfrak{B}_Q^* x_*)(0)$ for all $x_* \in \mathcal{D}(A_Q^*)$, and the transfer function coincides with the original transfer function $\widehat{\mathfrak{D}}$ on some right half-plane. We can now apply (21) and (22) with A, B, and C replaced by A_Q, B_Q, and C_Q, and with $\alpha \in \rho(A_Q)$ to recover the system operator S_Q. The semigroup, input map, output map, and input/output map of Σ_Q coincides with the maps given above. By construction, the conditions listed in Definition 4.1 are satisfied, i.e., Σ is pseudo-similar to Σ_Q with pseudo-similarity operator Q. Finally, it follows from condition (iii) in Definition 5.1 that Σ_Q is passive.

The explicit formula for the system operator S_Q given at the end of (ii) is contained in Corollary 4.8. □

Remark 5.3. Instead of appealing to the theory of well-posed linear systems it is possible to prove part (ii) of Theorem 5.2 by reducing it to the corresponding result in discrete time via the Cayley transform. The proof of Theorem 5.7 that we give below does not use part (ii) of Theorem 5.2. In that proof we use the Cayley transform to show that Σ is pseudo-similar to a passive system $\Sigma_{\sqrt{H}}$ with similarity operator $\sqrt{H}$. From this result we can get the general claim in part (ii) of Theorem 5.2 by using the polar factorization of Q.

We denote the set of all $H \in M_\Sigma$ for which the passive system node $\Sigma_{\sqrt{H}}$ defined in part (ii) *of Theorem* 5.2 *is minimal by $M_\Sigma^{\min}$.*

It is not difficult to show (using Lemma 2.7) that this minimality condition is equivalent to the following two conditions:

$$\begin{aligned} \bigvee_{\lambda \in \rho_\infty^+(A)} \mathcal{R}\left(\sqrt{H}(\lambda - \widehat{A})^{-1}B\right) &= \mathcal{X}, \\ \bigcap_{\lambda \in \rho_\infty^+(A)} \mathcal{N}\left(C(\lambda - A)^{-1}|_{\mathcal{D}(\sqrt{H})}\right) &= 0. \end{aligned} \tag{48}$$

For the formulation of our next theorem we recall the definition of the restricted Schur class $\mathcal{S}(\mathcal{U}, \mathcal{Y}; \Omega)$, where Ω is an open connected subset of $\mathbb{C}^+$: $\theta \in \mathcal{S}(\mathcal{U}, \mathcal{Y}; \Omega)$ means that θ is the restriction to Ω of a function in the Schur class $\mathcal{S}(\mathcal{U}, \mathcal{Y}, \mathbb{C}^+)$.

Theorem 5.4. *Let $\Sigma := (S; \mathcal{X}, \mathcal{U}, \mathcal{Y})$ be a system node with main operator A and transfer function $\widehat{\mathfrak{D}}$. Let $\rho_\infty^+(A)$ be the component of $\rho(A) \cap \mathbb{C}^+$ which contains some right half-plane.*

(i) *If Σ is H-passive, i.e., if $H \in M_\Sigma$, then $\widehat{\mathfrak{D}}|_{\rho_\infty^+(A)} \in \mathcal{S}(\mathcal{U}, \mathcal{Y}; \rho_\infty^+(A))$.*

(ii) *Conversely, suppose that Σ is minimal and that $\widehat{\mathfrak{D}}|_{\rho_\infty^+(A)} \in \mathcal{S}(\mathcal{U}, \mathcal{Y}; \rho_\infty^+(A))$. Then Σ is H-passive for some $H \in M_\Sigma^{\min}$.*

Proof. Proof of (i): Suppose generalized Σ is H-passive (see Theorem 5.7). By Theorem 5.2, Σ is pseudo-similar to a passive system $\Sigma_{\sqrt{H}}$, whose transfer function θ is a Schur function (see [AN96, Proposition 4.4] or [Sta05, Theorem 10.3.5 and Lemma 11.1.4]). By Theorem 4.11, the transfer functions of Σ and $\Sigma_{\sqrt{H}}$ coincide on the connected component of $\rho(A) \cap \mathbb{C}^+$. This proves (i).

Proof of (ii): Suppose that the transfer function coincides with some Schur function in some right half-plane. This Schur function has a minimal passive realization Σ_1; see, e.g., [AN96, Proposition 7.6] or [Sta05, Theorem 11.8.14]. Since the two transfer function coincides in some right-half plane, the input/output maps of the two minimal systems are the same, and consequently, by Theorem 4.11, Σ and Σ_1 are pseudo-similar with some pseudo-similarity Q. By Theorem 5.2, this implies that Σ is H-passive with $H = Q^*Q$. The system node $\Sigma_{\sqrt{H}}$ in part (ii) of Theorem 5.2 is unitarily similar to the system node Σ_1 with a similarity operator U obtained from the polar decomposition $Q = U\sqrt{H}$ of Q. Thus, $\Sigma_{\sqrt{H}}$ is minimal. □

Corollary 5.5. *If Σ is minimal, then $M_\Sigma^{\min}$ is nonempty if and only if M_Σ is nonempty.*

Proof. This follows directly from Theorem 5.4. □

In our next theorem we shall characterize the H-passivity of a system node Σ in terms of a solution of the *generalized* (*continuous time scattering*) *KYP inequality.*

Definition 5.6. Let $\Sigma := (S; \mathcal{X}, \mathcal{U}, \mathcal{Y})$ be a system node with system operator $S = \left[\begin{smallmatrix} A\&B \\ C\&D \end{smallmatrix}\right]$, main operator A, and control operator B, and let $\rho_\infty^+(A)$ be the component of $\rho(A) \cap \mathbb{C}^+$ which contains some right half-plane. By a solution of the *generalized* (*continuous time scattering*) *KYP inequality* induced by Σ we mean a linear operator H satisfying the following conditions.

(i) H is a positive operator on $\mathcal{X}$. Let $Q = \sqrt{H}$.
(ii) $(\lambda - A)^{-1}\mathcal{D}(Q) \subset \mathcal{D}(Q)$ for some $\lambda \in \rho_\infty^+(A)$.
(iii) $(\lambda - \widehat{A})^{-1}B\mathcal{U} \subset \mathcal{D}(Q)$ for some $\lambda \in \rho_\infty^+(A)$.
(iv) The operator QAQ^{-1}, defined on its natural domain consisting of those $x \in \mathcal{R}(Q)$ for which $Q^{-1}x \in \mathcal{D}(A)$ and $AQ^{-1}x \in \mathcal{D}(Q)$, is closable.
(v) For all $\left[\begin{smallmatrix} x_0 \\ u_0 \end{smallmatrix}\right] \in \mathcal{D}(S)$ with $x_0 \in \mathcal{D}(Q)$ and $A\&B \left[\begin{smallmatrix} x_0 \\ u_0 \end{smallmatrix}\right] \in \mathcal{D}(Q)$ we have

$$2\Re\langle Q[A\&B]\left[\begin{smallmatrix} x_0 \\ u_0 \end{smallmatrix}\right], Qx_0\rangle_{\mathcal{X}} + \|C\&D\left[\begin{smallmatrix} x_0 \\ u_0 \end{smallmatrix}\right]\|_{\mathcal{Y}}^2 \le \|u_0\|_{\mathcal{U}}^2. \tag{49}$$

If H is bounded with $\mathcal{D}(H) = \mathcal{X}$, then (ii) and (iii) are redundant, and if furthermore H^{-1} is bounded, then also (iv) is redundant. Thus, in this case H is a solution of the generalized KYP inequality if and only if (49) holds for all $\left[\begin{smallmatrix} x_0 \\ u_0 \end{smallmatrix}\right] \in \mathcal{D}(S)$. If $A\&B = \begin{bmatrix} A & B \end{bmatrix}$ and $C\&D = \begin{bmatrix} C & D \end{bmatrix}$, and if A, B, C, D, H and H^{-1} are bounded, then conditions (ii)–(iv) are satisfied and (49) reduces to the standard KYP inequality (7).

The significance of this definition is due to the following theorem.

Theorem 5.7. *Let* $\Sigma := (S; \mathcal{X}, \mathcal{U}, \mathcal{Y})$ *be a system node, and let* H *be a positive operator on* $\mathcal{X}$. *Then the following two conditions are equivalent:*

(i) Σ *is* H*-passive (i.e.,* $H \in M_\Sigma$*),*
(ii) H *is a solution of the generalized KYP-inequality induced by* Σ.

Moreover, Σ *is forward* H*-conservative if and only if condition* (v) *in Definition* 5.6 *holds with the inequality* (49) *replaced by the equality*

$$2\Re\langle Q[A\&B]\left[\begin{smallmatrix} x_0 \\ u_0 \end{smallmatrix}\right], Qx_0\rangle_{\mathcal{X}} + \|C\&D\left[\begin{smallmatrix} x_0 \\ u_0 \end{smallmatrix}\right]\|_{\mathcal{Y}}^2 = \|u_0\|_{\mathcal{U}}^2. \tag{50}$$

In particular, Σ is passive if and only if (49) holds with $Q = 1_{\mathcal{X}}$ for all $\left[\begin{smallmatrix} x_0 \\ u_0 \end{smallmatrix}\right] \in \mathcal{D}(S)$, and it is forward conservative if and only if (50) holds with $Q = 1_{\mathcal{X}}$ for all $\left[\begin{smallmatrix} x_0 \\ u_0 \end{smallmatrix}\right] \in \mathcal{D}(S)$.

As we shall see in a moment, one direction of the proof is fairly simple (the one which says that H-passivity of Σ implies that H is a solution of the generalized KYP-inequality). The proof of the converse is more difficult, especially the proof of the the validity of condition (ii) in Definition 5.1. For that part of the proof we shall need to study the H-passivity of the corresponding discrete time system obtained via a Cayley transform.

Following [AKP05], we call a discrete time system $\mathbf{\Sigma} := \left(\left[\begin{smallmatrix} \mathbf{A} & \mathbf{B} \\ \mathbf{C} & \mathbf{D} \end{smallmatrix}\right]; \mathcal{X}, \mathcal{U}, \mathcal{Y}\right)$ *H-passive* (or simply *passive* if $H = 1_{\mathcal{X}}$), where H is a positive operator on $\mathcal{X}$, if, with $Q := \sqrt{H}$,

$$\mathbf{A}\mathcal{D}(Q) \subset \mathcal{D}(Q), \qquad \mathbf{B}\mathcal{U} \subset \mathcal{D}(Q), \tag{51}$$

and if, for all $x_0 \in \mathcal{D}(Q)$ and $u_0 \in \mathcal{U}$,

$$\|Q(\mathbf{A}x_0 + \mathbf{B}u_0)\|_{\mathcal{X}}^2 + \|\mathbf{C}x_0 + \mathbf{D}u_0\|_{\mathcal{Y}}^2 \le \|Qx_0\|_{\mathcal{X}}^2 + \|u_0\|_{\mathcal{U}}^2. \tag{52}$$

In this case we also refer to H as a *solution of the discrete time (scattering) generalized KYP-inequality* induced by $\mathbf{\Sigma}$. If H is bounded with $\mathcal{D}(H) = \mathcal{X}$, then (51) is redundant, and (52) is equivalent to the discrete time scattering KYP inequality (11). In particular, passivity of $\mathbf{\Sigma}$ is equivalent to the requirement that $\left[\begin{smallmatrix} \mathbf{A} & \mathbf{B} \\ \mathbf{C} & \mathbf{D} \end{smallmatrix}\right]$ is a contraction from $\left[\begin{smallmatrix} \mathcal{X} \\ \mathcal{U} \end{smallmatrix}\right]$ to $\left[\begin{smallmatrix} \mathcal{X} \\ \mathcal{Y} \end{smallmatrix}\right]$.

Lemma 5.8. *Let* $\Sigma := (S; \mathcal{X}, \mathcal{U}, \mathcal{Y})$ *be a system node with main operator* A, *and let* Σ *and* H *satisfy conditions* (i)–(iii) *in Definition* 5.6, *with the same* $\lambda \in \rho_\infty^+(A)$ *in conditions* (ii) *and* (iii). *Then condition* (v) *in Definition* 5.6 *holds if and only if the Cayley transform* $\mathbf{\Sigma}(\lambda) := \left(\begin{bmatrix} \mathbf{A}(\lambda) & \mathbf{B}(\lambda) \\ \mathbf{C}(\lambda) & \mathbf{D}(\lambda) \end{bmatrix}; \mathcal{X}, \mathcal{U}, \mathcal{Y}\right)$ *of* Σ *(with the same parameter* λ *as in* (ii) *and* (iii)*) is* H*-passive.*

Proof. Clearly, by (28), (ii) and (iii) in Definition 5.6 imply (51). Thus, to prove the lemma it suffices to show that (49) is equivalent to (52).

According to Lemma 4.4, we have $\left[\begin{smallmatrix} x \\ u \end{smallmatrix}\right] \in \mathcal{D}(S)$ with $x \in \mathcal{D}(Q)$ and $A\&B\left[\begin{smallmatrix} x \\ u \end{smallmatrix}\right] \in \mathcal{D}(Q)$ if and only if $\left[\begin{smallmatrix} x \\ u \end{smallmatrix}\right] = F_\lambda \begin{bmatrix} \sqrt{2\Re\lambda} & 0 \\ 0 & 1_{\mathcal{U}} \end{bmatrix} \left[\begin{smallmatrix} x_0 \\ u_0 \end{smallmatrix}\right]$ for some $x_0 \in \mathcal{D}(Q)$ and some $u \in \mathcal{U}$.

Replacing $\left[\begin{smallmatrix} x_0 \\ u_0 \end{smallmatrix}\right]$ in (49) by $F_\lambda \left[\begin{smallmatrix} \sqrt{2\Re\lambda} & 0 \\ 0 & 1_{\mathcal{U}} \end{smallmatrix}\right] \left[\begin{smallmatrix} x_0 \\ u_0 \end{smallmatrix}\right]$ and using (21) and (22) we find that (49) is equivalent to the requirement that

$$\begin{aligned} 2\Re\big\langle Q\big[A(\lambda - A)^{-1}\sqrt{2\Re\lambda}\, x_0 + \lambda(\lambda - \widehat{A})^{-1} B u_0\big], Q(\lambda - A)^{-1}\sqrt{2\Re\lambda}\, x_0\big\rangle_{\mathcal{X}} \\ + \|C(\lambda - A)^{-1}\sqrt{2\Re\lambda}\, x_0 + \widehat{\mathfrak{D}}(\lambda) u_0\|^2_{\mathcal{Y}} \\ \le \|u_0\|^2_{\mathcal{U}} \end{aligned} \tag{53}$$

for all $x_0 \in \mathcal{D}(Q)$ and $u \in \mathcal{U}$. If we here replace $A(\lambda - A)^{-1}$ by $\lambda(\lambda - A)^{-1} - 1_{\mathcal{X}}$ and expand the resulting expression we get a large number of simple terms. A careful inspection shows that we get exactly the same terms by expanding (52) after replacing $\mathbf{A}(\lambda)$ by $2\Re\lambda(\lambda - A)^{-1} - 1_{\mathcal{X}}$ and replacing $\mathbf{B}(\lambda)$, $\mathbf{C}(\lambda)$, and $\mathbf{D}(\lambda)$ by the expressions given in (28). Thus, (49) and (52) are equivalent. □

Proof of Theorem 5.7. Suppose that Σ is H-passive. We must show that conditions (i)–(v) in Definition 5.6 hold. Condition (i) is the same as condition (i) in Definition 5.1. By Theorem 5.2, Σ is pseudo-similar to a system node $\Sigma_Q = (S_Q; \mathcal{X}, \mathcal{U}, \mathcal{Y})$, and (ii) and (iii) follow from Theorem 4.2 (for *all* $\lambda \in \rho^+_\infty(A)$; see (35)). By part (i) of Theorem 5.2, the operator QAQ^{-1} is closable (its closure is equal to the main operator of Σ_Q). Thus (i)–(iv) hold. Divide (46) by $t - s$, let $t - s \downarrow 0$, and use part (iii) of Definition 5.1 (and the closedness of Q) to get

$$2\Re\langle Q\dot{x}(t), Qx(t)\rangle_{\mathcal{X}} + \|y(t)\|_{\mathcal{Y}} \le \|u(t)\|_{\mathcal{U}}, \quad t \ge 0. \tag{54}$$

Here we substitute $\dot{x}(t) = A\&B \left[\begin{smallmatrix} x(t) \\ u(t) \end{smallmatrix}\right]$ and $y(t) = C\&D \left[\begin{smallmatrix} x(t) \\ u(t) \end{smallmatrix}\right]$ and take $t = 0$ to get (49) with $\left[\begin{smallmatrix} x_0 \\ u_0 \end{smallmatrix}\right]$ replaced by $\left[\begin{smallmatrix} x(0) \\ u(0) \end{smallmatrix}\right]$. Thus also (v) holds.

Conversely, suppose that H is a solution of the generalized KYP-inequality. Let us for the moment focus on the main operator A of S, and ignore the other parts of Σ. By Lemma 5.8, applied to a system node with main operator A but no input or output, the conditions (i) and (ii) imply that the Cayley transform $\mathbf{A}(\lambda)$ of A (with the same λ as in (ii)) satisfies $\mathbf{A}(\lambda)\mathcal{D}(Q) \subset \mathcal{D}(Q)$. In particular, we can define $\mathbf{A}_Q(\lambda) := Q\mathbf{A}(\lambda)Q^{-1}$ with $\mathcal{D}(\mathbf{A}_Q(\lambda)) = \mathcal{R}(Q)$. It follows from (v) that $\mathbf{A}_Q(\lambda)$ is a contraction from its domain (with the norm of $\mathcal{X}$) into $\mathcal{X}$. Thus, by density and continuity, $\mathbf{A}_Q(\lambda)$ can be extended to a contraction on $\mathcal{X}$, which we still denote by $\mathbf{A}_Q(\lambda)$.

We claim that $\mathbf{A}_Q(\lambda)$ does not have -1 as an eigenvalue. By the definition of $\mathbf{A}_Q(\lambda)$ as the closure of its restriction to $\mathcal{R}(Q)$, this is equivalent to the claim that if $x_n \in \mathcal{R}(Q)$, $x_n \to x$ in $\mathcal{X}$ and $y_n := (1_{\mathcal{X}} + \mathbf{A}_Q(\lambda))x_n \to 0$ in $\mathcal{X}$, then $x = 0$. Since $1_{\mathcal{X}} + \mathbf{A}_Q(\lambda) = 2\Re\lambda Q(\lambda - A)^{-1})Q^{-1}$, we have

$$2\Re\lambda x_n = (\lambda - QAQ^{-1})y_n.$$

By (iv), the operator $\lambda - QAQ^{-1}$ is closable. Now $y_n \to 0$ in $\mathcal{X}$ and $2\Re\lambda x_n \to 2\Re\lambda x$ in $\mathcal{X}$, so we must have $x = 0$. This proves that $\mathbf{A}_Q(\lambda)$ does not have -1 as an eigenvalue.

Since $\mathbf{A}_Q(\lambda)$ is a contraction which does not have -1 as an eigenvalue, it is the Cayley transform of the generator A_Q of a C_0 contraction semigroup $\mathfrak{A}_Q^t$, $t \geq 0$; see, e.g., [AN96, Theorem 5.2], [SF70, Theorem 8.1, p. 142], or [Sta05, Theorem 12.3.7]. By Theorem 4.10 (applied to the situation where there is no input or output), $\mathfrak{A}^t$, $t \geq 0$, is pseudo-similar to $\mathfrak{A}_Q^t$, $t \geq 0$, with pseudo-similarity operator Q. In particular, by Theorem 4.2, condition (ii) holds for *all* $\lambda \in \rho_\infty^+(A)$.

Since (ii) holds for all $\lambda \in \rho_\infty^+(A)$, we can use the same λ in (ii) as in (iii), and take the Cayley transform of the whole system node Σ. By Lemma 5.8, the Cayley transform $\mathbf{\Sigma}(\lambda) := \left(\left[\begin{smallmatrix} \mathbf{A}(\lambda) & \mathbf{B}(\lambda) \\ \mathbf{C}(\lambda) & \mathbf{D}(\lambda) \end{smallmatrix}\right]; \mathcal{X}, \mathcal{U}, \mathcal{Y}\right)$ is a discrete time scattering H-passive system. Therefore, by [AKP05, Proposition 4.2], this system is pseudo-similar to a passive system, with pseudo-similarity operator $Q = \sqrt{H}$. It is easy to see that the system operator of this contractive system must be the closure of $\left[\begin{smallmatrix} Q^{-1} & 0 \\ 0 & 1_{\mathcal{Y}} \end{smallmatrix}\right] \left[\begin{smallmatrix} \mathbf{A}(\lambda) & \mathbf{B}(\lambda) \\ \mathbf{C}(\lambda) & \mathbf{D}(\lambda) \end{smallmatrix}\right] \left[\begin{smallmatrix} Q & 0 \\ 0 & 1_{\mathcal{U}} \end{smallmatrix}\right]$ (cf. Corollary 4.8). Let us denote this system by $\mathbf{\Sigma}_Q(\lambda) := \left(\left[\begin{smallmatrix} \mathbf{A}_Q(\lambda) & \mathbf{B}_Q(\lambda) \\ \mathbf{C}_Q(\lambda) & \mathbf{D}_Q(\lambda) \end{smallmatrix}\right]; \mathcal{X}, \mathcal{U}, \mathcal{Y}\right)$. As we have shown above, $\mathbf{A}(\lambda)$ does not have -1 as an eigenvalue. This implies that $\mathbf{\Sigma}_Q(\lambda)$ is the Cayley transform with parameter λ of a scattering passive system node Σ_Q; see, e.g., [AN96, Theorem 5.2] or [Sta05, Theorem 12.3.7]. By Theorem 4.10, Σ and Σ_Q are pseudo-similar with pseudo-similarity operator Q. It then follows from Theorem 4.9 that condition (ii) in Definition 5.1 holds. Moreover, $\left[\begin{smallmatrix} x(t) \\ u(t) \end{smallmatrix}\right] \in \mathcal{D}(S)$ with $x(t) \in \mathcal{D}(Q)$ and $A\&B \left[\begin{smallmatrix} x(t) \\ u(t) \end{smallmatrix}\right] \in \mathcal{D}(Q)$ for all $t \geq 0$. Therefore, by (49), (54) holds for all $t \geq 0$. Integrating this inequality over the interval $[s, t]$ we get (46). □

It is possible to replace conditions (ii) and (iv) in Definition 5.6 by another equivalent condition, which can be formulated as follows.

Proposition 5.9. *The positive operator H is a solution of the generalized KYP-inequality if and only if, in addition to conditions* (i), (iii), *and* (v) *in Definition* 5.6, *the following condition holds:*

(ii′) *$\mathfrak{A}^t \mathcal{D}(Q) \subset \mathcal{D}(Q)$ for all $t \in \mathbb{R}^+$, and the function $t \mapsto Q\mathfrak{A}^t x_0$ is continuous on $\mathbb{R}^+$ (with values in $\mathcal{X}$) for all $x_0 \in \mathcal{D}(Q)$,*

where $\mathfrak{A}^t$, $t \geq 0$, is the semigroup on Σ.

Proof. The necessity of (ii′) follows from Theorem 5.7 and condition (ii) in Definition 5.1 (the trajectory x is given by $x(t) = \mathfrak{A}^t x_0$ when $u = 0$). Conversely, if (ii′) holds, then we obtain a C_0 semigroup $\mathfrak{A}_Q^t$, $t \geq 0$, in the same way as we did in the proof of the part (ii) of Theorem 5.2. By repeating the final part of the argument in the proof of the converse part of Theorem 5.7 we find that Σ is H-passive, and by the direct part of the same theorem, H is a solution of the generalized KYP-inequality. □

Corollary 5.10. *Let $\Sigma := (S; \mathcal{X}, \mathcal{U}, \mathcal{Y})$ be a system node, let H be a positive operator on $\mathcal{X}$, and let $Q = \sqrt{H}$. Then the following three conditions are equivalent:*

(i) *Σ is H-passive,*

(ii) *For* some $\lambda \in \rho_\infty^+(A)$, *the Cayley transform* $\left(\left[\begin{smallmatrix}\mathbf{A}(\lambda) & \mathbf{B}(\lambda)\\ \mathbf{C}(\lambda) & \mathbf{D}(\lambda)\end{smallmatrix}\right]; \mathcal{X},\mathcal{U},\mathcal{Y}\right)$ *of* Σ *with parameter* λ *is H-passive, and the closure of the operator* $Q^{-1}\mathbf{A}(\lambda)Q$ *does not have* -1 *as an eigenvalue.*

(iii) *For* all $\lambda \in \rho_\infty^+(A)$, *the Cayley transform* $\left(\left[\begin{smallmatrix}\mathbf{A}(\lambda) & \mathbf{B}(\lambda)\\ \mathbf{C}(\lambda) & \mathbf{D}(\lambda)\end{smallmatrix}\right]; \mathcal{X},\mathcal{U},\mathcal{Y}\right)$ *of* Σ *with parameter* λ *is H-passive, and the closure of the operator* $Q^{-1}\mathbf{A}(\lambda)Q$ *does not have* -1 *as an eigenvalue.*

In particular, when these conditions hold, then conditions (ii) *and* (iii) *in Definition* 5.6 *hold for* all $\lambda \in \rho_\infty^+(A)$.

Proof. As we saw in the first part of the proof of Theorem 5.7, if Σ is H-passive, then conditions (ii) and (iii) in Definition 5.6 hold for *all* $\lambda \in \rho_\infty^+(A)$. We also observed in the proof of Theorem 5.7 that condition (iv) in Definition 5.6 holds if and only if the closure of the operator $Q^{-1}\mathbf{A}(\lambda)Q$ does not have -1 as an eigenvalue. This, combined with Lemma 5.8, implies (iii). Trivially, (iii) $\Rightarrow$ (ii). That (ii) $\Rightarrow$ (i) was established in the proof of the converse part of Theorem 5.7. □

In our next theorem we compare solutions $H \in M_\Sigma^{\min}$ to each other by using the partial ordering of nonnegative self-adjoint operators on $\mathcal{X}$: if H_1 and H_2 are two nonnegative self-adjoint operators on the Hilbert space $\mathcal{X}$, then we write $H_1 \preceq H_2$ whenever $\mathcal{D}(H_2^{1/2}) \subset \mathcal{D}(H_1^{1/2})$ and $\|H_1^{1/2}x\| \le \|H_2^{1/2}x\|$ for all $x \in \mathcal{D}(H_2^{1/2})$. For bounded nonnegative operators H_1 and H_2 with $\mathcal{D}(H_2) = \mathcal{D}(H_1) = \mathcal{X}$ this ordering coincides with the standard ordering of bounded self-adjoint operators.

Theorem 5.11. *Let* $\Sigma := (S; \mathcal{X},\mathcal{U},\mathcal{Y})$ *be a minimal system node with transfer function* $\mathfrak{D}$ *satisfying the condition* $\mathfrak{D}|_{\rho_\infty^+(A)} \in \mathcal{S}(\mathcal{U},\mathcal{Y};\rho_\infty^+(A))$. *Then* $M_\Sigma^{\min}$ *is nonempty, and it contains a minimal element* $H_\circ$ *and a maximal element* $H_\bullet$, *i.e.,*

$$H_\circ \preceq H \preceq H_\bullet, \qquad H \in M_\Sigma^{\min}.$$

Proof. By Theorem 5.4, under the present assumption the set $M_\Sigma^{\min}$ is nonempty. We map both Σ and the pseudo-similar system $\Sigma_{\sqrt{H}}$ into discrete time via the Cayley transform with some parameter $\lambda \in \rho_\infty^+(A)$. By Proposition 5.10, H is a solution of the corresponding discrete time generalized KYP inequality, and by Lemma 3.1, the image $\boldsymbol{\Sigma}_{\sqrt{H}}$ of $\Sigma_{\sqrt{H}}$ under the Cayley transform is minimal. We denote the discrete version of $M_\Sigma^{\min}$ by $M_{\boldsymbol{\Sigma}}^{\min}$. According to [AKP05, Theorem 5.11 and Proposition 5.15], the set $M_{\boldsymbol{\Sigma}}^{\min}$ has a minimal solution $H_\circ$ and a maximal solution $H_\bullet$. The passivity and minimality of $\boldsymbol{\Sigma}_{\sqrt{H}}$ implies that the main operator of $\boldsymbol{\Sigma}_{\sqrt{H}}$ cannot have any eigenvalues with absolute value one, and in particular, it cannot have -1 as an eigenvalue. As we saw in the proof of Theorem 5.7, this condition is equivalent to condition (iv) in Definition 5.6 with $Q = \sqrt{H}$. Thus, due to the extra minimality condition on $\Sigma_{\sqrt{H}}$, there is a one-to-one correspondence between the solutions H of the continuous time generalized KYP-inequality and the discrete time generalized KYP-inequality, and the conclusion of Theorem 5.11 follows from [AKP05, Theorem 5.11 and Proposition 5.15]. □

The two extremal storage functions $E_{H_\circ}$ and $E_{H_\bullet}$ correspond to Willems' [Wil72a, Wil72b] *available storage* and *required supply*, respectively. See [Sta05, Remark 11.8.11] for details.

We remark that if $\Sigma = (S; \mathcal{X}, \mathcal{U}, \mathcal{Y})$ is a minimal passive system, then $M_\Sigma^{\min}$ is nonempty and $H_\circ \preceq 1_\mathcal{X} \preceq H_\bullet$ (since $1_\mathcal{X} \in M_\Sigma^{\min}$). In particular, both $H_\circ$ and $H_\bullet^{-1}$ are bounded.

We end this section by studying how H-passivity of a system is related to $\widetilde{H}$-passivity of its adjoint.

Theorem 5.12. *The system $\Sigma = (S; \mathcal{X}, \mathcal{U}, \mathcal{Y})$ is H-passive if and only if the adjoint system $\Sigma^* = (S^*; \mathcal{X}, \mathcal{Y}, \mathcal{U})$ is H^{-1}-passive.*

Proof. It suffices to prove this in one direction since $(\Sigma^*)^* = \Sigma$. Suppose that Σ is H-passive. Choose some $\alpha \in \rho(A)$, where A is the main operator of Σ. Then, by Proposition 5.10, the Cayley transform $\mathbf{\Sigma}(\alpha) := \left(\begin{bmatrix} \mathbf{A}(\alpha) & \mathbf{B}(\alpha) \\ \mathbf{C}(\alpha) & \mathbf{D}(\alpha) \end{bmatrix}; \mathcal{X}, \mathcal{U}, \mathcal{Y}\right)$ of Σ is H-passive, and -1 is not an eigenvalue of the closure $\mathbf{A}_Q(\lambda)$ of $Q^{-1}\mathbf{A}(\lambda)Q$. By [AKP05, Proposition 4.6], the adjoint system $\mathbf{\Sigma}(\alpha)^* := \left(\begin{bmatrix} \mathbf{A}(\alpha)^* & \mathbf{C}(\alpha)^* \\ \mathbf{B}(\alpha)^* & \mathbf{D}(\alpha)^* \end{bmatrix}; \mathcal{X}, \mathcal{Y}, \mathcal{U}\right)$ of Σ is H^{-1}-passive. The operator $\mathbf{A}_Q(\lambda)$ is a contraction which does not have -1 as an eigenvalue, and hence -1 is not an eigenvalue of $\mathbf{A}_Q(\lambda)^*$, which is the closure of $Q\mathbf{A}(\lambda)^*Q^{-1}$. The Cayley transform of Σ^* with parameter $\overline{\alpha} \in \rho(A^*)$ is equal to $\mathbf{\Sigma}(\alpha)^*$, and by Proposition 5.10, Σ^* is H^{-1}-passive. □

Theorem 5.13. *Let $\Sigma = (S; \mathcal{X}, \mathcal{U}, \mathcal{Y})$ be a system node. Then*

(i) *$H \in M_\Sigma$ if and only if $H^{-1} \in M_{\Sigma^*}$,*
(ii) *$H \in M_\Sigma^{\min}$ if and only if $H^{-1} \in M_{\Sigma^*}^{\min}$.*

Proof. Assertion (i) is a reformulation of Theorem 5.12. The second assertion follows from the fact that the system $\Sigma_{\sqrt{H}}$ is minimal if and only if $(\Sigma_{\sqrt{H}})^*$ is minimal (see Lemma 2.6), and $(\Sigma_{\sqrt{H}})^* = (\Sigma^*)_{\sqrt{H^{-1}}}$. □

Lemma 5.14. *Let $\Sigma = (S; \mathcal{X}, \mathcal{U}, \mathcal{Y})$ be a minimal system node which is self-adjoint in the sense that $\Sigma = \Sigma^* = (S^*; \mathcal{X}, \mathcal{Y}, \mathcal{U})$ (in particular, $\mathcal{U} = \mathcal{Y}$). If M_Σ is nonempty, then $H_\circ = H_\bullet^{-1}$.*

Proof. By Theorem 5.12 and the fact that Σ is self-adjoint, $H \in M_\Sigma^{\min}$ if and only if $H^{-1} \in M_\Sigma^{\min}$. The inequality $H^{-1} \preceq H_\bullet$ for all $H \in M_\Sigma^{\min}$ implies that $H_\bullet^{-1} \preceq H$ (see [AKP05, Proposition 5.4]). In particular $H_\bullet^{-1} \preceq H_\circ$. But we also have the converse inequality $H_\circ \preceq H_\bullet^{-1}$ since $H_\bullet^{-1} \in H_\Sigma^{\min}$. Thus, $H_\circ = H_\bullet^{-1}$. □

The identity $H_\circ = H_\bullet^{-1}$ implies, in particular, that $H_\circ \preceq H_\circ^{-1}$. It is not difficult to see that this implies that $H_\circ \preceq 1_\mathcal{X} \preceq H_\bullet$. However, we can say even more in this case.

Proposition 5.15. *Let $\Sigma = (S; \mathcal{X}, \mathcal{U}, \mathcal{Y})$ be a minimal system node for which M_Σ is nonempty and $H_\circ = H_\bullet^{-1}$. Then Σ is passive, i.e., $1_\mathcal{X} \in M_\Sigma^{\min}$.*

Proof. This follows from [Sta05, Theorem 11.8.14]. □

Definition 5.16. A minimal passive system Σ with the property that $H_\circ = H_\bullet^{-1}$ is called a *passive balanced system.*[9]

This is equivalent to [Sta05, Definition 11.8.13]. According to [Sta05, Theorem 11.8.14], every Schur function θ has a passive balanced realization, and it is unique up to unitary similarity.

We define $H_\odot \in M_\Sigma^{\min}$ to be a *balanced* solution of the generalized KYP inequality (49) if the system $\Sigma_{\sqrt{H_\odot}}$ constructed from $H_\odot$ is a passive balanced system in the sense of Definition 5.16. Thus, if Σ is minimal and M_Σ is nonempty, then *the generalized KYP inequality has a least one balanced solution $H_\odot$, and all the systems $\Sigma_{\sqrt{H_\odot}}$ obtained from these balanced solutions are unitarily similar.*

6. H-stability

The possible unboundedness of H and H^{-1} where H is a solution of the generalized KYP inequality (49) has important consequences for the stability analysis of Σ. Indeed, in the finite-dimensional setting it is sufficient to prove stability with respect to the storage function E_H defined in (3) in order to get stability with respect to the original norm in the state space, since all norms in a finite-dimensional space are equivalent. This is not true in the infinite-dimensional setting unless H and H^{-1} are bounded. Taking into account that H and H^{-1} may be unbounded we replace the definition of E_H given in (3) by

$$E_H(x) = \langle \sqrt{H}x, \sqrt{H}x \rangle, \quad x \in \mathcal{D}(\sqrt{H}). \tag{55}$$

In this more general setting stability with respect to one storage function E_{H_1} is not equivalent to stability with respect to another storage function E_{H_2}. Moreover, the natural norm to use for the adjoint system is the one obtained from $E_{H^{-1}}$ instead of E_H, taking into account that H is a solution of the generalized KYP inequality (49) if and only if $\tilde{H} = H^{-1}$ is a solution of the adjoint generalized KYP inequality.

Definition 6.1. Let H be a positive operator in a Hilbert space $\mathcal{X}$, and let $t \mapsto \mathfrak{A}^t$, $t \geq 0$, be a C_0 semigroup in $\mathcal{X}$. Then $t \mapsto \mathfrak{A}^t$, $t \geq 0$, is called

(i) strongly H-stable, if $\mathfrak{A}^t\mathcal{D}(H^{1/2}) \subset \mathcal{D}(H^{1/2})$ for all $t \geq 0$ and

$$\lim_{t\to\infty} \|H^{1/2}\mathfrak{A}^t x\| \to 0 \text{ for all } x \in \mathcal{D}(H^{1/2}),$$

(ii) strongly H-$*$-stable, if $(\mathfrak{A}^t)^*\mathcal{R}\left(H^{1/2}\right) \subset \mathcal{R}\left(H^{1/2}\right)$ for all $t \geq 0$ and

$$\lim_{t\to\infty} \|H^{-1/2}(\mathfrak{A}^t)^* x_*\| \to 0 \text{ for all } x_* \in \mathcal{R}\left(H^{1/2}\right),$$

(iii) strongly H-bistable if both (i) and (ii) above hold.

---

[9] We call this realization 'passive balanced' in order to distinguish it from other balanced realizations, such as Hankel balanced and LQG balanced realizations.

Theorem 6.2. *Let $\Sigma := (S; \mathcal{X}, \mathcal{U}, \mathcal{Y})$ be a minimal system node with transfer function $\mathfrak{D}$ satisfying the condition $\mathfrak{D}|_{\rho_\infty^+(A)} = \theta|_{\rho_\infty^+(A)}$ for some $\theta \in \mathcal{S}(\mathcal{U}, \mathcal{Y}; \mathbb{C}^+)$. Let $H_\circ$, $H_\bullet$, and $H_\odot$ be the minimal, the maximal, and a balanced solution in $M_\Sigma^{\min}$ of the generalized KYP inequality. Let $t \mapsto \mathfrak{A}^t$, $t \geq 0$, be the evolution semigroup of Σ. Then the following claims are true:*

(i) *$t \mapsto \mathfrak{A}^t$ is strongly $H_\circ$-stable if and only if the factorization problem*

$$\varphi(\lambda)^* \varphi(\lambda) = 1_{\mathcal{U}} - \theta(\lambda)^* \theta(\lambda) \quad \text{a.e. on } i\mathbb{R} \tag{56}$$

has a solution $\varphi \in \mathcal{S}(\mathcal{U}, \mathcal{Y}_\varphi; \mathbb{C}^+)$ for some Hilbert space $\mathcal{Y}_\varphi$.

(ii) *$t \mapsto \mathfrak{A}^t$ is strongly $H_\bullet$-$*$-stable if and only if the factorization problem*

$$\psi(\lambda)\psi(\lambda)^* = 1_{\mathcal{Y}} - \theta(\lambda)\theta(\lambda)^* \quad \text{a.e. on } i\mathbb{R} \tag{57}$$

has a solution $\psi \in \mathcal{S}(\mathcal{U}_\psi, \mathcal{Y}; \mathbb{C}^+)$ for some Hilbert space $\mathcal{U}_\psi$.

(iii) *$t \mapsto \mathfrak{A}^t$ is strongly $H_\odot$-bistable if and only if both the factorization problems in* (i) *and* (ii) *are solvable.*

In the case where H is the identity we simply call $t \mapsto \mathfrak{A}^t$ strongly stable, strongly $*$-stable, of strongly bi-stable.

Proof of Theorem 6.2. The proofs of all these claims are very similar to each other, so we only prove (i), and leave the analogous proofs of (ii) and (iii) to the reader.

We start by replacing the original system by the passive system $\Sigma_{\sqrt{H_\circ}}$. This system is strongly stable if and only if Σ is strongly $H_\circ$-stable. We map $\Sigma_{\sqrt{H_\circ}}$ into a discrete time system $\mathbf{\Sigma}$ by using the Cayley transform. It is easy to see that $\mathbf{\Sigma}$ is optimal in the sense of [AS05a] (i.e., it has the weakest norm among all passive minimal realizations of the same transfer function). By [SF70, Corollary, p. 149] or [Sta05, Theorem 12.3.10], the main operator $\mathbf{A}$ of $\mathbf{\Sigma}$ is strongly stable (i.e., $\mathbf{A} \in C_{0\bullet}$ in the terminology of [SF70]) if and only if the evolution semigroup of $\Sigma_{\sqrt{H_\circ}}$ is strongly stable, i.e., $t \mapsto \mathfrak{A}^t$ is strongly $H_\circ$-stable. By [AS05a, Lemma 4.4], $\mathbf{A}$ is strongly stable if and only if the discrete time analogue of (56) where $\mathbb{C}^+$ is replaced by the unit disk and θ is replaced by $\theta((\alpha - \overline{\alpha} z)/(1+z))$ has a solution (see (29)). But these two factorization problems are equivalent since $z \mapsto (\alpha - \overline{\alpha} z)/(1+z))$ is a conformal mapping of the unit disk onto the right half-plane. This proves (i). □

7. An example

In this section we present two examples based on the heat equation on a semi-infinite bar. Both of these are minimal systems with the same transfer function θ satisfying the conditions of Theorem 5.4 (so that the the KYP inequality has a generalized solution). The first example is exponentially stable, but $H_\bullet$ is unbounded and $H_\circ$ has an unbounded inverse. In the second example all $H \in M_\Sigma^{\min}$ are unbounded.

We consider a damped heat equation on $\mathbb{R}^+$ with Neumann control and Dirichlet observation, described by the system of equations

$$\begin{aligned} T_t(t,\xi) &= T_{\xi\xi}(t,\xi) - \alpha T(t,\xi), && t,\ \xi \ge 0, \\ T_\xi(t,0) &= -u(t), && t \ge 0, \\ T(t,0) &= y(t), && t \ge 0, \\ T(0,\xi) &= x_0(\xi), && \xi \ge 0. \end{aligned} \tag{58}$$

Here we suppose that the damping coefficient α satisfies $\alpha \ge 1$. The state space $\mathcal{X}$ of the standard realization $\Sigma(S;\mathcal{X},\mathbb{C},\mathbb{C})$ of this system is $\mathcal{X} = L^2(\mathbb{R}^+)$. We interpret $T(t,\xi)$ as a function $t \mapsto x(t)$, where $x(t) \in \mathcal{X}$ is the function $\xi \mapsto T(t,\xi)$, and define the system operator $S = \left[\begin{smallmatrix} A\&B \\ C\&D \end{smallmatrix}\right]$ as follows. We take the main operator to be $(Ax)(\xi) = x''(\xi) - \alpha x(\xi)$ for $x \in \mathcal{D}(A) := \{x \in W^{2,2}(\mathbb{R}^+) \mid x'(0) = 0\}$. We take the control operator to be $(Bc) = \delta_0 c$, $c \in \mathbb{C}$, where δ_0 is the Dirac delta at zero. We define $\mathcal{D}(S)$ to consist of those $\left[\begin{smallmatrix} x \\ c \end{smallmatrix}\right] \in \left[\begin{smallmatrix} \mathcal{X} \\ \mathbb{C} \end{smallmatrix}\right]$ for which x is of the form

$$x(\xi) = x(0) + c\xi + \int_0^\xi \int_0^\eta h(\nu)\,d\nu\,d\eta$$

for some $h \in L^2(\mathbb{R}^+)$, and define $[A\&B]\left[\begin{smallmatrix} x \\ c \end{smallmatrix}\right] = h - \alpha x$ and $[C\&D]\left[\begin{smallmatrix} x \\ c \end{smallmatrix}\right] = x(0)$.

This realization is unitarily similar to another one that we get by applying the Fourier cosine transform to all the vectors in the state space. The Fourier cosine transform is defined by $\tilde{x}(\omega) = \sqrt{2/\pi}\int_0^\infty \cos(\omega\xi)x(\xi)\,d\xi$ for $x \in L^1(\mathbb{R}^+) \cap L^2(\mathbb{R}^+)$, and it can be extended to a unitary and self-adjoint map of $L^2(\mathbb{R}^+)$ onto itself (so that it is its own inverse). Let us denote the Fourier cosine transform of $T(t,\xi)$ and $x_0(\xi)$ with respect to the ξ-variable by $\widetilde{T}(t,\omega)$ and $\tilde{x}_0(\omega)$, respectively. Then $\widetilde{T}(t,\omega)$ satisfies the following set of equations:

$$\begin{aligned} \widetilde{T}_t(t,\omega) &= -(\omega^2+\alpha)\widetilde{T}(t,\omega) + \sqrt{2/\pi}\,u(t), && t,\ \omega \ge 0, \\ y(t) &= \sqrt{2/\pi}\int_0^\infty \widetilde{T}(t,\omega)\,d\omega, && t \ge 0, \\ \widetilde{T}(0,\omega) &= \tilde{x}_0(\omega), && \omega \ge 0. \end{aligned} \tag{59}$$

The system operator $S_0 = \left[\begin{smallmatrix} [A\&B]_0 \\ [C\&D]_0 \end{smallmatrix}\right]$ of the similarity transformed system $\Sigma_0 = (S_0;\mathcal{X},\mathbb{C},\mathbb{C})$ is the following. The state space is still $\mathcal{X} = L^2(\mathbb{R}^+)$. The main operator is $(A_0\tilde{x})(\omega) = -(\omega^2+\alpha)\tilde{x}(\omega)$ for $\tilde{x} \in \mathcal{D}(A_0) := \{\tilde{x} \in \mathcal{X} \mid A_0\tilde{x} \in \mathcal{X}\}$, and the control operator is $(B_0 c)(\omega) = \sqrt{2/\pi}c$, $\omega \ge 0$, for $c \in \mathbb{C}$. The domain $\mathcal{D}(S_0)$ consists of those $\left[\begin{smallmatrix} \tilde{x} \\ c \end{smallmatrix}\right] \in \left[\begin{smallmatrix} \mathcal{X} \\ \mathbb{C} \end{smallmatrix}\right]$ for which $\left(\omega \mapsto -(\omega^2+\alpha)\tilde{x}(\omega) + \sqrt{2/\pi}\,c\right) \in \mathcal{X}$, and $[A\&B]_0$ and $[C\&D]_0$ are defined by $[A\&B]_0\left[\begin{smallmatrix} \tilde{x} \\ c \end{smallmatrix}\right](\omega) = -(\omega^2+\alpha)\tilde{x}(\omega) + \sqrt{2/\pi}\,c$, and $[C\&D]_0\left[\begin{smallmatrix} x \\ c \end{smallmatrix}\right] = \sqrt{2/\pi}\int_0^\infty \tilde{x}(\omega)\,d\omega$ for $\left[\begin{smallmatrix} \tilde{x} \\ c \end{smallmatrix}\right] \in \mathcal{D}(S_0)$. The evolution semigroup is given by $(\mathfrak{A}_0^t\tilde{x})(\xi) = \mathrm{e}^{-(\omega^2+\alpha)t}\tilde{x}(\omega)$, t, $\xi \ge 0$, and consequently, it is exponentially stable. From this representation it is easy to compute the transfer function: it is given for all $\lambda \in \rho(A_0) = \mathbb{C} \setminus (-\infty, -\alpha]$ by

$$\begin{aligned}\widehat{\mathfrak{D}}(\lambda) &= [C\&D]_0 \begin{bmatrix} (\lambda - \widehat{A}_0)^{-1} B_0 \\ 1_{\mathcal{U}} \end{bmatrix} \\ &= \frac{2}{\pi} \int_0^\infty \frac{d\omega}{\lambda + \alpha + \omega^2} = \frac{1}{\sqrt{\lambda + \alpha}}.\end{aligned}$$

In particular, $\widehat{\mathfrak{D}} \in \mathcal{S}(\mathbb{C}^+)$, since we assume that $\alpha \geq 1$. The corresponding impulse response is $b(t) = \frac{1}{\sqrt{\pi}} t^{-1/2} e^{-\alpha t}$, $t \geq 0$. It is easy to see that Σ_0 is minimal, hence so is Σ. Moreover, Σ_0 is exponentially stable, and it is self-adjoint in the sense that Σ_0 coincides with its adjoint Σ_0^*. Therefore, by Lemma 5.14 and Definition 5.16, Σ_0 *is passive balanced.* In particular, it is passive.

It is possible to apply Theorem 6.2 with $\theta(\lambda) = 1/\sqrt{\lambda + \alpha}$ to this example. In this case both factorization problems (i) and (ii) in that theorem coincide, and they are solvable. Consequently, the evolution semigroup $t \mapsto \mathfrak{A}^t$ is strongly $H_\circ$-stable, strongly $H_\bullet$-$*$-stable, and strongly $H_\odot$-bistable (and even exponentially $H_\odot$-stable in this case). Nevertheless, $t \mapsto \mathfrak{A}^t$ is *not* strongly $H_\circ$-$*$-stable or strongly $H_\bullet$-stable. This follows from the fact that θ does not have a meromorphic pseudo-continuation into the left half-plane (see [AS05a] for details).

A closer look at the preceding argument shows that in this example $H_\bullet = H_\circ^{-1}$ *must be unbounded.* This is equivalent to the claim that $\sqrt{H_\bullet}$ and $\sqrt{H_\circ}$ are *not* ordinary similarity transforms in $\mathcal{X}$ (since Σ_0 is passive $H_\circ = H_\bullet^{-1}$ must be bounded). Indeed, they can not be similarity transforms since the different semigroups have different stability properties.

In our second example we use a different method to realize the same impulse response $b(t) = \frac{1}{\sqrt{\pi}} t^{-1/2} e^{-\alpha t}$, $t \geq 0$, with transfer function $\theta(\lambda) = 1/\sqrt{\lambda + \alpha}$, $\lambda \in \mathbb{C}^+$, namely an exponentially weighted version of one of the standard Hankel realizations (we still take $\alpha \geq 1$ so that θ is a Schur function). We begin by first replacing θ by the shifted function $\theta_1(\lambda) := 1/\sqrt{\lambda + \alpha + 1}$, $\lambda \in \mathbb{C}^+$. The corresponding impulse response is $b_1(t) = \frac{1}{\sqrt{\pi}} t^{-1/2} e^{-(1+\alpha)t}$, $t \geq 0$. We realize θ_1 by means of the standard time domain output normalized shift realization described in, e.g., [Sta05, Example 2.6.5(ii)], and we denote this realization by $\Sigma_1 := (S_1; \mathcal{X}, \mathbb{C}, \mathbb{C})$. The state space of this realization is $\mathcal{X} = L^2(\mathbb{R}^+)$ and the system operator $S_1 = \left[\begin{smallmatrix} [A\&B]_1 \\ [C\&D]_1 \end{smallmatrix}\right]$ is defined as follows. We take the main operator to be $(A_1 x)(\xi) = x'(\xi)$ for $x \in \mathcal{D}(A_1) := W^{2,1}(\mathbb{R}^+)$. Then $\mathcal{X}^{-1} = W^{-1,2}(\mathbb{R}^+)$, and $\hat{A}_1 x$ is the distribution derivative of $x \in L^2(\mathbb{R}^+)$. We take the control operator to be $(B_1 c)(\xi) = b_1(\xi) c$ for $c \in \mathbb{C}$. We define $\mathcal{D}(S_1)$ to consist of those $\left[\begin{smallmatrix} x \\ c \end{smallmatrix}\right]$ for which $x \in L^2(\mathbb{R}^+)$ is of the form $x(\xi) = x(0) + \int_1^\xi h(\nu)\, d\nu - c \int_1^\xi b_1(\nu)\, d\nu$ for some $h \in L^2(\mathbb{R}^+)$, and define $[A\&B]_1 \left[\begin{smallmatrix} x \\ c \end{smallmatrix}\right] = h$ and $[C\&D]_1 \left[\begin{smallmatrix} x \\ c \end{smallmatrix}\right] = x(0)$. This realization is output normalized in the sense that the observability Gramian is the identity, and it is minimal because the range of the Hankel operator induced by b_1 is dense in $L^2(\mathbb{R}^+)$ (see [Fuh81, Theorem 3-5, p. 254]). The evolution semigroup $t \mapsto \mathfrak{A}_1^t$ is the left-shift semigroup on $L^2(\mathbb{R}^+)$, i.e., $(\mathfrak{A}_1^t x)(\xi) = x(t + \xi)$ for $t, \xi \geq 0$, and the spectrum of A_1 is the closed left half-plane $\{\Re\lambda \leq 0\}$. From this realization we get a minimal realization $\Sigma_2 := (S_2; \mathcal{X}, \mathbb{C}, \mathbb{C})$ of the original transfer function θ by taking $S_1 = S_1 + \left[\begin{smallmatrix} 1_{\mathcal{X}} & 0 \\ 0 & 0 \end{smallmatrix}\right]$. Clearly the spectrum of the main operator $A_2 := A_1 + 1_{\mathcal{X}}$ is the closed half-plane $\{\Re\lambda \leq 1\}$, the evolution semigroup $t \mapsto \mathfrak{A}_2^t$, given by $(\mathfrak{A}_2^t x)(\xi) = e^t x(t+\xi)$ for t, $\xi \geq 0$, is unbounded, and the transfer function $\mathfrak{D}_2$ is the restriction of θ to the half-plane $\Re\lambda > 1$.

Since $\theta|_{\mathbb{C}^+}$ is a Schur function, it follows from Theorem 5.4 that the generalized KYP inequality (49) has a solution H. Suppose that both H and H^{-1} are bounded. Then our original realization becomes passive if we replace the original norm by the norm induced by the storage function E_H. In particular, with respect to this norm the evolution semigroup is contractive. However, this is impossible since we known that the semigroup is unbounded with respect to the original norm, and the two norms are equivalent. This contradiction shows that H or H^{-1} is unbounded. In this particular case it follows from [Sta05, Theorems 9.4.7 and 9.5.2] that if $H \in M_\Sigma^{\min}$, then H^{-1} is bounded, hence H itself must be unbounded.

¿From the above example we can get another one where both H and H^{-1} must be unbounded for every $H \in M_\Sigma^{\min}$ as follows. We take two independent copies of the transfer function θ considered above, i.e, we look at the matrix-valued transfer function $\begin{bmatrix} \theta(\lambda) & 0 \\ 0 & \theta(\lambda) \end{bmatrix}$. We realize this transfer function by taking two independent realizations of the two blocks, so that we realize one of them with the exponentially weighted output normalized shift realization described above, and the other block with the adjoint of this realization. This will force both H and H^{-1} to be unbounded for every $H \in M_{\widetilde{\Sigma}}^{\min}$, where $\widetilde{\Sigma}$ is the combined system.

Acknowledgment

We gratefully acknowledge useful discussions with M.A. Kaashoek on the discrete time version of the generalized KYP inequality.

References

[AKP05] Damir Z. Arov, Marinus A. Kaashoek, and Derk R. Pik, *The Kalman–Yakubovich–Popov inequality and infinite dimensional discrete time dissipative systems*, J. Operator Theory (2005), 46 pages, To appear.

[AN96] Damir Z. Arov and Mark A. Nudelman, *Passive linear stationary dynamical scattering systems with continuous time*, Integral Equations Operator Theory **24** (1996), 1–45.

[Aro79] Damir Z. Arov, *Passive linear stationary dynamic systems*, Sibir. Mat. Zh. **20** (1979), 211–228, translation in Sib. Math. J. 20 (1979), 149–162.

[AS05a] Damir Z. Arov and Olof J. Staffans, *Bi-inner dilations and bi-stable passive scattering realizations of Schur class operator-valued functions*, Integral Equations Operator Theory (2005), 14 pages, To appear.

[AS05b] ______, *The infinite-dimensional continuous time Kalman–Yakubovich–Popov inequality (with scattering supply rate)*, Proceedings of CDC-ECC'05, 2005.

[AS05c] ______, *State/signal linear time-invariant systems theory. Part I: Discrete time*, The State Space Method, Generalizations and Applications (Basel Boston Berlin), Operator Theory: Advances and Applications, vol. 161, Birkhäuser-Verlag, 2005, pp. 115–177.

[AS05d] ______, *State/signal linear time-invariant systems theory. Part II: Passive discrete time systems*, Manuscript, 2005.

[Fuh81] Paul A. Fuhrmann, *Linear systems and operators in Hilbert space*, McGraw-Hill, New York, 1981.

[IW93] Vlad Ionescu and Martin Weiss, *Continuous and discrete-time Riccati theory: a Popov-function approach*, Linear Algebra Appl. **193** (1993), 173–209.

[Kal63] Rudolf E. Kalman, *Lyapunov functions for the problem of Lur'e in automatic control*, Proc. Nat. Acad. Sci. U.S.A. **49** (1963), 201–205.

[Kat80] Tosio Kato, *Perturbation theory for linear operators*, Springer-Verlag, Berlin Heidelberg New York, 1980.

[LR95] Peter Lancaster and Leiba Rodman, *Algebraic Riccati equations*, Oxford Science Publications, The Clarendon Press Oxford University Press, New York, 1995.

[LY76] Andrei L. Lihtarnikov and Vladimir A. Yakubovich, *A frequency theorem for equations of evolution type*, Sibirsk. Mat. Ž. **17** (1976), no. 5, 1069–1085, 1198, translation in Sib. Math. J. 17 (1976), 790–803 (1977).

[MSW05] Jarmo Malinen, Olof J. Staffans, and George Weiss, *When is a linear system conservative?*, Quart. Appl. Math. (2005), To appear.

[PAJ91] Ian R. Petersen, Brian D.O. Anderson, and Edmond A. Jonckheere, *A first principles solution to the non-singular H^∞ control problem*, Internat. J. Robust Nonlinear Control **1** (1991), 171–185.

[Pan99] Luciano Pandolfi, *The Kalman-Yakubovich-Popov theorem for stabilizable hyperbolic boundary control systems*, Integral Equations Operator Theory **34** (1999), no. 4, 478–493.

[Pop73] Vasile-Mihai Popov, *Hyperstability of control systems*, Editura Academiei, Bucharest, 1973, Translated from the Romanian by Radu Georgescu, Die Grundlehren der mathematischen Wissenschaften, Band 204.

[Sal87] Dietmar Salamon, *Infinite dimensional linear systems with unbounded control and observation: a functional analytic approach*, Trans. Amer. Math. Soc. **300** (1987), 383–431.

[Sal89] ______, *Realization theory in Hilbert space*, Math. Systems Theory **21** (1989), 147–164.

[SF70] Béla Sz.-Nagy and Ciprian Foiaş, *Harmonic analysis of operators on Hilbert space*, North-Holland, Amsterdam London, 1970.

[Šmu86] Yurii L. Šmuljan, *Invariant subspaces of semigroups and the Lax–Phillips scheme*, Deposited in VINITI, No. 8009-B86, Odessa, 49 pages, 1986.

[Sta02] Olof J. Staffans, *Passive and conservative infinite-dimensional impedance and scattering systems (from a personal point of view)*, Mathematical Systems Theory in Biology, Communication, Computation, and Finance (New York), IMA Volumes in Mathematics and its Applications, vol. 134, Springer-Verlag, 2002, pp. 375–414.

[Sta05] ———, *Well-posed linear systems*, Cambridge University Press, Cambridge and New York, 2005.

[Wei94a] George Weiss, *Regular linear systems with feedback*, Math. Control Signals Systems **7** (1994), 23–57.

[Wei94b] ———, *Transfer functions of regular linear systems. Part I: characterizations of regularity*, Trans. Amer. Math. Soc. **342** (1994), 827–854.

[Wil72a] Jan C. Willems, *Dissipative dynamical systems Part I: General theory*, Arch. Rational Mech. Anal. **45** (1972), 321–351.

[Wil72b] ———, *Dissipative dynamical systems Part II: Linear systems with quadratic supply rates*, Arch. Rational Mech. Anal. **45** (1972), 352–393.

[WT03] George Weiss and Marius Tucsnak, *How to get a conservative well-posed linear system out of thin air. I. Well-posedness and energy balance*, ESAIM. Control, Optim. Calc. Var. **9** (2003), 247–274.

[Yak62] Vladimir A. Yakubovich, *The solution of some matrix inequalities encountered in automatic control theory*, Dokl. Akad. Nauk SSSR **143** (1962), 1304–1307.

[Yak74] ———, *The frequency theorem for the case in which the state space and the control space are Hilbert spaces, and its application in certain problems in the synthesis of optimal control. I*, Sibirsk. Mat. Ž. **15** (1974), 639–668, 703, translation in Sib. Math. J. 15 (1974), 457–476 (1975).

[Yak75] ———, *The frequency theorem for the case in which the state space and the control space are Hilbert spaces, and its application in certain problems in the synthesis of optimal control. II*, Sibirsk. Mat. Ž. **16** (1975), no. 5, 1081–1102, 1132, translation in Sib. Math. J. 16 (1974), 828–845 (1976).

Damir Z. Arov
Division of Mathematical Analysis
Institute of Physics and Mathematics
South-Ukrainian Pedagogical University
65020 Odessa, Ukraine

Olof J. Staffans
Åbo Akademi University
Department of Mathematics
FIN-20500 Åbo, Finland
e-mail: `http://www.abo.fi/~staffans/`

Operator Theory:
Advances and Applications, Vol. 171, 73–87

From Toeplitz Eigenvalues through Green's Kernels to Higher-order Wirtinger–Sobolev Inequalities

Albrecht Böttcher and Harold Widom

Abstract. The paper is concerned with a sequence of constants which appear in several problems. These problems include the minimal eigenvalue of certain positive definite Toeplitz matrices, the minimal eigenvalue of some higher-order ordinary differential operators, the norm of the Green kernels of these operators, the best constant in a Wirtinger–Sobolev inequality, and the conditioning of a special least squares problem. The main result of the paper gives the asymptotics of this sequence.

Mathematics Subject Classification (2000). Primary 47B35; Secondary 26D10, 34B27, 39A12.

Keywords. Toeplitz matrix, minimal eigenvalue, Green's kernel, Wirtinger–Sobolev inequality.

1. Introduction and main result

There is a sequence $c_1, c_2, c_3, \dots$ of positive real numbers that emerges in various contexts. Here are five of them.

Minimal eigenvalues of Toeplitz matrices. Given a continuous function a on the complex unit circle $\mathbf{T}$, we denote by $\{a_k\}_{k=-\infty}^{\infty}$ the sequence of the Fourier coefficients,

$$a_k = \frac{1}{2\pi} \int_0^{2\pi} a(e^{i\theta}) e^{-ik\theta} d\theta,$$

and by $T_n(a)$ the $n \times n$ Toeplitz matrix $(a_{j-k})_{j,k=1}^n$. Suppose a is of the form $a(t) = |1-t|^{2\alpha} b(t)$ ($t \in \mathbf{T}$) where α is a natural number and b is a positive function on $\mathbf{T}$ whose Fourier coefficients are subject to the condition $\sum_{k=-\infty}^{\infty} |k|\,|b_k| < \infty$. Then the matrix $T_n(a)$ is positive definite and its smallest eigenvalue $\lambda_{\min}(T_n(a))$

satisfies

$$\lambda_{\min}(T_n(a)) \sim \frac{c_\alpha}{n^{2\alpha}}\, b(1) \quad \text{as} \quad n \to \infty \tag{1}$$

with a certain constant $c_\alpha \in (0, \infty)$ independent of b. Here and in what follows $x_n \sim y_n$ means that $x_n/y_n \to 1$. Kac, Murdock, and Szegő [8] proved that $c_1 = \pi^2$, and Parter [12] showed that $c_2 = 500.5467$, which is the fourth power of the smallest positive number μ satisfying $\cos\mu \cosh\mu = 1$.

Minimal eigenvalues of differential operators. For a natural number α, consider the boundary value problem

$$(-1)^\alpha\, u^{(2\alpha)}(x) = v(x) \quad \text{for} \quad x \in [0, 1], \tag{2}$$

$$u(0) = u'(0) = \cdots = u^{(\alpha-1)}(0) = 0, \quad u(1) = u'(1) = \cdots = u^{(\alpha-1)}(1) = 0. \tag{3}$$

The minimal eigenvalue of this boundary value problem can be shown to be just c_α. If $\alpha = 3$, then the equation $-u^{(6)} = \lambda u$ is satisfied by

$$u(x) = \sum_{k=0}^{5} A_k \exp\left(x\, \sqrt[6]{\lambda}\, \exp\left(\frac{(2k+1)\pi i}{6}\right)\right),$$

and the A_k's are the solution of a homogeneous linear 6×6 system with a matrix depending on λ. We found numerically that the smallest $\lambda > 0$ for which the determinant of this matrix is zero is approximately $\lambda = 61529$ and then computed the determinant for $\lambda = (2\pi)^6$ by hand and proved that it is zero. Thus, $c_3 = (2\pi)^6 = 61529$.

Norms of Green's kernels. Let $G_\alpha(x, y)$ be the Green kernel of problem (2), (3). The solution to (2), (3) is then given by

$$u(x) = \int_0^1 G_\alpha(x, y) v(y) dy. \tag{4}$$

It can be shown that $G_\alpha(x, y)$ is symmetric about the point $(\frac{1}{2}, \frac{1}{2})$ and that

$$G_\alpha(x, y) = \frac{x^\alpha y^\alpha}{[(\alpha-1)!]^2} \int_{\max(x,y)}^1 \frac{(t-x)^{\alpha-1}(t-y)^{\alpha-1}}{t^{2\alpha}}\, dt \tag{5}$$

for $x + y \geq 1$. Let K_α denote the integral operator defined by (4). It is clear that the minimal eigenvalue of (2), (3) equals the inverse of the maximal eigenvalue of the (compact and positive definite) operator K_α on $L^2(0, 1)$. As the maximal eigenvalue of K_α is its norm, we arrive at the equality $1/c_\alpha = \|K_\alpha\|$.

Best constants in Wirtinger–Sobolev inequalities. By a Wirtinger–Sobolev inequality one means an inequality of the form

$$\int_0^1 |u(x)|^2 dx \leq C \int_0^1 |u^{(\alpha)}(x)|^2 dx, \tag{6}$$

where u is required to satisfy certain additional (for example, boundary) conditions. It is well known that the best constant C for which (6) is true for all $u \in C^\alpha[0, 1]$ satisfying $\int_0^1 u(x) dx = 0$ and $u^{(j)}(0) = u^{(j)}(1)$ for $0 \leq j \leq \alpha - 1$

is equal to $1/(2\pi)^{2\alpha}$. However, problem (2), (3) leads to (6) with the additional constraints (3). In this case the best constant in (6) is $C = 1/c_\alpha$.

Conditioning of a least squares problem. Suppose we are given n complex numbers $y_1, \ldots, y_n$ and we want to know whether there exists a polynomial p of degree at most $\alpha - 1$ such that $p(j) = y_j$ for $1 \le j \le n$. Such a polynomial exists if and only if

$$\delta_k := y_k - \binom{\alpha}{1} y_{k+1} + \binom{\alpha}{2} y_{k+2} - \cdots + (-1)^\alpha y_{k+\alpha} = 0 \tag{7}$$

for $1 \le k \le n - \alpha$. Thus, to test the existence of p we may compute

$$D(y_1, \ldots, y_n) = \left(\sum_{k=1}^{n-\alpha} \delta_k^2 \right)^{1/2}$$

and ask whether this is a small number. Let $\mathcal{P}_\alpha$ denote the set of all polynomials of degree at most $\alpha - 1$ and put

$$E(y_1, \ldots, y_n) = \min_{p \in \mathcal{P}_\alpha} \left(\sum_{j=1}^{n} |y_j - p(j)|^2 \right)^{1/2}.$$

The question is whether $E(y_1, \ldots, y_n)$ may be large although $D(y_1, \ldots, y_n)$ is small. The answer to this question is (unfortunately) in the affirmative and is in precise form given by the formula

$$\max_{D(y_1,\ldots,y_n) \neq 0} \frac{E(y_1, \ldots, y_n)}{D(y_1, \ldots, y_n)} \sim \frac{n^\alpha}{\sqrt{c_\alpha}}. \tag{8}$$

Here is our main result on the constants c_α we have encountered in the five problems.

Theorem. *We have the asymptotics*

$$c_\alpha = \sqrt{8\pi\alpha} \left(\frac{4\alpha}{e} \right)^{2\alpha} \left[1 + O\left(\frac{1}{\sqrt{\alpha}} \right) \right] \quad \textit{as} \quad \alpha \to \infty \tag{9}$$

and the bounds

$$\frac{4\alpha - 2}{4\alpha^2 - \alpha} \frac{(4\alpha)! [\alpha!]^2}{[(2\alpha)!]^2} \le c_\alpha \le \frac{4\alpha + 1}{2\alpha + 1} \frac{(4\alpha)! [\alpha!]^2}{[(2\alpha)!]^2} \quad \textit{for every} \quad \alpha \ge 1. \tag{10}$$

In connection with (10), notice that

$$\frac{(4\alpha)! [\alpha!]^2}{[(2\alpha)!]^2} \sim \frac{1}{2} \sqrt{8\pi\alpha} \left(\frac{4\alpha}{e} \right)^{2\alpha}.$$

Thus, the upper bound in (10) is asymptotically exact, while the lower bound in (10) is asymptotically by the factor $1/(2\alpha)$ too small. This last defect is nasty, but on the other hand it is clear that $1/(2\alpha)$ is nothing in comparison with the astronomical growth of $(4\alpha/e)^{2\alpha}$.

We discuss the five problems quoted here in more detail in Section 2. The theorem will be proved in Section 3. Section 4 is devoted to an alternative approach to Wirtinger–Sobolev inequalities and gives a new proof of the coincidence of the constants in all the five problems.

2. Equivalence and history of the five problems

Toeplitz eigenvalues. For $\alpha = 1$, formula (1) goes back to Kac, Murdock, Szegő [8]. In the late 1950's, Seymour Parter and the second of the authors started tackling the general case, with Parter embarking on the Toeplitz case and the second of us on the Wiener-Hopf case. In [12] ($\alpha = 2$) and then in [11], [13] (general α), Parter established (1).

Subsequently, it turned out that the approach developed in [16], [17], [18] can also be used to derive (1). This approach is as follows. Let $[T_n^{-1}(a)]_{j,k}$ be the j,k entry of $T_n^{-1}(a) := (T_n(a))^{-1}$ and consider the functions

$$n\,[T_n^{-1}(a)]_{[nx],[ny]}, \quad (x,y) \in [0,1]^2, \tag{11}$$

where $[nz]$ is the smallest integer in $\{1,\dots,n\}$ that is greater than or equal to nz. Let $K^{(n)}$ denote the integral operator on $L^2(0,1)$ with the kernel (11). One can prove two things. First,

$$\left\| \frac{1}{n^{2\alpha}} K^{(n)} - \frac{1}{b(1)} V_\alpha \right\| \to 0 \quad \text{as} \quad n \to \infty, \tag{12}$$

where V_α is an integral operator with a certain completely identified kernel $F_\alpha(x,y)$. And secondly, the eigenvalues of $K^{(n)}$ are just the eigenvalues of $T_n^{-1}(a)$. These two insights imply that

$$\frac{1}{n^{2\alpha}} \frac{1}{\lambda_{\min}(T_n(a))} = \frac{1}{n^{2\alpha}} \lambda_{\max}(K^{(n)}) \to \frac{1}{b(1)} \lambda_{\max}(V_\alpha)$$

or equivalently,

$$\lambda_{\min}(T_n(a)) \sim \frac{1/\lambda_{\max}(V_\alpha)}{n^{2\alpha}} b(1).$$

The kernel $F_\alpha(x,y)$ is quite complicated, but it resembles the kernel $G_\alpha(x,y)$ given by (5).

Green's kernel. In [11] and [17] it was further established that $F_\alpha(x,y)$ is the Green kernel for the boundary problem (2), (3). This implies at once that actually $F_\alpha(x,y) = G_\alpha(x,y)$ and $V_\alpha = K_\alpha$. Thus, at this point it is clear that in the first three problems of the introduction we have to deal with one and the same constant c_α.

Expression (5) was found in [1]. That paper concentrates on the case where $b = 1$, that is, where $a(t) = |1-t|^{2\alpha}$ ($t \in \mathbf{T}$). Using a formula by Duduchava and Roch for the inverse of $T_n(|1-t|^{2\alpha})$, it is shown in a direct way that

$$n^{1-2\alpha}\,[T_n^{-1}(|1-t|^{2\alpha})]_{[nx],[ny]} \to G_\alpha(x,y) \quad \text{in} \quad L^\infty([0,1]^2).$$

Moreover, [1] has a short, self-contained, and elementary proof of the fact that $G_\alpha(x,y)$ is the Green kernel of (2), (3).

Rambour and Seghier [14] showed that

$$n^{1-2\alpha}\,[T_n^{-1}(|1-t|^{2\alpha}b(t))]_{[nx],[ny]} \to \frac{1}{b(1)}\,G_\alpha(x,y) \quad \text{in} \quad L^\infty([0,1]^2) \tag{13}$$

under the assumptions on b made in the introduction. Evidently, (13) implies (12) (but not vice versa). For $\alpha = 1$, result (13) was known from previous work of Courant, Friedrichs, and Lewy [4] and Spitzer and Stone [15]. The authors of [14] were obviously not aware of papers [11] and [17] and rediscovered again that $G_\alpha(x,y)$ is Green's kernel of (2), (3).

Wirtinger–Sobolev. The connection between the minimal eigenvalue of (2), (3) and the best constant in (6) with the boundary conditions (3) is nearly obvious. Indeed, we have

$$c_\alpha = \min \frac{((-1)^\alpha u^{(2\alpha)}, u)}{(u,u)},$$

where $(\cdot,\cdot)$ is the inner product in $L^2(0,1)$ and the minimum is over all nonzero and smooth functions u satisfying (3). Upon α times partially integrating and using the boundary conditions, one gets

$$c_\alpha = \min \frac{(u^{(\alpha)}, u^{(\alpha)})}{(u,u)} = \min \frac{\int_0^1 |u^{(\alpha)}(x)|^2 dx}{\int_0^1 |u(x)|^2 dx},$$

which is equivalent to saying that the best constant C in (6) with the boundary conditions (3) is $C = 1/c_\alpha$.

Numerous versions of inequalities of the Wirtinger–Sobolev type have been established for many decades. Chapter II of [10] is an excellent source for this topic, including the precise history and 218 references. The original inequality says that

$$\int_0^1 |u(x)|^2 dx - \left|\int_0^1 u(x)dx\right|^2 \le \frac{1}{(2\pi)^2}\int_0^1 |u'(x)|^2 dx \tag{14}$$

whenever $u \in C^1[0,1]$ and $u(0) = u(1)$. This inequality appears in different modifications, sometimes with the additional requirement that $\int_0^1 u(x)dx = 0$ and frequently over the interval $(0,2\pi)$, in which case the constant $1/(2\pi)^2$ becomes 1 (see, e.g., [6, pp. 184-187]). The proof of (14) is in fact very simple: take the Fourier expansion $u(x) = \sum u_k e^{2\pi i k x}$ and use Parseval's equality. We will say more on the matter in Section 4, which contains a direct proof of the fact that the best constant C in (6) with the boundary conditions (3) is the inverse of the constant c_α of (1).

The least squares problem. The least squares result is from [2]. Define the linear operator $\nabla : \mathbf{C}^n \to \mathbf{C}^n$ by $\nabla(y_1,\dots,y_n) = (\delta_1,\dots,\delta_{n-\alpha},0,\dots,0)$, where the δ_k's are given by (7), put $\operatorname{Ker}\nabla := \{y \in \mathbf{C}^n : \nabla y = 0\}$, and denote by $P_{\operatorname{Ker}\nabla}$ the orthogonal projection of $\mathbf{C}^n$ onto $\operatorname{Ker} \nabla$. The left-hand side of (8) is nothing but

$$\max_{y \notin \operatorname{Ker} \nabla} \frac{\|y - P_{\operatorname{Ker} \nabla}\, y\|_2}{\|\nabla y\|_2}, \tag{15}$$

where $\|\cdot\|_2$ is the ℓ^2 norm on $\mathbf{C}^n$. With ∇^+ denoting the Moore-Penrose inverse of ∇, we have the equality $I - P_{\operatorname{Ker} \nabla} = \nabla^+ \nabla$. This shows that (15) is the norm of ∇^+, that is, the inverse of the smallest nonzero singular value of ∇. But $\nabla\nabla^*$ can be shown to be of the form

$$J \begin{pmatrix} T_{n-\alpha}(|1-t|^{2\alpha}) & 0 \\ 0 & O_\alpha \end{pmatrix} J,$$

where J is a permutation matrix and O_α is the $\alpha \times \alpha$ zero matrix. Thus, the smallest nonzero singular value of ∇ is the square root of $\lambda_{\min}(T_{n-\alpha}(|1-t|^{2\alpha})) \sim c_\alpha / n^{2\alpha}$, which brings us back to the beginning.

A wrong conjecture. The first three values of c_α are

$$c_1 = \pi^2 = 9.8696, \quad c_2 = 500.5467, \quad c_3 = (2\pi)^6 = 61529,$$

and the first three values of $((\alpha+1)\pi/2)^{2\alpha}$ are

$$\pi^2 = 9.8696, \quad 493.1335, \quad (2\pi)^6 = 61529.$$

We all know that one should not guess the asymptotics of a sequence from its first three terms. But because of the amazing coincidence in the case $\alpha = 3$, it is indeed tempting to conjecture that $c_\alpha \sim ((\alpha+1)\pi/2)^{2\alpha}$. Our main result shows that this conjecture is wrong. The first three values of the correct asymptotics $c_\alpha \sim \sqrt{8\pi\alpha}\,(4\alpha/e)^{2\alpha}$ are

$$10.8555, \quad 531.8840, \quad 64269.$$

3. Proof of the main result

We employ the equality $1/c_\alpha = \|K_\alpha\|$, where K_α is the integral operator on $L^2(0,1)$ with kernel (5).

The kernel's peak. It will turn out that the main contribution to the kernel $G_\alpha(x,y)$ comes from a neighborhood of $(\frac{1}{2}, \frac{1}{2})$, and so for later convenience we consider instead the integral operator $\widetilde{K}_\alpha$ on $L^2(-1,1)$ whose kernel is

$$\widetilde{G}_\alpha(x,y) = \frac{1}{2}\, G_\alpha\left(\frac{1+x}{2}, \frac{1+y}{2}\right).$$

The operator $\widetilde{K}_\alpha$ has the same norm as K_α, its kernel is symmetric about $(0,0)$, and the main contribution to the kernel comes from a neighborhood of $(0,0)$. If we make the substitution $t \to (1+t)/2$ in the integral we see that

$$\widetilde{G}_\alpha(x,y) = \frac{1}{[(\alpha-1)!]^2}\, \frac{1}{4^\alpha}\, H_\alpha(x,y)$$

with

$$H_\alpha(x,y) = (1+x)^\alpha(1+y)^\alpha \int_{\max(x,y)}^1 \frac{(t-x)^{\alpha-1}(t-y)^{\alpha-1}}{((1+t)/2)^{2\alpha}}\, dt \tag{16}$$

when $x+y \geq 0$. We shall show that $H_\alpha(x,y)$ is equal to $(1/\alpha)(1-x^2)^\alpha(1-y^2)^\alpha$ plus a kernel whose norm is smaller by a factor $O(1/\sqrt{\alpha})$.

The logarithmic derivative in t of the function $(t-x)\,(t-y)/((1+t)/2)^2$ is

$$\frac{(2+x+y)t - x - y - 2xy}{(1+t)(t-x)(t-y)},$$

which is positive for $t > \max(x,y)$. (Recall that we are in the case $x+y \geq 0$.) Hence the function achieves its maximum $(1-x)(1-y)$ at $t=1$ and nowhere else. The function $(1+x)(1+y)(1-x)(1-y)$ achieves its maximum at $x=0$, $y=0$ and nowhere else. Putting these together we see that the function

$$(1+x)(1+y)\,\frac{(t-x)(t-y)}{((1+t)/2)^2}$$

achieves its maximum 1 at $t=1$, $x=0$, $y=0$, and outside a neighborhood of this point, say outside the set $t \geq 1-\varepsilon$, $|x| \leq \varepsilon$, $|y| \leq \varepsilon$, there is a bound

$$(1+x)(1+y)\,\frac{(t-x)(t-y)}{((1+t)/2)^2} < 1-\delta$$

for some $\delta > 0$. It follows that outside the same neighborhood the integrand in (16) with its outside factor is $O((1-\delta)^\alpha)$. This is also the bound after we integrate. We take any $\varepsilon < 1/2$, and have shown that

$$H_\alpha(x,y) = (1+x)^\alpha(1+y)^\alpha\chi_\varepsilon(x)\chi_\varepsilon(y) \int_{1-\varepsilon}^1 \frac{(t-x)^{\alpha-1}(t-y)^{\alpha-1}}{((1+t)/2)^{2\alpha}}\, dt + O((1-\delta)^\alpha),$$

where χ_ε is 1 on $[-\varepsilon,\varepsilon]$ and zero elsewhere. Substituting $t = 1-\tau$ we arrive at the formula

$$\begin{aligned} H_\alpha(x,y) &= \frac{(1-x^2)^\alpha(1-y^2)^\alpha}{(1-x)(1-y)}\,\chi_\varepsilon(x)\chi_\varepsilon(y) \int_0^\varepsilon \left[\frac{\left(1-\frac{\tau}{1-x}\right)\left(1-\frac{\tau}{1-y}\right)}{\left(1-\frac{\tau}{2}\right)^2}\right]^\alpha \times \\ &\quad \times \frac{d\tau}{\left(1-\frac{\tau}{1-x}\right)\left(1-\frac{\tau}{1-y}\right)} + O((1-\delta)^\alpha). \end{aligned} \tag{17}$$

The kernel's asymptotics. Let us compute the asymptotics of the kernel. The choice $\varepsilon < 1/2$ guarantees that $\tau/(1-x)$, $\tau/(1-y)$, $\tau/2$ belong to $(0,1)$. This implies that

$$\frac{\left(1-\frac{\tau}{1-x}\right)\left(1-\frac{\tau}{1-y}\right)}{\left(1-\frac{\tau}{2}\right)^2} = e^{-\tau\varphi(x,y)+O(\tau^2)}$$

with

$$\varphi(x,y) = \frac{1-xy}{(1-x)(1-y)}.$$

We split the integral in (17) into $\int_0^{1/\sqrt{\alpha}}$ and $\int_{1/\sqrt{\alpha}}^{\varepsilon}$. If $\varepsilon > 0$ is small enough, which we may assume, the term $O(\tau^2)$ is at most $\tau\varphi(x,y)/2$ in absolute value. Hence the integral $\int_{1/\sqrt{\alpha}}^{\varepsilon}$ is at most

$$\int_{1/\sqrt{\alpha}}^{\varepsilon} e^{-\alpha\tau\varphi(x,y)/2} O(1)d\tau = O\left(e^{-\gamma_1\sqrt{\alpha}}\right)$$

with some $\gamma_1 > 0$. For $\tau < 1/\sqrt{\alpha}$ we have $\alpha\tau^2 < 1$ and hence $e^{\alpha O(\tau^2)} = 1 + \alpha O(\tau^2)$. Consequently, the integral $\int_0^{1/\sqrt{\alpha}}$ is equal to

$$\begin{aligned} &\int_0^{1/\sqrt{\alpha}} e^{-\alpha\tau\varphi(x,y)} e^{\alpha O(\tau^2)}(1+O(\tau))d\tau \\ &= \int_0^{1/\sqrt{\alpha}} e^{-\alpha\tau\varphi(x,y)} \left(1+O(\tau)+\alpha O(\tau^2)\right) d\tau. \end{aligned} \tag{18}$$

Since, for $k = 0, 1, 2$,

$$\int_{1/\sqrt{\alpha}}^{\infty} \tau^k e^{-\alpha\tau\varphi(x,y)} d\tau = O\left(e^{-\gamma_2\sqrt{\alpha}}\right)$$

with $\gamma_2 > 0$ and

$$\int_0^{\infty} \tau^k e^{-\alpha\tau\varphi(x,y)} d\tau = O\left(\frac{1}{\alpha^{k+1}}\right),$$

it follows that (18) is

$$\begin{aligned} &\int_0^{\infty} e^{-\alpha\tau\varphi(x,y)} \left(1+O(\tau)+\alpha O(\tau^2)\right) d\tau + O\left(e^{-\gamma_2\sqrt{\alpha}}\right) \\ &= \left.\frac{e^{-\alpha\tau\varphi(x,y)}}{-\alpha\varphi(x,y)}\right|_0^{\infty} + O\left(\frac{1}{\alpha^2}\right) + \alpha\, O\left(\frac{1}{\alpha^3}\right) + O\left(e^{-\gamma_2\sqrt{\alpha}}\right) \\ &= \frac{1}{\alpha\varphi(x,y)} + O\left(\frac{1}{\alpha^2}\right). \end{aligned}$$

In summary,

$$\begin{aligned} H_\alpha(x,y) &= \frac{(1-x^2)^\alpha(1-y^2)^\alpha}{(1-x)(1-y)\,\alpha\,\varphi(x,y)}\,\chi_\varepsilon(x)\chi_\varepsilon(y) + O\left(\frac{1}{\alpha^2}\right) \\ &= \frac{(1-x^2)^\alpha(1-y^2)^\alpha}{\alpha\,(1-xy)}\,\chi_\varepsilon(x)\chi_\varepsilon(y) + O\left(\frac{1}{\alpha^2}\right), \end{aligned}$$

uniformly for $|x|, |y| \le \varepsilon$. Expanding near $x = y = 0$ we obtain

$$H_\alpha(x,y) = \frac{1}{\alpha}(1-x^2)^\alpha(1-y^2)^\alpha\,(1+O(xy)) + O\left(\frac{1}{\alpha^2}\right), \tag{19}$$

again uniformly. This was derived for $|x|, |y| \le \varepsilon$, but because of (17) we see that this holds uniformly for all x and y satisfying $x + y \ge 0$. This last condition can also be dropped by the symmetry of $H_\alpha(x,y)$.

The asymptotics of the norm. If an integral operator K is of the form

$$(Ku)(x) = \int_{-1}^{1} f(x)g(y)u(y)dy,$$

then $\|K\| = \|f\|_2\|g\|_2$, where $\|\cdot\|_2$ is the norm in $L^2(-1,1)$. Let us denote the integral operator with the kernel $H_\alpha(x,y)$ by M_α. Furthermore, in view of (19) we denote by M_α^0 and M_α^1 the integral operators with the kernels $(1-x^2)^\alpha(1-y^2)^\alpha$ and $O(xy)\,(1-x^2)^\alpha(1-y^2)^\alpha$, respectively. From (19) we infer that

$$\|M_\alpha\| = \frac{1}{\alpha}\,\|M_\alpha^0 + M_\alpha^1\| + O\left(\frac{1}{\alpha^2}\right) = \frac{1}{\alpha}\left[\|M_\alpha^0\| + O\left(\|M_\alpha^1\|\right) + O\left(\frac{1}{\alpha}\right)\right].$$

Since

$$\begin{aligned} \|M_\alpha^0\| &= \int_{-1}^{1}(1-x^2)^{2\alpha}dx = \sqrt{\frac{\pi}{2\alpha}}\left(1+O\left(\frac{1}{\alpha}\right)\right), \\ \|M_\alpha^1\| &= \int_{-1}^{1} O(x^2)\,(1-x^2)^{2\alpha}dx = O\left(\frac{1}{\alpha^{3/2}}\right), \end{aligned}$$

we finally get

$$\begin{aligned} \|K_\alpha\| &= \|\widetilde{K}_\alpha\| = \frac{1}{[(\alpha-1)!]^2}\,\frac{1}{4^{2\alpha}}\,\|M_\alpha\| \\ &= \frac{1}{[(\alpha-1)!]^2}\,\frac{1}{4^{2\alpha}\,\alpha}\left[\|M_\alpha^0\| + O\left(\|M_\alpha^1\|\right) + O\left(\frac{1}{\alpha}\right)\right] \\ &= \frac{\alpha^2}{\alpha^{2\alpha}e^{-2\alpha}2\pi\alpha}\left(1+O\left(\frac{1}{\alpha}\right)\right)\frac{1}{4^{2\alpha}\,\alpha}\left[\sqrt{\frac{\pi}{2\alpha}} + O\left(\frac{1}{\alpha}\right)\right] \\ &= \left(\frac{e}{4\alpha}\right)^{2\alpha}\frac{1}{\sqrt{8\pi\alpha}}\left(1+O\left(\frac{1}{\sqrt{\alpha}}\right)\right), \end{aligned}$$

which is the same as (9).

The lower bound. To prove the lower bound in (10), we start with (16) and the inequality

$$\frac{(t-x)(t-y)}{((1+t)/2)^2} \le (1-x)(1-y),$$

which was established in the course of the above proof. If $x+y \ge 0$, then $\max(x,y) \ge 0$ and consequently,

$$\begin{aligned} H_\alpha(x,y) &\le (1+x)^\alpha(1+y)^\alpha \int_0^1 (1-x)^{\alpha-1}(1-y)^{\alpha-1}\,\frac{dt}{((1+t)/2)^2} \\ &= 2(1+x)(1-x^2)^{\alpha-1}(1+y)(1-y^2)^{\alpha-1}. \end{aligned}$$

Hence

$$\|M_\alpha\| \le 2\int_{-1}^{1}(1+x)^2(1-x^2)^{2\alpha-2}dx = 4^{2\alpha}\,\frac{(2\alpha)!(2\alpha-2)!}{(4\alpha-1)!}$$

and thus

$$\|K_\alpha\| = \|\widetilde{K}_\alpha\| \le \frac{1}{[(\alpha-1)!]^2}\,\frac{1}{4^{2\alpha}}\,4^{2\alpha}\,\frac{(2\alpha)!(2\alpha-2)!}{(4\alpha-1)!} = \frac{\alpha^2}{\alpha!\alpha!}\,\frac{(2\alpha)!(2\alpha)!(4\alpha-1)}{(2\alpha-1)(2\alpha)(4\alpha)!},$$

which is equivalent to the assertion.

The upper bound. The proof of the upper bound in (10) is based on the observation that $1/c_\alpha$ is the best constant for which the inequality

$$\int_0^1 |u(x)|^2 dx \le \frac{1}{c_\alpha}\int_0^1 |u^{(\alpha)}(x)|^2 dx$$

is true for all $u \in C^\alpha[0,1]$ satisfying $u^{(j)}(0) = u^{(j)}(1) = 0$ for $0 \le j \le \alpha - 1$. If we insert $u(x) = x^\alpha(1-x)^\alpha$, the inequality becomes

$$\int_0^1 x^{2\alpha}(1-x)^{2\alpha}dx \le \frac{1}{c_\alpha}\int_0^1 \left[\frac{d^\alpha}{dx^\alpha}\left(x^\alpha(1-x)^\alpha\right)\right]^2 dx.$$

The integral on the left is $[(2\alpha)!]^2/(4\alpha+1)!$, and in the integral on the right we make the substitution $x = (1+y)/2$ to get

$$\int_0^1 \left[\frac{d^\alpha}{dx^\alpha}\left(x^\alpha(1-x)^\alpha\right)\right]^2 dx = \frac{1}{4^\alpha}\int_{-1}^1 \left[\frac{d^\alpha}{dy^\alpha}\left(y^2-1\right)^\alpha\right]^2 \frac{dy}{2}.$$

The function $(d^\alpha/dy^\alpha)(y^2-1)^\alpha$ is $2^\alpha\,\alpha!$ times the usual Legendre polynomial $P_\alpha(y)$ and it is well known that $\|P_\alpha\|_2^2 = 2/(2\alpha+1)$ (see, for example, [7]). Consequently, the integral on the right is

$$\frac{1}{4^\alpha}\,\frac{1}{2}\,2^{2\alpha}(\alpha!)^2\,\frac{2}{2\alpha+1} = \frac{(\alpha!)^2}{2\alpha+1}.$$

In summary,

$$\frac{[(2\alpha)!]^2}{(4\alpha+1)!} \le \frac{1}{c_\alpha}\,\frac{(\alpha!)^2}{2\alpha+1},$$

which is the asserted inequality.

Refinements. By carrying out the approximations further we could refine (17) to the form

$$H_\alpha(x,y) = \frac{1}{\alpha}(1-x^2)^\alpha(1-y^2)^\alpha\left(1 + \sum_{i\ge1,j\ge0}\frac{p_{ij}(x,y)}{\alpha^i}\right),$$

where each $p_{ij}(x,y)$ is a homogeneous polynomial of degree j. The operator with kernel $(1-x^2)^\alpha(1-y^2)^\alpha p_{ij}(x,y)$ has norm of the order $\alpha^{-(j+1)/2}$, so we get further approximations to H_α in this way, whence further approximations to the norm. (We do this by using the fact that the nonzero eigenvalues of a finite-rank kernel $\sum_{i=1}^m f_i(x)\,g_i(y)$ are the same as those of the $m \times m$ matrix whose i,j entry is the inner product $(f_i,\,g_j)$. One can see from this in particular that, because of evenness and oddness, with each approximation the power of α goes down by one.) However, these would probably not be of great interest.

4. Another approach to Wirtinger–Sobolev inequalities

We now show how Wirtinger–Sobolev integral inequalities can be derived from their discrete analogues, which, in dependence on the boundary conditions, are inequalities for circulant or Toeplitz matrices. In the Toeplitz case, we get in this way a new proof of the fact that the constants in the first and fourth problems are the same.

Discrete versions of Wirtinger–Sobolev type inequalities were first established by Fan, Taussky, and Todd [5], and the subject has been developed further since then (see, for example, [9] and the references therein). In particular, for circulant matrices the following is not terribly new, but it fits very well with the topic of this paper and perfectly illustrates the difference between the circulant and Toeplitz cases.

Circulant matrices. For a Laurent polynomial $a(t) = \sum_{k=-r}^{r} a_k t^k$ $(t \in \mathbf{T})$ and $n \geq 2r+1$, let $C_n(a)$ be the $n \times n$ circulant matrix whose first row is

$$(a_0, a_{-1}, \dots, a_{-r}, 0, \dots, 0, a_r, a_{r-1}, \dots, a_1).$$

Thus, $C_n(a)$ results from the Toeplitz matrix $T_n(a)$ by periodization. The singular values of $C_n(a)$ are $|a(\omega_n^j)|$ $(j = 1, \dots, n)$, where $\omega_n = e^{2\pi i/n}$.

Now let $a(t) = (1-t)^\alpha$ $(t \in \mathbf{T})$. One of the singular values of $C_n(a)$ is zero, which causes a slight complication. It is easily seen that $\operatorname{Ker} C_n(a) = \operatorname{span}\{(1, 1, \dots, 1)\}$. With notation as in Section 2, $I - P_{\operatorname{Ker} C_n(a)} = C_n^+(a) C_n(a)$ and hence

$$\|u - P_{\operatorname{Ker} C_n(a)} u\|_2 \leq \|C_n^+(a)\| \, \|C_n(a) u\|_2 \tag{20}$$

for all u in $\mathbf{C}^n$ with the ℓ^2 norm. The inverse of the (spectral) norm of the Moore-Penrose inverse $C_n^+(a)$ is the smallest nonzero singular value of $C_n(a)$ and consequently,

$$\frac{1}{\|C_n^+(a)\|} = |1 - \omega_n|^\alpha = \left(4 \sin^2 \frac{\pi}{n}\right)^{\alpha/2} \sim \frac{(2\pi)^\alpha}{n^\alpha} \tag{21}$$

The projection $P_{\operatorname{Ker} C_n(a)}$ acts by the rule

$$P_{\operatorname{Ker} C_n(a)} u = \left(\frac{1}{n} \sum_{j=1}^{n} u_j, \dots, \frac{1}{n} \sum_{j=1}^{n} u_j \right). \tag{22}$$

Inserting (21) and (22) in (20) we get

$$\begin{aligned} \|C_n(a) u\|_2^2 &\geq \left(4 \sin^2 \frac{\pi}{n}\right)^\alpha \sum_{i=1}^{n} \left| u_i - \frac{1}{n} \sum_{j=1}^{n} u_j \right|^2 \\ &= \left(4 \sin^2 \frac{\pi}{n}\right)^\alpha \left(\sum_{i=1}^{n} |u_i|^2 - \frac{1}{n} \left| \sum_{i=1}^{n} u_i \right|^2 \right). \end{aligned} \tag{23}$$

This is called a (higher-order) discrete Wirtinger–Sobolev inequality and was by different methods established in [9].

Periodic boundary conditions. As already said, the wanted inequality (24) follows almost immediately from Parseval's identity. So the following might seem unduly complicated. However, the analogue of (24) for zero boundary conditions is not straightforward from Parseval's identity, whereas just the following also works in that case.

Let u be a 1-periodic function in $C^\infty(\mathbf{R})$. We apply (23) to $u_n = (u(j/n))_{j=1}^n$. The jth component of $C_n(a)u_n$ is

$$u\left(\frac{j}{n}\right) - \binom{\alpha}{1} u\left(\frac{j+1}{n}\right) + \cdots + (-1)^\alpha u\left(\frac{j+\alpha}{n}\right) = u^{(\alpha)}\left(\frac{j}{n}\right)\frac{1}{n^\alpha} + O\left(\frac{1}{n^{\alpha+1}}\right),$$

the O being independent of j. It follows that

$$\|C_n(a)u\|_2^2 = \left(\sum_{j=1}^n \left|u^{(\alpha)}\left(\frac{j}{n}\right)\right|^2 \frac{1}{n^{2\alpha}}\right) + O\left(\frac{1}{n^{2\alpha}}\right).$$

Consequently, multiplying (23) by $n^{2\alpha-1}$ and passing to the limit $n \to \infty$ we arrive at the inequality

$$\int_0^1 |u^{(\alpha)}(x)|^2 dx \ge (2\pi)^{2\alpha}\left(\int_0^1 |u(x)|^2 dx - \left|\int_0^1 u(x)dx\right|^2\right). \tag{24}$$

Assume finally that $u \in C^\alpha[0,1]$ and $u^{(j)}(0) = u^{(j)}(1)$ for $0 \le j \le \alpha - 1$. We have $u(x) = \sum_{k=-\infty}^\infty e^{2\pi i k x}$ with

$$u_k = \int_0^1 u(x)e^{-2\pi i k x}dx.$$

We integrate the last equality α times partially and use the boundary conditions to obtain that

$$|u_k| = \frac{1}{(2\pi|k|)^\alpha}\left|\int_0^1 u^{(\alpha)}(x)e^{-2\pi i k x}dx\right|.$$

Since $u^{(\alpha)} \in L^2(0,1)$, we see that $|u_k| = v_k\, O(1/|k|^\alpha)$ with $\sum_{k=-\infty}^\infty v_k^2 < \infty$. This implies that

$$\sum_{k=-\infty}^\infty |k|^{2\alpha}|u_k|^2 < \infty. \tag{25}$$

We know that (24) is true with $u(x)$ replaced by $(S_N u)(x) = \sum_{k=-N}^N u_k e^{2\pi i k x}$,

$$\int_0^1 |(S_N u)^{(\alpha)}(x)|^2 dx \ge (2\pi)^{2\alpha}\left(\int_0^1 |(S_N u)(x)|^2 dx - \left|\int_0^1 (S_N u)(x)dx\right|^2\right). \tag{26}$$

From (25) we infer that

$$\int_0^1 |u^{(\alpha)}(x)|^2 dx - \int_0^1 |(S_N u)^{(\alpha)}(x)|^2 dx = \sum_{|k|>N} |k|^{2\alpha}|u_k|^2 = o(1),$$

$$\int_0^1 |u(x)|^2 dx - \int_0^1 |(S_N u)(x)|^2 dx = \sum_{|k|>N} |u_k|^2 = o(1),$$

and since $\int_0^1 (S_N u)(x)dx = \int_0^1 u(x)dx = u_0$, passage to the limit $N \to \infty$ in (26) yields (24) under the above assumptions on u.

Toeplitz matrices. Again let $a(t) = (1-t)^\alpha$ $(t \in \mathbf{T})$, but consider now the Toeplitz matrix $T_n(a)$ instead the circulant matrix $C_n(a)$. It can be easily verified or deduced from [3, formula (2.13)] or [19, formula (1.4)] that

$$T_n^*(a)T_n(a) = T_n(b) - R_\alpha$$

where $b(t) = |1-t|^{2\alpha}$ and R_α is a matrix of the form

$$R_\alpha = \begin{pmatrix} S_\alpha & 0 \\ 0 & O_{n-\alpha} \end{pmatrix}$$

with an $\alpha \times \alpha$ matrix S_α independent of n. Consequently,

$$\|T_n(a)u\|_2^2 = (T_n(a)u, T_n(a)u) = (T_n(b)u, u) - (R_\alpha u, u).$$

It follows that

$$\|T_n(a)u\|_2^2 \geq \lambda_{\min}(T_n(b)) \, \|u\|_2^2 - (R_\alpha u, u) \tag{27}$$

for all $u \in \mathbf{C}^n$. This is the Toeplitz analogue of (23).

Zero boundary conditions. Let $u \in C^\infty(\mathbf{R})$ be a function which vanishes identically outside $(0,1)$. As in the circulant case, we replace the u in (27) by $u_n = (u(j/n))_{j=1}^n$, multiply the result by $n^{2\alpha-1}$ and pass to the limit $n \to \infty$. Taking into account that $\lambda_{\min}(T_n(b)) \sim c_\alpha/n^{2\alpha}$, we obtain

$$\int_0^1 |u^{(\alpha)}(x)|^2 dx \geq c_\alpha \int_0^1 |u(x)|^2 - \lim_{n\to\infty} n^{2\alpha-1}(R_\alpha u_n, u_n). \tag{28}$$

By assumption, u and all its derivatives vanish at 0. This implies that

$$u\left(\frac{j}{n}\right) = \sum_{k=0}^{\alpha-1} \frac{u^{(k)}(0)}{k!}\frac{j^k}{n^k} + \frac{u^{(\alpha)}(\xi_{j,n})}{\alpha!}\frac{j^\alpha}{n^\alpha} = O\left(\frac{1}{n^\alpha}\right)$$

for each fixed j. Since $(R_\alpha u_n, u_n)$ is a bilinear form of $u(1/n), \ldots, u(\alpha/n)$, we arrive at the conclusion that $(R_\alpha u_n, u_n) = O(1/n^{2\alpha})$. Hence, (28) is actually the desired inequality

$$\int_0^1 |u^{(\alpha)}(x)|^2 dx \geq c_\alpha \int_0^1 |u(x)|^2 dx. \tag{29}$$

The approximation argument employed in the case of periodic boundary conditions is also applicable in the case at hand and allows us to relax the C^∞ assumption. It results that (29) is valid for every $u \in C^\alpha[0,1]$ satisfying $u^{(j)}(0) = u^{(j)}(1) = 0$ for $0 \le j \le \alpha - 1$.

References

[1] A. Böttcher, *The constants in the asymptotic formulas by Rambour and Seghier for inverses of Toeplitz matrices.* Integral Equations Operator Theory **50** (2004), 43–55.

[2] A. Böttcher, *On the problem of testing the structure of a matrix by displacement operations.* SIAM J. Numer. Analysis, to appear.

[3] A. Böttcher and B. Silbermann, *Introduction to Large Truncated Toeplitz Matrices.* Springer, New York 1999.

[4] R. Courant, K. Friedrichs, and H. Lewy, *Über die partiellen Differenzengleichungen der mathematischen Physik.* Math. Ann. **100** (1928), 32–74.

[5] K. Fan, O. Taussky, and J. Todd, *Discrete analogs of inequalities of Wirtinger.* Monatsh. Math. **59** (1955), 73–90.

[6] G.H. Hardy, J.E. Littlewood, and G. Pólya, *Inequalities.* 2nd ed., Cambridge University Press, Cambridge 1988.

[7] E. Jahnke, F. Emde, and F. Lösch, *Tables of Higher Functions.* 6th ed., McGraw-Hill, New York, Toronto, London and Teubner, Stuttgart 1960.

[8] M. Kac, W.L. Murdock, and G. Szegő, *On the eigenvalues of certain Hermitian forms.* J. Rational Mech. Anal. **2** (1953), 767–800.

[9] G.V. Milovanović and I. Ž. Milovanović, *Discrete inequalities of Wirtinger's type for higher differences.* J. Inequal. Appl. **1** (1997), 301–310.

[10] D.S. Mitrinović, J.E. Pečarić, and A.M. Fink, *Inequalities Involving Functions and Their Integrals and Derivatives.* Kluwer, Dordrecht 1991.

[11] S.V. Parter, *On the extreme eigenvalues of truncated Toeplitz matrices.* Bull. Amer. Math. Soc. **67** (1961), 191–196.

[12] S.V. Parter, *Extreme eigenvalues of Toeplitz forms and applications to elliptic difference equations.* Trans. Amer. Math. Soc. **99** (1961), 153–192.

[13] S.V. Parter, *On the extreme eigenvalues of Toeplitz matrices.* Trans. Amer. Math. Soc. **100** (1961), 263–276.

[14] P. Rambour and A. Seghier, *Formulas for the inverses of Toeplitz matrices with polynomially singular symbols.* Integral Equations Operator Theory **50** (2004), 83–114.

[15] F. Spitzer and C.J. Stone, *A class of Toeplitz forms and their application to probability theory.* Illinois J. Math. **4** (1960), 253–277.

[16] H. Widom, *On the eigenvalues of certain Hermitian operators.* Trans. Amer. Math. Soc. **88** (1958), 491–522.

[17] H. Widom, *Extreme eigenvalues of translation kernels.* Trans. Amer. Math. Soc. **100** (1961), 252–262.

[18] H. Widom, *Extreme eigenvalues of N-dimensional convolution operators.* Trans. Amer. Math. Soc. **106** (1963), 391–414.

[19] H. Widom, *Asymptotic behavior of block Toeplitz matrices and determinants. II.* Advances in Math. **21** (1976), 1–29.

Albrecht Böttcher
Fakultät für Mathematik
TU Chemnitz
D-09107 Chemnitz, Germany
e-mail: `aboettch@mathematik.tu-chemnitz.de`

Harold Widom
Department of Mathematics
University of California
Santa Cruz, CA 95064, USA
e-mail: `widom@math.ucsc.edu`

Operator Theory:
Advances and Applications, Vol. 171, 89–105

The Method of Minimal Vectors Applied to Weighted Composition Operators

Isabelle Chalendar, Antoine Flattot and Jonathan R. Partington

Abstract. We study the behavior of the sequence of minimal vectors corresponding to certain classes of operators on L^2 spaces, including weighted composition operators such as those induced by Möbius transformations. In conjunction with criteria for quasinilpotence, the convergence of sequences associated with the minimal vectors leads to the construction of hyperinvariant subspaces.

Mathematics Subject Classification (2000). Primary 41A29, 47A15; Secondary 47B33, 47B37.

Keywords. Minimal vectors, hyperinvariant subspaces, weighted composition operators, quasinilpotent operators.

1. Introduction

The construction of minimal vectors (y_n), corresponding to an operator T on a Hilbert space, was introduced by Enflo and his collaborators [2, 10, 11] as a method of constructing hyperinvariant subspaces for certain classes of linear operators. Further work in this area, extending the concept to more general Banach spaces, may be found in [1, 5, 7, 14, 15, 16]. As a result of this work, it has become apparent that it is important to be able to determine when the sequence $(T^n y_n)_n$ converges, since in many cases this gives an explicit construction of hyperinvariant subspaces.

In this paper we begin in Section 2 by considering the convergence in the case that $T^n T^{*n}$ is a multiplication operator on an $L^2(\mu)$ space for each n, extending the method initiated in [6] for normal operators. Some of the most important operators of this kind are weighted composition operators on L^2 of the unit interval or the closed unit disc. In Section 3 we obtain new criteria for the convergence of $(T^n y_n)_n$, which in Section 4 are combined with a characterization of quasinilpotence to give

The authors are grateful to the EPSRC for financial support under grant EP/C004418/1.

results on the existence of hyperinvariant subspaces. One case of particular interest is that of weighted composition operators induced by Möbius transformations, of which we give a detailed analysis. Some explicit examples on the unit interval are also presented. Finally, we discuss the applicability of the method to weighted composition operators induced by ergodic transformations, where a condition for quasinilpotence is already known.

All the operators considered in this paper will be bounded linear operators defined on (real or complex) Banach spaces. A nontrivial hyperinvariant subspace of an operator T acting on a Banach space $\mathcal{X}$ is a closed subspace $\mathcal{M}$ such that $\{0\} \neq \mathcal{M} \neq \mathcal{X}$ and $A\mathcal{M} \subseteq \mathcal{M}$ for all $A \in \{T\}' := \{A : \mathcal{X} \to \mathcal{X} : AT = TA\}$. Recall that an operator T is quasinilpotent if its spectral radius $r(T)$ is equal to 0 (or, equivalently if its spectrum is reduced to $\{0\}$).

Suppose that T is an operator on $\mathcal{X}$, with dense range, and that $f_0 \in \mathcal{X} \setminus \{0\}$. Take ϵ such that $\|f_0\| > \epsilon > 0$. For $n = 1, 2, \ldots$, a *(backward) minimal vector* y_n associated with T, n, f_0 and ϵ, is defined to be a vector of minimal norm such that $\|T^n y_n - f_0\| \leq \epsilon$. It is known that, if $\mathcal{X}$ is reflexive, then such minimal vectors exist and satisfy $\|T^n y_n - f_0\| = \epsilon$. If, in addition, $\mathcal{X}$ is strictly convex, then the minimal vectors are unique (see [7]).

Minimal vectors were introduced by Enflo in [2, 10] for $\mathcal{X}$ a Hilbert space. In [5, 7] they were defined in the context of a more general approximation problem, solved first in Hilbert spaces and then in general reflexive spaces.

From now on, let us fix an injective operator $T \in \mathcal{L}(\mathcal{X})$ with dense range, $f_0 \in \mathcal{X} \setminus \{0\}$, and $\epsilon > 0$ satisfying $\|f_0\| > \epsilon$.

2. Convergence of $(T^n y_n)_n$ for operators of normal type

We shall make use of the following expression:

$$f_0 - T^n y_n = (I + \mu_n A_n)^{-1} f_0,$$

where $A_n := T^n T^{*n}$ and μ_n is positive and uniquely determined by $\|f_0 - T^n y_n\| = \epsilon$. This follows easily from the formula

$$T^n y_n = \mu_n A_n (I + \mu_n A_n)^{-1} f_0,$$

which may be found in [2].

For $\Phi \in L^\infty(X, d\mu)$, M_Φ denotes the operator of multiplication by Φ on $L^2(X, d\mu)$, i.e., $M_\Phi(f) = \Phi f$ for all $f \in L^2(X, d\mu)$.

Definition 2.1. An operator $T \in \mathcal{L}(L^2(X, d\mu))$ is said to be of *normal type* if, for every positive integer n, the positive operator $A_n := T^n T^{*n}$ is equal to M_{Φ_n}, where $\Phi_n = F_n(\Phi)$ with $\Phi \in L^\infty(X, d\mu)$, $\Phi \geq 0$ and where $(F_n)_n$ is a sequence of nonnegative functions on $\mathbb{R}^+$ such that, whenever $t < r$,

$$\lim_{n\to\infty} \frac{F_n(t)}{F_n(r)} = 0.$$

Theorem 2.2. *Suppose that $T \in \mathcal{L}(L^2(X, d\mu))$ is of normal type or $A_n = M_{c_n}$, where c_n is a positive numerical constant. Then the sequence $(T^n y_n)_n$ converges in norm to a nonzero vector of $L^2(X, d\mu)$.*

Proof. First suppose that $A_n = M_{c_n}$ with $c_n \in \mathbb{R}^+$. Then $f_0 - T^n y_n = \frac{f_0}{1+\mu_n c_n}$ and since $\|f_0 - T^n y_n\| = \epsilon$, we have also $1 + \mu_n c_n = \frac{\|f_0\|}{\epsilon}$ and thus $T^n y_n$ is constant and equal to $(1 - \epsilon/\|f_0\|) f_0$.

Now suppose that T is of normal type. Then we have

$$f_0 - T^n y_n = J_n(\Phi) f_0,$$

where $J_n(t) = \frac{1}{1+\mu_n F_n(t)}$. Moreover, since $|J_n(z)| \leq |f_0(z)|$, it is sufficient to establish pointwise convergence almost everywhere, after which we obtain convergence in norm by the dominated convergence theorem.

For $r \geq 0$, we define $L_1(r) = \limsup J_n(r)$ and $L_2(r) = \liminf J_n(r)$. Clearly we have $0 \leq L_2(r) \leq L_1(r) \leq 1$ for all $r \geq 0$ and L_1 and L_2 are decreasing functions.

If $L_1(r) > 0$ for some value of r, then there is a sequence $(n_k)_k$ such that $(\mu_{n_k} F_{n_k}(r))_k$ remains bounded. So for $t < r$ we have $\mu_{n_k} F_{n_k}(t) \to 0$, because of the condition $\lim_{n\to+\infty} \frac{F_n(t)}{F_n(r)} = 0$. Therefore, $L_1(t) = 1$ for $t < r$. We conclude immediately that if $L_1(s) < 1$ for some s then $L_1(t) = 0$ for all $t > s$ (if not, we would have $L_1(s) = 1$).

Thus there is a number $r_1 \geq 0$ such that

$$L_1(t) = \begin{cases} 1 & \text{for } t < r_1, \\ 0 & \text{for } t > r_1 \end{cases} \tag{2.1}$$

and similarly we have a number $r_2 \geq 0$ with $r_2 \leq r_1$ such that

$$L_2(t) = \begin{cases} 1 & \text{for } t < r_2, \\ 0 & \text{for } t > r_2. \end{cases} \tag{2.2}$$

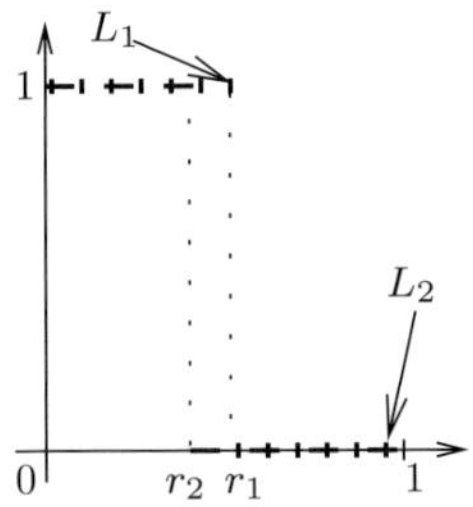

FIGURE 1. The graphs of L_1 and L_2

Now we have

$$\begin{aligned}\epsilon^2 &= \limsup_{n\to\infty} \|J_n(\Phi)f_0\|^2 \\ &\geq \int_{0\leq\phi(z)<r_1} |f_0(z)|^2\,d\mu(z) + \int_{\phi(z)=r_1} L_1(r_1)^2|f_0(z)|^2\,d\mu(z)\end{aligned}$$

and

$$\begin{aligned}\epsilon^2 &= \liminf_{n\to\infty} \|J_n(\Phi)f_0\|^2 \\ &\leq \int_{0\leq\phi(z)<r_2} |f_0(z)|^2 d\mu(z) + \int_{\phi(z)=r_2} L_2(r_2)^2|f_0(z)|^2\,d\mu(z).\end{aligned}$$

If $r_1 = r_2$, then either $L_1(r_1) = L_2(r_2)$ or $f_0(z) = 0$ a.e. on the set on which $\phi(z) = r_1$, and so the sequence $(J_n(\Phi)f_0)_n$ converges almost everywhere. Otherwise

$$\begin{aligned}(1 - L_2(r_2)^2)\int_{\phi(z)=r_2} |f_0(z)|^2\,d\mu(z) \quad &+ \quad \int_{r_2<\phi(z)<r_1} |f_0(z)|^2\,d\mu(z) \\ &+ \quad L_1(r_1)^2\int_{\phi(z)=r_1} |f_0(z)|^2\,d\mu(z) \leq 0,\end{aligned}$$

and hence each term is zero. It follows that $f_0(z) = 0$ almost everywhere on the set on which $L_1(\Phi(z)) \neq L_2(\Phi(z))$. Hence $(f_0 - T^n y_n)_n$ converges in norm to h where $h(z) = f_0(z)$ if $\Phi(z) < r_2$ and $h(z) = 0$ if $\Phi(z) > r_1$. □

3. Minimal vectors for weighted composition operators

3.1. Notation

Let X be either $[0, 1]$ or the closed unit disc $\overline{\mathbb{D}}$, with normalized Lebesgue measure μ. In this section we consider the case where $T \in \mathcal{L}(L^2(X))$ is of the form

$$Tf(t) = w(t)f(\gamma(t)),$$

with $w \in L^\infty(X)$, and $\gamma : X \to X$ an injective mapping, such that γ' exists and is piecewise continuous. In addition, we suppose that $w/(\gamma')^\alpha$ belongs to $L^\infty(X)$, where $\alpha = 1$ for $X = [0, 1]$ and $\alpha = 2$ for $X = \overline{\mathbb{D}}$.

The above conditions are sufficient to guarantee the continuity of T and we can easily check that:

$$T^*f(t) = \frac{\overline{w}(\gamma^{-1}(t))}{|\gamma'(\gamma^{-1}(t))|^\alpha}\chi_{\gamma(X)}(t)f(\gamma^{-1}(t)),$$

where χ_Ω denotes the characteristic function of a set Ω.

3.2. Convergence of $(T^n y_n)_n$

In order to determine whether $(T^n y_n)_n$ converges, we need to study the behavior of $A_n := T^n T^{*n}$. Since

$$T^n f(t) = w(t)w(\gamma(t))\cdots w(\gamma^{n-1}(t))f(\gamma^n(t))$$

and

$$T^{*n}f(t) = \frac{\overline{w}(\gamma^{-1}(t))}{|\gamma'(\gamma^{-1}(t))|^\alpha}\frac{\overline{w}(\gamma^{-2}(t))}{|\gamma'(\gamma^{-2}(t))|^\alpha}\cdots\frac{\overline{w}(\gamma^{-n}(t))}{|\gamma'(\gamma^{-n}(t))|^\alpha}\chi_{\gamma^n(X)}(t)f(\gamma^{-n}(t)),$$

we get:

$$A_n = M_{\Phi_n}, \text{ where } \Phi_n(t) = \frac{|w(t)|^2|w(\gamma(t))|^2\cdots|w(\gamma^{n-1}(t))|^2}{|\gamma'(t)\gamma'(\gamma(t))\cdots\gamma'(\gamma^{n-1}(t))|^\alpha}.$$

Therefore we have $A_n = M_{\Phi_n}$ with $\Phi_n = h\cdot h(\gamma)\cdots h(\gamma^{n-1})$ and h is the function of $L^\infty(X)$ defined by $h := |w|^2/|\gamma'|^\alpha$.

Obviously, as an immediate application of Theorem 2.2, we get the following corollary.

Corollary 3.1. *Let T be an operator on $L^2(X)$ defined as in Subsection* 3.1. *Suppose that h is constant. Then $(T^n y_n)_n$ converges in norm.*

Here are a few examples for $X = [0,1]$, which illustrate the previous corollary:

1. Let $m > 0$ and consider $w(t) = t^{\frac{m-1}{2}}$. Now, take $\theta \in [0,1]$ and consider $\gamma(t) = \{t^m + \theta\}$ where $\{x\}$ denotes the fractional part of x. Then $h(t) = \frac{1}{m}$ a.e.
2. Let $c > 0$ and then take $\gamma(t) = \frac{t+t^2}{2}$ and choose w such that $|w(t)|^2 = c(\frac{1}{2}+t)$. Then we get $h(t) = c$.

However, we can now give a fairly general context in which there is a sufficient condition implying the convergence of $(T^n y_n)_n$. To do this we define a partial order $\prec$ on X by saying that

$$t \prec r \quad \text{if and only if} \quad \limsup_{n\to\infty}\frac{h(\gamma^n(t))}{h(\gamma^n(r))} < 1.$$

We shall say that $\prec$ is *left-regular*, if every nonempty $Y \subseteq X$ such that $r \in Y$ and $t \prec r$ implies $t \in Y$, satisfies the following:

for each $\delta > 0$ there exists $r \in Y$ such that $\mu\{t \in Y : t \prec r\} > \mu(Y) - \delta$.

Likewise $\prec$ is *right-regular*, if every nonempty $Y \subseteq X$ such that $r \in Y$ and $r \prec t$ implies $t \in Y$, satisfies the following:

for each $\delta > 0$ there exists $r \in Y$ such that $\mu\{t \in Y : r \prec t\} > \mu(Y) - \delta$.

Later we shall give some examples of such partial orders $\prec$.

Theorem 3.2. *Let T be an operator on $L^2(X)$ defined as in Subsection* 3.1. *Suppose that h determines a partial order $\prec$ that is both left- and right-regular, such that for every $t \in X$ one has*

$$\mu\{r \in X : r \prec t \text{ or } t \prec r\} = 1. \tag{3.1}$$

Then $(T^n y_n)_n$ converges in norm.

Proof. The calculations above show that

$$(f_0 - T^n y_n)(t) = \frac{f_0(t)}{1 + \mu_n h(t) \dots h(\gamma^{n-1}(t))}.$$

Now set

$$\begin{cases} J_n(t) = \dfrac{1}{1 + \mu_n h(t) \dots h(\gamma^{n-1}(t))} \\ L_1(t) = \limsup\limits_{n\to\infty} J_n(t) \\ L_2(t) = \liminf\limits_{n\to\infty} J_n(t). \end{cases}$$

Obviously we have $0 \leq L_2(t) \leq L_1(t) \leq 1$.

It follows from (3.1) that for a fixed $t \in X$, almost all $r \in X$ are comparable to t in the sense that either $r \prec t$ or $t \prec r$.

Suppose that there exists $r \in X$ such that $L_1(r) > 0$. Then there exists a subsequence $(n_k)_k$ such that $(\mu_{n_k} h(r) \dots h(\gamma^{n_k-1}(r)))_k$ is bounded. Suppose that $t \prec r$; then it follows that the sequence $(\mu_{n_k} h(t) \dots h(\gamma^{n_k-1}(t)))_k$ tends to 0 and thus $L_1(t) = 1$. Hence, if in addition $L_1(r) < 1$, we must have $L_1(t) = 0$ for all t with $r \prec t$.

Thus there exist disjoint subsets $X_0^{(1)}$ and $X_1^{(1)}$ of X with total measure 1, such that

$$L_1(t) = \begin{cases} 0 & \text{if } t \in X_0^{(1)}, \\ 1 & \text{if } t \in X_1^{(1)}. \end{cases}$$

Similarly, suppose that there exists $r \in X$ such that $L_2(r) < 1$. Then there exists a subsequence $(n_k)_k$ such that $(\mu_{n_k} h(r) \dots h(\gamma^{n_k-1}(r)))_k$ is bounded below. Suppose that $r \prec t$; then it follows that the sequence $(\mu_{n_k} h(t) \dots h(\gamma^{n_k-1}(t)))_k$ tends to infinity, and thus $L_2(t) = 0$. Hence, if in addition $L_2(r) > 0$, we must have $L_2(t) = 1$ for all t with $t \prec r$.

Thus there exist disjoint subsets $X_0^{(2)}$ and $X_1^{(2)}$ of X with total measure 1, such that

$$L_2(t) = \begin{cases} 0 & \text{if } t \in X_0^{(2)}, \\ 1 & \text{if } t \in X_1^{(2)}. \end{cases}$$

Clearly, $X_0^{(1)} \subseteq X_0^{(2)}$ and $X_1^{(2)} \subseteq X_1^{(1)}$.

Since $\prec$ is left-regular, for each $\delta > 0$ we can choose $r \in X_1^{(1)}$ such that

$$\int_{\{t:\, t \prec r\}} |f_0(t)|^2 \, d\mu(t) \geq \int_{t \in X_1^{(1)}} |f_0(t)|^2 \, d\mu(t) - \delta.$$

Now we can find a subsequence $(n_k)_k$ such that $J_{n_k}(r) \to 1$, and so $J_{n_k}(t) \to 1$ for all $t \prec r$. Thus, for each $\delta > 0$,

$$\begin{aligned} \epsilon^2 &= \limsup_{n\to\infty} \|J_n f_0\|_2^2 \\ &\geq \lim_{k\to\infty} \int_{\{t:\, t \prec r\}} |J_{n_k}(t) f_0(t)|^2 \, d\mu(t) \geq \int_{t \in X_1^{(1)}} |f_0(t)|^2 \, d\mu(t) - \delta, \end{aligned}$$

and hence

$$\epsilon^2 \geq \int_{t\in X_1^{(1)}} |f_0(t)|^2 \, d\mu(t). \tag{3.2}$$

Similarly, since $\prec$ is right-regular, for each $\delta > 0$ we can choose $s \in X_0^{(2)}$ such that

$$\int_{\{t:\, s\prec t\}} |f_0(t)|^2 \, d\mu(t) \geq \int_{t\in X_0^{(2)}} |f_0(t)|^2 \, d\mu(t) - \delta,$$

and hence, using (3.1), we have

$$\int_{\{t:\, t\prec s\}} |f_0(t)|^2 \, d\mu(t) \leq \int_{t\in X_1^{(2)}} |f_0(t)|^2 \, d\mu(t) + \delta.$$

Now we can find a subsequence $(n_k)_k$ such that $J_{n_k}(s) \to 0$, and so $J_{n_k}(t) \to 0$ for all t with $s \prec t$. Thus, for each $\delta > 0$,

$$\begin{aligned} \epsilon^2 &= \liminf_{n\to\infty} \|J_n f_0\|_2^2 \\ &\leq \limsup_{k\to\infty} \int_{\{t:\, t\prec s\}} |J_{n_k}(t) f_0(t)|^2 \, d\mu(t) \leq \int_{t\in X_1^{(2)}} |f_0(t)|^2 \, d\mu(t) + \delta, \end{aligned}$$

and hence

$$\epsilon^2 \leq \int_{t\in X_1^{(2)}} |f_0(t)|^2 \, d\mu(t). \tag{3.3}$$

It follows from (3.2) and (3.3) that $f_0(t) = 0$ a.e. for $t \in X_1^{(1)} \backslash X_1^{(2)}$ and hence, using the dominated convergence theorem, we see that $(f_0 - T^n y_n)_n$ converges in norm to $f_0 \chi_{X_1^{(1)}}$. □

Corollary 3.3. *Let T be an operator on $L^2(X)$ as in Subsection* 3.1. *Suppose that there is a point $x_0 \in X$ such that whenever $|t - x_0| < |r - x_0|$ we have*

$$\limsup_{n\to\infty} \frac{h(\gamma^n(t))}{h(\gamma^n(r))} < 1. \tag{3.4}$$

Then $(T^n y_n)_n$ converges in norm.

Proof. This follows from Theorem 3.2, since $t \prec r$ means simply that $|t - x_0| < |r - x_0|$. □

Remark 3.4. A simple modification of the above arguments shows that for $X = [0, 1]$, if one has (3.4) whenever $x_0 < t < r$ or $r < t < x_0$, then $(T^n y_n)_n$ converges in norm. In this case there exist $r_1, r_1', r_2, r_2' \in [0, 1]$ such that the behavior of L_1 and L_2 is given by Figure 2.

Even in the particular case where T is a (dense range and injective) weighted composition operator on $L^2([0, 1])$, it is necessary to put conditions on T in order to guarantee the convergence of $(T^n y_n)_n$. Indeed, here is an explicit example where, for a suitable choice of f_0, the sequence of $(T^n y_n)_n$ does not converge.

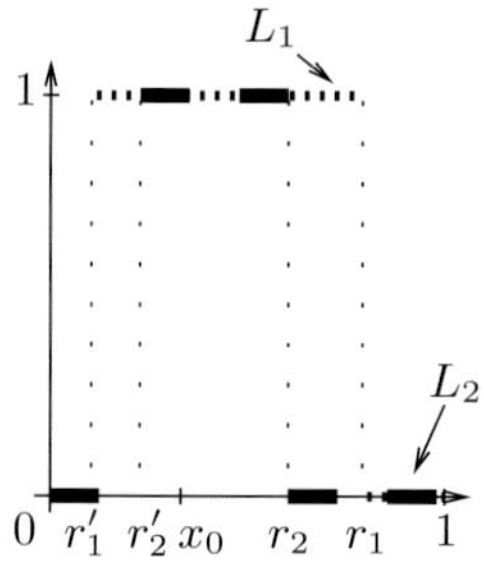

FIGURE 2. The graphs of L_1 and L_2 as in Remark 3.4

Example. Let $(\alpha_n)_{n\in\mathbb{Z}}$ be a strictly positive sequence such that $\sup_{n\in\mathbb{Z}}\alpha_n < \infty$ and let $(a_n)_{n\in\mathbb{Z}}$ be a subdivision of $[0,1]$ such that $\cdots < a_{-1} < a_0 < a_1 < \cdots$ and $\overline{\cup_{n\in\mathbb{Z}} I_n} = [0,1]$, with $I_n = [a_n, a_{n+1})$. Define γ be the piecewise linear mapping such that $\gamma(I_{k+1}) = I_k$ for all $k \in \mathbb{Z}$. Now set $e_k = \dfrac{\chi_{I_k}}{\mu(I_k)^{1/2}}$. Then $(e_k)_{k\in\mathbb{Z}}$ is an orthonormal sequence in $L^2([0,1], d\mu)$. Note that:

$$e_k(\gamma(t)) = \frac{\chi_{I_k}(\gamma(t))}{\mu(I_k)^{1/2}} = \frac{\chi_{I_{k+1}}(t)}{\mu(I_k)^{1/2}}.$$

Then we define w by:

$$w(t) = \sum_{k\in\mathbb{Z}} \alpha_k \frac{\mu(I_k)^{1/2}}{\mu(I_{k+1})^{1/2}} \chi_{I_{k+1}}(t).$$

If, in addition, $\sup_{k\in\mathbb{Z}} \alpha_k \dfrac{\mu(I_k)^{1/2}}{\mu(I_{k+1})^{1/2}} < \infty$, then $w \in L^\infty([0,1])$. The weighted composition operator T defined by $T(f(t)) = w(t)f(\gamma(t))$ satisfies $T(e_n) = \alpha_n e_{n+1}$.

We can show now that there exists f_0 such that $(T^n y_n)_n$ does not converge. First, we easily verify that $T^*(e_n) = \alpha_{n-1}e_{n-1}$. Let $f_0 = \sum_{k\in\mathbb{Z}} c_k e_k$ be in $L^2([0,1])$; from the equalities $f_0 - T^n y_n = (Id - \mu_n T^n T^{*n})^{-1} f_0$ and $\|f_0 - T^n y_n\| = \epsilon$, we get:

$$f_0 - T^n y_n = \sum_{k\in\mathbb{Z}} \frac{c_k}{1 + \mu_n(\alpha_{k-1}^2 \cdots \alpha_{k-n}^2)} e_k \tag{3.5}$$

and

$$\sum_{k\in\mathbb{Z}} \frac{|c_k|^2}{(1 + \mu_n(\alpha_{k-1}^2 \cdots \alpha_{k-n}^2))^2} = \epsilon^2.$$

Moreover, by the dominated convergence theorem, $(f_0 - T^n y_n)_n$ converges if and only if $\lim_{n\to\infty} \dfrac{1}{1 + \mu_n(\alpha_{k-1}^2 \cdots \alpha_{k-n}^2)}$ exists for all $k \in \mathbb{Z}$ such that $c_k \neq 0$. Take now $f_0 = e_0 + e_1$ and $0 < \epsilon < \sqrt{2}$. Then (3.5) gives:

$$\frac{1}{(1 + \mu_n(\alpha_{-1}^2 \cdots \alpha_{-n}^2))^2} + \frac{1}{(1 + \mu_n(\alpha_0^2 \cdots \alpha_{1-n}^2))^2} = \epsilon^2,$$

i.e.,

$$\frac{1}{(1+\lambda_n\alpha_{-n}^2)^2}+\frac{1}{(1+\lambda_n\alpha_0^2)^2}=\epsilon^2, \tag{3.6}$$

where $\lambda_n=\mu_n(\alpha_{-1}^2\dots\alpha_{1-n}^2)$.

Suppose that $\lim_{n\to\infty}\alpha_n$ does not exist. Then (3.6) implies that neither $(\lambda_n)_n$ nor the sequence $(T^n y_n)_n$ converges.

Remark 3.5. Another natural setting in which weighted composition operators have been much studied is the case of a space of analytic functions on the disc, such as the Hardy space H^2 or the Bergman space A^2 (see, for example [8]). The difficulty here is in the expression of T^{*n}, since it now involves an orthogonal projection, and thus T^nT^{*n} is no longer expressible as a multiplication. Even for the simple example $\gamma(z)=z$ and $w(z)=z-1$, it is still unknown whether the sequence $(T^n y_n)_n$ converges (see [5, 14]).

4. Hyperinvariant subspaces

4.1. Existence theorems

We first recall the following result, that appears in [2] in a Hilbert space context and that was generalized to an arbitrary Banach space in [15].

Theorem A *If there exists a subsequence $(y_{n_k})_k$ of minimal vectors such that $\lim_{k\to\infty}\frac{\|y_{n_k-1}\|}{\|y_{n_k}\|}=0$ and $(T^{n_k-1}y_{n_k-1})_k$ converges in norm, then T has a nontrivial hyperinvariant subspace. In particular, if T is quasinilpotent and $(T^n y_n)_n$ converges in norm, then T has a nontrivial hyperinvariant subspace.*

In order to provide concrete examples to illustrate Theorem A, we will first settle a useful lemma.

Lemma 4.1. *Let $(y_n)_n$ be a sequence of minimal vectors. Then we have*

$$\|y_n\|=\mu_n\|A_n^{1/2}(f_0-T^n y_n)\|,$$

where μ_n is a positive constant uniquely determined by the equality $\|f_0-T^n y_n\|=\epsilon$ and where $A_n=T^nT^{*n}$.

Proof. First let us recall that $y_n=\mu_nT^{*n}(f_0-T^n y_n)$, where μ_n is a positive constant uniquely determined by the equality $\|f_0-T^n y_n\|=\epsilon$. Therefore we have

$$\begin{aligned}\|y_n\|^2&=\langle y_n,y_n\rangle\\&=\mu_n^2\langle f_0-T^n y_n,(T^nT^{*n})(f_0-T^n y_n)\rangle\\&=\mu_n^2\langle f_0-T^n y_n,A_n(f_0-T^n y_n)\rangle\\&=\mu_n^2\|A_n^{1/2}(f_0-T^n y_n)\|^2.\end{aligned}$$

□

Remark 4.2. Unfortunately, under the hypothesis of Corollary 3.1, it is impossible to find a subsequence $(y_{n_k})_k$ such that $\lim_{k\to+\infty} \frac{\|y_{n_k-1}\|}{\|y_{n_k}\|} = 0$. Indeed, if $|w|^2/\gamma = c$, then $f_0 - T^n y_n = \frac{f_0}{1+\mu_n c^n}$ and thus $\mu_n = \left(\frac{\|f_0\|_2}{\epsilon} - 1\right)\frac{1}{c^n}$ and $\|y_n\|_2 = (\|f_0\|_2 - \epsilon)\frac{1}{c^{n/2}}$ since $A_n = M_{c^{n/2}}$. It follows that for all positive integers n we have:

$$\frac{\|y_{n-1}\|_2}{\|y_n\|_2} = \frac{1}{c^{1/2}}.$$

Here is an example of operator satisfying the hypothesis of Theorem A. Let $T \in \mathcal{L}(L^2([0,1]))$ defined by

$$Tf(t) = tf(t/2).$$

It is not difficult to check that

$$T^* f(t) = 4tf(2t)\chi_{(0,1/2)}(t).$$

Then, since

$$T^n f(t) = \frac{t^n}{2^{n(n-1)/2}} f\left(\frac{t}{2^n}\right) \quad \text{and} \quad T^{*n} f(t) = 4^n t^n 2^{n(n-1)/2} \chi_{[0,1/2^n]}(t) f(2^n t),$$

we get

$$A_n f(t) := T^n T^{*n} f(t) = 4^n t^{2n} 2^{-n^2} f(t).$$

Therefore $A_n = M_{\Phi_n}$ with $\Phi_n(t) = 4^n 2^{-n^2} t^{2n}$, and thus T is of normal type. Using Theorem 2.2, $(T^n y_n)_n$ converges in norm.

Moreover, T is quasinilpotent. Indeed, note that

$$\|T^n\| \leq \left\| \frac{t^n}{2^{n(n-1)/2}} \right\|_\infty = \frac{1}{2^{n(n-1)/2}},$$

and thus $\|T^n\|^{1/n} \leq \frac{1}{2^{(n-1)/2}}$, and then $\lim_{n\to+\infty} \|T^n\|^{1/n} = 0$.

Now, by Theorem A, T has a nontrivial hyperinvariant subspace.

It is clear that T has nontrivial invariant subspaces, although since a characterization of the the commutant $\{T\}'$ of T is not known, it is not otherwise obvious that T has nontrivial hyperinvariant subspaces.

The previous example is in fact a particular case of a much more general situation, which we now analyse.

Proposition 4.3. *Let T be an operator on $L^2(X)$ defined as in Subsection 3.1. Suppose that γ has a unique attractive fixed point x_0 such that $(\gamma^n(t))_n$ converges to x_0 uniformly on X, and that h is continuous at x_0. Then the spectral radius of T satisfies $r(T) = \sqrt{a}$, where $a = h(x_0)$. Therefore T is quasinilpotent if and only if $h(x_0) = 0$.*

Proof. For $f \in L^2(X)$ we have:

$$\|T^n f\|_2^2 = \int_{s\in\gamma^n(X)} \frac{|w(\gamma^{-n}(s))\dots w(\gamma^{-1}(s))|^2 |f(s)|^2}{|\gamma'(\gamma^{-n}(s))\dots\gamma'(\gamma^{-1}(s))|^\alpha}\, d\mu(s).$$

Suppose that $h(x_0) = a$ and take $\epsilon > 0$. Then there exists a positive integer K such that $k \geq K$ implies $\frac{|w(\gamma^k(t))|^2}{|\gamma'(\gamma^k(t))|^\alpha} < a + \epsilon$. So, for $n > K$, we have:

$$\|T^n f\|^2 \leq \|h\|_\infty^K (a+\epsilon)^{n-K} \|f\|_2^2 .$$

It follows that $\lim_{n\to\infty} \|T^n\|^{1/n} \leq a + \epsilon$ and thus $r(T) \leq \sqrt{a}$. If $a = 0$, then $r(T) = 0$. Suppose now that $a > 0$ and take $\epsilon < a$. Then there exists K' such that $k \geq K'$ implies $\frac{|w(\gamma^k(t))|^2}{|\gamma'(\gamma^k(t))|^\alpha} > a - \epsilon$. Let $a > \delta > 0$ and set $S_\delta = \{t : h(t) \geq \delta\}$. For $n > K'$ and $f = \chi_{\gamma^n(S_\delta)}$, we get:

$$\begin{aligned} \|T^n f\|_2^2 &= \int_{s\in\gamma^n(S_\delta)} \frac{|w(\gamma^{-n}(s))\dots w(\gamma^{-1}(s))|^2}{|\gamma'(\gamma^{-n}(s))\dots\gamma'(\gamma^{-1}(s))|^\alpha}\, ds \\ &\geq \delta^{K'} (a-\epsilon)^{n-K'} \|f\|_2^2 \end{aligned}$$

It follows that $\lim\limits_{n\to\infty} \|T^n\|^{1/n} \geq \sqrt{a-\epsilon}$ and thus $r(T) \geq \sqrt{a}$. Finally $r(T) = \sqrt{a}$ and obviously T is quasinilpotent if and only if $h(x_0) = 0$. □

Using Theorem 3.2, Proposition 4.3 and Theorem A, we get the following corollary.

Corollary 4.4. *Let T be an operator on $L^2(X)$ defined as in Subsection 3.1. Suppose that h determines a partial order $\prec$ that is both left- and right-regular, such that for every $t \in X$ one has (3.1) holding. Suppose also that γ has a unique attractive fixed point x_0 such that $(\gamma^n(t))_n$ converges to x_0 uniformly on X, and that h is continuous at x_0 with $h(x_0) = 0$. Then the operator T has nontrivial hyperinvariant subspaces.*

4.2. Möbius transformations on $\overline{\mathbb{D}}$

Now let $\gamma : \overline{\mathbb{D}} \to \overline{\mathbb{D}}$ be a Möbius transformation, i.e., a rational mapping of the form

$$\gamma(z) = \frac{az+b}{cz+d}, \quad \text{with} \quad ad - bc \neq 0 \quad \text{and} \quad |a|^2 + |b|^2 + 2|a\bar{b} - c\bar{d}| \leq |c|^2 + |d|^2.$$

From now on we exclude the case when $\gamma(z) = z$ for all z. As in [3], such transformations can be classified in terms of their two fixed points, which may coincide. By Schwarz's Lemma, it is not possible to have two fixed points in $\mathbb{D}$, nor two fixed points in $\mathbb{C} \cup \{\infty\} \setminus \overline{\mathbb{D}}$, and if there is a unique fixed point then it lies on the unit circle $\mathbb{T}$. The remaining cases are as follows:

1. Two fixed points, one in $\mathbb{D}$ and one outside $\overline{\mathbb{D}}$ (possibly at ∞).
2. Two fixed points, one on $\mathbb{T}$ and one outside $\overline{\mathbb{D}}$ (possibly at ∞).
3. Two fixed points, one in $\mathbb{D}$ and one on $\mathbb{T}$.
4. Two distinct fixed points on $\mathbb{T}$.
5. A unique fixed point on $\mathbb{T}$.

We shall see that the methods of this paper are most suited to cases 1 and 2 above.

Theorem 4.5. *Let $\gamma : \overline{\mathbb{D}} \to \overline{\mathbb{D}}$ be as in Case 1 or Case 2 above, and suppose that γ is not an automorphism. Let T be a weighted composition operator on $L^2(\overline{\mathbb{D}})$, as defined in Subsection 3.1. Suppose also that w is C^1 at the attractive fixed point $x_0 \in \overline{\mathbb{D}}$, with $w(x_0) = 0$ and $w'(x_0) \neq 0$. Then T has nontrivial hyperinvariant subspaces.*

Proof. By conjugating with an automorphism of $\overline{\mathbb{D}}$, we may suppose without loss of generality that $x_0 = 0$ or $x_0 = 1$.

Case 1: Without loss of generality the second fixed point is either real or infinite. Suppose first that it is at ∞. Then $\gamma(z) = kz$ with $|k| < 1$, since 0 is an attractive fixed point. Then

$$\frac{h(\gamma^n(t))}{h(\gamma^n(r))} = \left| \frac{w(k^n t)}{w(k^n r)} \right|^2 .$$

This converges to $|t/r|^2$ as $n \to \infty$, and thus $t \prec r$ if and only if $|t| < |r|$; clearly $\prec$ is left- and right-regular. It is easily seen that the remaining hypotheses of Corollary 4.4 are satisfied, and the result follows.

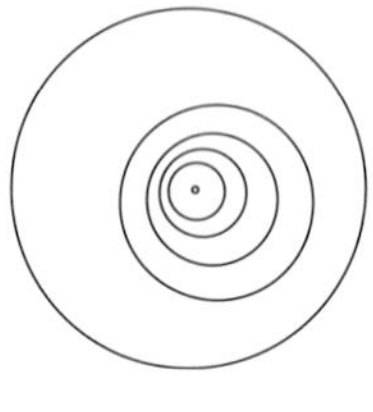

FIGURE 3. Representation of $\gamma^n(\overline{\mathbb{D}})$ in Case 1

If the second fixed point is finite, say x_1, then by conjugating with the mapping g, defined by $g(z) = z/(z - x_1)$, we deduce easily that γ^n has the form

$$\gamma^n(z) = \frac{k^n z x_1}{k^n z - z + x_1}, \qquad \text{with} \quad |k| < 1,$$

and asymptotically

$$|\gamma^n(z)| \sim |k|^n \left| \frac{z x_1}{z - x_1} \right| .$$

Thus we see that $t \prec r$ if and only if

$$\left| \frac{t x_1}{t - x_1} \right| < \left| \frac{r x_1}{r - x_1} \right| .$$

The argument is now completed as above.

Case 2: We begin with the case when the second fixed point is at ∞, and thus $\gamma(z) = 1 + k(z - 1)$ with $|k| < 1$. Thus

$$|\gamma^n(z) - 1| = |k|^n |z - 1|,$$

and so $t \prec r$ if and only if $|t - 1| < |r - 1|$; hence we may use the same arguments as before.

Finally, if the second fixed point is at $x_1 \in \mathbb{C}$, then by conjugating with the mapping g defined by $g(z) = \frac{z-1}{z-x_1}$, we obtain

$$\gamma^n(z) - 1 = \frac{k^n(z-1)(x_1-1)}{k^n(z-1) + x_1 - z}, \qquad \text{with} \quad |k| < 1,$$

and thus, asymptotically,

$$|\gamma^n(z) - 1| \sim |k|^n |x_1 - 1| \left| \frac{z-1}{z-x_1} \right|.$$

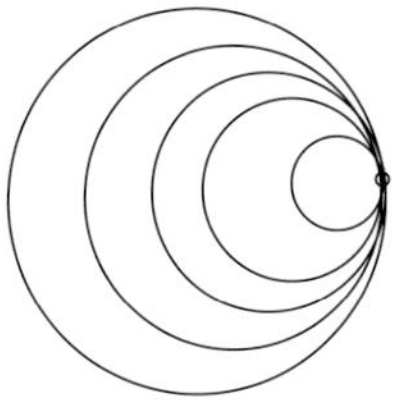

FIGURE 4. Representation of $\gamma^n(\overline{\mathbb{D}})$ in Case 2

We see that $t \prec r$ if and only if

$$\left| \frac{t-1}{t-x_1} \right| < \left| \frac{r-1}{r-x_1} \right|.$$

Once more the proof is completed by the same arguments. □

If w' vanishes at x_0, but w has a continuous higher derivative which is nonzero at x_0, then similar arguments can be applied. We omit the details.

It is necessary to exclude all cases when γ is an automorphism of $\overline{\mathbb{D}}$, since we no longer have the uniform convergence of its iterates. Moreover, in Cases 3 and 4 it is not possible to apply the methods presented so far, since the iterates of γ do not converge uniformly on $\overline{\mathbb{D}}$ (note that in Case 4 the mapping γ is necessarily an automorphism). There remains Case 5, which we now discuss separately.

Suppose now that γ is a parabolic mapping, that is, it has a unique fixed point x_0 on the unit circle. As in the proof of Theorem 4.5, we may assume without loss of generality that $x_0 = 1$. As in [3], one may conjugate γ by the mapping g, given by $g(z) = 1/(z-1)$, to obtain the mapping $z \mapsto z + \beta$ for some β with $\operatorname{Re}\beta < 0$. It follows that

$$\gamma^n(z) = 1 + \frac{z-1}{1 + n\beta(z-1)}.$$

Thus, asymptotically,

$$|1 - \gamma^n(z)| \sim \frac{1}{|n\beta|} \qquad \text{for all} \quad z \neq 1.$$

Although the iterates of γ do converge uniformly in this case, they do so at the same rate for all z, and thus the relation $\prec$ does not satisfy (3.1), and it is not possible to apply Corollary 4.4.

4.3. Mappings of the unit interval

On combining Remark 3.4 with Proposition 4.3 and Theorem A, we obtain a sufficient condition for the existence of nontrivial hyperinvariant subspaces, which is closely-related to Corollary 4.4. This is illustrated by the following examples.

Let $\gamma : [0,1] \to [0,1]$ defined by $\gamma(t) = \theta t + a$ with $0 \le a \le 1$ and θ satisfying both $-a \le \theta \le 1-a$ and $|\theta| < 1$. Then γ has a unique fixed point x_0 satisfying $x_0 = \frac{a}{1-\theta}$. Moreover $|\gamma(t_1) - \gamma(t_2)| = |\theta||t_1 - t_2|$ and thus γ is strictly contractive and the sequence $(\gamma^n(t))_n$ converges to x_0 uniformly on $[0,1]$. Then define w_1 on $[0,1]$ by:

$$w_1(t) = \begin{cases} \dfrac{x_0 - t}{x_0} & \text{if } 0 \le t \le x_0 \\[2ex] \dfrac{t - x_0}{1 - x_0} & \text{if } x_0 \le t \le 1 \end{cases}$$

and set $w = w_1^{1/2}$.

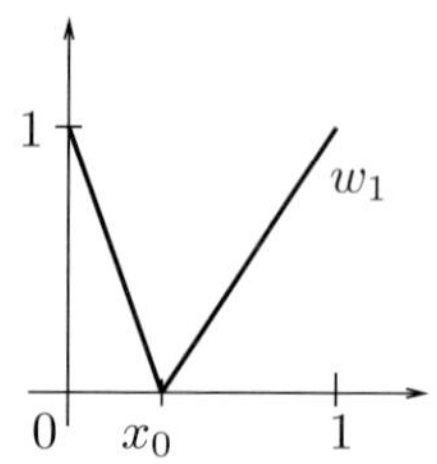

FIGURE 5. The graph of w_1

So, $w \in L^\infty([0,1])$; $h := \frac{w_1}{|\theta|} \in L^\infty([0,1])$ is continuous at x_0 and $h(x_0) = 0$.

We now have to study the quotient $\dfrac{h(\gamma^n(t))}{h(\gamma^n(r))}$ for some t, $r \in [0,1]$.

$$\frac{h(\gamma^n(t))}{h(\gamma^n(r))} = \begin{cases} \dfrac{x_0 - \gamma^n(t)}{x_0 - \gamma^n(r)} & \text{if } r < t < x_0 \\[2ex] \dfrac{\gamma^n(t) - x_0}{\gamma^n(r) - x_0} & \text{if } x_0 < t < r. \end{cases}$$

But

$$\gamma^n(t) = \theta^n t + a \sum_{k=0}^{n-1} \theta^k = \theta^n t + x_0(1 - \theta^n).$$

So $\gamma^n(t) - x_0 = \theta^n(t - x_0)$. Thus,

$$\frac{h(\gamma^n(t))}{h(\gamma^n(r))} = \begin{cases} \dfrac{x_0 - t}{x_0 - r} & \text{if } r < t < x_0 \\[2ex] \dfrac{t - x_0}{r - x_0} & \text{if } x_0 < t < r, \end{cases}$$

and in both cases we obtain:

$$\limsup_{n\to\infty}\frac{h(\gamma^n(t))}{h(\gamma^n(r))}<1.$$

We conclude that for these choices of γ and w, T has nontrivial hyperinvariant subspaces.

Now we take the same γ as before and $w=w_2$ is defined by:

$$w_2(t)=c(t-x_0)^{2m},\ \text{with } m\geq 1, c>0.$$

Then we have:

$$\begin{aligned}\frac{h(\gamma^n(t))}{h(\gamma^n(r))} &= \frac{(\gamma^n(t)-x_0)^{2m}}{(\gamma^n(r)-x_0)^{2m}}\\ &= \left(\frac{t-x_0}{r-x_0}\right)^{2m}\end{aligned}$$

If $x_0<t<r$ or if $r<t<x_0$ we obtain $0<\dfrac{t-x_0}{r-x_0}<1$. So, in both cases we have $\limsup_{n\to\infty}\frac{h(\gamma^n(t))}{h(\gamma^n(r))}<1$ and we conclude that the operator T has nontrivial hyperinvariant subspaces.

4.4. Bishop-type operators

The aim of this subsection is to discuss explicit examples of operators of the form

$$Tf(t)=w(t)f(\tau(t)),$$

where w is essentially bounded, and where τ is ergodic. Such operators include variations on the classical Bishop operator $Tf(t)=tf(\{t+\theta\})$, where θ is irrational and $\{\cdot\}$ denotes taking the remainder modulo 1, studied in [4, 9, 12, 13].

In view of trying to apply Theorem A, it is natural to characterize quasinilpotent Bishop-type operators. The answer to this is given in the next proposition.

Proposition 4.6 (Prop. 1.3 in [12]). *Let T be a Bishop-type operator defined by $Tf(t)=w(t)f(\tau(t))$. Let $w\in L^\infty([0,1])$ be such that $|w|$ is continuous a.e. Then $r(T)=e^{\int_0^1\log|w|d\mu}$ if $\log|w|\in L^1([0,1])$ and $r(T)=0$ otherwise. It follows that if $|w|$ is continuous a.e. then T is quasinilpotent if and only if $\log|w|\notin L^1([0,1])$)*

In order to be able to apply Theorem A, we need to prove the convergence of $(T^ny_n)_n$, but unfortunately, we are not able to decide this in general. Indeed, in this case our operator T is not of normal type: let $w\in L^\infty([0,1])$, $\frac{|w|^2}{|\gamma'|}\in L^\infty([0,1])$, and γ an ergodic and injective mapping.

Let $F_n=\prod_{k=0}^n h\circ\gamma^k$. We can prove that there exist $r<t$ such that $\left(\frac{F_n(r)}{F_n(t)}\right)_n$ does not tend to 0 when n tends to infinity in the following way. First, there exists r such that $\gamma(r)<r$. Indeed, otherwise $[r,1]$ would be invariant under γ and this is absurd since γ is ergodic. Now take $r=\gamma(t)$, so that $\frac{F_n(\gamma(t))}{F_n(t)}=\frac{h(\gamma^{n+1}(t))}{h(t)}$. Obviously we may assume that $h\neq 0$. Therefore there exists $\delta>0$ such that

$S_\delta = \{x \in (0,1) : h(x) > \delta\} \subset (0,1)$ is of positive measure. The ergodic property of γ implies that there exists $(n_k)_k$ such that $\gamma^{n_k}(t) \in S_\delta$ for all k. It follows that $\frac{F_{n_k}(\gamma(t))}{F_{n_k}(t)} \geq \frac{\delta}{h(t)}$, and thus cannot tend to 0.

Furthermore, it is not possible to apply Corollary 4.4 in this case, since condition (3.1) is never satisfied when γ is ergodic. To see this, we define, for any $\delta < \|h\|_\infty$ the set

$$R_\delta = \{x \in [0,1] : |h(x)| \geq \|h\|_\infty - \delta\}.$$

Then, by ergodicity, for each $t \in [0,1]$ there is a sequence $(n_k)_k$ with $\gamma^{n_k}(t) \in R_\delta$. Thus for every $r \in [0,1]$ we obtain

$$\limsup_{n\to\infty} \frac{h(\gamma^n(t))}{h(\gamma^n(r))} \geq \frac{\|h\|_\infty - \delta}{\|h\|_\infty} \qquad \text{for each} \quad \delta,$$

and hence the lim sup equals 1. Therefore in this case there are no points t with $t \prec r$.

Acknowledgement

The authors are grateful to the referee for reading the manuscript carefully and providing detailed comments.

References

[1] G. Androulakis. *A note on the method of minimal vectors.* Trends in Banach spaces and operator theory (Memphis, TN, 2001), volume 321 of Contemp. Math., Amer. Math. Soc., Providence, RI (2003), 29–36.

[2] S. Ansari and P. Enflo. *Extremal vectors and invariant subspaces.* Trans. Amer. Math. Soc. **350** (1998), 539–558.

[3] A.F. Beardon. *Iteration of rational functions.* Springer-Verlag, New York, 1991.

[4] D.P. Blecher and A.M. Davie. *Invariant subspaces for an operator on $L^2(\Pi)$ composed of a multiplication and a translation.* J. Operator Theory **23** (1990), 115–123.

[5] I. Chalendar and J.R. Partington. *Constrained approximation and invariant subspaces.* J. Math. Anal. Appl. **280** (2003), 176–187.

[6] I. Chalendar and J.R. Partington. *Convergence properties of minimal vectors for normal operators and weighted shifts.* Proc. Amer. Math. Soc. **133** (2005), 501–510.

[7] I. Chalendar, J.R. Partington, and M. Smith. *Approximation in reflexive Banach spaces and applications to the invariant subspace problem.* Proc. Amer. Math. Soc. **132** (2004), 1133–1142.

[8] C.C. Cowen and B.D. MacCluer. *Composition Operators on Spaces of Analytic Functions.* Studies in Advanced Mathematics, CRC Press, Boca Raton, FL, 1995.

[9] A.M. Davie. *Invariant subspaces for Bishop's operators.* Bull. London Math. Soc. **6** (1974), 343–348.

[10] P. Enflo. *Extremal vectors for a class of linear operators.* Functional analysis and economic theory (Samos, 1996), Springer, Berlin (1998), 61–64.

[11] P. Enflo and T. Hõim. *Some results on extremal vectors and invariant subspaces.* Proc. Amer. Math. Soc. **131** (2003), 379–387.

[12] G.W. MacDonald. *Invariant subspaces for Bishop-type operators.* J. Funct. Anal. **91** (1990), 287–311.

[13] H.A. Medina. *Connections between additive cocycles and Bishop operators.* Illinois J. Math. **40** (1996), 432–438.

[14] A. Spalsbury. *Vectors of minimal norm.* Proc. Amer. Math. Soc. **350** (1998), 2737–2745.

[15] V.G. Troitsky. *Minimal vectors in arbitrary Banach spaces.* Proc. Amer. Math. Soc. **132** (2004), 1177–1180.

[16] E. Wiesner. *Backward minimal points for bounded linear operators on finite-dimensional vector spaces.* Linear Algebra Appl. **338** (2001), 251–259.

Isabelle Chalendar
Institut Camille Jordan,
UFR de Mathématiques,
Université Claude Bernard Lyon 1,
F-69622 Villeurbanne Cedex, France
e-mail: `chalenda@math.univ-lyon1.fr`

Antoine Flattot
Institut Camille Jordan,
UFR de Mathématiques,
Université Claude Bernard Lyon 1,
F-69622 Villeurbanne Cedex, France
e-mail: `flattot@math.univ-lyon1.fr`

Jonathan R. Partington
School of Mathematics,
University of Leeds,
Leeds LS2 9JT, UK
e-mail: `J.R.Partington@leeds.ac.uk`

Operator Theory:
Advances and Applications, Vol. 171, 107–127

The Continuous Analogue of the Resultant and Related Convolution Operators

Israel Gohberg, Marinus A. Kaashoek and Leonid Lerer

Abstract. For a class of pairs of entire matrix functions the null space of the natural analogue of the classical resultant matrix is described in terms of the common Jordan chains of the defining entire matrix functions. The main theorem is applied to two inverse problems. The first concerns convolution integral operators on a finite interval with matrix valued kernel functions and complements earlier results of [6]. The second is the inverse problem for matrix-valued continuous analogues of Szegő orthogonal polynomials.

Mathematics Subject Classification (2000). Primary 47B35, 47B99, 45E10, 30D20; Secondary 33C47, 42C05.

Keywords. Resultant operator, continuous analogue of the resultant, Bezout operator, convolution integral operators on a finite interval, orthogonal matrix functions, inverse problems.

1. Introduction

We begin by stating the main theorem proved in this paper. This requires some preparations. Fix $\omega > 0$, and consider the $n \times n$ entire matrix functions

$$\mathcal{B}(\lambda) = I + \int_{-\omega}^{0} e^{i\lambda s} b(s)\, ds, \quad \mathcal{D}(\lambda) = I + \int_{0}^{\omega} e^{i\lambda s} d(s)\, ds. \tag{1.1}$$

Here $b \in L_1^{n\times n}[-\omega, 0]$ and $d \in L_1^{n\times n}[0, \omega]$. We define $R(\mathcal{B}, \mathcal{D})$ to be the operator on $L_1^n[-\omega, \omega]$ given by

$$\big(R(\mathcal{B}, \mathcal{D})f\big)(t) = \begin{cases} f(t) + \displaystyle\int_{-\omega}^{\omega} d(t-s)f(s)\, ds, & 0 \le t \le \omega, \\ f(t) + \displaystyle\int_{-\omega}^{\omega} b(t-s)f(s)\, ds, & -\omega \le t < 0. \end{cases} \tag{1.2}$$

The research of the third author was partially supported by a visitor fellowship of the Netherlands Organization for Scientific Research (NWO) and the Glasberg-Klein Research Fund at the Technion.

Here we follow the convention that $d(t)$ and $b(t)$ are zero whenever t does not belong to $[0, \omega]$ or $[-\omega, 0]$, respectively.

Our aim is to describe the null space of the operator $R(\mathcal{B}, \mathcal{D})$ in terms of the common spectral data of the two functions $\mathcal{B}$ and $\mathcal{D}$. Therefore we first review some elements of the spectral theory of entire matrix functions. Note that from the definitions in (1.1) it follows that

$$\lim_{\Im\lambda \leq 0,\, |\lambda|\to\infty} \mathcal{B}(\lambda) = I, \qquad \lim_{\Im\lambda \geq 0,\, |\lambda|\to\infty} \mathcal{D}(\lambda) = I.$$

Thus $\mathcal{B}$ has only a finite number of eigenvalues (zeros) in the closed lower half plane, and the same is true for $\mathcal{D}$ with respect to the closed upper half plane. We conclude that the number of common eigenvalues of $\mathcal{B}$ and $\mathcal{D}$ in $\mathbb{C}$ is finite. This allows us to define the *total common multiplicity* $\nu(\mathcal{B}, \mathcal{D})$ of $\mathcal{B}$ and $\mathcal{D}$, namely:

$$\nu(\mathcal{B}, \mathcal{D}) = \sum_{\lambda} \nu(\mathcal{B}, \mathcal{D}; \lambda),$$

where the sum is taken over the common eigenvalues, and $\nu(\mathcal{B}, \mathcal{D}; \lambda)$ is the common multiplicity of λ as a common eigenvalue of $\mathcal{B}$ and $\mathcal{D}$. For the definition of the latter notion we refer to Section 2 below which also presents more details about the notions used in the next paragraph.

Now let $\lambda_1, \dots, \lambda_{\ell_\circ}$ be the set of distinct common eigenvalues of $\mathcal{B}$ and $\mathcal{D}$ in $\mathbb{C}$. For each common eigenvalue λ_ℓ we let

$$\Xi_\ell = \{x_{jk,\ell} \mid k = 0, \dots, \kappa_{j,\ell} - 1,\ j = 1, \dots, p_\ell\}$$

stand for a canonical set of common Jordan chains of $\mathcal{B}$ and $\mathcal{D}$ at λ_ℓ. The set of vectors $\{\Xi_\ell\}_{\ell=1}^{\ell_\circ}$ generates the following set of $\mathbb{C}^n$-valued functions:

$$\widetilde{\Xi} = \{\widetilde{x}_{jk,\ell} \mid k = 0, \dots, \kappa_{j,\ell} - 1,\ j = 1, \dots, p_\ell,\ \ell = 1, \dots, \ell_\circ\}, \tag{1.3}$$

where

$$\widetilde{x}_{jk,\ell}(t) = e^{-i\lambda_\ell t} \sum_{\nu=0}^{k} \frac{(-it)^{k-\nu}}{(k-\nu)!} x_{j\nu,\ell}. \tag{1.4}$$

The functions $\widetilde{x}_{jk,\ell}$ will be considered on intervals $-\infty < \alpha < \beta < \infty$ for appropriate choices of α and β. Notice that for any interval $[\alpha, \beta]$ the system of functions $\widetilde{\Xi}$ is linearly independent, and

$$\dim \operatorname{span} \widetilde{\Xi} = \nu(\mathcal{B}, \mathcal{D}). \tag{1.5}$$

We can now state the main theorem of this paper.

Theorem 1.1. *Let $\mathcal{B}$ and $\mathcal{D}$ be the entire $n \times n$ matrix functions given by* (1.1), *and let $R(\mathcal{B}, \mathcal{D})$ be the operator on $L_1^n[-\omega, \omega]$ defined by* (1.2). *Assume that there exist entire $n \times n$ matrix functions $\mathcal{A}$ and $\mathcal{C}$,*

$$\mathcal{A}(\lambda) = I + \int_0^{\omega} e^{i\lambda s} a(s)\, ds, \quad \mathcal{C}(\lambda) = I + \int_{-\omega}^{0} e^{i\lambda s} c(s)\, ds, \tag{1.6}$$

where $a \in L_1^{n\times n}[0,\omega]$ *and* $c \in L_1^{n\times n}[-\omega,0]$, *such that*

$$\mathcal{A}(\lambda)\mathcal{B}(\lambda) = \mathcal{C}(\lambda)\mathcal{D}(\lambda), \quad \lambda \in \mathbb{C}. \tag{1.7}$$

Then the set of vector functions $\widetilde{\Xi}$ *in* (1.3), *considered in* $L_1^n[-\omega,\omega]$, *forms a basis of* Ker $R(\mathcal{B},\mathcal{D})$. *In particular,*

$$\dim \operatorname{Ker} R(\mathcal{B},\mathcal{D}) = \nu(\mathcal{B},\mathcal{D}). \tag{1.8}$$

The above theorem is new for the matrix case ($n > 1$); for the scalar case ($n = 1$) it has been proved in [7]. Note that in the scalar case condition (1.7) is redundant. In fact, more generally, if the matrix functions $\mathcal{B}$ and $\mathcal{D}$ commute, then one can always find matrix functions $\mathcal{A}$ and $\mathcal{C}$ of the form (1.6) such that (1.7) is fulfilled; the trivial choice $\mathcal{A} = \mathcal{D}$ and $\mathcal{C} = \mathcal{B}$ will do.

In [7], for $n = 1$, the operator $R(\mathcal{B},\mathcal{D})$ is introduced as the natural continuous analogue of the classical Sylvester resultant matrix for polynomials (for the latter see the survey article [20] or the book [23]). For this reason we call $R(\mathcal{B},\mathcal{D})$ the *resultant operator* associated with $\mathcal{B}$ and $\mathcal{D}$.

We shall refer to condition (1.7) as the *quasi commutativity property* of the quadruple $\{\mathcal{A},\mathcal{C};\mathcal{B},\mathcal{D}\}$. In general, without this property being satisfied, Theorem 1.1 does not remain true. This follows from a simple example with $n = 2$ given in [7]. On the other hand, in [7] it is also shown that the conclusion of Theorem 1.1, without the quasi commutativity property (1.7) being fulfilled, remains true provided one replaces the resultant operator $R(\mathcal{B},\mathcal{D})$ by a more complicated operator, acting between different L_1-spaces, namely by the operator $R_\varepsilon(\mathcal{B},\mathcal{D})$, with $\varepsilon > 0$, acting from $L_1^n[-\omega,\omega+\varepsilon]$ into $L_1^n[-\omega-\varepsilon,\omega+\varepsilon]$, defined by

$$(R_\varepsilon(\mathcal{B},\mathcal{D})f)(t) = \begin{cases} f(t) + \displaystyle\int_{-\omega}^{\omega+\varepsilon} d(t-s)f(s)\,ds, & 0 \le t \le \omega+\varepsilon, \\[2ex] f(t) + \displaystyle\int_{-\omega}^{\omega+\varepsilon} b(t-s)f(s)\,ds, & -\omega-\varepsilon \le t < 0. \end{cases}$$

In this paper we also present two applications of Theorem 1.1 to inverse problems for related convolution operators. The first application deals with the problem of reconstructing a matrix-valued kernel function of a convolution integral operator on a finite interval from four matrix functions satisfying integral equations with this kernel function. The second problem is the inverse problem for continuous analogues of orthogonal polynomials.

In our proof of Theorem 1.1 an essential role is played by Theorem 4.19 from the thesis [14]. The latter theorem describes the dimension of the null space of a Bezout type operator on $L_1^n[0,\omega]$ associated with entire matrix functions $\mathcal{A},\mathcal{B},\mathcal{C},\mathcal{D}$. The precise Bezout operator result needed in the present paper is stated in [5], where it is also proved by a new method. For the four matrix functions $\mathcal{A},\mathcal{B},\mathcal{C},\mathcal{D}$ appearing in Theorem 1.1 we show that after a simple extension the Bezout operator from [5] is equivalent to the resultant operator $R(\mathcal{B},\mathcal{D})$. To obtain this equivalence we rewrite the quasi commutativity property in operator form.

This is done by using a new method which is based on an analysis of related block Laurent operators.

The paper consists of five sections (including the present introduction). In Section 2 we review the elements of the spectral theory of entire $n \times n$ matrix functions that are used in this paper, which includes facts about common spectral data for pairs of such functions. In Section 3 we introduce the Bezout operator, state the result from [5] needed in the present paper, and establish the equivalence of the resultant operator with a simple extension of the Bezout operator. A more refined connection between the Bezout operator and the resultant operator, which turns out to be equivalent to the quasi commutativity property, is also given in this section. In Section 4 we prove Theorem 1.1. Section 5 presents the two applications of the main theorem to inverse problems.

Some words about notation. If $-\infty < \alpha < \beta < \infty$, then $L_1[\alpha, \beta]$ stands for the standard Banach spaces of Lebesgue integrable functions on $[\alpha, \beta]$. Any function $f \in L_1[\alpha, \beta]$ will be considered as a function on the entire real line by setting $f(t) = 0$ whenever $t < \alpha$ or $t > \beta$. Given a positive integer n we denote by $L_1^n[\alpha, \beta]$ the Banach space of all $n \times 1$ column vector functions with entries in $L_1[\alpha, \beta]$. Similarly, $L_1^{n\times n}[\alpha, \beta]$ consists of all $n \times n$ matrix functions with entries in $L_1[\alpha, \beta]$. As usual, we identify Lebesgue integrable functions which differ on a set of measure zero.

2. Preliminaries on the spectral theory of entire matrix functions

In this section we review the elements of the spectral theory of entire $n \times n$ matrix functions that are used in this paper.

2.1. Eigenvalues and Jordan chains

Let F be an entire $n \times n$ matrix function. We assume F to be *regular*. The latter means that $\det F(\lambda) \not\equiv 0$ on $\mathbb{C}$. As usual, the values of F are identified with their canonical action on $\mathbb{C}^n$. In what follows λ_0 is an arbitrary point in $\mathbb{C}$.

The point λ_0 is called an *eigenvalue* of F whenever there exists a vector $x_0 \neq 0$ in $\mathbb{C}^n$ such that $F(\lambda_0)x_0 = 0$. In that case the non-zero vector x_0 is called an *eigenvector* of F at λ_0. Note that λ_0 is an eigenvalue of F if and only if $\det F(\lambda_0) = 0$. In particular, in the scalar case, i.e., when n=1, the point λ_0 is an eigenvalue of F if and only if λ_0 is a zero of F. The *multiplicity* $\nu(\lambda_0)$ of the eigenvalue λ_0 of F is defined as the multiplicity of λ_0 as a zero of $\det F(\lambda)$. The set of eigenvectors of F at λ_0 together with the zero vector is equal to $\operatorname{Ker} F(\lambda_0)$.

An ordered sequence of vectors $x_0, x_1, \ldots, x_{r-1}$ in $\mathbb{C}^n$ is called a *Jordan chain of length* r of F at λ_0 if $x_0 \neq 0$ and

$$\sum_{j=0}^{k} \frac{1}{j!} F^{(j)}(\lambda_0) x_{k-j} = 0, \quad k = 0, \ldots, r-1. \tag{2.9}$$

Here $F^{(j)}(\lambda_0)$ is the jth derivative of F at λ_0. From $x_0 \neq 0$ and (2.9) it follows that λ_0 is an eigenvalue of F and x_0 is a corresponding eigenvector. The converse is also true, that is, x_0 is an eigenvector of F at λ_0 if and only if x_0 is the first vector in a Jordan chain for F at λ_0.

Given an eigenvector x_0 of F at λ_0 there are, in general, many Jordan chains for F at λ_0 which have x_0 as their first vector. However, the fact that F is regular implies that the lengths of these Jordan chains have a finite supremum which we shall call the *rank* of the eigenvector x_0.

To organize the Jordan chains corresponding to the eigenvalue λ_0 we proceed as follows. Choose an eigenvector x_{10} in $\operatorname{Ker} F(\lambda_0)$ such that the rank r_1 of x_{10} is maximal, and let $x_{10}, \ldots, x_{1r_1-1}$ be a corresponding Jordan chain. Next we choose among all vectors x in $\operatorname{Ker} F(\lambda_0)$, with x not a multiple of x_{10}, a vector x_{20} of maximal rank, r_2 say, and we choose a corresponding Jordan chain $x_{20}, \ldots, x_{2r_2-1}$. We proceed by induction. Assume

$$x_{10}, \ldots, x_{1r_1-1}, \ldots, x_{k0}, \ldots, x_{kr_k-1}$$

have been chosen. Then we choose x_{k+10} to be a vector in $\operatorname{Ker} F(\lambda_0)$ that does not belong to $\operatorname{span}\{x_{10}, \ldots, x_{k0}\}$ such that x_{k+10} is of maximal rank among all vectors in $\operatorname{Ker} F(\lambda_0)\backslash\operatorname{span}\{x_{10}, \ldots, x_{k0}\}$. In this way, in a finite number of steps, we obtain a basis $x_{10}, x_{20}, \ldots, x_{p0}$ of $\operatorname{Ker} F(\lambda_0)$ and corresponding Jordan chains

$$x_{10}, \ldots, x_{1r_1-1}, x_{20}, \ldots, x_{2r_2-1}, \ldots, x_{p0}, \ldots, x_{pr_p-1}. \tag{2.10}$$

The system (2.10) is called a *canonical system of Jordan chains* for F at λ_0. From the construction it follows that $p = \dim \operatorname{Ker} F(\lambda_0)$. The numbers $r_1 \geq r_2 \geq \cdots \geq r_p$ are uniquely determined by F and do not depend on the particular choices made above. In fact, these numbers coincide with the degrees of $\lambda - \lambda_0$ in the local Smith form of F at λ_0. The numbers $r_1, \ldots, r_p$ are called the *partial multiplicities* of F at λ_0. Their sum $r_1 + \cdots + r_p$ is equal to the multiplicity $\nu(\lambda_0)$.

The above definitions of eigenvalue, eigenvector and Jordan chain for F at λ_0 also make sense when F is non-regular or when F is a non-square entire matrix function on Ω. However, in that case it may happen that the supremum of the lengths of the Jordan chains with a given first vector is not finite. On the other hand, if for each non-zero vector x_0 in $\operatorname{Ker} F(\lambda_0)$ the supremum of the lengths of the Jordan chains with x_0 as first vector is finite, then we can define a canonical set of Jordan chains for F at λ_0 in the same way as it was done above for regular entire matrix functions.

More details on the above notions, including proofs, can be found in [13]; see also the book [10] or the appendix of [8].

2.2. Common eigenvalues and common Jordan chains

Throughout this section F_1 and F_2 are entire $n \times n$ matrix functions. Furthermore, $\lambda_0 \in \mathbb{C}$, and we assume that either F_1 or F_2 is regular on $\mathbb{C}$.

Let λ_0 be an arbitrary point in $\mathbb{C}$. We say that λ_0 is a *common eigenvalue* of F_1 and F_2 if there exists a vector $x_0 \neq 0$ such that $F_1(\lambda_0)x_0 = F_2(\lambda_0)x_0 = 0$. In this case we refer to x_0 as a *common eigenvector* of F_1 and F_2 at λ_0. Note that x_0 is a common eigenvector of F_1 and F_2 at λ_0 if and only if x_0 is a non-zero vector in

$$\operatorname{Ker} F_1(\lambda_0) \cap \operatorname{Ker} F_2(\lambda_0) = \operatorname{Ker} \begin{bmatrix} F_1(\lambda_0) \\ F_2(\lambda_0) \end{bmatrix}.$$

If an ordered sequence of vectors $x_0, x_1, \ldots, x_{r-1}$ is a Jordan chain for both F_1 and F_2 at λ_0, then we say that $x_0, x_1, \ldots, x_{r-1}$ is a *common Jordan chain* for F_1 and F_2 at λ_0. In other words, $x_0, x_1, \ldots, x_{r-1}$ is a common Jordan chain for F_1 and F_2 at λ_0 if and only if $x_0, x_1, \ldots, x_{r-1}$ is a Jordan chain for F at λ_0, where F is the non-square entire matrix function given by

$$F(\lambda) = \begin{bmatrix} F_1(\lambda) \\ F_2(\lambda) \end{bmatrix}, \quad \lambda \in \mathbb{C}. \tag{2.11}$$

Let x_0 be a common eigenvector of F_1 and F_2 at λ_0. Since F_1 or F_2 is regular, the lengths of the common Jordan chains of F_1 and F_2 at λ_0 with initial vector x_0 have a finite supremum. In other words, if x_0 is a non-zero vector in $\operatorname{Ker} F(\lambda_0)$, where F is the non-square analytic matrix function defined by (2.11), then the lengths of the Jordan chains of F at λ_0 with initial vector x_0 have a finite supremum. Hence (see the final paragraph of the previous section), for F in (2.11) a canonical set of Jordan chains of F at λ_0 is well defined. We say that

$$x_{10}, \ldots, x_{1r_1-1}, x_{20}, \ldots, x_{2r_2-1}, \ldots, x_{p0}, \ldots, x_{pr_p-1} \tag{2.12}$$

is a *canonical set of common Jordan chains* of F_1 and F_2 at λ_0 if the chains in (2.12) form a canonical set of Jordan chains for F at λ_0, where F is defined by (2.11). Furthermore, in that case the number

$$\nu(F_1, F_2; \lambda_0) := \sum_{j=1}^{p} r_j$$

is called the *common multiplicity* of λ_0 as a common eigenvalue of the analytic matrix functions F_1 and F_2.

For further details, including proofs, see [7].

3. Equivalence of Bezout and resultant operators

In this section we introduce the Bezout operator, recall some basic results from the paper [5], and clarify the connection between the resultant operator and the Bezout operator.

3.1. The Bezout operator and its main property

Throughout this section a, b, c, and d are $n \times n$ matrix functions, a and d belong to $L_1^{n\times n}[0,\omega]$, while b and c belong to $L_1^{n\times n}[-\omega,0]$. We denote by $\mathcal{A}$, $\mathcal{B}$, $\mathcal{C}$, $\mathcal{D}$ the entire $n \times n$ matrix functions given by

$$\mathcal{A}(\lambda) = I + \int_0^\omega e^{i\lambda s} a(s)\, ds, \quad \mathcal{B}(\lambda) = I + \int_{-\omega}^0 e^{i\lambda s} b(s)\, ds, \tag{3.1}$$

$$\mathcal{C}(\lambda) = I + \int_{-\omega}^0 e^{i\lambda s} c(s)\, ds, \quad \mathcal{D}(\lambda) = I + \int_0^\omega e^{i\lambda s} d(s)\, ds. \tag{3.2}$$

We shall assume that the functions a, b, c, and d satisfy the following additional condition

$$\mathcal{A}(\lambda)\mathcal{B}(\lambda) = \mathcal{C}(\lambda)\mathcal{D}(\lambda), \quad \lambda \in \mathbb{C}, \tag{3.3}$$

that is, the quadruple $\{\mathcal{A}, \mathcal{C}; \mathcal{B}, \mathcal{D}\}$ has the quasi commutativity property; cf. (1.7).

Given four functions as above, we let $T = I + \Gamma$, where Γ is the integral operator on $L_1^n[0, \omega]$ defined by

$$(\Gamma\varphi)(t) = \int_0^\omega \gamma(t, s)\varphi(s)\, ds, \quad 0 \le t \le \omega, \tag{3.4}$$

with

$$\begin{aligned} \gamma(t, s) &= a(t - s) + b(t - s) \\ &\quad + \int_0^{\min\{t,s\}} a(t - r)b(r - s) - c(t - \omega - r)d(r + \omega - s)\, dr. \end{aligned} \tag{3.5}$$

We refer to the operator

$$T = T\{\mathcal{A}, \mathcal{C}; \mathcal{B}, \mathcal{D}\} \tag{3.6}$$

as the *Bezout operator* associated with $\{\mathcal{A}, \mathcal{C}; \mathcal{B}, \mathcal{D}\}$. The main property of the Bezout operator used in this paper is given by the following theorem which is stated and proved in [5] and originates from [14].

Theorem 3.1. *Assume the quadruple $\{\mathcal{A}, \mathcal{C}; \mathcal{B}, \mathcal{D}\}$ satisfies the quasi commutativity property. Then the dimension of the null space of the Bezout operator T associated with $\{\mathcal{A}, \mathcal{C}; \mathcal{B}, \mathcal{D}\}$ is equal to the total common multiplicity of the entire matrix functions $\mathcal{B}$ and $\mathcal{D}$, that is,* $\dim \operatorname{Ker} T = \nu(\mathcal{B}, \mathcal{D})$.

For a comprehensive review of the history of Bezout matrices and Bezout operators we refer the reader to the final paragraphs in the Introduction of [5]; the early history can be found in [20]. For first results on Bezout operators for entire scalar functions see [7] and [21]; cf., Chapter 5 in [22].

3.2. The quasi commutativity property in operator form

In this section we restate in operator form the quasi commutativity property (3.3). The main result is Proposition 3.2 below. We begin with some preliminaries.

We shall adopt the following notations. If a lower case letter f denotes a function in $L_1^{n\times n}(\mathbb{R})$, then the corresponding bold face capital letter $\mathbf{F}$ denotes the convolution operator on $L_1^n(\mathbb{R})$ given by

$$(\mathbf{F}\varphi)(t) = \int_{-\infty}^\infty f(t - s)\varphi(s)\, ds, \quad -\infty < t < \infty. \tag{3.7}$$

We denote by $\mathcal{F}(\lambda)$ the symbol of the operator $I + \mathbf{F}$, that is,

$$\mathcal{F}(\lambda) = I + \int_{-\infty}^{\infty} e^{i\lambda s} f(s)\, ds. \tag{3.8}$$

Furthermore, for each $\nu \in \mathbb{Z}$ we define F_ν to be the convolution operator on $L_1^n[0,\omega]$ given by

$$(F_\nu \varphi)(t) = \int_0^\omega f(t - s + \nu\omega)\varphi(s)\, ds, \quad 0 \le t \le \omega.$$

We shall refer to the operators $F_\nu, \nu \in \mathbb{Z}$, as the *operators corresponding to* $\mathcal{F}$. With these operators F_ν we associate the block Laurent operator

$$L_{\mathbf{F}} = \begin{bmatrix} \ddots & & & & \\ & F_0 & F_{-1} & F_{-2} & \\ & F_1 & \boxed{F_0} & F_{-1} & \\ & F_2 & F_1 & F_0 & \\ & & & & \ddots \end{bmatrix}.$$

We consider $L_{\mathbf{F}}$ as a bounded linear operator on the space $\ell_{1,\mathbb{Z}}\big(L_1^n[0,\omega]\big)$. The latter space consists of all doubly infinite sequences $\varphi = (\varphi_j)_{j\in\mathbb{Z}}$ with $\varphi_j \in L_1^n[0,\omega]$ such that

$$\|\varphi\|_{\ell_{1,\mathbb{Z}}(L_1^n[0,\omega])} := \sum_{j=-\infty}^{\infty} \|\varphi_j\|_{L_1^n[0,\omega]} < \infty.$$

The spaces $L_1^n(R)$ and $\ell_{1,\mathbb{Z}}\big(L_1^n[0,\omega]\big)$ are isometrically equivalent, and for f and g in $L_1^{n\times n}(\mathbb{R})$ we have

$$L_{\mathbf{FG}} = L_{\mathbf{F}} L_{\mathbf{G}}. \tag{3.9}$$

Now let us return to the functions a, b, c, and d considered in the previous section. Our goal is to restate the quasi commutativity property (3.3) in operator form. We shall view a, b, c, and d as functions in $L_1^{n\times n}(\mathbb{R})$, with a and d having their support in $[0,\omega]$, while the support of b and c is in $[-\omega,0]$. Thus, using the terminology of the previous paragraph, the functions $\mathcal{A}$, $\mathcal{B}$, $\mathcal{C}$, $\mathcal{D}$ in (3.1), (3.2) are the symbols of the operators $I + \mathbf{A}$, $I + \mathbf{B}$, $I + \mathbf{C}$ and $I + \mathbf{D}$, respectively. Let us consider the operators $A_\nu, \nu \in \mathbb{Z}$, corresponding to $\mathcal{A}$. Since a has its support in $[0,\omega]$, we have the following properties:

(i) $(A_0\varphi)(t) = \displaystyle\int_0^t a(t-s)\varphi(s)\, ds, \quad 0 \le t \le \omega,$

(ii) $(A_1\varphi)(t) = \displaystyle\int_t^\omega a(t+\omega-s)\varphi(s)\, ds, \quad 0 \le t \le \omega,$

(iii) $A_\nu = 0$ for $\nu \neq 0, \nu \neq 1$.

Similarly, since b has its support in $[-\omega,0]$, the operators $B_\nu, \nu \in \mathbb{Z}$, corresponding to $\mathcal{B}$ have the following properties:

(j) $(B_0\varphi)(t) = \displaystyle\int_t^\omega b(t-s)\varphi(s)\, ds, \quad 0 \le t \le \omega,$

(jj) $(B_{-1}\varphi)(t) = \int_0^t b(t-s-\omega)\varphi(s)\,ds, \quad 0 \le t \le \omega,$

(jjj) $B_\nu = 0$ for $\nu \ne 0, \nu \ne -1$.

Analogous results hold for the operators $C_\nu, \nu \in \mathbb{Z}$, corresponding to $\mathcal{C}$ and for $D_\nu, \nu \in \mathbb{Z}$, corresponding to $\mathcal{D}$. In particular, the above formulas show that all the operators

$$I + A_0, \quad I + B_0, \quad I + C_0, \quad I + D_0$$

are invertible operators on $L_1^n[0,\omega]$.

The next proposition is the main result of this section.

Proposition 3.2. *The quasi commutativity property* (3.3) *is equivalent to the following two conditions:*

$$(I + A_0)B_{-1} = C_{-1}(I + D_0), \quad (I + C_0)D_1 = A_1(I + B_0). \tag{3.10}$$

Proof. Since $\mathcal{A}$, $\mathcal{B}$, $\mathcal{C}$, and $\mathcal{D}$ are the symbols of the convolution operators $I + \mathbf{A}$, $I + \mathbf{B}$, $I + \mathbf{C}$, and $I + \mathbf{D}$, respectively, condition (3.3) is equivalent to

$$(I + \mathbf{A})(I + \mathbf{B}) = (I + \mathbf{C})(I + \mathbf{D}), \tag{3.11}$$

which according to (3.9) can be rewritten as

$$L_{\mathbf{A}} + L_{\mathbf{B}} + L_{\mathbf{A}}L_{\mathbf{B}} = L_{\mathbf{C}} + L_{\mathbf{D}} + L_{\mathbf{C}}L_{\mathbf{D}}. \tag{3.12}$$

Now recall the properties (i)–(iii) for the operator A_ν, the properties (j)–(jjj) for the operators B_ν, and the analogous properties for the operators C_ν and D_ν ($\nu \in \mathbb{Z}$). By comparing the entries in the infinite operator matrices determined by the left- and right-hand sides of (3.12) we see that (3.3) is equivalent to

(α) $B_{-1} + A_0B_{-1} = C_{-1} + C_{-1}D_0$,

(β) $A_0 + B_0 + A_0B_0 + A_1B_{-1} = C_0 + D_0 + C_0D_0 + C_{-1}D_1$

(γ) $A_1 + A_1B_0 = D_1 + C_0D_1$.

Obviously, (α) is the same as the first part of (3.10), and (γ) is the same as the second part of (3.10). Thus to complete the proof we have to show that (3.10) implies condition (β).

Consider the functions $f = a + b + a * b$ and $g = c + d + c * d$, where $*$ denotes the convolution product in $L_1^{n\times n}(\mathbb{R})$. Then $L_{\mathbf{F}}$ is equal to the left-hand side of (3.12), and $L_{\mathbf{G}}$ to the right-hand side of (3.12). The first part of (3.10) yields $F_{-1} = G_{-1}$, and the second part of (3.10) implies $F_1 = G_1$. Now notice that $F_{-1} = G_{-1}$ is equivalent to $f(t) = g(t)$ for each $-2\omega \le t \le 0$, and $F_1 = G_1$ is equivalent to $f(t) = g(t)$ for each $0 \le t \le 2\omega$, i.e.,

$$F_{-1} = G_{-1} \iff f|_{[-2\omega,\,0]} = g|_{[-2\omega,\,0]},$$
$$F_1 = G_1 \iff f|_{[0,\,2\omega]} = g|_{[0,\,2\omega]}.$$

In particular, if (3.10) holds, then $f|_{[-\omega,\,\omega]} = g|_{[-\omega,\,\omega]}$, which is equivalent to $F_0 = G_0$. But $F_0 = G_0$ is equivalent to (β). This completes the proof. $\square$

3.3. The equivalence theorem

Let $\mathcal{A}$, $\mathcal{B}$, $\mathcal{C}$, $\mathcal{D}$ be the four entire $n \times n$ matrix functions given by (3.1), (3.2), and assume that the quasi commutativity property (3.3) is satisfied. In this section we show that after an appropriate extension the Bezout operator T associated with the quadruple $\{\mathcal{A}, \mathcal{C}; \mathcal{B}, \mathcal{D}\}$ is equivalent to the resultant operator $R(\mathcal{B}, \mathcal{D})$. To state the precise result we need some additional notation.

First recall that the Bezout operator T is given by $T = I + \Gamma$, where Γ is the integral operator with kernel function (3.5). Using this formula and notations from the previous section we see that

$$T = (I + A_0)(I + B_0) - C_{-1}D_1. \tag{3.13}$$

Here A_0, B_0, C_{-1} and D_1 are the operators on $L_1^n[0, \omega]$ introduced in the previous section. Next, we consider the operators

$$H^+ : L_1^n[-\omega, 0] \to L_1^n[0, \omega], \quad (H^+ f)(t) = f(t - \omega) \quad (0 \le t \le \omega), \tag{3.14}$$

$$H^- : L_1^n[0, \omega] \to L_1^n[-\omega, 0], \quad (H^- f)(t) = f(t + \omega) \quad (\omega \le t \le 0). \tag{3.15}$$

Note that H^+ and H^- are both invertible and $(H^+)^{-1} = H^-$. Representing the space $L_1^n[-\omega, \omega]$ as the natural direct sum of the spaces $L_1^n[-\omega, 0]$ and $L_1^n[0, \omega]$, the resultant operator $R(\mathcal{B}, \mathcal{D})$ can be written as a 2×2 operator matrix as follows:

$$R(\mathcal{B}, \mathcal{D}) = \begin{bmatrix} I + H^- B_0 H^+ & H^- B_{-1} \\ D_1 H^+ & I + D_0 \end{bmatrix} \quad \text{on} \quad L_1^n[-\omega, 0] \oplus L_1^n[0, \omega]. \tag{3.16}$$

Here B_{-1} D_0, and D_1 are operators on $L_1^n[0, \omega]$ of which the definition can be found in the previous section. Since the operator $I + D_0$ is invertible, formula (3.16) allows us to factor $R(\mathcal{B}, \mathcal{D})$ as follows:

$$\begin{aligned} R(\mathcal{B}, \mathcal{D}) &= \begin{bmatrix} I & H^- B_{-1}(I + D_0)^{-1} \\ 0 & I \end{bmatrix} \\ &\quad \times \begin{bmatrix} H^- \Upsilon H^+ & 0 \\ 0 & I + D_0 \end{bmatrix} \begin{bmatrix} I & 0 \\ (I + D_0)^{-1} D_1 H^+ & I \end{bmatrix}, \end{aligned} \tag{3.17}$$

where

$$\Upsilon = I + B_0 - B_{-1}(I + D_0)^{-1} D_1. \tag{3.18}$$

Since the quasi commutativity property is satisfied, we know from Proposition 3.2 that $B_{-1}(I + D_0)^{-1} = (I + A_0)^{-1} C_{-1}$, and therefore we obtain from (3.13) and (3.18) that

$$\begin{aligned} \Upsilon &= I + B_0 - (I + A_0)^{-1} C_{-1} \\ &= (I + A_0)^{-1}\{(I + A_0)(I + B_0) - C_{-1} D_{-1}\} = (I + A_0)^{-1} T. \end{aligned}$$

We conclude that the resultant operator $R(\mathcal{B},\mathcal{D})$ and the Bezout operator T are related in the following way:

$$R(\mathcal{B},\mathcal{D}) = \begin{bmatrix} H^-(I+A_0)^{-1} & H^-B_{-1}(I+D_0)^{-1} \\ 0 & I \end{bmatrix} \tag{3.19}$$

$$\times \begin{bmatrix} T & 0 \\ 0 & I \end{bmatrix} \begin{bmatrix} H^+ & 0 \\ D_1H^+ & I+D_0 \end{bmatrix}.$$

Since the first and the third factor in the right-hand side of (3.19) are invertible, we arrive at the following result.

Theorem 3.3. *Let the entire $n \times n$ matrix functions $\mathcal{A},\mathcal{B},\mathcal{C},\mathcal{D}$ be given by (3.1), (3.2), and assume that the quasi commutativity property (3.3) is fulfilled. Then the Bezout operator T associated with the quadruple $\{\mathcal{A},\mathcal{B},\mathcal{C},\mathcal{D}\}$ is equivalent after extension to the resultant operator $R(\mathcal{B},\mathcal{D})$, with the equivalence after extension relation being given by (3.19). In particular,*

$$\dim \operatorname{Ker} R(\mathcal{B},\mathcal{D}) = \dim \operatorname{Ker} T. \tag{3.20}$$

3.4. A Kravitsky type formula

In this subsection we present a proposition which gives some further insight in the use of the quasi commutativity property.

Let $\mathcal{A}$, $\mathcal{B}$, $\mathcal{C}$, $\mathcal{D}$ be four entire $n \times n$ matrix functions given by (3.1), (3.2). With the pair $\mathcal{A}$, $\mathcal{C}$ we associate an operator $S(\mathcal{A},\mathcal{C})$ on $L_1^n[-\omega,\omega]$ by setting

$$(S(\mathcal{A},\mathcal{C})f)(t) = f(t) + \int_{-\omega}^{0} a(t-s)f(s)\,ds + \int_{0}^{\omega} c(t-s)f(s)\,ds, \qquad -\omega \le t \le \omega.$$

We call $S(\mathcal{A},\mathcal{C})$ the *pair operator* associated with $\mathcal{A}$, $\mathcal{C}$. This operator can be viewed as a transpose to the left resultant operator of $\mathcal{A}$ and $\mathcal{C}$.

As in the previous section we write the space $L_1^n[-\omega,\omega]$ as the natural direct sum of the spaces $L_1^n[-\omega,0]$ and $L_1^n[0,\omega]$. This allows us to represent the pair operator $S(\mathcal{A},\mathcal{C})$ by a 2×2 operator matrix,

$$S(\mathcal{A},\mathcal{C}) = \begin{bmatrix} I+H^-A_0H^+ & H^-C_{-1} \\ A_1H^+ & I+C_0 \end{bmatrix} \quad \text{on} \quad L_1^n[-\omega,0] \oplus L_1^n[0,\omega]. \tag{3.21}$$

Here H^+ and H^- are the operators defined by (3.14) and (3.15), respectively.

Proposition 3.4. *Let the entire $n \times n$ matrix functions $\mathcal{A},\mathcal{B},\mathcal{C},\mathcal{D}$ be given by (3.1), (3.2). If the quadruple $\{\mathcal{A},\mathcal{C};\mathcal{B},\mathcal{D}\}$ has the quasi commutativity property, then*

$$S(\mathcal{A},\mathcal{C}) \begin{bmatrix} I_{L_1^n[-\omega,0]} & 0 \\ 0 & -I_{L_1^n[0,\omega]} \end{bmatrix} R(\mathcal{B},\mathcal{D}) = \begin{bmatrix} H^-TH^+ & 0 \\ 0 & T \end{bmatrix}, \tag{3.22}$$

where T is the Bezout operator associated with $\{\mathcal{A},\mathcal{C};\mathcal{B},\mathcal{D}\}$. Conversely, if the left-hand side of (3.22) is equal to a block diagonal 2×2 operator matrix relative to *the partitioning* $L_1^n[-\omega,0]\oplus L_1^n[0,\omega]$, *then* $\{\mathcal{A},\mathcal{C};\mathcal{B},\mathcal{D}\}$ *has the quasi commutativity property and formula* (3.22) *holds true.*

Proof. Using the operator matrix representation of $R(\mathcal{B},\mathcal{D})$ and $S(\mathcal{A},\mathcal{C})$ given by (3.16) and (3.21), respectively, we have

$$S(\mathcal{A},\mathcal{C})\begin{bmatrix} I_{L_1^n[-\omega,0]} & 0 \\ 0 & -I_{L_1^n[0,\omega]}\end{bmatrix}R(\mathcal{B},\mathcal{D})$$

$$=\begin{bmatrix} H^-\big((I+A_0)(I+B_0)-C_{-1}D_1\big)H^+ & H^-\big((I+A_0)B_- - C_{-1}(I+D_0)\big) \\ \big(A_1(I+B_0)-(I+C_0)D_1\big)H^+ & -(I+C_0)(I+D_0)+A_+B_{-1}\end{bmatrix},$$

Since the operators H^+ and H^- are invertible, we see that the left-hand side of (3.22) is a block diagonal 2×2 operator matrix relative to the partitioning $L_1^n[-\omega,0]\oplus L_1^n[0,\omega]$ if and only if

$$(I+A_0)B_- - C_{-1}(I+D_0)=0,\quad A_1(I+B_0)-(I+C_0)D_1=0.$$

By Proposition 3.2 the latter two identities are equivalent to the requirement that $\{\mathcal{A},\mathcal{C};\mathcal{B},\mathcal{D}\}$ has the quasi commutativity property.

Next assume that $\{\mathcal{A},\mathcal{C};\mathcal{B},\mathcal{D}\}$ has the quasi commutativity property. As we have seen in the proof of Proposition 3.2, the quasi commutativity property implies that

$$A_0+B_0+A_0B_0+A_1B_{-1}=C_0+D_0+C_0D_0+C_{-1}D_1.$$

It follows that the Bezout operator associated with $\{\mathcal{A},\mathcal{C};\mathcal{B},\mathcal{D}\}$ is not only given by (3.13) but also by

$$T=(I+C_0)(I+D_0)-A_1B_{-1}. \tag{3.23}$$

The formulas (3.13) and (3.23), together with the result of the preceding paragraph, prove (3.22). □

Formula (3.22) has been proved in [16] for the classical resultant matrix and the classical Bezout matrix corresponding to two scalar polynomials.

4. Proof of the main theorem

In this section we prove Theorem 1.1. To do this it suffices to show that any vector function from the system $\widetilde{\Xi}$ defined by (1.3) belongs to Ker $R(\mathcal{B},\mathcal{D})$. Indeed, when this result has been established, then we know that

$$\dim\operatorname{Ker}R(\mathcal{B},\mathcal{D})\geq\nu(\mathcal{B},\mathcal{D}). \tag{4.1}$$

On the other hand, from (3.20) and the fact (see Theorem 3.1) that $\dim\operatorname{Ker}T=\nu(\mathcal{B},\mathcal{D})$, we see that we have equality in (4.1), which proves Theorem 1.1.

To prove that any vector function from the system $\widetilde{\Xi}$ defined by (1.3) belongs to Ker $R(\mathcal{B},\mathcal{D})$ we follow the lead of [7] (on the way correcting some misprints in the statement of Theorem 1.1 and in computations on page 198 of [7]).

Let $x_0, x_1, \ldots, x_{r-1}$ in $\mathbb{C}^n$ be a common Jordan chain of the entire $n \times n$ matrix functions $\mathcal{B}$ and $\mathcal{D}$ corresponding to the common eigenvalue λ_0. As in Section 1 we associate with the chain $x_0, x_1, \ldots, x_{r-1}$ the functions

$$\widetilde{x}_k(t) = e^{-i\lambda_0 t} \sum_{\nu=0}^{k} \frac{(-it)^\nu}{\nu!} x_{k-\nu}, \quad k = 0, \ldots, r-1. \tag{4.2}$$

In order to prove (4.1) it suffices to show that the functions $\widetilde{x}_0, \widetilde{x}_1, \ldots, \widetilde{x}_{r-1}$ belong to the null space of $R(\mathcal{B}, \mathcal{D})$. To do this we introduce operators $R(\mathcal{B})$ and $R(\mathcal{D})$ from $L_1^n[-\omega, \omega]$ into $L_1^n[-\omega, 0]$ and $L_1^n[0, \omega]$, respectively, by setting

$$\begin{aligned}
(R(\mathcal{B})f)(t) &= f(t) + \int_{-\omega}^{\omega} b(t-s) f(s)\, ds, \quad -\omega \le t \le 0, \\
(R(\mathcal{D})f)(t) &= f(t) + \int_{-\omega}^{\omega} d(t-s) f(s)\, ds, \quad 0 \le t \le \omega.
\end{aligned}$$

Obviously,

$$\operatorname{Ker} R(\mathcal{B}, \mathcal{D}) = \operatorname{Ker} R(\mathcal{B}) \cap \operatorname{Ker} R(\mathcal{D}).$$

Thus (4.1) will be proved when we establish the following lemma.

Lemma 4.1. *Fix $\lambda_0 \in \mathbb{C}$, let $x_0, x_1, \ldots, x_{r-1}$ be vectors in $\mathbb{C}^n$, and consider the associated functions $\widetilde{x}_0, \widetilde{x}_1, \ldots, \widetilde{x}_{r-1}$ defined by formula (4.2). Then the vectors $x_0, x_1, \ldots, x_{r-1}$ form a Jordan chain of $\mathcal{B}$ at λ_0 (of $\mathcal{D}$ at λ_0) if and only if the functions $\widetilde{x}_0, \widetilde{x}_1, \ldots, \widetilde{x}_{r-1}$ belong to* Ker $R(\mathcal{B})$ *(belong to* Ker $R(\mathcal{D})$*).*

Proof. We prove the statement concerning $R(\mathcal{B})$; the result for $R(\mathcal{D})$ follows in a similar way.

Recall (see (2.9)) that $x_0, x_1, \ldots, x_{r-1}$ is a Jordan chain of $\mathcal{B}$ at λ_0 if and only if

$$\sum_{j=0}^{k} \frac{1}{j!} \mathcal{B}^{(j)}(\lambda_0) x_{k-j} = 0, \quad k = 0, \ldots, r-1. \tag{4.3}$$

Since $\mathcal{B}(\lambda)$ is given by the right-hand side of (3.1), we have

$$\mathcal{B}^{(j)}(\lambda) = \int_{-\omega}^{0} (is)^j e^{i\lambda s} b(s)\, ds, \quad j = 1, 2, \ldots. \tag{4.4}$$

Next, note that for each $f \in L_1^n[-\omega, \omega]$ we have

$$\begin{aligned}
(R(\mathcal{B})f)(t) &= f(t) + \int_{t-\omega}^{t+\omega} b(s) f(t-s)\, ds \\
&= f(t) + \int_{-\omega}^{0} b(s) f(t-s)\, ds, \quad -\omega \le t \le 0.
\end{aligned}$$

Fix $0 \le \nu \le r-1$, $x \in \mathbb{C}^n$, and consider the functions

$$\varphi_\nu(t) = e^{-i\lambda_0 t} \frac{(-it)^\nu}{\nu!}, \quad \widetilde{\varphi}_\nu(t) = \varphi_\nu(t) x \quad (-\omega \le t \le \omega).$$

Then for $-\omega \le t \le 0$ we have

$$\begin{aligned}
(R(\mathcal{B})\widetilde{\varphi}_\nu)(t) &= \widetilde{\varphi}_\nu(t) + \int_{-\omega}^{0} b(s)\widetilde{\varphi}_\nu(t-s)\,ds \\
&= \varphi_\nu(t)x + \int_{-\omega}^{0} \frac{(is-it)^\nu}{\nu!} e^{-i\lambda_0(t-s)} b(s)x\,ds \\
&= \varphi_\nu(t)x + \sum_{p=0}^{\nu} \varphi_p(t)\frac{1}{(\nu-p)!}\int_{-\omega}^{0} (is)^{\nu-p} e^{i\lambda_0 s} b(s)x\,ds.
\end{aligned}$$

Using the definition of $\mathcal{B}$, given in (3.1), and formula (4.4) we obtain

$$\begin{aligned}
(R(\mathcal{B})\widetilde{\varphi}_\nu)(t) &= \varphi_\nu(t)\mathcal{B}(\lambda_0)x + \sum_{p=0}^{\nu-1} \varphi_p(t)\frac{1}{(\nu-p)!}\mathcal{B}^{(\nu-p)}(\lambda_0)x \\
&= \sum_{p=0}^{\nu} \varphi_p(t)\frac{1}{(\nu-p)!}\mathcal{B}^{(\nu-p)}(\lambda_0)x.
\end{aligned}$$

Next notice that the function $\widetilde{x}_k$ defined by (4.2) is also given by

$$\widetilde{x}_k(t) = \sum_{\nu=0}^{k} \varphi_\nu(t)x_{k-\nu}, \qquad -\omega \le t \le \omega.$$

Thus the calculation in the previous paragraph yields

$$\begin{aligned}
(R(\mathcal{B})\widetilde{x}_k)(t) &= \sum_{\nu=0}^{k}\sum_{p=0}^{\nu} \varphi_p(t)\frac{1}{(\nu-p)!}\mathcal{B}^{(\nu-p)}(\lambda_0)x_{k-\nu} \\
&= \sum_{p=0}^{k} \varphi_p(t)\Big(\sum_{\nu=p}^{k} \frac{1}{(\nu-p)!}\mathcal{B}^{(\nu-p)}(\lambda_0)x_{k-\nu}\Big) \\
&= \sum_{p=0}^{k} \varphi_p(t)\Big(\sum_{\nu=0}^{k-p} \frac{1}{\nu!}\mathcal{B}^{(\nu)}(\lambda_0)x_{(k-p)-\nu}\Big).
\end{aligned}$$

Now assume that $x_0, x_1, \ldots, x_{r-1}$ is a Jordan chain for $\mathcal{B}$ at λ_0. Then in the formula of the preceding paragraph the last term between parentheses is equal to zero because of (4.3). We conclude that $\widetilde{x}_k$ belongs to the null space of $R(\mathcal{B})$ for $k = 0, \ldots, r-1$. Conversely, assume that $\widetilde{x}_0, \widetilde{x}_1, \ldots, \widetilde{x}_{r-1}$ belong to $\operatorname{Ker} R(\mathcal{B})$. Then

$$\sum_{p=0}^{k} \varphi_p(\cdot)\Big(\sum_{\nu=0}^{k-p} \frac{1}{\nu!}\mathcal{B}^{(\nu)}(\lambda_0)x_{(k-p)-\nu}\Big) = 0, \quad k = 0, \ldots, r-1.$$

Since the functions $\varphi_0, \ldots, \varphi_{r-1}$ are linearly independent in $L_1[-\omega, \omega]$, it follows that the vectors defined by the sum between parentheses are zero. This holds for $m = k - p$ running from 0 to $r-1$. Hence, again using (4.3), we see that $x_0, x_1, \ldots, x_{r-1}$ is a Jordan chain for $\mathcal{B}$ at λ_0. □

5. Applications to inverse problems

In this section we present applications of the main theorem to the inverse problem for convolution operators on a finite interval and to the inverse problem for continuous analogues of orthogonal polynomials.

5.1. The inverse problem for convolution operators on a finite interval

First we recall the results of [6] concerning inversion of convolution operators on a finite interval. Let $k \in L_1^{n\times n}[-\omega,\omega]$, and on $L_1^n[0,\omega]$ consider the operator $M = I - K$, where K is the integral operator on $L_1^n[0,\omega]$ given by

$$(Kf)(t) = \int_0^\omega k(t-s)f(s)\,ds, \quad 0 \le t \le \omega. \tag{5.1}$$

With the kernel function k we also associate the following four integral equations:

$$a(t) - \int_0^\omega k(t-s)a(s)\,ds = k(t), \quad 0 \le t \le \omega, \tag{5.2}$$

$$b(t) - \int_{-\omega}^0 b(s)k(t-s)\,ds = k(t), \quad -\omega \le t \le 0, \tag{5.3}$$

$$c(t) - \int_{-\omega}^0 k(t-s)c(s)\,ds = k(t), \quad -\omega \le t \le 0, \tag{5.4}$$

$$d(t) - \int_0^{-\omega} d(s)k(t-s)\,ds = k(t), \quad 0 \le t \le \omega. \tag{5.5}$$

The inversion theorem proved in [6] reads as follows.

Theorem 5.1. *The operator $M = I - K$ is invertible on $L_1^n[0,\omega]$ if and only equations (5.2) and (5.3) have solutions in $L_1^{n\times n}[0,\omega]$ and $L_1^{n\times n}[-\omega,0]$, respectively, or equations (5.4) and (5.5) have solutions in $L_1^{n\times n}[0,\omega]$ and $L_1^{n\times n}[-\omega,0]$, respectively. Moreover, in this case M^{-1} is the integral operator given by*

$$(M^{-1}f)(t) = f(t) + \int_0^\omega \gamma(t,s)f(s)\,ds, \quad 0 \le t \le \omega, \tag{5.6}$$

with the kernel function γ being given by

$$\begin{aligned}\gamma(t,s) &= a(t-s) + b(t-s)\\ &\quad + \int_0^{\min\{t,s\}} [a(t-r)b(r-s) - c(t-r-\omega)d(r-s+\omega)]\,dr,\end{aligned}$$

where a, b, c and d are the (unique) solutions of (5.2)–(5.5).

Note that the operator M^{-1} in the above theorem is of the same form as the Bezout operator defined in Section 3. Theorem 5.1 remains true if the operator $M = I - K$, where K is given by (5.1), is considered on $L_p^n[0,\omega]$ with $1 \le p \le \infty$; see [2], Section 7.2.

In the scalar case, that is when $n = 1$, equations (5.3) and (5.5) are redundant. For this scalar case Theorem 5.1 has been proved in [12] (see also [4], Section III.8). For a recent proof of Theorem 5.1, in a L_2-setting, using ideas from mathematical system theory, see [9].

The following inverse problem is related to Theorem 5.1. Given four matrix functions a, d in $L_1^{n\times n}[0,\omega]$ and b, c in $L_1^{n\times n}[-\omega,0]$, find $k \in L_1^{n\times n}[-\omega,\omega]$ such that the corresponding operator $M = I - K$ is invertible and the given functions are solutions of the equations (5.2)–(5.5) for this kernel function k. The solution to this problem is also given in the paper [6]. To state this solution introduce the following operator acting on $L_1^n[0,\omega] \oplus L_1^n[0,\omega]$:

$$\Lambda = \begin{bmatrix} I + B_0 & B_{-1} \\ D_1 & I + D_0 \end{bmatrix}. \tag{5.7}$$

Theorem 5.2. *Given $a, d \in L_1^{n\times n}[0,\omega]$ and $b, c \in L_1^{n\times n}[-\omega,0]$, there exists a matrix functions $k \in L_1^{n\times n}[-\omega,\omega]$ such that the given matrix functions are solutions of the equations (5.2)–(5.5) if and only if the following conditions are fulfilled:*

$$C_{-1}(I + D_0) = (I + A_0)B_{-1}, \quad (I + C_0)D_{+1} = A_{+1}(I + B_0), \tag{5.8}$$

and the operator Λ is invertible. Moreover, if these conditions are satisfied, then the matrix function k is uniquely determined by

$$k(t) = \begin{cases} (\Lambda^{-1}h)(t+\omega) & \text{for } -\omega \le t \le 0, \\ (\Lambda^{-1}h)(t) & \text{for } \ 0 \le t \le \omega, \end{cases} \tag{5.9}$$

where

$$h(t) = \begin{bmatrix} h_1(t) \\ h_2(t) \end{bmatrix}, \quad h_1(t) = b(t-\omega), \quad h_2(t) = d(t) \quad (0 \le t \le \omega) \tag{5.10}$$

The aim of the present subsection is to present a more transparent version of Theorem 5.2. For this purpose we associate with the given functions $a, d \in L_1^{n\times n}[0,\omega]$ and $b, c \in L_1^{n\times n}[-\omega,0]$ the entire $n \times n$ matrix functions $\mathcal{A}$, $\mathcal{B}$, $\mathcal{C}$ and $\mathcal{D}$ defined by formulas (3.1), (3.2). As already shown in Proposition 3.2, condition (5.8) in Theorem 5.2 is equivalent to the quasi commutativity property:

$$\mathcal{A}(\lambda)\mathcal{B}(\lambda) = \mathcal{C}(\lambda)\mathcal{D}(\lambda), \quad \lambda \in \mathbb{C}.$$

Next we claim that the operator Λ in (5.7) is similar to the resultant operator $R(\mathcal{B},\mathcal{D})$ acting on $L_1^{n\times n}[-\omega,\omega]$. Indeed, in view of (3.16) we have

$$\begin{bmatrix} H_- & 0 \\ 0 & I \end{bmatrix} \Lambda \begin{bmatrix} H_+ & 0 \\ 0 & I \end{bmatrix} = \begin{bmatrix} I + H^-BH^+ & H^-B_- \\ D_+H^+ & I + D \end{bmatrix} = R(\mathcal{B},\mathcal{D}).$$

Thus the invertibility of the operator Λ in Theorem 5.2 is equivalent to the invertibility of the resultant operator $R(\mathcal{B},\mathcal{D})$. In other words, using Theorem 1.1, the invertibility of Λ is equivalent to $\nu(\mathcal{B},\mathcal{D}) = 0$. Notice that $\nu(\mathcal{B},\mathcal{D}) = 0$ is equivalent to $\operatorname{Ker}\mathcal{B}(\lambda) \cap \operatorname{Ker}\mathcal{D}(\lambda) = 0$ for each $\lambda \in \mathbb{C}$.

The above considerations lead us to the following new version of Theorem 5.2.

Theorem 5.3. *Let* $a, d \in L_1^{n\times n}[0,\omega]$ *and* $b, c \in L_1^{n\times n}[-\omega, 0]$ *be given, and let* $\mathcal{A}, \mathcal{B}, \mathcal{C}, \mathcal{D}$ *be the corresponding entire* $n \times n$ *matrix functions defined by* (3.1), (3.2). *There exists a matrix function* $k \in L_1^{n\times n}[-\omega,\omega]$ *such that the given matrix functions* a, b, c, d *are solutions of the equations* (5.2)–(5.5) *if and only if for each* $\lambda \in \mathbb{C}$ *the following two conditions are satisfied:*

$$\mathcal{A}(\lambda)\mathcal{B}(\lambda) = \mathcal{C}(\lambda)\mathcal{D}(\lambda), \quad \operatorname{Ker}\mathcal{B}(\lambda) \cap \operatorname{Ker}\mathcal{D}(\lambda) = 0.$$

In this case the function k *is uniquely determined by the formulas*

$$k = [R(\mathcal{B},\mathcal{D})]^{-1} f, \quad f(t) = \begin{cases} b(t) & \text{for } -\omega \le t \le 0, \\ d(t) & \text{for } \ 0 \le t \le \omega, \end{cases}$$

In the scalar case Theorem 5.3 easily follows from [12] by making use of the resultant operator for scalar entire functions (see [7]). For the case of finite block Toeplitz matrices the analogous result can be found in [11].

5.2. The inverse problem for orthogonal matrix functions

A matrix function k in $L_1^{n\times n}[-\omega,\omega]$ is said to be *hermitian* if $k = k^*$, where $k^*(t) = k(-t)^*$. Given such a matrix function, consider on $L_1^{n\times n}[0,\omega]$ the equation

$$\varphi(t) - \int_0^\omega k(t-s)\varphi(s)\,ds = k(t), \quad 0 \le t \le \omega, \tag{5.11}$$

and set

$$\Phi(\lambda) = I + \int_0^\omega e^{i\lambda t}\varphi(t)\,dt. \tag{5.12}$$

We shall refer to the matrix function Φ as the *orthogonal matrix function generated by* k (related to ω). With k we also associate the integral operator K defined by (5.1), which will now be considered on $L_2^n[0,\omega]$.

For $n = 1$ functions of this type have been introduced by M.G. Krein in [17], who showed that for the case when $I - K$ is strictly positive on $L_2[0,\omega]$, the function Φ exhibit properties, with respect to the real line, that are analogous to those of the classical Szegő orthogonal polynomials with respect to the unit circle. For that reason in [17] these functions Φ are called “continuous analogues of orthogonal polynomials.” Later, M.G. Krein and H. Langer [18], [19] considered the case of a non-definite $I - K$ (still in case $n = 1$). They proved two remarkable results on orthogonal functions for this case. Their first result relates the number of zeroes of Φ in the upper half plane to the spectrum of the operator $I - K$. This result has been generalized to the matrix case ($n > 1$) in [3] and [1] (see [2] for a detailed exposition and related developments). The second Krein-Langer result deals with the following inverse problem: given a matrix function

$$F(\lambda) = I + \int_0^\omega e^{i\lambda t} f(t)\,dt, \quad f \in L_1^{n\times n}[0,\omega], \tag{5.13}$$

find criteria such that F is an orthogonal matrix function generated by some hermitian $k \in L_1^{n\times n}[-\omega,\omega]$ (i.e., $F = \Psi$), and if possible give a formula for k. For the case $n = 1$ the following result is proved in [19].

Theorem 5.4. *Let F be a scalar function of the form* (5.13). *There exists a hermitian function $k \in L_1[-\omega,\omega]$, i.e., $k(t) = \overline{k(-t)}$, such that F is the orthogonal function generated by k if and only if F has no real zeroes and no conjugate pairs λ_0, $\overline{\lambda}_0$ of zeroes.*

To the best of our knowledge there is no generalization of this result to the matrix case ($n > 1$). The present paper provides a first step in this direction. In this subsection we reduce this inverse problem for the matrix case to a certain factorization problem, and we derive an explicit formula for the kernel function k.

First consider Theorems 5.1 and 5.3 for the case of a hermitian kernel k. Note that if $k^* = k \in L_1^{n\times n}[-\omega,\omega]$, then the operator M in Theorem 5.1, considered on $L_2^n[0,\omega]$, is selfadjoint, and in this case to obtain the inversion formula of Theorem 5.1 one has to solve only one of the equations (5.2), (5.3) and one of the equations (5.4), (5.5). Indeed, in this hermitian case $b^* = a$ and $c^* = d$. Moreover, the corresponding inverse theorem for a selfadjoint convolution operator on a finite interval should be based on two given functions instead of the four functions used in the general case. To be more specific in the selfadjoint case Theorem 5.3 takes the following form.

Theorem 5.5. *Given φ, $\psi \in L_1^{n\times n}[0,\omega]$, put*

$$\Phi(\lambda) = I + \int_0^\omega e^{i\lambda t}\varphi(t)\,dt, \qquad \Psi(\lambda) = I + \int_0^\omega e^{i\lambda t}\psi(t)\,dt,$$

and let $\Phi^*(\lambda) = \Phi(\bar{\lambda})^*$ and $\Psi^*(\lambda) = \Psi(\bar{\lambda})^*$. Then there is a hermitian matrix function $k \in L_1^{n\times n}[-\omega,\omega]$ such that

$$\varphi(t) - \int_0^\omega k(t-s)\varphi(s)\,ds = k(t), \quad 0 \le t \le \omega, \tag{5.14}$$

$$\psi(t) - \int_0^\omega \psi(s)k(t-s)\,ds = k(t), \quad 0 \le t \le \omega, \tag{5.15}$$

if and only if for each $\lambda \in \mathbb{C}$ the following two conditions are satisfied:

$$\Phi(\lambda)\Phi^*(\lambda) = \Psi^*(\lambda)\Psi(\lambda), \qquad \operatorname{Ker}\Phi^*(\lambda) \cap \operatorname{Ker}\Psi(\lambda) = \{0\}.$$

In this case the function k is uniquely determined by the formulas

$$k = [R(\Phi^*,\Psi)]^{-1}f, \quad f(t) = \begin{cases} \varphi^*(t) & \text{for } -\omega \le t \le 0, \\ \psi(t) & \text{for } 0 \le t \le \omega, \end{cases}$$

The connection between Theorem 5.5 and the inverse problem for orthogonal matrix functions is now transparent: we have to produce a hermitian matrix function $k \in L_1^{n\times n}[-\omega,\omega]$ such that (5.14) holds, given one matrix function φ only, while in Theorem 5.4 we have given φ along with ψ (which has to satisfy (5.15)).

Thus Theorem 5.4 leads to the following result.

Theorem 5.6. *Given*

$$F(\lambda) = I + \int_0^\omega e^{i\lambda t} f(t)\, dt, \quad f \in L_1^{n\times n}[0,\omega],$$

there exists a hermitian matrix function k in $L_1^{n\times n}[-\omega,\omega]$ such that F is the orthogonal matrix function generated by k if and only if $F(\cdot)F^*(\cdot)$ admits a factorization

$$F(\lambda)F^*(\lambda) = \Psi^*(\lambda)\Psi(\lambda), \quad \lambda \in \mathbb{C}, \tag{5.16}$$

where Ψ is an entire $n \times n$ matrix function of the form

$$\Psi(\lambda) = I + \int_0^\omega e^{i\lambda t} \psi(t)\, dt, \quad \psi \in L_1^{n\times n}[0,\omega],$$

and

$$\operatorname{Ker} F^*(\lambda) \cap \operatorname{Ker} \Psi(\lambda) = \{0\}, \quad \lambda \in \mathbb{C}. \tag{5.17}$$

In this case the kernel function k is given by

$$k = [R(F^*, \Psi)]^{-1} g, \quad g(t) = \begin{cases} f^*(t) & \textit{for} -\omega \le t \le 0, \\ \psi(t) & \textit{for} \quad 0 \le t \le \omega, \end{cases}$$

It is interesting to specify the above theorem for the scalar case. To do this, we first note that for $n = 1$ a factorization (5.16), (5.17) implies that F has no real zero and no conjugate pairs $\lambda_0, \overline{\lambda_0}$ of zeroes. Indeed, assume $F(\lambda_0) = 0$. Then (5.16) implies that $\Psi(\lambda) = 0$ and/or $\Psi(\overline{\lambda_0}) = 0$. However, since $F(\lambda_0) = 0$, we have $F^*(\overline{\lambda_0}) = 0$, and hence (5.17) excludes $\Psi(\overline{\lambda_0}) = 0$. Thus $\Psi(\lambda) = 0$. But then (5.17) shows that $F^*(\overline{\lambda_0}) \neq 0$, and hence $F(\lambda_0) \neq 0$, which is a contradiction. Thus F has no real zero and no conjugate pairs $\lambda_0, \overline{\lambda_0}$ of zeroes. Conversely, if $n = 1$ and F has no real zero and no conjugate pairs $\lambda_0, \overline{\lambda_0}$ of zeroes, then (5.16), (5.17) hold with $\Psi = F$. Furthermore in that case, the kernel function k is obtained by

$$k = [R(F^*, F)]^{-1} g, \quad g(t) = \begin{cases} f^*(t) & \text{for} -\omega \le t \le 0, \\ f(t) & \text{for} \quad 0 \le t \le \omega, \end{cases} \tag{5.18}$$

We conclude that in the scalar case Theorem 5.6 implies Theorem 5.4. Moreover, the formula for k in (5.18) is new in this scalar case.

In the general matrix case the meaning of the factorization (5.16), (5.17) in terms of the zero data of $F(\lambda)$ is much more involved and will be the topic of a future publication.

References

[1] H. Dym, On the zeros of some continuous analogues of matrix orthogonal polynomials and a related extension problem with negative squares, *Comm. Pure Appl. Math* **47** (1994), 207–256.

[2] R.L. Ellis and I. Gohberg, *Orthogonal Systems and Convolution Operators*, Operator Theory: Advances and Applications, **140**, Birkhäuser Verlag, Basel, 2003.

[3] R.L. Ellis, I. Gohberg and D.C. Lay, Distribution of zeros of matrix-valued continuous analogues of orthogonal polynomials, *Oper. Theory: Adv. Appl.* **58** (1992), 26–70.

[4] I.C. Gohberg, and I.A. Fel'dman, *Convolution equations and projection methods for their solution*, Transl. Math. Monographs **41**, Amer. Math. Soc., Providence RI, 1974.

[5] I. Gohberg, I. Haimovici, M.A. Kaashoek, and L. Lerer, The Bezout integral operator: main property and underlying abstract scheme, Operator Theory: Advances and Applications, to appear.

[6] I. Gohberg, and G. Heinig, On matrix valued integral operators on a finite interval with matrix kernels that depend on the difference of arguments, *Rev. Roumaine Math. Pures Appl.*, **20** (1975), 1. 55–73. (in Russian)

[7] I. Gohberg and G. Heinig, The resultant matrix and its generalizations, II. Continual analog of resultant matrix, *Acta Math. Acad. Sci. Hungar* **28** (1976), 198–209, [in Russian].

[8] I. Gohberg, M.A. Kaashoek, and F. van Schagen, *Partially specified matrices and operators: classification, completion, applications*, OT **79** Birkhäuser Verlag, Basel, 1995.

[9] I. Gohberg, M.A. Kaashoek and F. van Schagen, On inversion of convolution integral operators on a finite interval, in: *Operator Theoretical Methods and Applications to Mathematical Physics. The Erhard Meister Memorial Volume*, OT **147**, Birkhäuser Verlag, Basel, 2004, pp. 277–285.

[10] I. Gohberg, P. Lancaster and L. Rodman, *Matrix Polynomials*, Academic Press, New York, 1982.

[11] I. Gohberg and L. Lerer, Matrix generalizations of M.G. Krein theorems on orthogonal polynomials, *Oper. Theory: Adv. Appl.* **34** (1988), 137–202.

[12] I. Gohberg and A.A. Semencul, On the inversion of finite Toeplitz matrices and their continuous analogues, *Mat. Issled.* **7**, No. 2 (1972), 201–223 [in Russian].

[13] I.C. Gohberg, and E.I. Sigal, An operator generalization of the logarithmic residue theorem and the theorem of Rouché. *Mat.Sbornik* 84(126) (1971), 607–629 [in Russian]; English transl. *Math. USSR, Sbornik* 13 (1971), 603–625.

[14] I. Haimovici, *Operator equations and Bezout operators for analytic operator functions*, Ph.D. thesis, Technion Haifa, Israel, 1991 [in Hebrew].

[15] I. Haimovici, and L. Lerer, Bezout operators for analytic operator functions, I. A general concept of Bezout operator, *Integral Equations Oper. Theory* **21** (1995), 33–70.

[16] N. Kravitsky, On the discriminant function of two commuting nonselfadjoint operators, *Integral Equations Oper. Theory* **3** (1980), 97–125.

[17] M.G. Krein, Continuous analogues of propositions about orthogonal polynomials on the unit circle, *Dokl. Akad. Nauk SSSR* **105:4** (1955), 637–640 [in Russian].

[18] M.G. Krein and H. Langer, Continuous analogues of orthogonal polynomials on the unit circle with respect to an indefinite weight and related extension problems, *Soviet Math. Dokl.* **32** (1983), 553–557.

[19] M.G. Krein and H. Langer, On some continuation problems which are closely related to the theory of operators in spaces Π_κ. IV: Continuous analogues of orthogonal polynomials on the unit circle with respect to an indefinite weight and related continuation problems for some classes of functions, *J. Oper. Theory* **13** (1985), 299–417.

[20] M.G. Krein and M.A. Naimark, The method of symmetric and hermitian forms in theory of separation of the roots of algebraic equations, GNTI, Kharkov, 1936 [in Russian], English translation in *Linear and Multilinear Algebra* **10**, (1981), 265–308.

[21] L.A. Sakhnovich, Operatorial Bezoutiant in the theory of separation of roots of entire functions, *Functional Anal. Appl.* **10** (1976), 45–51 [in Russian].

[22] L.A. Sakhnovich, *Integral equations with difference kernels on finite intervals*, OT **84**, Birkhäuser Verlag, Basel, 1996.

[23] B.L. van der Waerden, *Moderne Algebra I, II*, Springer, Berlin, 1937.

Israel Gohberg
School of Mathematical Sciences
Raymond and Beverly Faculty of Exact Sciences
Tel-Aviv University
Ramat Aviv 69978, Israel
e-mail: `gohberg@math.tau.ac.il`

Marinus A. Kaashoek
Afdeling Wiskunde
Faculteit der Exacte Wetenschappen, Vrije Universiteit
De Boelelaan 1081a
NL-1081 HV Amsterdam, The Netherlands
e-mail: `ma.kaashoek@few.vu.nl`

Leonid Lerer
Department of Mathematics
Technion – Israel Institute of Technology
Haifa 32000, Israel
e-mail: `llerer@techunix.technion.ac.il`

Operator Theory:
Advances and Applications, Vol. 171, 129–146

Split Algorithms for Centrosymmetric Toeplitz-plus-Hankel Matrices with Arbitrary Rank Profile

Georg Heinig and Karla Rost

Abstract. Split Levinson and Schur algorithms for the inversion of centrosymmetric Toeplitz-plus-Hankel matrices are designed that work, in contrast to previous algorithms, for matrices with any rank profile. Furthermore, it is shown that the algorithms are related to generalized ZW-factorizations of the matrix and its inverse.

Mathematics Subject Classification (2000). Primary 65F05; Secondary 15A06, 15A23, 15A09.

Keywords. Toeplitz-plus-Hankel matrix, split algorithm, ZW-factorization, centrosymmetric matrix.

1. Introduction

This paper is dedicated to the solution of a linear system $A_n\mathbf{f} = \mathbf{b}$ with an $n \times n$ nonsingular centrosymmetric Toeplitz-plus-Hankel (T+H) coefficient matrix

$$A_n = [\, a_{i-j} + s_{i+j-1} \,]_{i,j=1}^{n} \tag{1.1}$$

the entries of which are in a given field $\mathbb{F}$ of characteristics different from 2.

Recall that an $n \times n$ matrix A is called centrosymmetric if $A = J_n A J_n$, where J_n stands for the $n \times n$ matrix of the counteridentity,

$$J_n = \begin{bmatrix} 0 & & 1 \\ & \cdot^{\cdot^{\cdot}} & \\ 1 & & 0 \end{bmatrix}.$$

The T+H matrix A_n is centrosymmetric if it has a representation (1.1) with a symmetric Toeplitz matrix $[\, a_{i-j} \,]$ and a persymmetric Hankel matrix $[\, s_{i+j-1} \,]$. It is a surprising conclusion that centrosymmetric T+H matrices are also symmetric. Clearly, symmetric Toeplitz matrices are special cases of centrosymmetric T+H matrices.

For $n \times n$ T+H matrices algorithms with $O(n^2)$ computational complexity were designed in [24], [25], [12], [7], [21], and in many other papers. In our paper [20] it was shown that the complexity of solution algorithms can be essentially reduced (at least by the factor 4) if the centrosymmetry of the matrix is taken into account. The algorithms proposed in [20] are developed in the spirit of the "split" algorithms of P. Delsarte and Y. Genin in [4], [5] for the solution of symmetric Toeplitz systems and of our paper [19] for skewsymmetric Toeplitz systems. However, the algorithms in [20] work only under the condition that all central submatrices of A_n are nonsingular.

The main aim of the present paper is to overcome this additional restriction and so design fast algorithms that work for any nonsingular centrosymmetric T+H matrix. The corresponding algorithms generalize not only those in [20] but also those in our recent papers [22] and [23], in which split algorithms for symmetric and skewsymmetric Toeplitz matrices with arbitrary rank profile are presented.

The secondary aim is to discuss the relations between the algorithms and factorizations of the matrix A_n and its inverse A_n^{-1}. More precisely, we will show that the Levinson-type algorithm is related to a generalized WZ-factorization of A_n^{-1} and the Schur-type algorithm to a ZW-factorization of A_n. Note that WZ-factorization was originally introduced by D.J. Evans and his coworkers for the parallel solution of tridiagonal systems (see [6], [26] and references therein).

The rest of the paper is built as follows. First we recall some observations from [18] and [20] about representations of centrosymmetric T+H matrices in Sections 2. Section 3 starts with an inversion formula for centrosymmetric T+H matrices from [20]. Then a Gohberg-Semencul-type representation is derived, which is new and particularly convenient if $\mathbb{F}$ is not the field of real or complex numbers or if the entries of the matrix are integers.

In Section 4 we present a split Levinson-type algorithm for the computation of the data in the inversion formula and in Section 5 a split Schur-type algorithm for computing the residuals. The algorithms, in principle, coincide with those developed in [22] for the solution of symmetric Toeplitz systems. Therefore we refrain from presenting all details.

Section 6 is dedicated to a discussion of how linear systems with coefficient matrix A_n can be solved without using inversion formulas.

In Section 7 we show that the Levinson-type algorithm produces a generalized WZ-factorization of A_n^{-1} and the Schur-type algorithm a generalized ZW-factorization of A_n.

Let us agree upon some notations. Occasionally we will use polynomial language. For a matrix $A = [\,a_{ij}\,]$, $A(t,s)$ will denote the bivariate polynomial $A(t,s) = \sum_{i,j} a_{ij} t^{i-1} s^{j-1}$, which is called the *generating function* of A. In the same spirit the polynomial $\mathbf{x}(t)$ is defined for a vector $\mathbf{x}$.

For a vector $\mathbf{u} = (u_i)_{i=1}^{l}$, let $M_k(\mathbf{u})$ denote the $(k+l-1) \times k$ matrix

$$M_k(\mathbf{u}) = \left.\begin{bmatrix} u_1 & & 0 \\ \vdots & \ddots & \\ u_l & & u_1 \\ & \ddots & \vdots \\ 0 & & u_l \end{bmatrix}\right\} k+l-1.$$

It is easily checked that, for $\mathbf{x} \in \mathbb{F}^k$, $(M_k(\mathbf{u})\mathbf{x})(t) = \mathbf{u}(t)\mathbf{x}(t)$.

We let $\mathbf{e}_k \in \mathbb{F}^n$ stand for the kth vector in the standard basis of $\mathbb{F}^n$, and $\mathbf{0}_k$ will be the zero vector of length k.

2. Representations

The starting point is to use a representation different from (1.1) for a centrosymmetric T+H matrix A_n.

Let $\mathbb{F}^n_+$, $\mathbb{F}^n_-$ be the subspaces of $\mathbb{F}^n$ consisting of all symmetric or skewsymmetric vectors, respectively. A vector $\mathbf{u} \in \mathbb{F}^n$ is called symmetric if $\mathbf{u} = J_n\mathbf{u}$ and skewsymmetric if $\mathbf{u} = -J_n\mathbf{u}$. Then $P_n^{\pm} = \frac{1}{2}(I_n \pm J_n)$ are the projections onto $\mathbb{F}^n_{\pm}$, respectively. Since A_n is assumed to be centrosymmetric, the subspaces $\mathbb{F}^n_{\pm}$ are invariant under A_n.

A centrosymmetric T+H matrix (1.1) can be represented in the form

$$A_n = T_n^+ P_n^+ + T_n^- P_n^-, \tag{2.1}$$

with symmetric Toeplitz matrices $T_n^{\pm} = [\,c^{\pm}_{|i-j|}\,]_{i,j=1}^n$, $c_i^{\pm} = a_i \pm s_{n-i}$ and conversely, each matrix of this form is a centrosymmetric T+H matrix. For details we refer to [18]. Thus a linear system $A_n\mathbf{f} = \mathbf{b}$ is equivalent to the two symmetric Toeplitz systems $T_n^{\pm}\mathbf{f}_{\pm} = P_n^{\pm}\mathbf{b}$, where $\mathbf{f} = \mathbf{f}_+ + \mathbf{f}_-$, $\mathbf{f}_{\pm} \in \mathbb{F}^n_{\pm}$.

Note that the matrices $T_n^{\pm}$ in the representation (2.1) might be singular, but they can have nullity at most 1 (for a discussion of this point see [18]).

Besides the matrix $A_n = [\,r_{ij}\,]_{i,j=1}^n$ we consider its central submatrices $A_k = [\,r_{ij}\,]_{i,j=l+1}^{n-l}$ for $k = n-2l$ and $l = 0, 1, \ldots, [n/2]$, where $[\,\cdot\,]$ denotes the integer part. These matrices inherit the centrosymmetry from A_n. Furthermore, the following is obvious.

Proposition 2.1. *If A_n is of the form (2.1), then for $k = n-2l$, $l = 0, 1, \ldots, [n/2]$, the central submatrices A_k are given by*

$$A_k = P_k^+ T_k^+ P_k^+ + P_k^- T_k^- P_k^-, \tag{2.2}$$

where $T_k^{\pm} = [\,c^{\pm}_{|i-j|}\,]_{i,j=1}^k$, $c_i^{\pm} = a_i \pm s_{i+n}$.

Moreover we consider a nonsingular, centrosymmetric extension A_{n+2} of A_n,

$$A_{n+2} = T_{n+2}^+ P_{n+2}^+ + T_{n+2}^- P_{n+2}^-,$$

where $T^{\pm}_{n+2} = [c^{\pm}_{|i-j|}]^{n+2}_{i,j=1}$. It can easily be checked that if A_n is nonsingular, then for almost all choices of the numbers $c^{\pm}_n$ and $c^{\pm}_{n+1}$ the matrix A_{n+2} is also nonsingular.

3. Inversion formulas

In [18] it was shown that the inverse of a centrosymmetric T+H matrix can be represented in terms of special T+H Bezoutians which are defined next. It is convenient to give the definition in polynomial language.

Let $\mathbf{u}, \mathbf{v} \in \mathbb{F}^{n+2}$ be either symmetric or skewsymmetric vectors. The T+H Bezoutian of $\mathbf{u}$ and $\mathbf{v}$ is, by definition, the $n \times n$ matrix $B = B(\mathbf{u}, \mathbf{v})$ with the generating function

$$B(t,s) = \frac{\mathbf{u}(t)\mathbf{v}(s) - \mathbf{v}(t)\mathbf{u}(s)}{(t-s)(1-ts)}\,.$$

Note that these *T+H Bezoutians* are special cases of more general T+H Bezoutians, which were introduced in [14].

Obviously, B is a symmetric matrix and in the case, where $\mathbf{u}, \mathbf{v}$ are both symmetric or both skewsymmetric vectors, B is also centrosymmetric. Moreover, it is easily checked that the columns and rows of B are symmetric if $\mathbf{u}$ and $\mathbf{v}$ are symmetric, and they are skewsymmetric if $\mathbf{u}$ and $\mathbf{v}$ are skewsymmetric. The entries of the matrix B can be constructed recursively from $\mathbf{u}$ and $\mathbf{v}$ with $O(n^2)$ operations (see [18]).

Now we recall Theorem 3.1 of [20], which is a slight modification of Theorem 3.3 in [18].

Theorem 3.1. *Let A_{n+2} be an $(n+2) \times (n+2)$ nonsingular centrosymmetric extension of A_n. Then the equations*

$$\begin{aligned} T^+_n \mathbf{x}^+_n &= P^+_n \mathbf{e}_n\,, \quad T^+_{n+2}\mathbf{x}^+_{n+2} = P^+_{n+2}\mathbf{e}_{n+2}\,,\\ T^-_n \mathbf{x}^-_n &= P^-_n \mathbf{e}_n\,, \quad T^-_{n+2}\mathbf{x}^-_{n+2} = P^-_{n+2}\mathbf{e}_{n+2} \end{aligned} \tag{3.1}$$

have unique symmetric or skewsymmetric solutions $\mathbf{x}^{\pm}_n$ and $\mathbf{x}^{\pm}_{n+2}$, respectively[1]*, and*

$$A_n^{-1} = \frac{1}{r_+} B(\mathbf{x}^+_{n+2}, \widetilde{\mathbf{x}}^+_n) + \frac{1}{r_-} B(\mathbf{x}^-_{n+2}, \widetilde{\mathbf{x}}^-_n), \tag{3.2}$$

where $r_{\pm}$ is the last component of $\mathbf{x}^{\pm}_{n+2}$, and $\widetilde{\mathbf{x}}^{\pm}_n \in \mathbb{F}^{n+2}$ is the vector obtained from $\mathbf{x}^{\pm}_n \in \mathbb{F}^n_{\pm}$ by adding a zero at the top and the bottom.

In order to solve a linear system with coefficient matrix A_n one needs an efficient way for matrix-vector multiplication by $B(\mathbf{x}^{\pm}_{n+2}, \widetilde{\mathbf{x}}^{\pm}_n)$, which are both symmetric and centrosymmetric matrices. There are several possibilities for this.

[1]Here the superscript + at a vector indicates that the vector is symmetric and − that the vector is skewsymmetric.

If $\mathbb{F}$ is the field of real or complex numbers, then these matrices have matrix representations that include only discrete Fourier matrices or matrices of other trigonometric transformations and diagonal matrices. This allows one to carry out matrix-vector multiplication with computational complexity $O(n \log n)$. Concerning the complex case we refer to [15], and concerning the real case to [16] and [17]. However, this approach is not applicable for all fields $\mathbb{F}$. It is also inconvenient for matrices with integer entries or medium size problems.

A second possibility is to use the splitting approach in [11], where it is shown that, in principle, the T+H Bezoutians introduced here are related to "Chebyshev-Hankel Bezoutians" which are Hankel Bezoutians with respect to bases of Chebyshev polynomials. The Chebyshev-Hankel Bezoutians can be represented with the help of elements of matrix algebras related to the algebra of τ-matrices (see [2], [3]).

One can also use results from [18] (Section 5) concerning the relation between T+H Bezoutians of odd order $n = 2l - 1$ and classical Hankel Bezoutians of order l via a transformation S_l that is generated by Pascal's triangle, together with matrix representations of Hankel Bezoutians. However, if $\mathbb{F}$ is the field of real or complex numbers this approach cannot be recommended since the transformation S_l is ill-conditioned.

In this paper we present a different possibility that is based on the relation between T+H Bezoutians and Toeplitz Bezoutians. Using this relation, a matrix representation of Toeplitz Bezoutians, like the Gohberg-Semencul formula [8], can be applied.

The Toeplitz Bezoutian of $\mathbf{p}, \mathbf{q} \in \mathbb{F}^{n+1}$ is, by definition, the matrix $B_T(\mathbf{p}, \mathbf{q})$ with generating function

$$B_T(\mathbf{p}, \mathbf{q})(t, s) = \frac{\mathbf{p}(t)\mathbf{q}(s^{-1})s^n - \mathbf{q}(t)\mathbf{p}(s^{-1})s^n}{1 - ts} . \tag{3.3}$$

Notice that

$$(J_{n+1}\mathbf{p})(t) = \mathbf{p}(t^{-1})t^n .$$

Obviously, $B_T(\mathbf{p}, \mathbf{q}) = -B_T(\mathbf{q}, \mathbf{p})$. It is well known that a nonsingular matrix is the inverse of a Toeplitz matrix if and only if it is a Toeplitz Bezoutian. In the case of a symmetric Toeplitz Bezoutian $\mathbf{p}$ can be chosen as symmetric and $\mathbf{q}$ as skewsymmetric.

Proposition 3.2. *Let* $\mathbf{p} \in \mathbb{F}_+^{n+1}$ *and* $\mathbf{q} \in \mathbb{F}_-^{n+1}$,

$$\mathbf{u}_\pm(t) = (1 \pm t)\mathbf{p}(t),$$
$$\mathbf{v}_\pm(t) = (1 \mp t)\mathbf{q}(t).$$

Then

$$B(\mathbf{u}_\mp, \mathbf{v}_\mp) = \pm 2\, B_T(\mathbf{p}, \mathbf{q}) P_n^\pm .$$

Proof. From (3.3) we obtain

$$\begin{aligned}
&2(B_T(\mathbf{p},\mathbf{q})P_n^+)(t,s)\\
=&\frac{(s-t)(\mathbf{p}(t)\mathbf{q}(s)+\mathbf{q}(t)\mathbf{p}(s))+(1-ts)(\mathbf{p}(t)\mathbf{q}(s)-\mathbf{q}(t)\mathbf{p}(s))}{(t-s)(1-ts)}\\
=&\frac{\mathbf{p}(t)\mathbf{q}(s)(1+s-t-ts)-\mathbf{q}(t)\mathbf{p}(s)(1+t-s-ts)}{(t-s)(1-ts)}\\
=&\frac{\mathbf{p}(t)\mathbf{q}(s)(1-t)(1+s)-\mathbf{q}(t)\mathbf{p}(s)(1+t)(1-s)}{(t-s)(1-ts)}\\
=&B(\mathbf{u}_-,\mathbf{v}_-)(t,s).
\end{aligned}$$

Analogously, $2(B_T(\mathbf{p},\mathbf{q})P_n^-)(t,s)=-B(\mathbf{u}_+,\mathbf{v}_+)(t,s)$ is shown. □

In our situation, the given $\mathbf{u}_\pm(t)$ and $\mathbf{v}_\pm(t)$ might be not divisible by $t\pm 1$. But nevertheless, the following is true.

Proposition 3.3. *For given* $\mathbf{u}_+,\ \mathbf{v}_+\in\mathbb{F}_+^{n+2}$ *there exist* $\mathbf{p}\in\mathbb{F}_+^{n+1}$ *and* $\mathbf{q}\in\mathbb{F}_-^{n+1}$ *such that*

$$B(\mathbf{u}_+,\mathbf{v}_+)=-2\,B_T(\mathbf{p},\mathbf{q})P_n^-\,. \tag{3.4}$$

Proof. It is immediately verified that, for $a,b,c,d\in\mathbb{F}$,

$$B(a\mathbf{u}_++b\mathbf{v}_+,c\mathbf{u}_++d\mathbf{v}_+)=(ad-bc)B(\mathbf{u}_+,\mathbf{v}_+)$$

We choose a,b,c,d such that $ad-bc=1$, $a\mathbf{u}_+(-1)+b\mathbf{v}_+(-1)=0$ and $c\mathbf{u}_+(1)+d\mathbf{v}_+(1)=0$. It can easily be checked that this is possible. Then we set

$$\mathbf{p}(t)=(a\mathbf{u}_+(t)+b\mathbf{v}_+(t))(1+t)^{-1},\quad \mathbf{q}(t)=(c\mathbf{u}_+(t)+d\mathbf{v}_+(t))(1-t)^{-1}\,.$$

Now $\mathbf{p}\in\mathbb{F}_+^{n+1}$, $\mathbf{q}\in\mathbb{F}_-^{n+1}$ and (3.4) holds. □

A similar proposition can be shown for $\mathbf{u}_-,\mathbf{v}_-\in\mathbb{F}_-^{n+2}$. We arrived at the following.

Theorem 3.4. *The inverse of a centrosymmetric T+H matrix* A_n *admits a representation*

$$A_n^{-1}=B_T(\mathbf{p},\mathbf{q})P_n^-+B_T(\widetilde{\mathbf{p}},\widetilde{\mathbf{q}})P_n^+\,, \tag{3.5}$$

for some $\mathbf{p},\widetilde{\mathbf{p}}\in\mathbb{F}_+^{n+1}$, $\mathbf{q},\widetilde{\mathbf{q}}\in\mathbb{F}_-^{n+1}$.

The vectors $\mathbf{p},\widetilde{\mathbf{p}},\mathbf{q},\widetilde{\mathbf{q}}$ can be constructed from the solutions of (3.1) in $O(n)$ operations, as is shown in the proof of Proposition 3.3.

Note that Theorem 3.4 is trivial if T_n^+ and T_n^- are both nonsingular. However, there are cases in which one or more of the matrices $T_n^\pm$ are singular and A_n is nonsingular (see [18]).

Theorem 3.4 can be used in connection with matrix representations of Toeplitz Bezoutians. The most familiar one is the Gohberg-Semencul formula

$$B_T(\mathbf{p},\mathbf{q}) = \begin{bmatrix} p_0 & & \\ \vdots & \ddots & \\ p_{n-1} & \cdots & p_0 \end{bmatrix}\begin{bmatrix} q_n & \cdots & q_1 \\ & \ddots & \vdots \\ & & q_n \end{bmatrix} - \begin{bmatrix} q_0 & & \\ \vdots & \ddots & \\ q_{n-1} & \cdots & q_0 \end{bmatrix}\begin{bmatrix} p_n & \cdots & p_1 \\ & \ddots & \vdots \\ & & p_n \end{bmatrix}, \tag{3.6}$$

where $\mathbf{p} = (p_i)_{i=0}^n$, $\mathbf{q} = (q_i)_{i=0}^n$. For other representations we refer to [13], [1], [9], [10].

Applying this formula to the matrix-vector multiplication $A_n^{-1}\mathbf{b}$, the centrosymmetry of the two Toeplitz Bezoutians in (3.5) can be exploited, which implies that $B_T(\mathbf{p},\mathbf{q})P_n^-\mathbf{b}$ is skewsymmetric and $B_T(\widetilde{\mathbf{p}},\widetilde{\mathbf{q}})P_n^+\mathbf{b}$ is symmetric. Hence only the upper half of the vectors $B_T(\mathbf{p},\mathbf{q})P_n^-\mathbf{b}$ and $B_T(\widetilde{\mathbf{p}},\widetilde{\mathbf{q}})P_n^+\mathbf{b}$ have to be computed. This leads to another formula for A_n^{-1} involving matrices of size about $n/2$.

We derive this formula for the case where n is even, $n = 2m$. The case of odd n can be handled analogously.

We have

$$P_n^{\pm}\mathbf{b} = \begin{bmatrix} \mathbf{c}_{\pm} \\ \pm J_m\mathbf{c}_{\pm} \end{bmatrix}$$

where

$$\begin{bmatrix} \mathbf{c}_- \\ \mathbf{c}_+ \end{bmatrix} = \frac{1}{2}\,Q\,\mathbf{b} \text{ and } Q = \begin{bmatrix} I_m & -J_m \\ I_m & J_m \end{bmatrix}.$$

Since $B_T(\mathbf{p},\mathbf{q})$ and $B_T(\widetilde{\mathbf{p}},\widetilde{\mathbf{q}})$ are centrosymmetric, we have

$$B_T(\mathbf{p},\mathbf{q})P_n^-\mathbf{b} = \begin{bmatrix} \mathbf{d}_- \\ -J_m\mathbf{d}_- \end{bmatrix}, \quad B_T(\widetilde{\mathbf{p}},\widetilde{\mathbf{q}})P_n^+\mathbf{b} = \begin{bmatrix} \mathbf{d}_+ \\ J_m\mathbf{d}_+ \end{bmatrix}$$

for some $\mathbf{d}_{\pm} \in \mathbb{F}^m$. Thus

$$A_n^{-1}\mathbf{b} = \begin{bmatrix} \mathbf{d}_+ + \mathbf{d}_- \\ J_m(\mathbf{d}_+ - \mathbf{d}_-) \end{bmatrix} = Q^T \begin{bmatrix} \mathbf{d}_- \\ \mathbf{d}_+ \end{bmatrix}.$$

It remains to show how the vectors $\mathbf{d}_{\pm}$ can be obtained from $\mathbf{c}_{\pm}$. For this we introduce some notation. We associate a vector $\mathbf{u} = (u_i)_{i=0}^n$ with the $m \times m$ Hankel matrix $H(\mathbf{u})$ and the $m \times m$ lower triangular Toeplitz matrix $L(\mathbf{u})$ defined by

$$H(\mathbf{u}) = \begin{bmatrix} u_1 & u_2 & \cdots & u_m \\ u_2 & & \unicode{x22F0} & \vdots \\ \vdots & \unicode{x22F0} & & \vdots \\ u_m & \cdots & \cdots & u_{n-1} \end{bmatrix}, \quad L(\mathbf{u}) = \begin{bmatrix} u_0 & & & \\ u_1 & u_0 & & \\ \vdots & & \ddots & \\ u_{m-1} & \cdots & \cdots & u_0 \end{bmatrix}.$$

Now we obtain from (3.6) that $\mathbf{d}_\pm = C_\pm \mathbf{c}_\pm$ where

$$\begin{aligned} C_- &= -L(\mathbf{p})(H(\mathbf{q}) + L(\mathbf{q})^T) - L(\mathbf{q})(L(\mathbf{p})^T - H(\mathbf{p})) \\ C_+ &= L(\widetilde{\mathbf{p}})(H(\widetilde{\mathbf{q}}) - L(\widetilde{\mathbf{q}})^T) - L(\widetilde{\mathbf{q}})(L(\widetilde{\mathbf{p}})^T + H(\widetilde{\mathbf{p}})) \end{aligned}$$

We arrived at the following.

Corollary 3.5. *The inverse of A_n admits the representation*

$$A_n^{-1} = \frac{1}{2} Q^T \begin{bmatrix} C_- & 0 \\ 0 & C_+ \end{bmatrix} Q\,. \tag{3.7}$$

4. Split Levinson-type algorithm

We show in this section how the solutions of (3.1) can be computed recursively. Since the recursions for the + and the − vectors are the same, we omit this sign as a subscript or superscript. So T_k stands for T_k^+ or T_k^-, $\mathbf{x}_k$ stands for $\mathbf{x}_k^+$ or $\mathbf{x}_k^-$ and so on. For more details concerning the derivation and background material we refer to [22].

Let $\{n_1, \dots, n_r\}$, $n_1 < \cdots < n_r = n$, be the set of all integers $j \in \{1, \dots, n\}$ for which the restriction of T_j to $\mathbb{F}_\pm^j$ (as a linear operator in $\mathbb{F}_\pm^j$) is invertible and $n - j$ is even, and $n_{r+1} = n + 2$. We set $d_k = \frac{n_k - n_{k-1}}{2}$. Instead of T_{n_k} we write $T^{(k)}$.

Let $\mathbf{x}^{(k)} \in \mathbb{F}_\pm^{n_k}$ be the solution of

$$T^{(k)}\mathbf{x}^{(k)} = 2\, P_k \mathbf{e}_1\,. \tag{4.1}$$

Then the solutions of (3.1) are given by $\frac{1}{2}\mathbf{x}^{(r)}$ and $\frac{1}{2}\mathbf{x}^{(r+1)}$. It can be shown (see [22]) that $\mathbf{x}^{(k)}$ has the form

$$\mathbf{x}^{(k)} = \begin{bmatrix} \mathbf{0}_{d_k-1} \\ \mathbf{u}^{(k)} \\ \mathbf{0}_{d_k-1} \end{bmatrix}$$

for some $\mathbf{u}^{(k)} \in \mathbb{F}_\pm^{n_{k-1}+2}$ with nonvanishing first component. In polynomial language this means that $\mathbf{x}^{(k)}(t) = t^{d_k-1}\mathbf{u}^{(k)}(t)$.

We show how $(\mathbf{x}^{(k+1)}, d_{k+1})$ can be found from $(\mathbf{x}^{(k)}, d_k)$ and $(\mathbf{x}^{(k-1)}, d_{k-1})$. First we compute the residuals

$$[\,c_i \;\dots\; c_{n_k-1+i}\,]\,\mathbf{x}^{(k)} = r_i^{(k)} \tag{4.2}$$

for $i = 1, \dots, d_k$ and, if $d_k > d_{k-1}$, also $r_i^{(k-1)}$ for $i = d_{k-1}+1, \dots, d_k$. The numbers $r_i^{(k-1)}$ for $i = 1, \dots, d_{k-1}$ are given from the previous step. Then we form a $(d_k+1) \times (d_k+1)$ triangular Toeplitz matrix $C^{(k)}$ and a vector $\mathbf{r}_0^{(k-1)}$ according to

$$C^{(k)} = \begin{bmatrix} r_0^{(k)} & & 0 \\ \vdots & \ddots & \\ r_{d_k}^{(k)} & \dots & r_0^{(k)} \end{bmatrix}, \quad \mathbf{r}_0^{(k-1)} = \begin{bmatrix} r_0^{(k-1)} \\ \vdots \\ r_{d_k}^{(k-1)} \end{bmatrix}, \tag{4.3}$$

where $r_0^{(k)} = r_0^{(k-1)} = 1$ and solve the system $C^{(k)}\mathbf{c} = \mathbf{r}_0^{(k-1)}$.

Let

$$\widetilde{\mathbf{c}} = \begin{bmatrix} \mathbf{c} \\ \mathbf{c}' \end{bmatrix} \in \mathbb{F}_+^{2d_k+1}$$

the symmetric extension of $\mathbf{c}$. We define the vector

$$\mathbf{w}^{(k)} = M_{2d_k+1}(\mathbf{u}^{(k)})\,\widetilde{\mathbf{c}} - \begin{bmatrix} \mathbf{0}_{d_k+1} \\ \mathbf{x}^{(k-1)} \\ \mathbf{0}_{d_k+1} \end{bmatrix} \in \mathbb{F}_\pm^{n_k+2}.$$

Next we compute residuals for $\mathbf{w}^{(k)}$ via

$$\alpha_i = \begin{bmatrix} c_{i-1} & \dots & c_{n_k+i} \end{bmatrix} \mathbf{w}^{(k)}$$

starting with $i = 1$ until we arrive at the first nonzero α_i. Let $\alpha_\mu \neq 0$ and $\alpha_i = 0$ for $1 \leq i < \mu$. Then we have $d_{k+1} = \mu$, i.e., $n_{k+1} = n_k + 2\mu$, and $\mathbf{u}^{(k+1)} = \frac{1}{\alpha_\mu}\mathbf{w}^{(k)}$. Finally, we set $\mathbf{p}^{(k)} = \frac{1}{\alpha_\mu}\widetilde{\mathbf{c}} \in \mathbb{F}_+^{2d_k+1}$ and $q^{(k)} = \frac{1}{\alpha_\mu}$. For convenience, we write the recursions for $\mathbf{u}^{(k+1)}$ and $\mathbf{x}^{(k+1)}$ in polynomial form. Recall that $\mathbf{x}^{(k+1)}(t) = t^{d_{k+1}-1}\,\mathbf{u}^{(k+1)}(t)$. Let us summarize.

Theorem 4.1. *The polynomials $\mathbf{u}^{(k)}(t)$ satisfy the recursion*

$$\mathbf{u}^{(k+1)}(t) = \mathbf{p}^{(k)}(t)\,\mathbf{u}^{(k)}(t) - q^{(k)}\,t^{d_{k-1}+d_k}\,\mathbf{u}^{(k-1)}(t),$$

and the polynomials $\mathbf{x}^{(k)}(t)$ the recursion

$$\mathbf{x}^{(k+1)}(t) = t^{d_{k+1}-d_k}\mathbf{p}^{(k)}(t)\,\mathbf{x}^{(k)}(t) - q^{(k)}\,t^{d_{k+1}+d_k}\,\mathbf{x}^{(k-1)}(t),$$

where $k = 1, \dots, r$.

The computation of the initial vectors $\mathbf{x}^{(1)}$ and $\mathbf{x}^{(2)}$ requires the solution of corresponding symmetric Toeplitz systems of order n_1 and n_2.

After applying the recursion in Theorem 4.1 r times for both the + and the − vectors we arrive at the data which appear in the inversion formula (3.2).

5. Split Schur-type algorithms

The Schur-type algorithm presented below computes the complete set of residuals $r_i^{(k)}$ defined by (4.2) recursively. These residuals can be used in two ways. Firstly, the computation of them can replace the inner product calculation in the Levinson-type algorithm, which leads to a reduction of the parallel complexity from $O(n \log n)$ to $O(n)$ (for n processors). Secondly, the Schur-type algorithm implies a factorization, which will be discussed in Section 7.

Theorem 4.1 immediately leads to the relations

$$r_i^{(k+1)} = \sum_{j=0}^{2d_k} p_j^{(k)} r_{i+d_{k+1}-d_k+j}^{(k)} - q^{(k)} r_{i+d_{k+1}+d_k}^{(k-1)},$$

where $p_j^{(k)}$ are the coefficients of $\mathbf{p}^{(k)}(t)$. Introducing the polynomials

$$\mathbf{r}^{(k)}(t) = \sum_{i=0}^{n+2-n_k} r_i^{(k)} t^i$$

this can be written in the following form.

Theorem 5.1. *The polynomials $\mathbf{r}^{(k)}(t)$ satisfy the recursion*

$$\mathbf{r}^{(k+1)}(t) = \mathbf{p}^{(k)}(t^{-1})\mathbf{r}^{(k)}(t)t^{-d_{k+1}+d_k} - q^{(k)}\,\mathbf{r}^{(k-1)}(t)t^{-d_k-d_{k+1}}.$$

The initializations for this algorithm is obtained from the initializations of the $\mathbf{x}^{(k)}$.

6. Solution of linear systems

In this section we show how to solve a linear system $A_n\mathbf{f} = \mathbf{b}$ without using the inversion formula. As mentioned in Section 2, this system is equivalent to the two systems $T_n^{\pm}\mathbf{f}_{\pm} = \mathbf{b}_{\pm}$, where $\mathbf{b}_{\pm} = P_n^{\pm}\mathbf{b}$. We use the notation introduced in Section 4 and again we omit the subscripts and superscripts $+$ and $-$, for simplicity.

Suppose that $\mathbf{b} = (b_i)_{i=1}^n$, where $b_{n+1-i} = \pm b_i$ $(i = 1, \dots, n)$. Let $\mathbf{b}^{(k)}$ be the vector of the central n_k components of $\mathbf{b}$, $\mathbf{b}^{(k)} = (b_i)_{i=\nu_k+1}^{\nu_k+n_k}$, $\nu_k = \frac{1}{2}(n - n_k)$, $k = 1, \dots, r$. Then $\mathbf{b}^{(k)}$ is also symmetric or skewsymmetric.

We consider the systems

$$T^{(k)}\mathbf{f}^{(k)} = \mathbf{b}^{(k)}.$$

Our aim is to compute $\mathbf{f}^{(k)}$ from $\mathbf{f}^{(k-1)}$ and $\mathbf{u}^{(k)}$.

The vector $\mathbf{b}^{(k)}$ and and the matrix $T^{(k)}$ are of the form

$$\mathbf{b}^{(k)} = \begin{bmatrix} \pm\widehat{\mathbf{b}}_1^{(k)} \\ \mathbf{b}^{(k-1)} \\ \mathbf{b}_1^{(k)} \end{bmatrix}, \quad T^{(k)} = \begin{bmatrix} * & \widehat{B}^{(k)} & * \\ * & T^{(k-1)} & * \\ * & B^{(k)} & * \end{bmatrix},$$

where $\widehat{\mathbf{b}}_1^{(k)} = J_{d_k}\mathbf{b}_1^{(k)}$, and

$$B^{(k)} = \begin{bmatrix} c_{n_{k-1}} & \cdots & c_1 \\ \vdots & \ddots & \vdots \\ c_{n_{k-1}+d_k-1} & \cdots & c_{d_k} \end{bmatrix}, \qquad \widehat{B}^{(k)} = J_{d_k}B^{(k)}J_{n_{k-1}}.$$

We compute the vectors $\mathbf{s}^{(k)} = B^{(k)}\mathbf{f}^{(k-1)}$. Then

$$T^{(k)} \begin{bmatrix} \mathbf{0} \\ \mathbf{f}^{(k-1)} \\ \mathbf{0} \end{bmatrix} = \begin{bmatrix} \pm\widehat{\mathbf{s}}^{(k)} \\ \mathbf{b}^{(k-1)} \\ \mathbf{s}^{(k)} \end{bmatrix},$$

where $\widehat{\mathbf{s}}^{(k)} = J_{d_k}\mathbf{s}^{(k)}$.

We have

$$T^{(k)} M_{2d_k-1}(\mathbf{u}^{(k)}) = \begin{bmatrix} 0 & \pm r_0^{(k)} & \dots & \pm r_{d_k-1}^{(k)} \\ \vdots & & \ddots & \vdots \\ 0 & 0 & & \pm r_0^{(k)} \\ \mathbf{0} & \mathbf{0} & & \mathbf{0} \\ r_0^{(k)} & 0 & & 0 \\ \vdots & \ddots & & \vdots \\ r_{d_k-1}^{(k)} & \dots & r_0^{(k)} & 0 \end{bmatrix}. \tag{6.1}$$

We introduce the $d_k \times d_k$ matrix

$$C_1^{(k)} = \begin{bmatrix} r_0^{(k)} & & 0 \\ \vdots & \ddots & \\ r_{d_k-1}^{(k)} & \dots & r_0^{(k)} \end{bmatrix}$$

which is a submatrix of the matrix $C^{(k)}$ in (4.3). Now the following is easily checked.

Theorem 6.1. *Let $\mathbf{d}^{(k)} \in \mathbb{F}^{d_k}$ be the solutions of*

$$C_1^{(k)}\mathbf{d}^{(k)} = \mathbf{b}_1^{(k)} - \mathbf{s}^{(k)}. \tag{6.2}$$

Then the solution $\mathbf{f}^{(k)}$ is given by

$$\mathbf{f}^{(k)} = \begin{bmatrix} \mathbf{0} \\ \mathbf{f}^{(k-1)} \\ \mathbf{0} \end{bmatrix} + M_{2d_k-1}(\mathbf{u}^{(k)})\,\mathbf{z}^{(k)} \tag{6.3}$$

where $\mathbf{z}^{(k)} = \begin{bmatrix} \mathbf{d}^{(k)} \\ * \end{bmatrix} \in \mathbb{F}_+^{2d_k-1}$ is the symmetric extension of $\mathbf{d}^{(k)}$.

The computation of the initial vector $\mathbf{f}^{(1)}$ requires the solution of a symmetric Toeplitz system of order n_1.

For one step of the recursion one has to first compute the vector $\mathbf{s}^{(k)}$ which consists in d_k inner product calculations, then to solve a $d_k \times d_k$ triangular Toeplitz system with the coefficient matrix $C_1^{(k)}$ to get $\mathbf{z}^{(k)}$, and finally to apply formula (6.3).

Inner product calculations can be avoided if the full residual vector is updated at each step.

7. Block ZW-factorization

We show that the algorithm described in the Section 5 can be used to compute a block ZW-factorization of A_n. First we recall some concepts.

A matrix $A = [\,a_{ij}\,]_{i,j=1}^n$ is called *W-matrix* if $a_{ij} = 0$ for all (i,j) for which $i > j$ and $i + j > n + 1$, or $i < j$ and $i + j \leq n$. The matrix A will be called *unit W-matrix* if in addition $a_{ii} = 1$ for $i = 1, \dots, n$ and $a_{i,n+1-i} = 0$ for $i \neq \frac{n+1}{2}$. The transpose of a W-matrix is called a *Z-matrix.* A matrix which is both a Z- and a W-matrix will be called *X-matrix.* The names arise from the shapes of the set of all possible positions for nonzero entries, which are as follows:

$$W = \begin{bmatrix} \bullet & & & & & \bullet \\ \bullet & \circ & & & \circ & \bullet \\ \bullet & \circ & \circ & \circ & \circ & \bullet \\ \bullet & \circ & \bullet & \bullet & \circ & \bullet \\ \bullet & \bullet & & & \bullet & \bullet \\ \bullet & & & & & \bullet \end{bmatrix}, \qquad Z = \begin{bmatrix} \bullet & \bullet & \bullet & \bullet & \bullet & \bullet \\ & \circ & \circ & \circ & \bullet & \\ & & \circ & \bullet & & \\ & & \bullet & \circ & & \\ & \bullet & \circ & \circ & \circ & \\ \bullet & \bullet & \bullet & \bullet & \bullet & \bullet \end{bmatrix},$$

$$X = \begin{bmatrix} \bullet & & & & & \bullet \\ & \bullet & & & \bullet & \\ & & \bullet & \bullet & & \\ & & \bullet & \bullet & & \\ & \bullet & & & \bullet & \\ \bullet & & & & & \bullet \end{bmatrix}.$$

A unit W- or Z-matrix A is obviously nonsingular and a system $A\mathbf{f} = \mathbf{b}$ can be solved by back substitution with $\frac{n^2}{2}$ additions and $\frac{n^2}{2}$ multiplications.

A representation $A = ZXW$ of a nonsingular matrix A in which Z is a Z-matrix, W is a W-matrix, and X an X-matrix is called *ZW-factorization.* WZ-factorization is analogously defined.

A necessary and sufficient condition for a matrix $A = [\,a_{jk}\,]_{j,k=1}^n$ to admit a ZW-factorization is that the central submatrices $A_{n+2-2l}^c = [\,a_{jk}\,]_{j,k=l}^{n+1-l}$ are nonsingular for all natural numbers $l = 1, \dots, [\,\frac{n}{2}\,]$. A matrix with this property will be called *centro-nonsingular.* Under the same condition A^{-1} admits a WZ-factorization. Among all ZW-factorizations of A there is a unique one in which the factors are unit.

Symmetry properties of the matrix are inherited in the factors of the unit ZW-factorization. If A is symmetric, then $W = Z^T$. If A is centrosymmetric, then all factors Z, X and W are also centrosymmetric. All this follows from the uniqueness of the unit ZW-factorization.

The matrix A_n is both symmetric and centrosymmetric, so it admits a factorization $A_n = ZXZ^T$, in which Z is centrosymmetric and X is centrosymmetric and symmetric, provided that A_n is centro-nonsingular.

If A_n is not centro-nonsingular, then one might look for block ZW-factorizations. We show that such a factorization exists and can be found using the split

Schur algorithm. However, the block factorization we present will not be a generalization of the unit ZW-factorization but rather a generalization of a modification of it as described next.

For sake of simplicity of notation we assume that n is even, $n = 2m$. The case of odd n is similar. We introduce the $n \times n$ X-matrix

$$\Theta_n = \begin{bmatrix} -1 & & & & & 1 \\ & \ddots & & & \iddots & \\ & & -1 & 1 & & \\ & & 1 & 1 & & \\ & \iddots & & & \ddots & \\ 1 & & & & & 1 \end{bmatrix}.$$

Obviously, $\Theta_n^{-1} = \frac{1}{2}\Theta_n$.

If Z_0 is an $n \times n$ centrosymmetric Z-matrix, then the matrix $Z_1 = Z_0\Theta_n$ has the property $J_n Z_1 = Z_0 J_n \Theta_n$. That means that the first m columns of Z_1 are skewsymmetric, while the last m columns are symmetric. Let us call a matrix with this property *column-symmetric*. If moreover the X-matrix built from the diagonal and antidiagonal of Z_1 is equal to Θ_n, then Z_1 will be referred to as *unit*.

The unit ZW-factorization $A = Z_0 X_0 Z_0^T$ of a centrosymmetric, symmetric matrix A can be transformed into a ZW-factorization $A = ZXZ^T$ in which Z is unit column-symmetric. We will call this factorization *unit column-symmetric ZW-factorization*. Since the product

$$\begin{bmatrix} -1 & 1 \\ 1 & 1 \end{bmatrix}\begin{bmatrix} a & b \\ b & a \end{bmatrix}\begin{bmatrix} -1 & 1 \\ 1 & 1 \end{bmatrix}$$

is a diagonal matrix, the X-factor in the unit column-symmetric ZW-factorization is actually a diagonal matrix. Thus, provided that A_n is centro-nonsingular, A_n admits a factorization $A_n = ZDZ^T$ in which Z is unit column-symmetric and D is diagonal.

We show now that for the general case there is a *block* unit column-symmetric ZW-factorization of A_n. This is a representation $A_n = ZDZ^T$ in which D is a block diagonal matrix and Z is a column-symmetric Z-matrix. The diagonal blocks of D are assumed to have size $d_k^{\pm} \times d_k^{\pm}$ and the corresponding diagonal blocks of Z are $\pm I_{d_k^{\pm}}$. In this case we call the Z-matrix *block unit* column-symmetric. Clearly, such a factorization is, if it exists, unique.

Let Σ_d denote the $(2d-1) \times d$ matrix

$$\Sigma_d = \left.\begin{bmatrix} & & & 1 & & & \\ & & 1 & & 1 & & \\ & \iddots & & & & \ddots & \\ 1 & & & & & & 1 \end{bmatrix}\right\} d\,.$$

The $n_k \times (2d_k - 1)$ matrix $M_{2d_k-1}(\mathbf{u}^{(k)})\Sigma_{d_k}^T$ has symmetric or skewsymmetric columns, depending on whether we have the + or − case, and has the form

$$\begin{bmatrix} \pm J_{d_k} U^{(k)} \\ * \\ U^{(k)} \end{bmatrix},$$

where $U^{(k)}$ is the (nonsingular) upper triangular Toeplitz matrix

$$U^{(k)} = \begin{bmatrix} u_0^{(k)} & \dots & u_{d_k-1}^{(k)} \\ & \ddots & \vdots \\ & & u_0^{(k)} \end{bmatrix}$$

and $u_j^{(k)}$ $(k = 0, \dots, d_k - 1)$ are the first components of $\mathbf{u}^{(k)} \in \mathbb{F}^{n_{k-1}+2}$ possibly extended by zeros in case where $n_{k-1} + 2 < d_k$. We evaluate, for the + and − cases, the matrices $M_{2d_k-1}(\mathbf{u}^{(k)})\Sigma_{d_k}^T (U^{(k)})^{-1}$. These matrices will be extended to $n \times d_k$ matrices by adding symmetrically zeros at the top and the bottom. The resulting matrices will be denoted by $W_\pm^{(k)}$. Clearly, $W_\pm^{(k)}$ has again symmetric or skewsymmetric columns. From now on we indicate by a subscript or superscript the + or − case.

We form the block matrix

$$W = [\, W_-^{(r_-)} J_{d_{r_-}^-} \,, \dots , W_-^{(1)} J_{d_1^-} \,, W_+^{(1)} \,, \dots , W_+^{(r_+)} \,], \tag{7.1}$$

which is a block unit column-symmetric W-matrix.

Next we evaluate $A_n W$. According to (6.1) we have

$$T_n^\pm W_\pm^{(k)} = \begin{bmatrix} \pm J_{\nu_k + d_k^\pm} R_\pm^{(k)} \\ O \\ R_\pm^{(k)} \end{bmatrix} (U_\pm^{(k)})^{-1},$$

where

$$R_\pm^{(k)} = \begin{bmatrix} & & r_0^{\pm(k)} \\ & \iddots & \vdots \\ r_0^{\pm(k)} & \dots & r_{d_k^\pm - 1}^{\pm(k)} \\ r_1^{\pm(k)} & \dots & r_{d_k^\pm}^{\pm(k)} \\ \vdots & & \vdots \\ r_{\nu_k^\pm}^{\pm(k)} & \dots & r_{\nu_k^\pm + d_k^\pm - 1}^{\pm(k)} \end{bmatrix},$$

$\nu_k^\pm = \frac{1}{2}(n - n_k^\pm)$ and O denotes the zero matrix of size $(n - n_{k-1}^\pm) \times d_k^\pm$.

We introduce matrices $V_\pm^{(k)}$ by

$$V_\pm^{(k)} = \begin{bmatrix} r_0^{\pm(k)} & \cdots & r_{d_k^\pm - 1}^{\pm(k)} \\ & \ddots & \vdots \\ & & r_0^{\pm(k)} \end{bmatrix}$$

and the matrices $Z_\pm^{(k)}$ by

$$Z_\pm^{(k)} = T_n^\pm W_\pm^{(k)} U_\pm^{(k)} (V_\pm^{(k)})^{-1} J_{d_k^\pm} = \begin{bmatrix} \pm J_{\nu_k^\pm + d_k^\pm} R_\pm^{(k)} \\ O \\ R_\pm^{(k)} \end{bmatrix} (V_\pm^{(k)})^{-1} J_{d_k^\pm} \,.$$

We arrange the matrices $Z_\pm^{(k)}$ to the $n \times n$ matrix

$$Z = [\, Z_-^{(r_-)} J_{d_{r_-}^-} \,,\; \dots \,,\; Z_-^{(1)} J_{d_1^-} \,,\; Z_+^{(1)} \,,\; \dots \,,\; Z_+^{(r_+)} \,] \,.$$

This matrix is a block unit column-symmetric Z-matrix.

We have now the relation

$$A_n W = ZD \,, \tag{7.2}$$

where $D = \operatorname{diag}(D_-^{(r_-)}, \dots, D_-^{(1)}, D_+^{(1)}, \dots, D_+^{(r_+)})$ and

$$D_+^{(k)} = J_{d_k^+} V_+^{(k)} (U_+^{(k)})^{-1}, \quad D_-^{(k)} = V_-^{(k)} (U_-^{(k)})^{-1} J_{d_k^-} \,.$$

Note that the matrices $D_+^{(k)}$ are lower triangular and the matrices $D_-^{(k)}$ are upper triangular Hankel matrices.

Since the inverse of a W-matrix is again a W-matrix we have a block ZW-factorization $A_n = ZDW^{-1}$. The matrix $2\,W^{-1}$ is a block unit row-symmetric W-matrix. Taking the uniqueness of unit ZW-factorization into account we conclude that $Z^T = 2\,W^{-1}$, thus

$$A_n = \frac{1}{2} ZDZ^T. \tag{7.3}$$

Moreover we have obtained a WZ-factorization of A_n^{-1} as

$$A_n^{-1} = \frac{1}{2} WD^{-1}W^T. \tag{7.4}$$

Summing up, we arrived at the following.

Theorem 7.1. *Any nonsingular centrosymmetric T+H matrix A_n admits a representation $A_n = ZEZ^T$ in which Z is a block unit column-symmetric Z-matrix and E is a block diagonal matrix the diagonal blocks of which are triangular Hankel matrices.*

In order to find the block ZW-factorization of A_n one has to run the split Schur algorithm from Section 5. This gives the factor Z. In order to evaluate the block diagonal factor one has to find the first $d_k^\pm$ components of the vectors $\mathbf{u}_\pm^{(k)}$. This can be done running partly the split Levinson algorithm described in

Section 4. Unfortunately the numbers d_k are not known a priori, so some updating might be necessary during the procedure.

8. Concluding remarks

The split Levinson and Schur algorithms developed in this paper can be used in different ways for fast solution of a linear system $A_n\mathbf{f} = \mathbf{b}$ with a nonsingular, centrosymmetric T+H coefficient matrix A_n. There are basically three approaches, namely (a) via an inversion formula, (b) via factorization, and (c) by direct recursion.

Approach (a) means that first the data in an inversion formula, e.g., in (3.2) or (3.5), are computed and then this formula is applied for fast matrix-vector multiplication. This can be done in a standard way with $O(n^2)$ operations. In the cases $\mathbb{F} = \mathbb{R}$ and $\mathbb{F} = \mathbb{C}$ it is more efficient to use representations of T+H Bezoutians that involve only diagonal matrices and Fourier or real trigonometric transformations (see [15], [16], [17]). The application of these formulas reduces the amount for matrix-vector multiplication to $O(n \log n)$ operations.

The data in the inversion formula can be computed by means of the split Levinson algorithm alone or, in order to avoid inner products, together with the split Schur algorithm. The second method is more expensive in sequential but preferable in parallel processing. Another advantage of the second method is that it can be speeded up with the help of a divide and conquer strategy and FFT to an $O(n \log^2 n)$-complexity algorithm, at least in the cases $\mathbb{F} = \mathbb{R}$ and $\mathbb{F} = \mathbb{C}$ (see [11]).

Approach (b) is to use a factorization of the matrix A_n, of its inverse or formula (7.2). To obtain the block ZW-factorization (7.3) of A_n one has to run the split Schur algorithm for the full vectors and the split Levinson algorithm for a few components depending on the rank profile. In the centro-nonsingular case this reduces to the recursion of just one number. Finding the solution $\mathbf{f}$ requires finally the solution of a Z^T and Z-system.

Running only the split Levinson algorithm will provide a block WZ-factorization (7.4) of A_n^{-1}. The solution of the system is then obtained after two matrix-vector multiplications.

To evaluate the factors W, Z and D in formula (7.2) one has to run both the split Levinson and Schur algorithms. After that one has to solve a system with the coefficient matrix ZD by back substitution and to multiply the result by W.

Approach (c) is to use the direct recursion formula from Section 6 together with the split Levinson algorithm. Recall that there is also a Schur-type version of the recursion in Section 6, so that inner products can be avoided. Approach (c) is preferable if only a single system has to be solved.

The comparison of the complexities of the different algorithms will be similar to that for the centro-nonsingular case discussed in [20]. Of course, the complexities depend on the rank profile of the central submatrices of A_n, but it is remarkable that the presence of singular central submatrices does not essentially increase the complexity. On the contrary, in many cases the complexity is smaller than in the centro-nonsingular case (see [22], [23]). Let us also point out that complexity is not the only criterion to select an algorithm. If finite precision arithmetic is used, then another important issue is stability. As practical experience and theoretical results indicate, Schur-type algorithms have better stability behavior than their Levinson counterparts.

References

[1] G. Ammar, P. Gader, *A variant of the Gohberg-Semencul formula involving circulant matrices*, SIAM J. Matrix Analysis Appl., **12**, 3 (1991), 534–540.

[2] D. Bini, V. Pan, *Polynomial and Matrix Computations*, Birkhäuser Verlag, Basel, Boston, Berlin, 1994.

[3] E. Bozzo, C. di Fiore, *On the use of certain matrix algebras associated with discrete trigonometric transforms in matrix displacement decompositions*, SIAM J. Matrix Analysis Appl., **16** (1995), 312–326.

[4] P. Delsarte, Y. Genin, *The split Levinson algorithm, IEEE Transactions on Acoustics Speech, and Signal Processing* ASSP-34 (1986), 470–477.

[5] P. Delsarte, Y. Genin, *On the splitting of classical algorithms in linear prediction theory*, IEEE Transactions on Acoustics Speech, and Signal Processing ASSP-35 (1987), 645-653.

[6] D.J. Evans, M. Hatzopoulos, *A parallel linear systems solver*, Internat. J. Comput. Math., **7**, 3 (1979), 227–238.

[7] I. Gohberg, I. Koltracht, *Efficient algorithm for Toeplitz plus Hankel matrices*, Integral Equations Operator Theory, **12**, 1 (1989), 136–142.

[8] I. Gohberg, A.A. Semencul, *On the inversion of finite Toeplitz matrices and their continuous analogs* (in Russian), Matemat. Issledovanya, **7**, 2 (1972), 201–223.

[9] I. Gohberg, V. Olshevsky, *Circulants, displacements and decompositions of matrices*, Integral Equations Operator Theory, **15**, 5 (1992), 730–743.

[10] G. Heinig, *Matrix representations of Bezoutians*, Linear Algebra Appl., **223/224** (1995), 337–354.

[11] G. Heinig, *Chebyshev-Hankel matrices and the splitting approach for centrosymmetric Toeplitz-plus-Hankel matrices*, Linear Algebra Appl., **327**, 1-3 (2001), 181–196.

[12] G. Heinig, P. Jankowski, K. Rost, *Fast inversion algorithms of Toeplitz-plus-Hankel matrices*, Numerische Mathematik, **52** (1988), 665–682.

[13] G. Heinig, K. Rost, *Algebraic Methods for Toeplitz-like Matrices and Operators*, Birkhäuser Verlag: Basel, Boston, Stuttgart, 1984.

[14] G. Heinig, K. Rost, *On the inverses of Toeplitz-plus-Hankel matrices*, Linear Algebra Appl., **106** (1988), 39–52.

[15] G. Heinig, K. Rost, *DFT representations of Toeplitz-plus-Hankel Bezoutians with application to fast matrix-vector multiplication*, Linear Algebra Appl., **284** (1998), 157–175.

[16] G. Heinig, K. Rost, *Hartley transform representations of inverses of real Toeplitz-plus-Hankel matrices*, Numerical Functional Analysis and Optimization, **21** (2000), 175–189.

[17] G. Heinig, K. Rost, *Efficient inversion formulas for Toeplitz-plus-Hankel matrices using trigonometric transformations*, In: V. Olshevsky (Ed.), *Structured Matrices in Mathematics, Computer Science, and Engineering*, vol. 281, AMS-Series *Contemporary Mathematics* (2001), 247–264.

[18] G. Heinig, K. Rost, *Centrosymmetric and centro-skewsymmetric Toeplitz-plus-Hankel matrices and Bezoutians*, Linear Algebra Appl., **366** (2003), 257–281.

[19] G. Heinig, K. Rost, *Fast algorithms for skewsymmetric Toeplitz matrices*, Operator Theory: Advances and Applications, Birkhäuser Verlag, Basel, Boston, Berlin, 135 (2002), 193–208.

[20] G. Heinig, K. Rost, *Fast algorithms for centro-symmetric and centro-skewsymmetric Toeplitz-plus-Hankel matrices*, Numerical Algorithms, **33** (2003), 305–317.

[21] G. Heinig, K. Rost, *New fast algorithms for Toeplitz-plus-Hankel matrices*, SIAM J. Matrix Analysis Appl., **25**, 3 (2004), 842–857.

[22] G. Heinig, K. Rost, *Split algorithms for symmetric Toeplitz matrices with arbitrary rank profile*, Numer. Linear Algebra Appl., in print.

[23] G. Heinig, K. Rost, *Split algorithms for skewsymmetric Toeplitz matrices with arbitrary rank profile*, Theoretical Computer Science, **315**, 2-3 (2004), 453–468.

[24] G.A. Merchant, T.W. Parks, *Efficient solution of a Toeplitz-plus-Hankel coefficient matrix system of equations*, IEEE Transact. on ASSP, **30**, 1 (1982), 40–44.

[25] A.B. Nersesjan, A.A. Papoyan, *Construction of the matrix inverse to the sum of Toeplitz and Hankel matrices* (Russian), Izv AN Arm. SSR, Matematika, **8**, 2 (1983), 150–160.

[26] S. Chandra Sekhara Rao, *Existence and uniqueness of WZ factorization*, Parallel Comp., **23**, 8 (1997), 1129–1139.

Georg Heinig
Dept. of Math. and Comp. Sci.
Kuwait University
P.O.Box 5969, Safat 13060, Kuwait

Karla Rost
Dept. of Mathematics
University of Chemnitz
D-09107 Chemnitz, Germany
e-mail: krost@mathematik.tu-chemnitz.de

Operator Theory:
Advances and Applications, Vol. 171, 147–170

Schmidt-Representation of Difference Quotient Operators

Michael Kaltenbäck and Harald Woracek

Abstract. We consider difference quotient operators in de Branges Hilbert spaces of entire functions. We give a description of the spectrum and a formula for the spectral subspaces. The question of completeness of the system of eigenvectors and generalized eigenvectors is discussed. For certain cases the s-numbers and the Schmidt-representation of the operator under discussion is explicitly determined.

Mathematics Subject Classification (2000). Primary 46E20; Secondary 47B06.

Keywords. Hilbert space, entire function, compact operator, s-number.

1. Introduction and Preliminaries

Let $\mathcal{H}$ be a Hilbert space whose elements are entire functions. We call $\mathcal{H}$ a *de Branges Hilbert space*, or *dB-space* for short, if it satisfies the following axioms (cf. [dB]):

(dB1) For each $w \in \mathbb{C}$ the functional $F \mapsto F(w)$ is continuous.

(dB2) If $F \in \mathcal{H}$, then also $F^{\#}(z) := \overline{F(\overline{z})}$ belongs to $\mathcal{H}$. For all $F, G \in \mathcal{H}$

$$(F^{\#}, G^{\#}) = (G, F) .$$

(dB3) If $w \in \mathbb{C} \setminus \mathbb{R}$ and $F \in \mathcal{H}$ with $F(w) = 0$, then also

$$\frac{z - \overline{w}}{z - w} F(z) \in \mathcal{H} .$$

For all $F, G \in \mathcal{H}$ with $F(w) = G(w) = 0$

$$\left(\frac{z - \overline{w}}{z - w} F(z), \frac{z - \overline{w}}{z - w} G(z) \right) = (F, G) .$$

For a dB-space $\mathcal{H}$ the linear space of *associated functions* can be defined as

$$\operatorname{Assoc} \mathcal{H} := \mathcal{H} + z\mathcal{H}.$$

Consider the operator $\mathcal{S}$ of multiplication by the independent variable in the dB-space $\mathcal{H}$:

$$\operatorname{dom}\mathcal{S} := \{F \in \mathcal{H} :\ zF(z) \in \mathcal{H}\}$$
$$(\mathcal{S}F)(z) := zF(z),\ \ F \in \operatorname{dom}\mathcal{S}.$$

By **(dB1)**–**(dB3)** the operator $\mathcal{S}$ is a closed symmetric operator with defect index $(1,1)$, is real with respect to the involution $F \mapsto F^{\#}$, and the set of regular points of $\mathcal{S}$ equals $\mathbb{C}$.

There is a natural bijection between the set $\operatorname{Assoc}\mathcal{H}$ and the set of all relational extensions $\mathcal{A}$ of $\mathcal{S}$ with nonempty resolvent set, see, e.g., [KW1, Proposition 4.6]. It is established by the formula

$$(\mathcal{A} - w)^{-1}F(z) = \frac{F(z) - \frac{S(z)}{S(w)}F(w)}{z - w},\ \ w \in \mathbb{C}, S(w) \neq 0. \tag{1.1}$$

Throughout this paper we will denote the relation corresponding to a function $S \in \operatorname{Assoc}\mathcal{H}$ via (1.1) by $\mathcal{A}_S$ and will put $\mathcal{R}_S := \mathcal{A}_S^{-1}$. Note that

$$\ker(\mathcal{A}_S - w)^{-1} = \operatorname{span}\{S\} \cap \mathcal{H}\,,$$

and therefore $\mathcal{A}_S$ is a proper relation if and only if $S \in \mathcal{H}$. Moreover, it is a consequence of the formula (1.1) that the finite spectrum of the relation $\mathcal{A}_S$ is given by the zeros of the function S. Hence

$$\sigma(\mathcal{R}_S) \setminus \{0\} = \left\{\lambda \in \mathbb{C} :\ S\left(\frac{1}{\lambda}\right) = 0\right\}.$$

If $S(0) \neq 0$ the relation $\mathcal{R}_S$ has no multivalued part, i.e., is an operator, and is given by (1.1) with $w = 0$. As is seen by a perturbation argument (cf. Lemma 2.1) it is in fact a compact operator.

In this note we give some results on the completeness of the system of eigenvalues and generalized eigenvalues (Theorem 3.3) and determine the s-numbers and the Schmidt-representation of $\mathcal{R}_S$ (Theorem 4.5) when S belongs to a certain subclass of $\operatorname{Assoc}\mathcal{H}$. In fact, we are mainly interested in the operator $\mathcal{R}_E$ where $\mathcal{H} = \mathcal{H}(E)$ in the sense explained further below in this introduction. However, our results are valid, and thus stated, for a slightly bigger subclass of $\operatorname{Assoc}\mathcal{H}$. As a preliminary result, in Section 2, we give a self-contained proof of the explicit form of spectral subspaces of $\mathcal{R}_S$ at nonzero eigenvalues.

In the final Section 5 we add a discussion of the operator $\mathcal{R}_E$ in the case of a space which is symmetric about the origin, for the definition see (5.1). This notion was introduced by de Branges, and originates in the classical theory of Fourier transforms, i.e., the theory of Paley-Wiener spaces. In our context it turns out that in this case $\mathcal{R}_E$ is selfadjoint with respect to a canonical Krein space inner product on $\mathcal{H}(E)$. An investigation of the case of symmetry is of particular interest for several reasons. Firstly, in spaces symmetric about the origin a rich structure theory is available and thus much stronger results can be expected, cf. [KWW2], [B]. Secondly, it appears in many applications, as for example in the spectral theory of strings, cf. [LW], [KWW1], the theory of Hamiltonian systems with semibounded spectrum, cf. [W], or in the theory of functions of classical analysis like Gauss' hypergeometric functions, cf. [dB], or the Riemann Zeta-function, cf. [KW2]. Moreover, by comparing Lemma 4.3 with Theorem 1 of [OS] more operator theoretic methods could be brought into the theory of sampling sequences in de Branges spaces symmetric about the origin. It is a possible future direction of research to investigate these subjects.

The present note should also be viewed as a possible starting point with connections to several areas of research. For example in Corollary 4.6 we actually apply the present results to the field of growth of entire function. The proofs given are often elementary, which is due to the fact that we (basically) deal with those operators $\mathcal{R}_S$ having one-dimensional imaginary part. Thus many notions are accessible to explicit computation.

Let us collect some necessary preliminaries. A function f analytic in the open upper half-plane $\mathbb{C}^+$ is said to be of *bounded type*, $f \in N(\mathbb{C}^+)$, if it can be written as a quotient of two bounded analytic functions. If the assumption that f is analytic is weakened to f being merely meromorphic, we speak of functions of *bounded characteristic*, $f \in \tilde{N}(\mathbb{C}^+)$. If $f \in \tilde{N}(\mathbb{C}^+)$ there exists a real number $\operatorname{mt} f$, the *mean type* of f, such that for all $\theta \in (0, \pi)$ with possible exception of a set of measure zero

$$\lim_{r\to\infty} \frac{\log|f(re^{i\theta})|}{r} = \operatorname{mt} f \cdot \sin\theta .$$

For $f \in N(\mathbb{C}^+)$ the mean type can be obtained as

$$\operatorname{mt} f = \limsup_{y\to+\infty} \frac{\log|f(iy)|}{y} .$$

An entire function E is said to be long to the *Hermite-Biehler class*, $E \in \mathcal{HB}$, if it has no zeros in $\mathbb{C}^+$ and satisfies

$$\left|E^{\#}(z)\right| \le \left|E(z)\right|, \ z \in \mathbb{C}^+ .$$

If, additionally, E has no real zeros we shall write $E \in \mathcal{HB}^\times$.

Recall that the notions of dB-spaces and Hermite-Biehler functions are intimately related: For given $E \in \mathcal{HB}$ define $\mathcal{H}(E)$ to be the set of all entire functions F such that $E^{-1}F$ and $E^{-1}F^{\#}$ are of bounded type and nonpositive mean type in $\mathbb{C}^+$ and, moreover, belong to $L^2(\mathbb{R})$. If $\mathcal{H}(E)$ is equipped with the inner product

$$(F, G) := \int_{\mathbb{R}} F(t)\overline{G(t)} \, \frac{dt}{|E(t)|^2} ,$$

it becomes a dB-space. Conversely, for any given dB-space $\mathcal{H}$ there exists a function $E \in \mathcal{HB}$ such that $\mathcal{H} = \mathcal{H}(E)$. In fact the function E is in essence uniquely determined by $\mathcal{H}$: Let $E_1, E_2 \in \mathcal{HB}$ and write $E_1 = A_1 - iB_1$, $E_2 = A_2 - iB_2$, with $A_1 = A_1^{\#}$, $A_2 = A_2^{\#}$, etc. Then we have $\mathcal{H}(E_1) = \mathcal{H}(E_2)$ if and only if there exists a 2×2-matrix M whose entries are real numbers and which has determinant 1 such that $(A_2, B_2) = (A_1, B_1)M$.

By its definition a dB-space $\mathcal{H}$ is a reproducing kernel Hilbert space; denote its reproducing kernel by $K(w,z)$, i.e.,

$$F(w) = (F(.), K(w,.)),\ F \in \mathcal{H}, w \in \mathbb{C}.$$

If $\mathcal{H}$ is written as $\mathcal{H} = \mathcal{H}(E)$, the kernel $K(w,z)$ can be represented in terms of E:

$$\begin{aligned} K(w,z) &= \frac{E(z)E^{\#}(\overline{w}) - E(\overline{w})E^{\#}(z)}{2\pi i(\overline{w} - z)},\ z \neq \overline{w}, \\ K(\overline{z},z) &= \frac{-1}{2\pi i}\left(E'(z)E^{\#}(z) - E(z)E^{\#}(z)'\right). \end{aligned}$$

The function E, and thus also $E^{\#}$ as well as any linear combination of those functions, belongs to $\operatorname{Assoc}\mathcal{H}$ and henceforth gives rise to an extension of the operator $\mathcal{S}$. Thereby the functions ($E = A - iB$)

$$S_\phi(z) := \frac{1}{2}e^{i(\phi-\frac{\pi}{2})}E(z) + \frac{1}{2}e^{-i(\phi-\frac{\pi}{2})}E^{\#}(z) = \sin\phi\, A(z) - \cos\phi\, B(z),\ \phi \in \mathbb{R},$$

play a special role: The set

$$\{\mathcal{A}_{S_\phi} : \phi \in [0,\pi)\}$$

is equal to the set of selfadjoint extensions of $\mathcal{S}$, cf. [KW1, Proposition 6.1]. Note that there exists $\phi \in [0,\pi)$ such that $S_\phi \in \mathcal{H}$ if and only if $\overline{\operatorname{dom}\mathcal{S}} \neq \mathcal{H}$ in which case ϕ is unique and $\overline{\operatorname{dom}\mathcal{S}} \oplus \operatorname{span}\{S_\phi\} = \mathcal{H}$, cf. [dB, Theorem 29, Problem 46]. Let us note for later reference that the reproducing kernel K can be expressed in terms of the functions S_ϕ as

$$\begin{aligned} K(w,z) &= \frac{S_\phi(z)S_{\phi+\frac{\pi}{2}}(\overline{w}) - S_{\phi+\frac{\pi}{2}}(z)S_\phi(z)}{\pi i(\overline{w} - z)},\ z \neq \overline{w}, \\ K(\overline{z},z) &= \frac{1}{\pi}\left(S_\phi(z)S'_{\phi+\frac{\pi}{2}}(z) - S'_\phi(z)S_{\phi+\frac{\pi}{2}}(z)\right). \end{aligned} \tag{1.2}$$

If f is analytic at a point w we denote by $\operatorname{Ord}_w f \in \mathbb{N} \cup \{0\}$ the order of w as a zero of f. Note that by the definition of $\mathcal{H}(E)$ we have

$$(\mathfrak{d}\mathcal{H})(w) := \min_{F \in \mathcal{H}} \operatorname{Ord}_w F = \begin{cases} \operatorname{Ord}_w E & , w \in \mathbb{R} \\ 0 & , w \notin \mathbb{R} \end{cases}. \tag{1.3}$$

We will confine our attention to dB-spaces $\mathcal{H}$ with $\mathfrak{d}\mathcal{H} = 0$ which means, by virtue of (1.3), to restrict to dB-spaces that can be written as $\mathcal{H} = \mathcal{H}(E)$ with $E \in \mathcal{H}B^\times$. This assumption is no essential restriction since, if $E \in \mathcal{H}B$ and C denotes a Weierstraß product formed with the real zeros of E, we have $C^{-1}E \in \mathcal{H}B$ and the mapping $F \mapsto C^{-1}F$ is an isometry of $\mathcal{H}(E)$ onto $\mathcal{H}(C^{-1}E)$, cf. [KW3, Lemma 2.4].

2. Spectral subspaces

Let us start with the following observation:

2.1. Lemma. *Let $S \in \operatorname{Assoc} \mathcal{H}$, $S(0) \neq 0$. Then the operator $\mathcal{R}_S$ is compact.*

Proof. Let $S, T \in \operatorname{Assoc} \mathcal{H}$, $S(0), T(0) \neq 0$, then $\mathcal{R}_S$ and $\mathcal{R}_T$ differ only by a one-dimensional operator:

$$\begin{aligned}(\mathcal{R}_S - \mathcal{R}_T) F(z) &= \frac{F(z) - \frac{S(z)}{S(0)} F(0)}{z} - \frac{F(z) - \frac{T(z)}{T(0)} F(0)}{z} \\ &= \frac{1}{z} \left[\frac{T(z)}{T(0)} - \frac{S(z)}{S(0)} \right] (F(.), K(0,.)) \,.\end{aligned} \tag{2.1}$$

Choose $T = S_\phi$ where ϕ is such that $S_\phi(0) \neq 0$. For this choice the operator $\mathcal{R}_T$ is a bounded selfadjoint operator whose nonzero spectrum is discrete, cf. [KW1, Proposition 4.6] and hence $\mathcal{R}_T$ is compact. It follows that $\mathcal{R}_S$ is compact for any $S \in \operatorname{Assoc} \mathcal{H}$, $S(0) \neq 0$. □

2.2. *Remark.* Assume that $\mathcal{H} = \mathcal{H}(E)$ with a function $E \in \mathcal{H}B$ of finite order ρ. Let $S \in \operatorname{Assoc} \mathcal{H}$, $S(0) \neq 0$, be given. Then for any $\rho' > \rho$ the operator $\mathcal{R}_S$ belongs to the symmetrically-normed ideal $\mathfrak{S}_{\rho'}$ (cf. [GK]).

To see this recall, e.g., from [dB], that the nonzero spectrum of the selfadjoint operator $\mathcal{R}_{S_\phi}$ consists of the simple eigenvalues $\{\lambda \in \mathbb{R} : S_\phi(\frac{1}{\lambda}) = 0\}$. Since E is of order ρ, also every function S_ϕ possesses this growth, cf. [KW3, Theorem 3.4]. Thus, for every $\rho' > \rho$, its zeros μ_k satisfy

$$\sum \frac{1}{{\mu_k}^{\rho'}} < \infty \,.$$

We start with determining the spectral subspaces of $\mathcal{R}_S$. The following result is standard, however, since it is a basic tool for the following and no explicit reference is known to us, we provide a complete proof.

If $\sigma(\mathcal{R}_S) \cap M$ is an isolated component of the spectrum denote by $\mathcal{P}_M$ the corresponding Riesz-projection.

2.3. Proposition. *Let $\mathcal{H}$ be a dB-space, $\mathfrak{d}\mathcal{H} = 0$, and let $S \in \operatorname{Assoc} \mathcal{H}$, $S(0) \neq 0$, be given. Unless $\mathcal{H}$ is finite-dimensional and $S \in (\operatorname{Assoc} \mathcal{H}) \setminus \mathcal{H}$, we have*

$$\sigma(\mathcal{R}_S) = \left\{ \lambda \in \mathbb{C} : S\left(\frac{1}{\lambda}\right) = 0 \right\} \cup \{0\}.$$

In the case $\dim \mathcal{H} < \infty$, $S \in (\operatorname{Assoc} \mathcal{H}) \setminus \mathcal{H}$,

$$\sigma(\mathcal{R}_S) = \left\{ \lambda \in \mathbb{C} : S\left(\frac{1}{\lambda}\right) = 0 \right\}.$$

If $\lambda \in \sigma(\mathcal{R}_S) \setminus \{0\}$, the Riesz-projection $\mathcal{P}_{\{\lambda\}}$ is given as ($n := \operatorname{Ord}_{\lambda^{-1}} S$)

$$\left(\mathcal{P}_{\{\lambda\}} F\right)(z) = \sum_{l=1}^{n} \frac{1}{(n-l)!} \frac{d^{n-l}}{dz^{n-l}} \left[\left(z - \frac{1}{\lambda}\right)^n \frac{F(z)}{S(z)} \right]_{z=\frac{1}{\lambda}} \frac{S(z)}{(z - \frac{1}{\lambda})^l} \,. \tag{2.2}$$

The spectral subspace $\operatorname{ran}\mathcal{P}_{\{\lambda\}}$ *is spanned by the Jordan chain*

$$\left\{ \begin{pmatrix} \frac{S(z)}{1-\lambda z} \\ \vdots \\ \frac{S(z)}{(1-\lambda z)^n} \end{pmatrix}^T (\lambda M)^k \begin{pmatrix} 0 \\ \vdots \\ 0 \\ 1 \end{pmatrix}, k = 0, \dots, n-1 \right\} \tag{2.3}$$

where we have put

$$M := \begin{pmatrix} 0 & 1 & \cdots & 1 \\ 0 & 0 & \ddots & \vdots \\ \vdots & & \ddots & 1 \\ 0 & 0 & \cdots & 0 \end{pmatrix}$$

If $S \in \operatorname{dom}(\mathcal{S}^k)$ *for some* $k \in \mathbb{N} \cup \{0\}$, *then*

$$\{ S(z), (1+z)S(z), \dots, (1+\dots+z^k)S(z) \} \tag{2.4}$$

is a Jordan chain of $\mathcal{R}_S$ *at* 0. *Moreover, any Jordan-chain at* 0 *is of the form* $\{p(z)S(z), \mathcal{R}_S(p(z)S(z)), \mathcal{R}_S^2(p(z)S(z)), \dots, \alpha S(z)\}$ *where* p *is a polynomial of degree at most* k *and* α *is constant.*

Proof. Assume that $0 \notin \sigma(\mathcal{R}_S)$. Then $|\sigma(\mathcal{R}_S)| < \infty$ and since every nonzero spectral point is an eigenvalue of finite type we conclude that $\dim \mathcal{H} < \infty$. Since

$$\ker \mathcal{R}_S = \operatorname{span}\{S\} \cap \mathcal{H}, \tag{2.5}$$

we must have $S \notin \mathcal{H}$. Conversely assume that $\dim \mathcal{H} < \infty$, $S \notin \mathcal{H}$. Then $\sigma(\mathcal{R}_S) = \sigma_p(\mathcal{R}_S)$ and by (2.5) we have $0 \notin \sigma(\mathcal{R}_S)$.

Let $\lambda \in \sigma(\mathcal{R}_S) \setminus \{0\}$ be given. In order to compute the Riesz-projection $\mathcal{P}_{\{\lambda\}}$ we use the relation ($\mu \in \rho(\mathcal{R}_S)$, $\mu \neq 0$)

$$(\mathcal{R}_S - \mu)^{-1} = -\frac{1}{\mu} - \frac{1}{\mu^2}\left(\mathcal{A}_S - \frac{1}{\mu}\right)^{-1}.$$

Choose a sufficiently small circle Γ around λ so that neither 0 nor any spectral point other than λ is contained in the interior of Γ. Then

$$\begin{aligned}
\left(\mathcal{P}_{\{\lambda\}} F\right)(z) &= \frac{-1}{2\pi i} \oint_\Gamma \left((\mathcal{R}_S - \mu)^{-1} F \right)(z)\, d\mu \\
&= \frac{-1}{2\pi i} \oint_\Gamma \left(\left(-\frac{1}{\mu} - \frac{1}{\mu^2}\left(\mathcal{A}_S - \frac{1}{\mu}\right)^{-1} \right) F \right)(z)\, d\mu \\
&= \frac{-1}{2\pi i} \oint_\Gamma \left(\left(\mathcal{A}_S - \frac{1}{\mu}\right)^{-1} F \right)(z) \left(-\frac{1}{\mu^2} d\mu \right) \\
&= \frac{-1}{2\pi i} \oint_{\frac{1}{\Gamma}} \left((\mathcal{A}_S - \nu)^{-1} F \right)(z)\, d\nu
\end{aligned}$$

$$= \frac{-1}{2\pi i} \oint_{\frac{1}{\Gamma}} \frac{F(z) - \frac{S(z)}{S(\nu)} F(\nu)}{z - \nu} \, d\nu$$

$$= \frac{F(z)}{2\pi i} \oint_{\frac{1}{\Gamma}} \frac{d\nu}{\nu - z} - \frac{S(z)}{2\pi i} \oint_{\frac{1}{\Gamma}} \frac{F(\nu)}{(\nu - z)S(\nu)} \, d\nu \,. \tag{2.6}$$

Assume that z is located in the exterior of the circle Γ^{-1}, put $\xi := \lambda^{-1}$ and let $n := \operatorname{Ord}_\xi S$. Then the first integral in (2.6) vanishes and the integrand in the second term is analytic with exception of a pole at ξ with order n. Thus

$$\left(\mathcal{P}_{\{\lambda\}} F\right)(z) = -S(z) \operatorname{Res}_{\mu=\xi} \frac{F(\mu)}{(\mu - z)S(\mu)}$$
$$= -\frac{S(z)}{(n-1)!} \frac{d^{n-1}}{d\mu^{n-1}} \left[\frac{(\mu - \xi)^n F(\mu)}{(\mu - z)S(\mu)} \right]_{\mu=\xi}$$

and we compute

$$\frac{d^{n-1}}{d\mu^{n-1}} \left[\frac{(\mu - \xi)^n F(\mu)}{(\mu - z)S(\mu)} \right]_{\mu=\xi}$$
$$= \sum_{k=0}^{n-1} \binom{n-1}{k} \frac{d^k}{d\mu^k} \left[\frac{1}{\mu - z} \right] \cdot \frac{d^{n-1-k}}{d\mu^{n-1-k}} \left[\frac{(\mu - \xi)^n F(\mu)}{S(\mu)} \right]_{\mu=\xi}$$
$$= \sum_{k=0}^{n-1} \frac{(n-1)!}{(n-1-k)!} \frac{d^{n-1-k}}{d\mu^{n-1-k}} \left[\frac{(\mu - \xi)^n F(\mu)}{S(\mu)} \right]_{\mu=\xi} \frac{(-1)^k}{(\mu - z)^{k+1}} \,.$$

Then $\mathcal{P}_{\{\lambda\}} F$ coincides for z in the exterior of Γ^{-1} with the function on the right-hand side of (2.2). Since both functions are entire this establishes (2.2).

Put $\Phi_l(z) := (1 - \lambda z)^{-l} S(z)$, $l = 1, 2, \ldots, n$. Then, by the already proved formula (2.2), we have $\operatorname{ran} \mathcal{P}_S \subseteq \operatorname{span}\{\Phi_1, \ldots, \Phi_n\}$. We compute

$$(\mathcal{R}_S - \lambda) \, \Phi_k(z) = \frac{1}{z} \left[\frac{S(z)}{(1 - \lambda z)^k} - \frac{S(z)}{S(0)} \Phi_k(0) \right] - \lambda \frac{S(z)}{(1 - \lambda z)^k}$$
$$= \frac{S(z) - (1 - \lambda z)^k S(z) - \lambda z S(z)}{z(1 - \lambda z)^k} = \frac{S(z)}{(1 - \lambda z)^{k-1}} \cdot \frac{1 - (1 - \lambda z)^{k-1}}{z}$$
$$= \frac{S(z)}{(1 - \lambda z)^{k-1}} \lambda \sum_{l=0}^{k-2} (1 - \lambda z)^l = \sum_{j=1}^{k-1} \lambda \Phi_j(z) \,.$$

Hence $\operatorname{span}\{\Phi_1, \ldots, \Phi_n\}$ is an invariant subspace for $\mathcal{R}_S$ and with respect to the basis $\{\Phi_1, \ldots, \Phi_n\}$ the operator $\mathcal{R}_S - \lambda$ has the matrix representation

$$\mathcal{R}_S - \lambda = \lambda M \,.$$

The only eigenvalue of this matrix is 0 and therefore $\operatorname{ran} \mathcal{P}_S = \operatorname{span}\{\Phi_1, \ldots, \Phi_n\}$. Moreover, this space is spanned by the Jordan chain (2.3).

Assume that $S \in \operatorname{dom}(\mathcal{S}^k)$ and put $\tau_l(z) := (1 + z + \cdots + z^l)S(z)$, $l = 0, 1, \ldots, k$. Then

$$\mathcal{R}_S\, \tau_0(z) = \mathcal{R}_S\, S(z) = 0\,,$$

and for $l \geq 1$

$$\mathcal{R}_S\, \tau_l(z) = \frac{1}{z}\left[(1 + z + \cdots + z^l)S(z) - \frac{S(z)}{S(0)} \cdot S(0)\right] = \tau_{l-1}(z).$$

We see that (2.4) is a Jordan chain at 0. □

3. Completeness of eigenvectors

In general the system $\mathcal{E}$ of eigenvectors and generalized eigenvectors of $\mathcal{R}_S$ need not be complete. For example consider a dB-space $\mathcal{H}$ with $1 \in (\operatorname{Assoc}\mathcal{H}) \setminus \mathcal{H}$. By Proposition 2.3 the operator $\mathcal{R}_1$ has no eigenvectors. However, in two special situations a completeness result holds. The following statements answer the question on completeness of eigenvectors in our particular situation. They complement classical results on completeness of eigenvectors such as [KL, K, L, M].

The first case is easily explained, it follows immediately from Proposition 2.3. Denote by $\mathbb{C}[z]$ the set of all polynomials with complex coefficients.

3.1. Proposition. *Let $\mathcal{H}$ be a dB-space. Assume that $\mathbb{C}[z] \subseteq \mathcal{H}$ and that $S \in \mathbb{C}[z]$, $S(0) \neq 0$. Then*

$$\begin{aligned} \operatorname{span}\mathcal{E} &= \mathbb{C}[z]\,, \\ \operatorname{ran}\mathcal{P}_{\{0\}} &\supseteq S(z)\mathbb{C}[z]\,, \\ \operatorname{ran}\mathcal{P}_{\mathbb{C}\setminus\{0\}} &= \{p \in \mathbb{C}[z] : \deg p < \deg S\}\,. \end{aligned} \tag{3.1}$$

The following are equivalent:

(i) $\operatorname{cls}\mathcal{E} = \mathcal{H}$,

(ii) $\overline{\mathbb{C}[z]} = \mathcal{H}$,

(iii) $\operatorname{ran}\mathcal{P}_{\{0\}} = \operatorname{cls}\{z^k S(z) : k = 0, 1, 2, \ldots\}$.

Proof. Let $n := \deg S$. Since S has exactly n zeros (taking into account multiplicities) and $S(0) \neq 0$, we conclude from Proposition 2.3 that

$$\operatorname{span} \bigcup_{\{\lambda:\, S(\lambda)=0\}} \operatorname{ran}\mathcal{P}_{\{\lambda\}}$$

is an n-dimensional linear space which contains only polynomials with degree less than n. Hence

$$\operatorname{ran}\mathcal{P}_{\mathbb{C}\setminus\{0\}} = \{p \in \mathbb{C}[z] : \deg p < \deg S\}\,.$$

Moreover, since $S \in \operatorname{dom}(\mathcal{S}^k)$ for all $k \in \mathbb{N}$, Proposition 2.3 implies that

$$S(z), (1 + z)S(z), (1 + z + z^2)S(z), \ldots$$

is a Jordan chain of infinite length of $\mathcal{R}_S$ at 0. We have

$$\operatorname{span}\{S(z), (1 + z)S(z), (1 + z + z^2)S(z), \ldots\} = S(z)\mathbb{C}[z].$$

We have proved all relations (3.1) and henceforth also the equivalence of (i), (ii) and (iii). □

Let $\mathcal{H}$ be a dB-space and write $\mathcal{H} = \mathcal{H}(E)$ for some $E \in \mathcal{H}B$. The next result, Theorem 3.3, which is the first main result of this note, deals with functions $S \in \operatorname{span}\{E, E^\#\} =: \mathcal{D}$. It will be proved that generically for such S a completeness result holds true. However, let us first clarify the meaning of the (two-dimensional) space $\mathcal{D}$.

3.2. Lemma. *Let a dB-space $\mathcal{H}$ be given and write $\mathcal{H} = \mathcal{H}(E)$ for some $E \in \mathcal{H}B$. The space $\mathcal{D} = \operatorname{span}\{E, E^\#\}$ does not depend on the choice of E. We can write $\mathcal{D}$ as the disjoint union*

$$\mathcal{D} = \mathcal{G} \cup \mathcal{C} \cup \mathcal{G}^\#,$$

with

$$\mathcal{C} := \{\, \alpha T : \alpha \in \mathbb{C}, T \in \operatorname{Assoc}\mathcal{H}, \mathcal{A}_T \text{ selfadjoint} \,\},$$

$$\mathcal{G} := \{\, \rho H : \rho > 0, H \in \mathcal{H}B, \mathcal{H}(H) = \mathcal{H} \,\}.$$

We have $\mathcal{D} \cap \mathcal{H} \subseteq \mathcal{C}$, $\dim(\mathcal{D} \cap \mathcal{H}) \leq 1$, and $\{S \in \mathcal{D} : \operatorname{Ord}_0 S > \operatorname{Ord}_0 E\} = \operatorname{span}\{S_\phi\}$ for an appropriate $\phi \in [0, \pi)$.

Proof. Let $E, H \in \mathcal{H}B$ and write $E = A - iB$, $H = K - iL$, with A, B, K, L real. Then, by [KW1, Corollary 6.2], we have $\mathcal{H}(E) = \mathcal{H}(H)$ if and only if for some 2×2-matrix M with real entries and $\det M = 1$ the relation $(K, L) = (A, B)M$ holds. Hence

$$\operatorname{span}\{H, H^\#\} = \operatorname{span}\{K, L\} = \operatorname{span}\{A, B\} = \operatorname{span}\{E, E^\#\}.$$

Choose $E \in \mathcal{H}B$ such that $\mathcal{H} = \mathcal{H}(E)$ and write $E = A - iB$. Let $S \in \mathcal{D}$, then $S = uA + vB$ for some appropriate $u, v \in \mathbb{C}$. Then

$$\left(\frac{S + S^\#}{2}, i\frac{S - S^\#}{2} \right) = (A, B) \begin{pmatrix} \operatorname{Re} u & -\operatorname{Im} u \\ \operatorname{Re} v & -\operatorname{Im} v \end{pmatrix}. \tag{3.2}$$

Consider the determinant Δ of the matrix on the right-hand side of (3.2). We have $\Delta = 0$ if and only if $u\overline{v} \in \mathbb{R}$ and hence $S = \alpha S_\phi$ for certain $\alpha \in \mathbb{C}$ and $\phi \in [0, \pi)$. If $\Delta > 0$, the function

$$H(z) := \frac{1}{\sqrt{\Delta}} S(z)$$

belongs to $\mathcal{H}B$ and $\mathcal{H}(H) = \mathcal{H}(E)$. Thus $S \in \mathcal{G}$. In case $\Delta < 0$ apply this argument to $S^\#$ instead of S to conclude that $S \in \mathcal{G}^\#$.

The fact that $\mathcal{D} \cap \mathcal{H}$ is at most one-dimensional and is a subset of $\mathcal{C}$ was proved in [dB, Problem 85]. Since we have $E \in \mathcal{D}$, the set of all functions of $\mathcal{D}$ which vanish at the origin with higher order than E is a at most one-dimensional subspace of $\mathcal{D}$. The present assertion follows since there exists a (unique) value $\phi \in [0, \pi)$ such that $\operatorname{Ord}_0 S_\phi > \operatorname{Ord}_0 E$. □

For a dB-space $\mathcal{H}$ and numbers $\alpha, \beta \leq 0$ denote by $\mathcal{H}_{(\alpha,\beta)}$ the closed linear subspace (cf. [KW3, Lemma 2.6])

$$\mathcal{H}_{(\alpha,\beta)} := \left\{ F \in \mathcal{H} : \operatorname{mt} \frac{F}{E} \leq \alpha, \operatorname{mt} \frac{F^\#}{E} \leq \beta \right\}.$$

3.3. Theorem. *Let $\mathcal{H}$ be a dB-space, $\mathfrak{d}\mathcal{H} = 0$. Assume that $S \in \mathcal{D}$, $S(0) \neq 0$, and put $\tau := \frac{1}{2} \operatorname{mt} S^{-1} S^\#$. Then*

$$\mathcal{E}^\perp = \left\{ F \in \mathcal{H} : \operatorname{Ord}_w F \geq \operatorname{Ord}_w S^\# \text{ for all } w \in \mathbb{C} \right\}. \tag{3.3}$$

We have $\operatorname{cls} \mathcal{E} = \mathcal{H}$ if and only if $\tau = 0$. In the case $\tau \neq 0$

$$\operatorname{cls} \mathcal{E} = \begin{cases} \mathcal{H}_{(0,2\tau)} & , \tau < 0 \\ \mathcal{H}_{(-2\tau,0)} & , \tau > 0 \end{cases},$$

and

$$\mathcal{E}^\perp = S^\#(z) e^{i\tau z} \mathcal{H}(e^{-i|\tau|z}).$$

Proof. If $S \in \mathcal{C}$, we have $\tau = 0$ and $\operatorname{cls} \mathcal{E} = \mathcal{H}$ by [dB, Theorem 22]. In the case $S \in \mathcal{G}$ we may assume without loss of generality that $\mathcal{H} = \mathcal{H}(S)$.

Let $\mathcal{H} = \mathcal{H}(E)$ be given, then Proposition 2.3 implies that

$$\begin{aligned} &(\lambda \in \mathbb{C},\ E(\lambda^{-1}) = 0), \\ &\operatorname{ran} \mathcal{P}_{\{\lambda\}} = \operatorname{span} \left\{ \frac{\partial^k}{\partial z^k} K(\frac{1}{\overline{\lambda}}, z) : 0 \leq k < \operatorname{Ord}_{\frac{1}{\overline{\lambda}}} E \right\}, \end{aligned}$$

and we conclude that (3.3) holds.

Let $F \in \mathcal{H}(E)$ and consider the inner-outer factorizations of $E^{-1}F, E^{-1}E^\# \in N(\mathbb{C}^+)$:

$$\frac{F}{E}(z) = B(z)U(z), \quad \frac{E^\#}{E}(z) = B_1(z) e^{-iz \operatorname{mt}(E^{-1}E^\#)},$$

where B and B_1 denote the Blaschke products to the zeros of F and $E^\#$, respectively, and U is an outer function. By the already proved relation (3.3) we have $F \perp \operatorname{cls} \mathcal{E}$ if and only if $B_1 | B$ in $H^2(\mathbb{C}^+)$. Under the hypothesis $\tau = 0$ this tells us that $F(E^\#)^{-1}$ belongs to $H^2(\mathbb{C}^+)$, and hence that $F = 0$ since

$$\frac{1}{E}\mathcal{H}(E) = H^2(\mathbb{C}^+) \ominus \frac{E^\#}{E} H^2(\mathbb{C}^+).$$

Next assume that $\tau < 0$ and put $E_0(z) := E(z) \exp[-i\tau z]$. Then $\mathcal{H}(E_0) = \exp[-i\tau z] \mathcal{H}_{(2\tau,0)}$ and (cf. [KW3, Theorem 2.7])

$$\mathcal{H}(E) = \mathcal{H}_{(2\tau,0)} \oplus E_0(z) \mathcal{H}(e^{i\tau z}).$$

Hence also

$$\mathcal{H}(E) = \mathcal{H}_{(0,2\tau)} \oplus E^\#(z) e^{i\tau z} \mathcal{H}(e^{i\tau z}).$$

From what we have proved in the previous paragraph ($\mathrm{mt}(E_0^{-1}E_0^\#) = 0$, $\mathrm{Ord}_w E_0 = \mathrm{Ord}_w E$), we know that a function $F \in \mathcal{H}_{(0,2\tau)} = (\mathcal{H}(E_0)\exp[i\tau z])^\#$ with $\mathrm{Ord}_w F \geq \mathrm{Ord}_w E^\#$ for all $w \in \mathbb{C}$ must vanish identically. Hence

$$\mathcal{E}^\perp = E^\#(z)\exp[i\tau z]\mathcal{H}(\exp[i\tau z])$$

and therefore $\mathrm{cls}\,\mathcal{E} = \mathcal{H}_{(0,2\tau)}$.

Finally let us turn to the case that $S \in \mathcal{G}^\#$. Applying the already proved result to the function $S^\#$ we obtain the assertion of the theorem also in this case.

□

3.4. *Remark.* We would like to mention that the part of Theorem 3.3 which states that $\mathrm{cls}\,\mathcal{E} = \mathcal{H}$ if and only if $\tau = 0$, could also be approached differently. In fact it can be deduced from a statement which is asserted without a proof in [GT]. However, the approach chosen here is self-contained and gives a more detailed result.

The case $\tau \neq 0$ in Theorem 3.3 is actually exceptional: Put $\mathcal{D}_0 := \{S \in \mathcal{D} : \mathrm{mt}(S^{-1}S^\#) \neq 0\} \cup \{0\}$, then

3.5. Lemma. *We have either* $\mathcal{D}_0 = \{0\}$ *or*

$$\mathcal{D}_0 = \mathrm{span}\{H_0\} \cup \mathrm{span}\{H_0^\#\},$$

for some $H_0 \in \mathcal{G}$.

Proof. Assume that $H_0 \in \mathcal{H}B$ generates the space $\mathcal{H}$ and satisfies $\mathrm{mt}(H_0^{-1}H_0^\#) < 0$. All other functions $H \in \mathcal{H}B$ with $\mathcal{H}(H) = \mathcal{H}$ are obtained as ($H = K - iL$, $H_0 = K_0 - iL_0$)

$$(K, L) = (K_0, L_0)M,$$

where M runs through the group of all real 2×2-matrices with determinant 1.

Every such matrix M can be factorized uniquely as

$$M = \begin{pmatrix} \cos\gamma & \sin\gamma \\ -\sin\gamma & \cos\gamma \end{pmatrix} \begin{pmatrix} \lambda & t \\ 0 & \frac{1}{\lambda} \end{pmatrix}, \tag{3.4}$$

with $\gamma \in [0, 2\pi)$, $\lambda > 0$ and $t \in \mathbb{R}$.

If the second factor in (3.4) is not present, i.e., if $(\lambda, t) = (1, 0)$, we have $H(z) = H_0(z)e^{-i\gamma}$. We see that $H \in \mathrm{span}\{H_0\}$ and thus as well $\mathrm{mt}(H^{-1}H^\#) < 0$.

Assume that $(\lambda, t) \neq (1, 0)$. Put $H_1 := K_1 - iL_1$ where

$$(K_1, L_1) := (K_0, L_0)\begin{pmatrix} \cos\gamma & \sin\gamma \\ -\sin\gamma & \cos\gamma \end{pmatrix}.$$

We already saw that $\mathrm{mt}(H_1^{-1}H_1^\#) < 0$. The functions H_1 and H are connected by

$$(K, L) = (K_1, L_1)\begin{pmatrix} \lambda & t \\ 0 & \frac{1}{\lambda} \end{pmatrix}.$$

We compute

$$\frac{H^\#}{H_1^\#} = \frac{\lambda K_1 + i(\frac{1}{\lambda}L_1 + tK_1)}{K_1 + iL_1} = \lambda + i\frac{(\frac{1}{\lambda} - \lambda)L_1 + tK_1}{K_1 + iL_1}$$

$$= \lambda + \frac{i}{\sqrt{(\frac{1}{\lambda} - \lambda)^2 + t^2}} \cdot \frac{S_{\phi,1}}{H_1^\#},$$

for a certain $\phi \in [0, 2\pi)$. Since

$$\operatorname{mt} \frac{S_{\phi,1}}{H_1^\#} = \operatorname{mt} \frac{H_1}{H_1^\#} > 0,$$

it follows that

$$\operatorname{mt} \frac{H^\#}{H_1^\#} = \operatorname{mt} \frac{H_1}{H_1^\#}.$$

Since both, H and H_1, generate the same dB-Hilbert space, we must have $\operatorname{mt}(H_1^{-1}H) = 0$ and conclude that

$$\operatorname{mt} \frac{H^\#}{H} = \operatorname{mt} \left[\frac{H^\#}{H_1^\#} \cdot \frac{H_1^\#}{H_1} \cdot \frac{H_1}{H}\right] = 0.$$

We have proved that the set of all functions $H \in \mathcal{HB}$, $\mathcal{H}(H) = \mathcal{H}$, with $\operatorname{mt}(H^{-1}H^\#) < 0$ is either empty or of the form $H_0(z)e^{-i\gamma}$, $\gamma \in [0, 2\pi)$. From this knowledge the assertion of the lemma can be easily deduced: First note that, by $S_\phi = S_\phi^\#$, we have $\operatorname{mt}(S^{-1}S^\#) = 0$ for all functions $S \in \mathcal{C} \setminus \{0\}$ and hence $\mathcal{D}_0 \cap \mathcal{C} = \{0\}$. Moreover, by $\operatorname{mt}(S^{-1}S^\#) = -\operatorname{mt}[(S^\#)^{-1}S]$, it suffices to determine $\mathcal{D}_0 \cap \mathcal{G}$. Finally, since for $\rho > 0$ we have $\operatorname{mt}[(\rho S)^{-1}(\rho S)^\#] = \operatorname{mt}(S^{-1}S^\#)$, we are in the case $H \in \mathcal{HB}$, $\mathcal{H}(H) = \mathcal{H}$. □

4. s-numbers

In this section we investigate more closely the operators $\mathcal{R}_S$ for $S \in \mathcal{D}$. The next lemma is immediate from [S, §4], and will therefore stated without a proof:

4.1. Lemma. *Assume that $S \in \mathcal{D}$, $S(0) \neq 0$. Then*

$$\left[(\mathcal{A}_S - \overline{w})^{-1}\right]^* = (\mathcal{A}_{S^\#} - w)^{-1},\ w \in \mathbb{C},\ S(\overline{w}) \neq 0. \tag{4.1}$$

In the case $S \in \mathcal{C} \setminus \{0\}$, $S(0) \neq 0$, the operator $\mathcal{R}_S$ is selfadjoint. Hence the Schmidt-representation of $\mathcal{R}_S$ can be read off its spectral representation:

$$\mathcal{R}_S = \sum_{\{\lambda \in \mathbb{C}:\, S(\lambda) = 0\}} \frac{1}{\lambda}(., \Phi_\lambda)\Phi_\lambda\,,$$

with $\Phi_\lambda = K(\lambda, z)K(\lambda, \lambda)^{-\frac{1}{2}}$. If $S \in \mathcal{D} \setminus \mathcal{C}$, then $\mathcal{R}_S$ is no longer selfadjoint. However, let us remark the following (cf. [dB, Theorem 27])

4.2. Lemma. *Assume that $S \in \mathcal{D} \setminus \mathcal{C}$. Then either $\mathcal{R}_S$ or $-\mathcal{R}_S$ is dissipative, depending on whether $S \in \mathcal{G}$ or $S \in \mathcal{G}^\#$.*

Proof. It suffices to prove that $\mathcal{R}_E$ is dissipative whenever $\mathcal{H} = \mathcal{H}(E)$. We compute (cf. (2.1))

$$\begin{aligned}\operatorname{Im}(\mathcal{R}_E F, F) &= \left(\frac{\mathcal{R}_E - \mathcal{R}_E^*}{2i} F, F\right) = \left(\frac{\mathcal{R}_E - \mathcal{R}_{E^\#}}{2i} F, F\right)\\ &= \left(\frac{E^\#(z)E(0) - E(z)E^\#(0)}{2iE(0)E^\#(0)z} F(0), F(z)\right)\\ &= \pi \frac{F(0)}{|E(0)|^2}\left(K(0,z), F(z)\right) = \pi \frac{|F(0)|^2}{|E(0)|^2}\,.\end{aligned}$$

□

In order to compute the Schmidt-representation of $\mathcal{R}_S$ we need some information about the spectrum of $\mathcal{R}_S^* \mathcal{R}_S$.

4.3. Lemma. *Let $S, T \in (\operatorname{Assoc}\mathcal{H}) \setminus \mathcal{H}$, $S(0), T(0) \neq 0$, and put*

$$U_{S,T}(z) := T(z)S(-z) + T(-z)S(z)\,.$$

Then

$$\sigma(\mathcal{R}_S \mathcal{R}_T) \setminus \{0\} = \left\{ \lambda \in \mathbb{C} : U_{S,T}\left(\frac{1}{\sqrt{\lambda}}\right) = 0 \right\}. \tag{4.2}$$

We have $0 \notin \sigma_p(\mathcal{R}_S \mathcal{R}_T)$. Denote by $\mathcal{E}_\lambda$ the geometric eigenspace at a nonzero eigenvalue $\lambda \in \sigma(\mathcal{R}_S \mathcal{R}_T) \setminus \{0\}$. Then $\dim \mathcal{E}_\lambda = 1, 2$ where the latter case appears if and only if

$$T\left(\pm\frac{1}{\sqrt{\lambda}}\right) = S\left(\pm\frac{1}{\sqrt{\lambda}}\right) = 0\,.$$

If $\dim \mathcal{E}_\lambda = 2$, then

$$\mathcal{E}_\lambda = \operatorname{span}\left\{ \frac{T(z)}{1 - \lambda z^2}, \frac{zS(z)}{1 - \lambda z^2} \right\}. \tag{4.3}$$

Let $\dim \mathcal{E}_\lambda = 1$. If $\left(T(\frac{1}{\sqrt{\lambda}}), S(\frac{1}{\sqrt{\lambda}})\right) \neq (0,0)$, then

$$\mathcal{E}_\lambda = \operatorname{span}\left\{ \frac{1}{1 - \lambda z^2}\left[T(z)\frac{1}{\sqrt{\lambda}} S\left(\frac{1}{\sqrt{\lambda}}\right) - zS(z)T\left(\frac{1}{\sqrt{\lambda}}\right)\right]\right\},$$

if $\left(T(-\frac{1}{\sqrt{\lambda}}), S(-\frac{1}{\sqrt{\lambda}})\right) \neq (0,0)$, then

$$\mathcal{E}_\lambda = \operatorname{span}\left\{ \frac{1}{1 - \lambda z^2}\left[T(z)\frac{1}{\sqrt{\lambda}} S\left(-\frac{1}{\sqrt{\lambda}}\right) - zS(z)T\left(-\frac{1}{\sqrt{\lambda}}\right)\right]\right\}. \tag{4.4}$$

Proof. Since for any nonzero constants a, b we have $\mathcal{R}_{aS} = \mathcal{R}_S$, $\mathcal{R}_{bT} = \mathcal{R}_T$, and $U_{aS,bT}(z) = abU_{S,T}(z)$, we can assume without loss of generality that $S(0) =$

$T(0) = 1$. We compute $\mathcal{R}_S \mathcal{R}_T$:

$$\begin{aligned}\mathcal{R}_S \mathcal{R}_T F(z) &= \frac{1}{z}\left[\frac{F(z) - T(z)F(0)}{z} - S(z)\left(F'(0) - T'(0)F(0)\right)\right] \\ &= \frac{1}{z^2}\left[F(z) - T(z)F(0) - zS(z)\left(F'(0) - T'(0)F(0)\right)\right].\end{aligned}$$

It is readily seen from this formula that $\ker \mathcal{R}_S \mathcal{R}_T = \{0\}$, since the hypothesis $S, T \in (\operatorname{Assoc} \mathcal{H}) \setminus \mathcal{H}$ implies that the functions $zS(z)$ and $T(z)$ are linearly independent and $\operatorname{span}\{zS(z), T(z)\} \cap \mathcal{H} = \{0\}$.

Let $\mu \in \mathbb{C}$ and assume that $F \in \ker(\mathcal{R}_S \mathcal{R}_T - \mu)$, $F \not\equiv 0$. Then we must have

$$\begin{aligned}F(z)(1 - \mu z^2) &= T(z)F(0) + zS(z)\left(F'(0) - T'(0)F(0)\right) \\ &= T(z)\phi_0 + zS(z)\phi_1,\end{aligned} \tag{4.5}$$

with certain $\phi_0, \phi_1 \in \mathbb{C}$, not both zero. In the sequel always put $\nu := (\sqrt{\mu})^{-1}$ where we choose, e.g., the square root lying in the right half-plane. From (4.5) we see that

$$\begin{array}{rcrl} T(\nu)\phi_0 & + & \nu S(\nu)\phi_1 & = 0 \\ T(-\nu)\phi_0 & - & \nu S(-\nu)\phi_1 & = 0 \end{array}. \tag{4.6}$$

Hence,

$$0 = \det\begin{pmatrix} T(\nu) & \nu S(\nu) \\ T(-\nu) & -\nu S(-\nu)\end{pmatrix} = (-\nu) U_{S,T}(\nu),$$

and we conclude that the inclusion "$\subseteq$" in (4.2) holds.

Conversely, assume that $\mu \in \mathbb{C} \setminus \{0\}$ and that $U_{S,T}(\nu) = 0$ where again $\nu = (\sqrt{\mu})^{-1}$. Then the system (4.6) of linear equations has nontrivial solutions. If (ϕ_0, ϕ_1) is any such nontrivial solution of (4.6), the function

$$F(z) := \frac{1}{1 - \mu z^2}\left[T(z)\phi_0 + zS(z)\phi_1\right]$$

is entire and belongs to the space $\mathcal{H}$. We have $F(0) = \phi_0$ and

$$F'(0) = \lim_{z \to 0} \frac{F(z) - \phi_0}{z} = \lim_{z \to 0} \frac{1}{1 - \mu z^2}\left[\frac{T(z) - 1}{z}\phi_0 + S(z)\phi_1\right] = T'(0)\phi_0 + \phi_1.$$

Hence F satisfies the first equality in (4.5) and thus belongs to $\ker(\mathcal{R}_S \mathcal{R}_T - \mu)$.

The formulas (4.3) and (4.4) for $\mathcal{E}_\lambda$ now follow on solving the linear system (4.6). □

4.4. Lemma. *Assume that $E \in \mathcal{H}B$, $E(0) = 1$, and denote by $K(w, z)$ the reproducing kernel of the space $\mathcal{H}(E)$. Let $z, w \in \mathbb{C}$, $w \neq 0$, $z \neq \overline{w}$. Then*

$$\mathcal{R}_E K(w, z) = \frac{1}{2\pi i z \overline{w}(\overline{w} - z)}\left[\overline{w} E(\overline{w})\left(E(z) - E^{\#}(z)\right) - zE(z)\left(E(\overline{w}) - E^{\#}(\overline{w})\right)\right].$$

Proof. We substitute in the definition of $\mathcal{R}_E$:

$$\mathcal{R}_E K(w,z) = \frac{1}{z}\left[\frac{E(z)E^\#(\overline{w}) - E(\overline{w})E^\#(z)}{2\pi i(\overline{w}-z)} - E(z)\frac{E(\overline{w}) - E^\#(\overline{w})}{2\pi i\overline{w}}\right]$$

$$= \frac{1}{2\pi i z\overline{w}(\overline{w}-z)}\left[\overline{w}E(z)E^\#(\overline{w}) - \overline{w}E(\overline{w})E^\#(z)\right.$$
$$\left. -(\overline{w}-z)\left(E(z)E^\#(\overline{w}) - E(z)E(\overline{w})\right)\right]$$

$$= \frac{1}{2\pi i z\overline{w}(\overline{w}-z)}\left[-\overline{w}E(\overline{w})E^\#(z) + \overline{w}E(z)E(\overline{w}) + zE(z)E^\#(\overline{w}) - zE(z)E(\overline{w})\right]$$

$$= \frac{1}{2\pi i z\overline{w}(\overline{w}-z)}\left[\overline{w}E(\overline{w})\left(E(z) - E^\#(z)\right) - zE(z)\left(E(\overline{w}) - E^\#(\overline{w})\right)\right]. \qquad \square$$

4.5. Theorem. *Let $\mathcal{H}$ be a dB-space, $\mathfrak{d}\mathcal{H} = 0$, let $S \in \mathcal{D}\setminus\mathcal{C}$ and let $\rho > 0$ be such that $\mathcal{H} = \mathcal{H}(\rho^{-1}S)$ ($\mathcal{H} = \mathcal{H}(\rho^{-1}S^\#)$, respectively). The zeros of the function*

$$U_{S,S^\#}(z) = S(z)S^\#(-z) + S(-z)S^\#(z)$$

are real, simple, nonzero, and symmetric with respect to 0. Denote the sequence of positive zeros of $U_{S,S^\#}$ by

$$0 < \mu_1 < \mu_2 < \cdots.$$

Then the s-numbers of the operator $\mathcal{R}_S$ are given as

$$s_j(\mathcal{R}_S) = \frac{1}{\mu_j},\ j = 1, 2, \ldots$$

In the Schmidt-representation $\mathcal{R}_S = \sum s_j(.,\phi_j)\psi_j$ we have

$$\phi_j(z) = \rho\sqrt{2\pi}\frac{S(-\mu_j)K(\mu_j,z) + S(\mu_j)K(-\mu_j,z)}{|S(\mu_j)S^\#(-\mu_j)U'_{S,S^\#}(\mu_j)|^{\frac{1}{2}}}, \tag{4.7}$$

$$\psi_j(z) = \rho\sqrt{2\pi}\frac{S(-\mu_j)K(\mu_j,z) - S(\mu_j)K(-\mu_j,z)}{|S(\mu_j)S^\#(-\mu_j)U'_{S,S^\#}(\mu_j)|^{\frac{1}{2}}}. \tag{4.8}$$

Proof. Let us first assume that $\mathcal{H} = \mathcal{H}(E)$ and determine the Schmidt-representation of $\mathcal{R}_E$. By Lemma 4.3 the zeros of $U := U_{E,E^\#}$ ($= U_{E^\#,E}$) are exactly the eigenvalues of the selfadjoint operator $\mathcal{R}_E\mathcal{R}_{E^\#}$ and, hence, are all real. The fact that all zeros of U are simple will follow from the following relation which holds for all λ with $U(\lambda) = 0$:

$$\frac{1}{2\pi i}U'(\lambda)E(\lambda)E^\#(-\lambda) = \|E(-\lambda)K(\lambda,z) \pm E(\lambda)K(-\lambda,z)\|^2. \tag{4.9}$$

In order to establish this formula we compute the norm on the right-hand side of (4.9):

$$\|E(-\lambda)K(\lambda,z) \pm E(\lambda)K(-\lambda,z)\|^2$$
$$= |E(-\lambda)|^2K(\lambda,\lambda) + |E(\lambda)|^2K(-\lambda,-\lambda)$$
$$\pm\left[E(-\lambda)\overline{E(\lambda)}K(\lambda,-\lambda) + E(\lambda)\overline{E(-\lambda)}K(-\lambda,\lambda)\right].$$

Since $\lambda \in \mathbb{R}$ and $E(\lambda)E^{\#}(-\lambda) = -E(-\lambda)E^{\#}(\lambda)$ the term in the square bracket vanishes:

$$E(-\lambda)E^{\#}(\lambda)\left(\frac{E(-\lambda)E^{\#}(\lambda) - E(\lambda)E^{\#}(-\lambda)}{2\pi i(2\lambda)}\right) + E(\lambda)E^{\#}(-\lambda)\left(\frac{E(\lambda)E^{\#}(-\lambda) - E(-\lambda)E^{\#}(\lambda)}{2\pi i(-2\lambda)}\right)$$
$$= \frac{1}{2\pi i\lambda}\left[\left(E(-\lambda)E^{\#}(\lambda)\right)^2 - \left(E(\lambda)E^{\#}(-\lambda)\right)^2\right] = 0\,.$$

The first two summands compute as

$$E(-\lambda)E^{\#}(-\lambda)\left(\frac{E'(\lambda)E^{\#}(\lambda) - E(\lambda)E'(\lambda)^{\#}}{-2\pi i}\right) + E(\lambda)E^{\#}(\lambda)\left(\frac{E'(-\lambda)E^{\#}(-\lambda) - E(-\lambda)E'(-\lambda)^{\#}}{-2\pi i}\right)$$
$$= \frac{E(-\lambda)E^{\#}(\lambda)}{-2\pi i}\left(E'(\lambda)E^{\#}(-\lambda) - E(\lambda)E'(-\lambda)^{\#}\right) + \frac{E(\lambda)E^{\#}(-\lambda)}{-2\pi i}\left(-E(-\lambda)E'(\lambda)^{\#} + E'(-\lambda)E^{\#}(\lambda)\right)$$
$$= \frac{E(-\lambda)E^{\#}(\lambda)}{-2\pi i}\left[E'(\lambda)E^{\#}(-\lambda) - E(\lambda)E'(-\lambda)^{\#} + E(-\lambda)E'(\lambda)^{\#} - E'(-\lambda)E^{\#}(\lambda)\right] = \frac{E(-\lambda)E^{\#}(\lambda)}{-2\pi i}U'(\lambda)\,.$$

By definition the singular values of $\mathcal{R}_E$ are the positive roots of the eigenvalues of $\mathcal{R}_E^*\mathcal{R}_E = \mathcal{R}_{E^{\#}}\mathcal{R}_E$ and hence $s_j(\mathcal{R}_E) = \mu_j^{-1}$.

The elements ϕ_j in the Schmidt-representation for $\mathcal{R}_E$ are the members of the orthogonal system of eigenvectors of $\mathcal{R}_E^*\mathcal{R}_E$. Note here that by Lemma 4.3 the operator $\mathcal{R}_E^*\mathcal{R}_E$ has only simple eigenvalues. Hence ϕ_j must be a scalar multiple of the function

$$\frac{1}{\mu_j^2 - z^2}\left[E(z)\mu_j E^{\#}(\mu_j) - E^{\#}(z)zE(\mu_j)\right]\,.$$

We compute ($U(\lambda) = 0$)

$$E(-\lambda)K(\lambda, z) + E(\lambda)K(-\lambda, z)$$
$$= E(-\lambda)\frac{E(z)E^{\#}(\lambda) - E(\lambda)E^{\#}(z)}{2\pi i(\lambda - z)} + E(\lambda)\frac{E(z)E^{\#}(-\lambda) - E(-\lambda)E^{\#}(z)}{2\pi i(-\lambda - z)}$$
$$= \frac{E(-\lambda)}{2\pi i}\left[\frac{E(z)E^{\#}(\lambda) - E(\lambda)E^{\#}(z)}{\lambda - z} + \frac{E(z)E^{\#}(\lambda) + E(\lambda)E^{\#}(z)}{\lambda + z}\right]$$
$$= \frac{E(-\lambda)}{\pi i}\,\frac{E(z)E^{\#}(\lambda)\lambda - E(\lambda)E^{\#}(z)z}{\lambda^2 - z^2}\,.$$

By virtue of (4.9) we may take ϕ_j as stated in (4.7).

It remains to identify the elements ψ_j. First note that they must form an orthonormal system of eigenvalues of $\mathcal{R}_E\mathcal{R}_E^*$. Hence ψ_j is a scalar multiple of

$$\frac{1}{\mu_j^2 - z^2}\left[E(z)zE^\#(\mu_j) - E^\#(z)\mu_j E(\mu_j)\right] .$$

A similar computation as in the previous paragraph shows that ψ_j henceforth is a scalar multiple of the function

$$E(-\lambda)K(\lambda, z) - E(\lambda)K(-\lambda, z) .$$

In order to prove (4.8) it is therefore sufficient to evaluate at a point $z = z_0$ with $\psi_j(z_0) \neq 0$ in the equation

$$\mathcal{R}_E\phi_j = \frac{1}{\mu_j}\psi_j . \tag{4.10}$$

We have $(U(\lambda) = 0)$

$$\begin{aligned} &E(-\lambda)K(\lambda, 0) - E(\lambda)K(-\lambda, 0) \\ &= E(-\lambda)\frac{E^\#(\lambda) - E(\lambda)}{2\pi i\lambda} - E(\lambda)\frac{E^\#(-\lambda) - E(-\lambda)}{2\pi i(-\lambda)} = \frac{-E(-\lambda)E(\lambda)}{\pi i\lambda} . \end{aligned}$$

Note that this value is nonzero. In order to evaluate the left-hand side of (4.10) we use Lemma 4.4:

$$\begin{aligned} &\mathcal{R}_E\left(E(-\lambda)K(\lambda, z) + E(\lambda)K(-\lambda, z)\right) \\ &= \frac{E(-\lambda)}{2\pi i\lambda(\lambda - z)}\left[\lambda E(\lambda)\frac{E(z) - E^\#(z)}{z} - E(z)\left(E(\lambda) - E^\#(\lambda)\right)\right] \\ &\quad + \frac{E(\lambda)}{2\pi i\lambda(\lambda + z)}\left[-\lambda E(-\lambda)\frac{E(z) - E^\#(z)}{z} - E(z)\left(E(-\lambda) - E^\#(-\lambda)\right)\right] . \end{aligned}$$

Letting z tend to 0 we see that the first summands of each square bracket cancel. For λ with $U(\lambda) = 0$, therefore,

$$\mathcal{R}_E\left(E(-\lambda)K(\lambda, 0) + E(\lambda)K(-\lambda, 0)\right) = \frac{-E(-\lambda)E(\lambda)}{\pi\lambda^2} ,$$

and we obtain the desired representation of $\mathcal{R}_E$.

Let $S \in \mathcal{D}\setminus\mathcal{C}$ be given. We reduce the assertion to the already proved case.

Assume that $S \in \mathcal{G}$ and write $S = \rho E$ with $\rho > 0$ and $\mathcal{H} = \mathcal{H}(E)$. We saw that

$$\mathcal{R}_S = \mathcal{R}_E = \sum s_j\left(., \phi_j(E)\right)\psi_j(E) ,$$

where $\phi_j(E)$ and $\psi_j(E)$ denote the elements (4.7) and (4.8) with E instead of S. Then

$$\begin{aligned} \phi_j(E) &= \sqrt{2\pi}\frac{E(-\mu_j)K(\mu_j, z) + E(\mu_j)K(-\mu_j, z)}{|E(\mu_j)E^\#(-\mu_j)U'_{E,E^\#}(\mu_j)|^{\frac{1}{2}}} \\ &= \sqrt{2\pi}\frac{\rho^{-1}\left(S(-\mu_j)K(\mu_j, z) + S(\mu_j)K(-\mu_j, z)\right)}{\rho^{-2}|S(\mu_j)S^\#(-\mu_j)U'_{S,S^\#}(\mu_j)|^{\frac{1}{2}}} . \end{aligned}$$

The same computation applies to $\psi_j(E)$ and hence the assertion of the theorem follows for $S \in \mathcal{G}$.

Assume finally that $S \in \mathcal{G}^\#$ and write $S = \rho E^\#$ with $\rho > 0$ and $\mathcal{H} = \mathcal{H}(E)$. We have

$$\mathcal{R}_S = \mathcal{R}_{E^\#} = (\mathcal{R}_E)^* = \sum s_j\,(.,\psi_j(E))\,\phi_j(E).$$

The element $\phi_j(E)$ computes as

$$\phi_j(E) = \rho\sqrt{2\pi}\,\frac{\overline{S(-\mu_j)}K(\mu_j,z) + \overline{S(\mu_j)}K(-\mu_j,z)}{|S^\#(\mu_j)S(-\mu_j)U'_{S^\#,S}(\mu_j)|^{\frac{1}{2}}}.$$

Since $U_{S^\#,S}(\mu_j) = 0$, it follows that this expression is equal to

$$\rho\sqrt{2\pi}\,\frac{S(-\mu_j)K(\mu_j,z) - S(\mu_j)K(-\mu_j,z)}{|S(\mu_j)S^\#(-\mu_j)U'_{S,S^\#}(\mu_j)|^{\frac{1}{2}}}\cdot\frac{\overline{S(-\mu_j)}}{S(-\mu_j)}.$$

Analogously

$$\psi_j(E) = \rho\sqrt{2\pi}\,\frac{S(-\mu_j)K(\mu_j,z) + S(\mu_j)K(-\mu_j,z)}{|S(\mu_j)S^\#(-\mu_j)U'_{S,S^\#}(\mu_j)|^{\frac{1}{2}}}\cdot\frac{\overline{S(-\mu_j)}}{S(-\mu_j)},$$

and henceforth also in the case $S \in \mathcal{G}^\#$ the assertion of the theorem follows. □

4.6. Corollary. *Let E be a Hermite-Biehler function of finite order $\rho > 1$ and assume that E can be written as*

$$E(z) = e^{-iaz}\prod_{n\in\mathbb{N}}\left(1 - \frac{z}{z_n}\right)\exp\left[z\,\mathrm{Re}\,\frac{1}{z_n} + \cdots + \frac{z^p}{p}\,\mathrm{Re}\,\frac{1}{z_n^p}\right], \tag{4.11}$$

with $a \geq 0$ and $z_n \in \mathbb{C}^-$, compare [KW3, Lemma 3.12]. *Then the order of the function*

$$F(z) := E(z)E^\#(-z) + E(-z)E^\#(z)$$

is also equal to ρ.

Proof. First of all let us note that, since E is of the form (4.11) and we assume that $\rho > 1$, the order of the product $\prod_{n\in\mathbb{N}}(1 - \frac{z}{z_n})\exp[z\,\mathrm{Re}\,\frac{1}{z_n} + \cdots + \frac{z^p}{p}\,\mathrm{Re}\,\frac{1}{z_n^p}]$ must also be equal to ρ.

The fact that the order of F does not exceed the order of E is clear. Assume that the order of F is $\rho' < \rho$ and choose $\epsilon > 0$ such that $1 < \rho' + \epsilon < \rho' + 2\epsilon < \rho$.

Denote by μ_k the sequence of zeros of F, then the series $\sum_k |\mu_k|^{-(\rho'+\epsilon)}$ is convergent. By the above theorem the operator $\mathcal{R}_E$ in the space $\mathcal{H}(E)$ belongs to the class $\mathfrak{S}_{\rho'+\epsilon}$. By the proof of Lemma 2.1 for every $S \in \mathrm{Assoc}\,\mathcal{H}(E)$ we have $\mathcal{R}_S \in \mathfrak{S}_{\rho'+\epsilon}$.

Consider the function $A(z) := \frac{1}{2}(E(z) + E^\#(z))$, and denote by (λ_k) the sequence of its zeros. Since $\mathcal{R}_A$ is a selfadjoint operator of the class $\mathfrak{S}_{\rho'+\epsilon}$ and its spectrum coincides with $\{\lambda_k\}$, we know that $\sum_k |\lambda_k|^{-(\rho'+\epsilon)}$ converges. By [KW3, Theorem 3.17], applied with the growth function $\lambda(r) := r^{\rho'+2\epsilon}$, we obtain that there exists a Hermite-Biehler function E_1 of order $\rho'' \leq \rho + 2\epsilon$ and a real and zero free function C such that $\mathcal{H}(E) = C \cdot \mathcal{H}(E_1)$. By [KW3, Lemma 2.4], it follows that

$$\mathcal{H}(E_1) = \frac{1}{C}\mathcal{H}(E) = \mathcal{H}(\frac{1}{C}E),$$

and in particular $C^{-1}E \in \operatorname{Assoc}\mathcal{H}(E_1)$. Thus the order of $C^{-1}E$ cannot exceed the order of ρ'' of E_1. However, since C is zero free and E is of the form (4.11), certainly the order of $C^{-1}E$ is at least equal to ρ and we have reached a contradiction. □

The sequence of zeros of the function $U_{S,S^\#}$ can be obtained from the knowledge of a phase function. Recall from [dB, Problem 48] that, if $E \in \mathcal{H}B$, a *phase function* is a continuous function $\varphi : \mathbb{R} \to \mathbb{R}$ such that

$$E(t)e^{i\varphi(t)} \in \mathbb{R}.$$

By this relation the function φ is uniquely determined up to integer multiples of π.

4.7. Lemma. *Let $E \in \mathcal{H}B$ be given and let φ be a phase function of E. Then $U_{E,E^\#}(t) = 0$ if and only if*

$$\varphi(t) - \varphi(-t) \equiv \frac{\pi}{2} \mod \pi\,.$$

Proof. We have for $t \in \mathbb{R}$

$$U_{E,E^\#}(t) = E^\#(-t)E^\#(t)\left[\frac{E(t)}{E^\#(t)} + \frac{E(-t)}{E^\#(-t)}\right].$$

Both summands in the square bracket are complex numbers of modulus 1. Hence their sum vanishes if and only if their arguments differ by an odd multiple of π, i.e.,

$$\arg\frac{E(t)}{E^\#(t)} \equiv \arg\frac{E(-t)}{E^\#(-t)} + \pi \mod 2\pi.$$

Since

$$\arg\frac{E(t)}{E^\#(t)} \equiv -2\varphi(t) \mod 2\pi, \quad \arg\frac{E(-t)}{E^\#(-t)} \equiv -2\varphi(-t) \mod 2\pi,$$

we obtain

$$-2\varphi(t) \equiv -2\varphi(-t) + \pi \mod 2\pi.$$ □

4.8. *Remark.* Assume that E satisfies the functional equation $E^\#(z) = E(-z)$. Then $E(0) \in \mathbb{R}$, hence we may choose a phase function φ such that $\varphi(0) = 0$. Then φ is an odd function and hence $U_{E,E^\#}(t) = 0$ if and only if

$$\varphi(t) \equiv \frac{\pi}{4} \mod \frac{\pi}{2}.$$

This observation is explained by the fact that in the present case the function $U_{E,E^\#}$ can be factorized as

$$U_{E,E\#}(z) = 4S_{\frac{\pi}{4}}(z)S_{\frac{3\pi}{4}}(z).$$

5. Spaces symmetric about the origin

Let us consider the situation that the dB-space $\mathcal{H}$ is *symmetric with respect to the origin* (cf. [dB]), i.e., has the property that the mapping $F(z) \mapsto F(-z)$ is an isometry of $\mathcal{H}$ into itself. By [dB, Theorem 47] an equivalent property is that $\mathcal{H}$ can be written as $\mathcal{H} = \mathcal{H}(E)$ with some $E \in \mathcal{H}B$ satisfying

$$E^{\#}(z) = E(-z), \ z \in \mathbb{C}. \tag{5.1}$$

This symmetry property can also be read off the reproducing kernel $K(w, z)$ of the space $\mathcal{H}$: In order that $\mathcal{H}$ is symmetric about the origin it is necessary and sufficient that

$$K(w, z) = K(-w, -z), \ w, z \in \mathbb{C}\,. \tag{5.2}$$

A space $\mathcal{H}$ being symmetric about the origin can be decomposed orthogonally as

$$\mathcal{H} = \mathcal{H}^g \oplus \mathcal{H}^u\,, \tag{5.3}$$

where

$$\mathcal{H}^g := \{\, F \in \mathcal{H} : \, F(-z) = F(z) \,\}\,,$$
$$\mathcal{H}^u := \{\, F \in \mathcal{H} : \, F(-z) = -F(z) \,\}\,.$$

The orthogonal projections $P^g : \mathcal{H} \to \mathcal{H}^g$ and $P^u : \mathcal{H} \to \mathcal{H}^u$ are given by

$$(P^g F)(z) = \frac{F(z) + F(-z)}{2}, \ (P^u F)(z) = \frac{F(z) - F(-z)}{2}\,. \tag{5.4}$$

In particular the reproducing kernel K^g of $\mathcal{H}^g$ (K^u of $\mathcal{H}^u$, respectively) is given by

$$\begin{aligned} K^g(w, z) &= \frac{1}{2}\left(K(w, z) + K(w, -z)\right) = \frac{1}{2}\left(K(w, z) + K(-w, z)\right)\,, \\ K^u(w, z) &= \frac{1}{2}\left(K(w, z) - K(w, -z)\right) = \frac{1}{2}\left(K(w, z) - K(-w, z)\right)\,. \end{aligned} \tag{5.5}$$

Taking (5.3) as a fundamental decomposition $\mathcal{H}$ can be regarded as a Krein space $\langle \mathcal{H}, [.,.] \rangle$, $[.,.] = (\mathcal{J}.,.)$ where the fundamental symmetry $\mathcal{J}$ is given by

$$\mathcal{J} = \begin{pmatrix} I & 0 \\ 0 & -I \end{pmatrix} : \begin{matrix} \mathcal{H}^g \\ \oplus \\ \mathcal{H}^u \end{matrix} \longrightarrow \begin{matrix} \mathcal{H}^g \\ \oplus \\ \mathcal{H}^u \end{matrix}\,.$$

Note that, by the formulas (5.4), $\mathcal{J}$ is nothing else but the isometry $(\mathcal{J}F)(z) = F(-z)$.

5.1. Theorem. *Assume that $\mathcal{H}$, $\mathfrak{d}\mathcal{H} = 0$, is symmetric with respect to the origin and write $\mathcal{H} = \mathcal{H}(E)$ where $E \in \mathcal{H}B^{\times}$ satisfies* (5.1) *and $E(0) > 0$. Then the operator $-i\mathcal{R}_E$ is selfadjoint in the Krein space $\langle \mathcal{H}, [.,.] \rangle$. In fact,*

$$-i\mathcal{J}\mathcal{R}_E = \sum_j (-1)^{j+1} s_j \,(.,\phi_j)\,\phi_j\,, \tag{5.6}$$

where

$$\{(-1)^{j+1}s_j : j=1,2,\dots\} = \left\{w\in\mathbb{C}: S_{\frac{\pi}{2}}(\frac{1}{w})=0\right\}.$$

Let $\lambda\in\sigma(-i\mathcal{R}_E)\backslash\{0\}$ and let e_λ be a corresponding eigenvector. Then e_λ is neutral if and only if either $\lambda\notin\mathbb{R}$ or $-i\lambda^{-1}\in i\mathbb{R}^-$ is a multiple root of E. Denote by $\hat{n}(t)$ the number of zeros of E lying in $[0,-it]$ counted according to their multiplicities. If $-i\lambda^{-1}\in i\mathbb{R}^-$ is a simple root of E, then

$$\operatorname{sgn}[e_\lambda,e_\lambda]=(-1)^{\hat{n}(\lambda^{-1})+1}. \tag{5.7}$$

Proof. Let $\mathcal{R}_E=\sum_j s_j(.,\phi_j)\phi_j$ be the Schmidt-representation of $\mathcal{R}_E$. Then

$$-i\mathcal{J}\mathcal{R}_E=\sum_j(-i)s_j(.,\phi_j)\mathcal{J}\psi_j\,,$$

and hence establishing (5.6) amounts to show that

$$\mathcal{J}\psi_j=i(-1)^{j+1}\phi_j\,. \tag{5.8}$$

From (5.6) selfadjointness with respect to $[.,.]$ follows immediately.

The function ψ_j is given by (4.8) and we obtain from (5.2)

$$\frac{1}{\sqrt{\pi}}|E(\mu_j)|\cdot\left|U'_{E,E^\#}(\mu_j)\right|^{\frac12}\cdot\mathcal{J}\psi_j=\mathcal{J}\left[E(-\mu_j)K(\mu_j,z)-E(\mu_j)K(-\mu_j,z)\right]$$
$$=E(-\mu_j)K(-\mu_j,z)-E(\mu_j)K(\mu_j,z)=\overline{E(\mu_j)}K(\mu_j,z)-\overline{E(\mu_j)}K(-\mu_j,z)\,.$$

Let φ be the phase function with $\varphi(0)=0$, then by Remark 4.8 the numbers $\mu_j=s_j^{-1}$ are such that

$$\varphi(\mu_j)=\frac{\pi}{4}+(j-1)\frac{\pi}{2}\,, \tag{5.9}$$

which means that

$$\arg E(\mu_j)=i\left(\frac{\pi}{4}-j\frac{\pi}{2}\right).$$

From this we obtain

$$\overline{E(\mu_j)}=|E(\mu_j)|\,e^{-i(\frac{\pi}{4}-j\frac{\pi}{2})}=E(\mu_j)e^{-2i(\frac{\pi}{4}-j\frac{\pi}{2})}=E(\mu_j)(-i)(-1)^j\,,$$

and by symmetry $\overline{E(-\mu_j)}=E(-\mu_j)i(-1)^j$. Thus we have

$$\overline{E(\mu_j)}K(\mu_j,z)-\overline{E(\mu_j)}K(-\mu_j,z)$$
$$=i(-1)^{j+1}\left[E(-\mu_j)K(\mu_j,z)+E(\mu_j)K(-\mu_j,z)\right],$$

and (5.8) follows. Next note that, since φ is odd, (5.9) implies

$$\varphi\left((-1)^{j+1}\mu_j\right)=\begin{cases}\frac{\pi}{4}+\frac{j-1}{2}\pi & ,\ j\text{ odd}\\ \frac{\pi}{4}-\frac{j}{2}\pi & ,\ j\text{ even}\end{cases},$$

and hence $(-1)^{j+1}\mu_j$, $j=1,3,5,\dots$, enumerates the positive zeros of $S_{\frac{\pi}{4}}$ and $(-1)^{j+1}\mu_j$, $j=2,4,6,\dots$, the negative zeros of this function.

Let $\lambda\in\sigma(-i\mathcal{R}_E)$ be given. By Proposition 2.3 the geometric eigenspace at λ is one-dimensional and spanned by $e_\lambda:=K(i\overline{\lambda}^{-1},z)$. The assertion concerning neutrality of eigenvectors follows from the selfadjointness of $\mathcal{R}_E$: If $\lambda \notin \mathbb{R}$ the eigenvector must be neutral. If $\lambda \in \mathbb{R}$ and $-i\lambda^{-1}$ is a multiple root of E then by Proposition 2.3 there exists a Jordan chain at λ and, hence, the eigenvector must be neutral. It remains to consider the case that $-i\lambda^{-1}$ is a simple root of E. To this end we compute (put $w := i\overline{\lambda}^{-1}$)

$$[e_\lambda, e_\lambda] = [K(w,z), K(w,z)] = (K(-w,z), K(w,z)) = K(-w,w)\,.$$

Since $w \in i\mathbb{R}^-$, we have

$$K(-w,w) = K(\overline{w},w) = \frac{i}{2\pi}\left(E'(w)E^\#(w) - E(w)E^\#(w)'\right) = iE'(w)\frac{E^\#(w)}{2\pi}\,.$$

From the symmetry relation (5.1) we find that $E(i\mathbb{R}) \subseteq \mathbb{R}$. Since $E(0) > 0$ and E has no zeros in $\mathbb{C}^+$, we have $E(i\mathbb{R}^+) \subseteq \mathbb{R}^+$ and therefore

$$\operatorname{sgn}[e_\lambda, e_\lambda] = \operatorname{sgn} iE'(w)\,.$$

Consider the function $f(t) := E(-it) : [0,\infty) \to \mathbb{R}$. Then $f(0) > 0$ and hence at a simple zero t_0 of f we have

$$\operatorname{sgn} f'(t_0) = (-1)^{\hat{n}(f,t_0)}\,,$$

where $\hat{n}(f,t_0)$ denotes the number of zeros of f in $[0,t_0]$ counted according to their multiplicities. Since $f'(t) = -iE'(-it)$ the relation (5.7) follows. □

Let us conclude with giving the matrix representation of $\mathcal{R}_E$ with respect to the decomposition (5.3).

5.2. Lemma. *With respect to* (5.3) *the operator* $\mathcal{R}_E$ *has the representation*

$$\mathcal{R}_E = \sum_j \frac{4\pi i(-1)^{j+1}s_j}{|U'_{E,E^\#}(\mu_j)|} M_j\,,$$

with

$$M_j = \begin{pmatrix} (\,.\,, K^g(\mu_j,z))\,K^g(\mu_j,z) & i(-1)^j\,(\,.\,, K^u(\mu_j,z))\,K^g(\mu_j,z) \\ i(-1)^j\,(\,.\,, K^g(\mu_j,z))\,K^u(\mu_j,z) & -(\,.\,, K^u(\mu_j,z))\,K^u(\mu_j,z) \end{pmatrix}\,.$$

Proof. We determine the decomposition of ϕ_j with respect to (5.3). Using (5.9) we obtain

$$\begin{aligned}\phi_j &= \frac{\sqrt{2\pi}}{|E(\mu_j)|\cdot|U'_{E,E^\#}(\mu_j)|^{\frac12}}\cdot\frac{|E(\mu_j)|}{\sqrt{2}} \\ &\quad\cdot\left[\left((-1)^{[\frac{j}{2}]} - i(-1)^{[\frac{j+1}{2}]}\right)K(\mu_j,z) + \left((-1)^{[\frac{j}{2}]} + i(-1)^{[\frac{j+1}{2}]}\right)K(-\mu_j,z)\right] \\ &= \frac{2\sqrt{\pi}}{|U'_{E,E^\#}(\mu_j)|^{\frac12}}\left[(-1)^{[\frac{j}{2}]}K^g(\mu_j,z) - i(-1)^{[\frac{j+1}{2}]}K^u(\mu_j,z)\right]\,.\end{aligned}$$

Substituting into (5.6) the assertion of the lemma follows. □

5.3. *Remark.* The value of $|U'_{E,E^\#}(\mu_j)|$ can be determined from A and B: Making use of (5.9) we obtain from $U_{E,E^\#}(z) = E(z)^2 + E^\#(z)^2$ that

$$U'_{E,E^\#}(\mu_j) = 2\sqrt{2}|E(\mu_j)|(-1)^{[\frac{j}{2}]}\left(A'(\mu_j) + (-1)^j B'(\mu_j)\right) .$$

Since $\operatorname{sgn}(-i)E(\mu_j)^2 = (-1)^j$, we obtain from (4.9) that $\operatorname{sgn} U'_{E,E^\#}(\mu_j) = (-1)^j$ and, hence, conclude that

$$\left|U'_{E,E^\#}(\mu_j)\right| = 2\sqrt{2}|E(\mu_j)|(-1)^{[\frac{j+1}{2}]}\left(A'(\mu_j) + (-1)^j B'(\mu_j)\right) .$$

References

[B] A. Baranov: *Polynomials in the de Branges spaces of entire functions,* Arkiv för Matematik, to appear.

[dB] L. de Branges: *Hilbert spaces of entire functions,* Prentice-Hall, London 1968.

[GGK] I. Gohberg, S. Goldberg, M.A. Kaashoek: *Classes of linear operators Vol.I,* Oper. Theory Adv. Appl. 49, Birkhäuser Verlag, Basel 1990.

[GK] I. Gohberg, M.G. Krein: *Introduction to the theory of linear nonselfadjoint operators,* Translations of mathematical monographs 18, Providence, Rhode Island 1969.

[GT] G. Gubreev, A. Tarasenko: *Representability of a de Branges matrix in the form of a Blaschke-Potapov product and completeness of some function families,* Math. Zametki (Russian) 73 (2003), 796–801.

[KWW1] M. Kaltenbäck, H. Winkler, H. Woracek: *Strings, dual strings, and related canonical systems,* submitted.

[KWW2] M. Kaltenbäck, H. Winkler, H. Woracek: *DeBranges spaces of entire functions symmetric about the origin,* submitted.

[KW1] M. Kaltenbäck, H. Woracek: *Pontryagin spaces of entire functions I,* Integral Equations Operator Theory 33 (1999), 34–97.

[KW2] M. Kaltenbäck, H. Woracek: *Hermite-Biehler Functions with zeros close to the imaginary axis,* Proc. Amer. Math. Soc. 133 (2005), 245–255.

[KW3] M. Kaltenbäck, H. Woracek: *De Branges space of exponential type: General theory of growth,* Acta Sci. Math. (Szeged), to appear.

[KL] M.V. Keldyš, V.B. Lidskiĭ: *On the spectral theory of non-selfadjoint operators,* Proc. Fourth All-Union Math. Congr. (Leningrad, 1961), Vol. I, 101–120, Izdat. Akad. Nauk SSSR (Russian).

[K] M.G. Kreĭn: *A contribution to the theory of linear non-selfadjoint operators,* Dokl. Akad. Nauk SSSR (Russian) 130 (1960), 254–256.

[LW] H. Langer, H. Winkler: *Direct and inverse spectral problems for generalized strings,* Integral Equations Operator Theory **30** (1998), 409–431.

[L] V.B. Lidskiĭ: *Conditions for completeness of a system of root subspaces for non-selfadjoint operators with discrete spectrum,* Trudy Moskov. Mat. Obšč. (Russian) 8 (1959), 83–120.

[M] V.I. Macaev: *Several theorems on completeness of root subspaces of completely continuous operators,* Dokl. Akad. Nauk SSSR (Russian) 155 (1964), 273–276.

[OS] J. Ortega-Cerda, K. Seip: *Fourier frames,* Annals of Mathematics 155 (2002), 789–806.

[S] L. Sakhnovic: *Method of operator identities and problems of analysis,* St. Petersburg Math. Journal 5(1) (1993), 3–80.

[W] H. Winkler: *Canonical systems with a semibounded spectrum,* Operator Theory Adv.Appl. **106** (1998), 397–417.

Michael Kaltenbäck and Harald Woracek
Institut für Analysis und Scientific Computing
Technische Universität Wien
Wiedner Hauptstraße 8–10
A-1040 Wien, Austria
e-mail: michael.kaltenbaeck@tuwien.ac.at
e-mail: harald.woracek@tuwien.ac.at

Operator Theory:
Advances and Applications, Vol. 171, 171–188

Algebras of Singular Integral Operators with Piecewise Continuous Coefficients on Weighted Nakano Spaces

Alexei Yu. Karlovich

Abstract. We find Fredholm criteria and a formula for the index of an arbitrary operator in the Banach algebra of singular integral operators with piecewise continuous coefficients on Nakano spaces (generalized Lebesgue spaces with variable exponent) with Khvedelidze weights over either Lyapunov curves or Radon curves without cusps. These results "localize" the Gohberg-Krupnik Fredholm theory with respect to the variable exponent.

Mathematics Subject Classification (2000). Primary 45E05; Secondary 46E30, 47B35.

Keywords. Weighted Nakano space, Khvedelidze weight, one-dimensional singular integral operator, Lyapunov curve, Radon curve, Fredholmness.

1. Introduction

The study of one-dimensional singular integral operators (SIOs) with piecewise continuous (PC) coefficients on weighted Lebesgue spaces was started by Khvedelidze in the fifties and then was continued in the sixties by Widom, Simonenko, Gohberg, Krupnik, and others. The starting point for those investigations was the sufficient conditions for the boundedness of the Cauchy singular integral operator S on Lebesgue spaces with power weights over Lyapunov curves proved in 1956 by Khvedelidze [27]. Gohberg and Krupnik constructed the Fredholm theory for SIOs with PC coefficients under the assumptions of the Khvedelidze theorem and this theory is the heart of their monograph [16] first published in Russian in 1973 (see also the monographs [6, 20, 33, 35, 36]). In the same year Hunt, Muckenhoupt, and Wheeden proved that for the boundedness of S on $L^p(\mathbb{T}, w)$ it is necessary and sufficient that the weight w belongs to the so-called Muckenhoupt class $A_p(\mathbb{T})$, here $\mathbb{T}$ denotes the unit circle. In 1982 David proved that S is bounded on L^2

This work is partially supported by an F.C.T. (Portugal) grant.

over a rectifiable curve if and only if the curve is a Carleson curve. After some hard analysis one can conclude, finally, that S is bounded on a weighted Lebesgue space over a rectifiable curve if and only the weight belongs to a Carleson curve analog of the Muckenhoupt class (see [11], [2] and also [36]). In 1992 Spitkovsky [43] made the next significant step after Gohberg and Krupnik (20 years later!): he proved Fredholm criteria for an individual SIO with PC coefficients on Lebesgue spaces with Muckenhoupt weights over Lyapunov curves. Finally, Böttcher and Yu. Karlovich extended Spitkovsky's result to the case of arbitrary Carleson curves and Banach algebras of SIOs with PC coefficients. With their work the Fredholm theory of SIOs with PC coefficients is available in the maximal generality (that, is, when the Cauchy singular integral operator S is bounded on weighted Lebesgue spaces). We recommend the nice paper [3] for a first reading about this topic and [2] for a complete and self-contained analysis (see also [4]).

It is quite natural to consider the same problems in other, more general, spaces of measurable functions on which the operator S is bounded. Good candidates for this role are rearrangement-invariant spaces (that is, spaces with the property that norms of equimeasurable functions are equal). These spaces have nice interpolation properties and boundedness results can be extracted from known results for Lebesgue spaces applying interpolation theorems. The author extended (some parts of) the Böttcher-Yu. Karlovich Fredholm theory of SIOs with PC coefficients to the case of rearrangement-invariant spaces with Muckenhoupt weights [22, 24]. Notice that necessary conditions for the Fredholmness of an individual singular integral operator with PC coefficients are obtained in [25] for weighted reflexive Banach function spaces (see [1, Ch. 1]) on which the operator S is bounded.

Nakano spaces $L^{p(\cdot)}$ (generalized Lebesgue spaces with variable exponent) are a nontrivial example of Banach function spaces which are not rearrangement-invariant, in general. Many results about the behavior of some classical operators on these spaces have important applications to fluid dynamics (see [10] and the references therein). Recently Kokilashvili and S. Samko proved [29] that the operator S is bounded on weighted Nakano spaces for the case of nice curves, nice weights, and nice (but variable!) exponents. They also extended the Gohberg-Krupnik Fredholm criteria for an individual SIO with PC coefficients to this situation [30]. So, Nakano spaces are a natural context for the "localization" of the Gohberg-Krupnik theory with respect to the variable exponent. In this paper we proved Fredholm criteria and a formula for the index of an arbitrary operator in the Banach algebra of SIOs with PC coefficients on Nakano spaces (generalized Lebesgue spaces with variable exponent) with Khvedelidze weights over either Lyapunov curves or Radon curves without cusps. These results generalize [30] (see also [25]) to the case of Banach algebras and the results of [15] (see also [14]) to the case of variable exponents (notice also that Radon curves were not considered in [15]). Basically, under the assumptions of the theorem of Kokilashvili and Samko, we can replace the constant exponent p by the value of the variable exponent $p(t)$ at each point t of the contour of integration in the Gohberg-Krupnik Fredholm theory [15].

The paper is organized as follows. In Section 2 we define weighted Nakano spaces and discuss the boundedness of the Cauchy singular integral operator S on weighted Nakano spaces. Section 3 contains Fredholm criteria for an individual SIO with PC coefficients on weighted Nakano spaces. In Section 4 we formulate the Allan-Douglas local principle and the two projections theorem. The results of Section 4 are the main tools allowing us to construct the symbols calculus for the Banach algebra of SIOs with PC coefficients in Section 5. Finally, in Section 6, we prove an index formula for an arbitrary operator in the Banach algebra of SIOs with PC coefficients acting on a Nakano space with a Khvedelidze weight over either a Lyapunov curve or a Radon curve without cusps.

2. Preliminaries

2.1. The Cauchy singular integral

Let Γ be a Jordan (i.e., homeomorphic to a circle) rectifiable curve. We equip Γ with the Lebesgue length measure $|d\tau|$ and the counter-clockwise orientation. The *Cauchy singular integral* of a measurable function $f : \Gamma \to \mathbb{C}$ is defined by

$$(Sf)(t) := \lim_{R\to 0} \frac{1}{\pi i} \int_{\Gamma\setminus\Gamma(t,R)} \frac{f(\tau)}{\tau - t} d\tau \quad (t \in \Gamma),$$

where the "portion" $\Gamma(t, R)$ is

$$\Gamma(t, R) := \{\tau \in \Gamma : |\tau - t| < R\} \quad (R > 0).$$

It is well known that $(Sf)(t)$ exists almost everywhere on Γ whenever f is integrable (see [11, Theorem 2.22]).

2.2. Weighted Nakano spaces $L^{p(\cdot)}$

Function spaces $L^{p(\cdot)}$ of Lebesgue type with variable exponent p were studied for the first time by Orlicz [42] in 1931, but notice that other kind of Banach spaces are named after him. Inspired by the successful theory of Orlicz spaces, Nakano defined in the late forties [40, 41] so-called *modular spaces*. He considered the space $L^{p(\cdot)}$ as an example of modular spaces. Musielak and Orlicz [38] in 1959 extended the definition of modular spaces by Nakano. Actually, that paper was the starting point for the theory of Musielak-Orlicz spaces (generalized Orlicz spaces generated by Young functions with a parameter), see [37].

Let $p : \Gamma \to [1, \infty)$ be a measurable function. Consider the convex modular (see [37, Ch. 1] for definitions and properties)

$$m(f, p) := \int_\Gamma |f(\tau)|^{p(\tau)} |d\tau|.$$

Denote by $L^{p(\cdot)}$ the set of all measurable complex-valued functions f on Γ such that $m(\lambda f, p) < \infty$ for some $\lambda = \lambda(f) > 0$. This set becomes a Banach space with respect to the *Luxemburg-Nakano norm*

$$\|f\|_{L^{p(\cdot)}} := \inf\left\{\lambda > 0 : \ m(f/\lambda, p) \le 1\right\}$$

(see, e.g., [37, Ch. 2]). So, the spaces $L^{p(\cdot)}$ are a special case of Musielak-Orlicz spaces. Sometimes the spaces $L^{p(\cdot)}$ are referred to as Nakano spaces (see, e.g., [13, p. 151], [19, p. 179]). We will follow this tradition. Clearly, if $p(\cdot) = p$ is constant, then the Nakano space $L^{p(\cdot)}$ is isometrically isomorphic to the Lebesgue space L^p. Therefore, sometimes $L^{p(\cdot)}$ are called generalized Lebesgue spaces with variable exponent or, simply, variable L^p spaces.

A nonnegative measurable function w on the curve Γ is referred to as a *weight* if $0 < w(t) < \infty$ almost everywhere on Γ. The *weighted Nakano space* is defined by

$$L^{p(\cdot)}_w = \left\{f \text{ is measurable on } \Gamma \text{ and } fw \in L^{p(\cdot)}\right\}.$$

The norm in this space is defined as usual by $\|f\|_{L^{p(\cdot)}_w} = \|fw\|_{L^{p(\cdot)}}$.

2.3. Carleson, Lyapunov, and Radon curves

A rectifiable Jordan curve Γ is said to be a *Carleson* (or *Ahlfors-David regular*) *curve* if

$$\sup_{t\in\Gamma}\sup_{R>0}\frac{|\Gamma(t,R)|}{R} < \infty,$$

where $|\Omega|$ denotes the measure of a measurable set $\Omega \subset \Gamma$. Much information about Carleson curves can be found in [2].

On a rectifiable Jordan curve we have $d\tau = e^{i\theta_\Gamma(\tau)}|d\tau|$ where $\theta_\Gamma(\tau)$ is the angle between the positively oriented real axis and the naturally oriented tangent of Γ at τ (which exists almost everywhere). A rectifiable Jordan curve Γ is said to be a *Lyapunov curve* if

$$|\theta_\Gamma(\tau) - \theta_\Gamma(t)| \le c|\tau - t|^\mu$$

for some constants $c > 0, \mu \in (0, 1)$ and for all $\tau, t \in \Gamma$. If θ_Γ is a function of bounded variation on Γ, then the curve gamma Γ is called a *Radon curve* (or a *curve of bounded rotation*). It is well known that Lyapunov curves are smooth, while Radon curves may have at most countable set of corner points or cusps. All Lyapunov curves and Radon curves without cusps are Carleson curves (see, e.g., [28, Section 2.3]).

2.4. Boundedness of the Cauchy singular integral operator

We shall assume that

$$1 < \operatorname*{ess\,inf}_{t\in\Gamma} p(t), \quad \operatorname*{ess\,sup}_{t\in\Gamma} p(t) < \infty. \tag{1}$$

In this case the conjugate exponent

$$q(t) := \frac{p(t)}{p(t)-1} \quad (t \in \Gamma)$$

has the same property.

Not so much is known about the boundedness of the Cauchy singular integral operator S on weighted Nakano spaces $L_w^{p(\cdot)}$ for general curves, general weights, and general exponents $p(\cdot)$. From [25, Theorem 6.1] we immediately get the following.

Theorem 2.1. *Let Γ be a rectifiable Jordan curve, let $w : \Gamma \to [0, \infty]$ be a weight, and let $p : \Gamma \to [0, \infty)$ be a measurable function satisfying* (1). *If the Cauchy singular integral generates a bounded operator S on the weighted Nakano space $L_w^{p(\cdot)}$, then*

$$\sup_{t\in\Gamma}\sup_{R>0}\frac{1}{R}\|w\chi_{\Gamma(t,R)}\|_{L^{p(\cdot)}}\|\chi_{\Gamma(t,R)}/w\|_{L^{q(\cdot)}}<\infty. \tag{2}$$

From the Hölder inequality for Nakano spaces (see, e.g., [37] or [32]) and (2) we deduce that if S is bounded on $L_w^{p(\cdot)}$, then Γ is necessarily a Carleson curve. If the exponent $p(\cdot) = p \in (1, \infty)$ is constant, then (2) is simply the famous Muckenhoupt condition A_p (written in the symmetric form):

$$\sup_{t\in\Gamma}\sup_{R>0}\frac{1}{R}\left(\int_{\Gamma(t,R)}w^p(\tau)|d\tau|\right)^{1/p}\left(\int_{\Gamma(t,R)}w^{-q}(\tau)|d\tau|\right)^{-1/q}<\infty,$$

where $1/p + 1/q = 1$. It is well known that for classical Lebesgue spaces L^p this condition is not only necessary, but also sufficient for the boundedness of the Cauchy singular integral operator S. A detailed proof of this result can be found in [2, Theorem 4.15].

Consider now a power weight of the form

$$\varrho(t) := \prod_{k=1}^{N}|t-\tau_k|^{\lambda_k}, \quad \tau_k\in\Gamma, \quad k\in\{1,\dots,N\}, \quad N\in\mathbb{N}, \tag{3}$$

where all λ_k are real numbers. Introduce the class $\mathcal{P}$ of exponents $p : \Gamma \to [1, \infty)$ satisfying (1) and

$$|p(\tau)-p(t)|\le\frac{A}{-\log|\tau-t|} \tag{4}$$

for some $A \in (0, \infty)$ and all $\tau, t \in \Gamma$ such that $|\tau - t| < 1/2$.

Criteria for the boundedness of the Cauchy singular integral operator on Nakano spaces with power weights (3) were recently proved by Kokilashvili and Samko [29] under the condition that the curve Γ and the variable exponent $p(\cdot)$ are sufficiently nice.

Theorem 2.2. (see [29, Theorem 2]). *Let Γ be either a Lyapunov Jordan curve or a Radon Jordan curve without cusps, let ϱ be a power weight of the form* (3), *and let $p \in \mathcal{P}$. The Cauchy singular integral operator S is bounded on the weighted Nakano space $L_\varrho^{p(\cdot)}$ if and only if*

$$0<\frac{1}{p(\tau_k)}+\lambda_k<1 \quad \textit{for all} \quad k\in\{1,\dots,N\}. \tag{5}$$

For weighted Lebesgue spaces this result is classic, for Lyapunov curves it was proved by Khvedelidze [27]. Therefore the weights of the form (3) are often called *Khvedelidze weights*. We shall follow this tradition. For Lebesgue spaces over Radon curves without cusps the above result was proved by Danilyuk and Shelepov [8, Theorem 2]. The proofs and history can be found in [7, 16, 28, 36].

Notice that if p is constant and Γ is a Carleson curve, then (5) is equivalent to the fact that ϱ is a Muckenhoupt weight (see, e.g., [2, Chapter 2]). Analogously one can prove that if the exponent p belong to the class $\mathcal{P}$ and the curve Γ is Carleson, then the power weight (3) satisfies the condition (2) if and only if (5) is fulfilled. The proof of this fact is essentially based on the possibility of estimation of the norms of power functions in Nakano spaces with exponents in the class $\mathcal{P}$ (see also [25, Lemmas 5.7 and 5.8] and [29], [31]).

2.5. Is the condition $p \in \mathcal{P}$ necessary for the boundedness?

What can be said about the necessity of the condition $p \in \mathcal{P}$ in Theorem 2.2? We conjecture that this condition is not necessary, that is, the Cauchy singular integral operator can be bounded on $L^{p(\cdot)}_\varrho$, but p does not satisfy (4). This conjecture is supported by the following observation made by Andrei Lerner [34].

It is well known that, roughly speaking, singular integrals can be controlled by maximal functions. Denote by $\mathcal{M}(\mathbb{R}^n)$ the class of exponents $p : \mathbb{R}^n \to [1, \infty)$ which are essentially bounded and bounded away from 1 and such that the Hardy-Littlewood maximal operator is bounded on $L^{p(\cdot)}(\mathbb{R}^n)$. Diening and Růžička [10, Theorem 4.8] proved that if $p \in \mathcal{M}(\mathbb{R}^n)$ and there exists $s \in (0, 1)$ such that $s/p(t) + 1/\widetilde{q}(t) = 1$ and $\widetilde{q} \in \mathcal{M}(\mathbb{R}^n)$, then the Calderón-Zygmund singular integral operator is bounded on $L^{p(\cdot)}(\mathbb{R}^n)$. A weighted analog of this theorem was used by Kokilashvili and Samko (see [29] and also [31]) to prove Theorem 2.2. Notice also that the author and Lerner [26, Theorem 2.7] proved that if $p, q \in \mathcal{M}(\mathbb{R}^n)$, then the Calderón-Zygmund singular integral operator is bounded on $L^{p(\cdot)}(\mathbb{R}^n)$. On the other hand, Diening [9] showed that the following conditions are equivalent:

(i) $p \in \mathcal{M}(\mathbb{R}^n)$;
(ii) $q \in \mathcal{M}(\mathbb{R}^n)$;
(iii) there exists $s \in (0, 1)$ such that $s/p(t) + 1/\widetilde{q}(t) = 1$ and $\widetilde{q} \in \mathcal{M}(\mathbb{R}^n)$.

So, $p \in \mathcal{M}(\mathbb{R}^n)$ implies the boundedness of the Calderón-Zygmund singular integral operator on $L^{p(\cdot)}(\mathbb{R}^n)$.

Lerner [34], among other things, observed that

$$p(x) = \alpha + \sin(\log\log(1/|x|)\chi_E(x)),$$

where $\alpha > 2$ is some constant and χ_E is the characteristic function of the ball $E := \{x \in \mathbb{R}^n : |x| \le 1/e\}$, belongs to $\mathcal{M}(\mathbb{R}^n)$. Clearly, the exponent p in this example is discontinuous at the origin, so it does not satisfy (an $\mathbb{R}^n$ analog of) the condition (4). This exponent belongs to the class of pointwise multipliers for BMO (the space of functions of bounded mean oscillation). For descriptions of pointwise multipliers for BMO, see Stegenga [44], Janson [18] (local case) and Nakai, Yabuta [39] (global case). So, we strongly believe that necessary and sufficient conditions for the boundedness of the Cauchy singular integral operator (and other singular integrals and maximal functions) on Nakano spaces $L^{p(\cdot)}$ should be formulated in terms of integral means of the exponent p (i.e., in BMO terms), but not in pointwise terms like (4).

3. Fredholm criteria

3.1. Fredholm operators

A bounded linear operator A on a Banach space is said to be Fredholm if its image is closed and both so-called defect numbers

$$n(A) := \dim \ker A, \quad d(A) := \dim \ker A^*$$

are finite. In this case the difference $n(A) - d(A)$ is referred to as the index of the operator A and is denoted by $\operatorname{Ind} A$. Basic properties of Fredholm operators are discussed in [5, 16, 20, 35, 36] and in many other monographs.

3.2. Singular integral operators with piecewise continuous coefficients

In the following we shall suppose that Γ is either a Lyapunov Jordan curve or a Radon Jordan curve without cusps, the variable exponent p belongs to the class $\mathcal{P}$, and the Khvedelidze weight (3) satisfies the conditions (5). Then, by Theorem 2.2, the operator S is bounded on the weighted Nakano space $L^{p(\cdot)}_{\varrho}$. Let I be the identity operator on $L^{p(\cdot)}_{\varrho}$. Put

$$P := (I + S)/2, \quad Q := (I - S)/2.$$

Let L^∞ denote the space of all measurable essentially bounded functions on Γ. We denote by PC the Banach algebra of all piecewise continuous functions on Γ: a function $a \in L^\infty$ belongs to PC if and only if the finite one-sided limits

$$a(t \pm 0) := \lim_{\tau \to t \pm 0} a(\tau)$$

exist for every $t \in \Gamma$.

For $a \in PC$ denote by aI the operator of multiplication by a. Obviously, it is bounded on $L^{p(\cdot)}_{\varrho}$. If B is a bounded operator, then we will simply write aB for the product $aI \cdot B$. The operators of the form $aP + bQ$ with $a, b \in PC$ are called *singular integral operators (SIOs) with piecewise continuous (PC) coefficients.*

Theorem 3.1. *The operator $aP+bQ$, where $a, b \in PC$, is Fredholm on the weighted Nakano space $L^{p(\cdot)}_{\varrho}$ if and only if*

$$a(t \pm 0) \neq 0, \quad b(t \pm 0) \neq 0, \quad -\frac{1}{2\pi} \arg \frac{g(t-0)}{g(t+0)} + \frac{1}{p(t)} + \lambda(t) \notin \mathbb{Z}$$

for all $t \in \Gamma$, where $g = a/b$ and

$$\lambda(t) := \begin{cases} \lambda_k, & \text{if } \ t = \tau_k, \quad k \in \{1, \dots, N\}, \\ 0, & \text{if } \ t \notin \Gamma \setminus \{\tau_1, \dots, \tau_N\}. \end{cases}$$

If a, b have only finite numbers of jumps and $\varrho = 1$, this result was obtained in [30, Theorem A] (as well as a formula for the index of the operator $aP + bQ$). In the present form this result is contained in [25, Theorem 8.3]. For Lebesgue spaces with Khvedelidze weights over Lyapunov curves the corresponding result was obtained in the late sixties by Gohberg and Krupnik [16, Ch. 9].

3.3. Widom-Gohberg-Krupnik arcs

Given $z_1, z_2 \in \mathbb{C}$ and $r \in (0,1)$, put

$$\mathcal{A}(z_1, z_2; r) := \{z_1, z_2\} \cup \left\{ z \in \mathbb{C} \setminus \{z_1, z_2\} : \arg \frac{z - z_1}{z - z_2} \in 2\pi r + 2\pi\mathbb{Z} \right\}.$$

This is a circular arc between z_1 and z_2 (which contains its endpoints z_1 and z_2). Clearly, $\mathcal{A}(z, z; \nu)$ degenerates to the point $\{z\}$ and $\mathcal{A}(z_1, z_2; 1/2)$ is the line segment between z_1 and z_2. A connection of these arcs to Fredholm properties of singular integral operators with piecewise continuous coefficients on $L^p(\mathbb{R})$ was first observed by Widom in 1960. Gohberg and Krupnik expressed their Fredholm theory of SIOs with PC coefficients on Lebesgue spaces with Khvedelidze weights over piecewise Lyapunov curves in terms of these arcs. For more about this topic we refer to the books [5, 16, 20, 36], where the Gohberg-Krupnik Fredholm theory is presented; see also more recent monographs [2, 4], where generalizations of Widom-Gohberg-Krupnik arcs play an essential role in the Fredholm theory of Toeplitz operators with PC symbols on Hardy spaces with Muckenhoupt weights.

Fix $t \in \Gamma$ and consider a function $\chi_t \in PC$ which is continuous on $\Gamma \setminus \{t\}$ and satisfies $\chi_t(t-0) = 0$ and $\chi_t(t+0) = 1$.

From Theorem 3.1 we immediately get the following.

Corollary 3.2. *We have*

$$\{\lambda \in \mathbb{C} : (\chi_t - \lambda)P + Q \text{ is not Fredholm on } L^{p(\cdot)}_\varrho\} = \mathcal{A}\big(0, 1; 1/p(t) + \lambda(t)\big).$$

4. Tools for the construction of the symbol calculus

4.1. The Allan-Douglas local principle

Let B be a Banach algebra with identity. A subalgebra Z of B is said to be a central subalgebra if $zb = bz$ for all $z \in Z$ and all $b \in B$.

Theorem 4.1. (see [5, Theorem 1.34(a)]). *Let B be a Banach algebra with unit e and let Z be closed central subalgebra of B containing e. Let $M(Z)$ be the maximal ideal space of Z, and for $\omega \in M(Z)$, let J_ω refer to the smallest closed two-sided ideal of B containing the ideal ω. Then an element b is invertible in B if and only if $b + J_\omega$ is invertible in the quotient algebra B/J_ω for all $\omega \in M(Z)$.*

4.2. The two projections theorem

The following two projections theorem was obtained by Finck, Roch, Silbermann [12] and Gohberg, Krupnik [17].

Theorem 4.2. *Let F be a Banach algebra with identity e, let $\mathcal{C} = \mathbb{C}^{n\times n}$ be a Banach subalgebra of F which contains e, and let p and q be two projections in F such that $cp = pc$ and $cq = qc$ for all $c \in \mathcal{C}$. Let $W = \operatorname{alg}(\mathcal{C}, p, q)$ be the smallest closed subalgebra of F containing $\mathcal{C}, p, q$. Put*

$$x = pqp + (e-p)(e-q)(e-p),$$

denote by $\operatorname{sp} x$ the spectrum of x in F, and suppose the points 0 and 1 are not isolated points of $\operatorname{sp} x$. Then

(a) *for each $\mu \in \operatorname{sp} x$ the map σ_μ of $\mathcal{C} \cup \{p, q\}$ into the algebra $\mathbb{C}^{2n\times 2n}$ of all complex $2n \times 2n$ matrices defined by*

$$\sigma_\mu c = \begin{pmatrix} c & 0 \\ 0 & c \end{pmatrix}, \quad \sigma_\mu p = \begin{pmatrix} E & 0 \\ 0 & 0 \end{pmatrix}, \tag{6}$$

$$\sigma_\mu q = \begin{pmatrix} \mu E & \sqrt{\mu(1-\mu)}E \\ \sqrt{\mu(1-\mu)}E & (1-\mu)E \end{pmatrix}, \tag{7}$$

where $c \in \mathcal{C}$, E denotes the $n \times n$ unit matrix and $\sqrt{\mu(1-\mu)}$ denotes any complex number whose square is $\mu(1-\mu)$, extends to a Banach algebra homomorphism $\sigma_\mu : W \to \mathbb{C}^{2n\times 2n}$;

(b) *an element $a \in W$ is invertible in F if and only if $\det \sigma_\mu a \neq 0$ for all $\mu \in \operatorname{sp} x$;*

(c) *the algebra W is inverse closed in F if and only if the spectrum of x in W coincides with the spectrum of x in F.*

A further generalization of the above result to the case of N projections is contained in [2].

5. Algebra of singular integral operators

5.1. The ideal of compact operators

The curve Γ divides the complex plane $\mathbb{C}$ into the bounded simply connected domain D^+ and the unbounded domain D^-. Without loss of generality we assume that $0 \in D^+$. Let $X_n := [L^{p(\cdot)}_\varrho]_n$ be a direct sum of n copies of weighted Nakano spaces $X := L^{p(\cdot)}_\varrho$, let $\mathcal{B} := \mathcal{B}(X_n)$ be the Banach algebra of all bounded linear operators on X_n, and let $\mathcal{K} := \mathcal{K}(X_n)$ be the closed two-sided ideal of all compact operators on X_n. We denote by $C^{n\times n}$ (resp. $PC^{n\times n}$) the collection of all continuous (resp. piecewise continuous) $n \times n$ matrix functions, that is, matrix-valued functions with entries in C (resp. PC). Put $I^{(n)} := \operatorname{diag}\{I, \dots, I\}$ and $S^{(n)} := \operatorname{diag}\{S, \dots, S\}$. Our aim is to get Fredholm criteria for an operator $A \in \mathcal{U} := \operatorname{alg}(PC^{n\times n}, S^{(n)})$, the smallest Banach subalgebra of $\mathcal{B}$ which contains all operators of multiplication by matrix-valued functions in $PC^{n\times n}$ and the operator $S^{(n)}$.

Lemma 5.1. *$\mathcal{K}$ is contained in* $\operatorname{alg}(C^{n\times n}, S^{(n)})$, *the smallest closed subalgebra of $\mathcal{B}$ which contains the operators of multiplication by continuous matrix-valued functions and the operator $S^{(n)}$.*

Proof. The proof of this statement is standard, here we follow the presentation in [21, Lemma 9.1]. First, notice that it is sufficient to prove the statement for $n = 1$. By [32, Theorem 2.3 and Corollary 2.7] (see also [37]), (1) is equivalent to the reflexivity of the Nakano space $L^{p(\cdot)}$. Then, in view of [25, Proposition 2.11], the set of all rational functions without poles on Γ is dense in both weighted spaces $L^{p(\cdot)}_\varrho$ and $L^{q(\cdot)}_{1/\varrho}$. Hence $\{t^k\}_{k=-\infty}^{\infty}$ is a basis in $L^{p(\cdot)}_\varrho$ (we assumed that $0 \in D^+$), whence $L^{p(\cdot)}_\varrho$ has the approximating property: each compact operator on $L^{p(\cdot)}_\varrho$ can be approximated in the operator norm by finite-rank operators as closely as desired. So, it is sufficient to show that a finite-rank operator on $L^{p(\cdot)}_\varrho$ belongs to $\operatorname{alg}(C, S)$. Since $[L^{p(\cdot)}_\varrho]^* = L^{q(\cdot)}_{1/\varrho}$ (again see [32] or [37]), a finite-rank operator on $L^{p(\cdot)}_\varrho$ is of the form

$$(Kf)(t) = \sum_{j=1}^{m} a_j(t) \int_\Gamma b_j(\tau) f(\tau)d\tau, \quad t \in \Gamma, \tag{8}$$

where $a_j \in L^{p(\cdot)}_\varrho$ and $b_j \in L^{q(\cdot)}_{1/\varrho}$. Since C is dense in $L^{p(\cdot)}_\varrho$ and in $L^{q(\cdot)}_{1/\varrho}$, one can approximate in the operator norm every operator of the form (8) by operators of the same form but with $a_j, b_j \in C$. Therefore it is sufficient to prove that the operator (8) with $a_j, b_j \in C$ belongs to $\operatorname{alg}(C, S)$. But the latter fact is obvious because

$$K = \sum_{j=1}^{m} a_j (S\chi I - \chi S) b_j I,$$

where $\chi(\tau) = \tau$ for $\tau \in \Gamma$. □

5.2. Operators of local type

We shall denote by $\mathcal{B}^\pi$ the Calkin algebra $\mathcal{B}/\mathcal{K}$ and by A^π the coset $A+\mathcal{K}$ for any operator $A \in \mathcal{B}$. An operator $A \in \mathcal{B}$ is said to be of local type if $AcI^{(n)} - cA$ is compact for all $c \in C$, where $cI^{(n)}$ denotes the operator of multiplication by the diagonal matrix-valued function $\operatorname{diag}\{c, \dots, c\}$. It easy to see that the set $\mathcal{L}$ of all operators of local type is a closed subalgebra of $\mathcal{B}$.

Proposition 5.2.

(a) *We have $\mathcal{K} \subset \mathcal{U} \subset \mathcal{L}$.*
(b) *An operator $A \in \mathcal{L}$ is Fredholm if and only if the coset A^π is invertible in the quotient algebra $\mathcal{L}^\pi := \mathcal{L}/\mathcal{K}$.*

Proof. (a) The embedding $\mathcal{K} \subset \mathcal{U}$ follows from Lemma 5.1, the embedding $\mathcal{U} \subset \mathcal{L}$ follows from the fact that $cS - ScI$ is a compact operator on $L^{p(\cdot)}_\varrho$ for $c \in C$ (see, e.g., [25, Lemma 6.5]).

(b) Straightforward. □

5.3. Localization

From Proposition 5.2(a) we deduce that the quotient algebras $\mathcal{U}^\pi := \mathcal{U}/\mathcal{K}$ and $\mathcal{L}^\pi := \mathcal{L}/\mathcal{K}$ are well defined. We shall study the invertibility of an element A^π of $\mathcal{U}^\pi$ in the larger algebra $\mathcal{L}^\pi$ by using the localization techniques (more precisely, Theorem 4.1). To this end, consider

$$\mathcal{Z}^\pi := \{(cI^{(n)})^\pi : c \in C\}.$$

From the definition of $\mathcal{L}$ it follows that $\mathcal{Z}^\pi$ is a central subalgebra of $\mathcal{L}^\pi$. The maximal ideal space $M(\mathcal{Z}^\pi)$ of $\mathcal{Z}^\pi$ may be identified with the curve Γ via the Gelfand map $\mathcal{G}$ given by

$$\mathcal{G} : \mathcal{Z}^\pi \to C, \quad (\mathcal{G}(cI^{(n)})^\pi)(t) = c(t) \quad (t \in \Gamma).$$

In accordance with Theorem 4.1, for every $t \in \Gamma$ we define $\mathcal{J}_t \subset \mathcal{L}^\pi$ as the smallest closed two-sided ideal of $\mathcal{L}^\pi$ containing the set

$$\{(cI^{(n)})^\pi : \quad c \in C, \quad c(t) = 0\}.$$

Consider a function $\chi_t \in PC$ which is continuous on $\Gamma \setminus \{t\}$ and satisfies $\chi_t(t-0) = 0$ and $\chi_t(t+0) = 1$. For $a \in PC^{n\times n}$ define the function $a_t \in PC^{n\times n}$ by

$$a_t := a(t-0)(1-\chi_t) + a(t+0)\chi_t. \tag{9}$$

Clearly $(aI^{(n)})^\pi - (a_tI^{(n)})^\pi \in \mathcal{J}_t$. Hence, for any operator $A \in \mathcal{U}$, the coset $A^\pi + \mathcal{J}_t$ belongs to the smallest closed subalgebra $\mathcal{W}_t$ of $\mathcal{L}^\pi/\mathcal{J}_t$ containing the cosets

$$p := \big((I^{(n)} + S^{(n)})/2\big)^\pi + \mathcal{J}_t, \quad q := (\chi_t I^{(n)})^\pi + \mathcal{J}_t, \tag{10}$$

where $\chi_t I^{(n)}$ denotes the operator of multiplication by the diagonal matrix-valued function $\operatorname{diag}\{\chi_t, \dots, \chi_t\}$ and the algebra

$$\mathcal{C} := \{(cI^{(n)})^\pi + \mathcal{J}_t : \quad c \in \mathbb{C}^{n\times n}\}. \tag{11}$$

The latter algebra is obviously isomorphic to $\mathbb{C}^{n\times n}$, so $\mathcal{C}$ and $\mathbb{C}^{n\times n}$ can be identified to each other.

5.4. The spectrum of $pqp + (e-p)(e-q)(e-p)$

Since $P^2 = P$ on $L^{p(\cdot)}_\varrho$ (see, e.g., [25, Lemma 6.4]) and $\chi_t^2 - \chi_t \in C$, $(\chi_t^2 - \chi_t)(t) = 0$, it is easy to see that

$$p^2 = p, \quad q^2 = q, \quad pc = cp, \quad qc = cq \tag{12}$$

for every $c \in \mathcal{C}$, where p, q and $\mathcal{C}$ are given by (10) and (11). To apply Theorem 4.2 to the algebras $F = \mathcal{L}^\pi/\mathcal{J}_t$ and $W = \mathcal{W}_t = \operatorname{alg}(\mathcal{C}, p, q)$, we have to identify the spectrum of

$$pqp + (e-p)(e-q)(e-p) = \big(P^{(n)}\chi_t P^{(n)} + Q^{(n)}(1-\chi_t)Q^{(n)}\big)^\pi + \mathcal{J}_t \tag{13}$$

in the algebra $F = \mathcal{L}^\pi/\mathcal{J}_\tau$, here $P^{(n)} := (I^{(n)} + S^{(n)})/2$ and $Q^{(n)} := (I^{(n)} - S^{(n)})/2$.

Lemma 5.3. *Let $\chi_t \in PC$ be a continuous function on $\Gamma \backslash \{t\}$ such that $\chi_t(t-0) = 0$, $\chi_t(\tau + 0) = 1$ and $\chi_t(\Gamma \setminus \{t\}) \cap \mathcal{A}(0, 1; 1/p(t) + \lambda(t)) = \emptyset$. Then the spectrum of (13) in the algebra $\mathcal{L}^\pi/\mathcal{J}_t$ coincides with $\mathcal{A}(0, 1; 1/p(t) + \lambda(t))$.*

Proof. Once we have at hand Corollary 3.2, the proof of this lemma can be developed by a literal repetition of the proof of [21, Lemma 9.4]. It is only necessary to replace the spiralic horn $\mathcal{S}(0, 1; \delta_t; \alpha_M, \beta_M)$ in that proof by the Widom-Gohberg-Krupnik circular arc $\mathcal{A}(0, 1; 1/p(t) + \lambda(t))$. A nice discussion of the relations between (spiralic) horns and circular arcs and their role in the Fredholm theory of SIOs can be found in [2] and [3]. □

5.5. Symbol calculus

Now we are in a position to prove the main result of this paper.

Theorem 5.4. *Define the "arcs bundle"*

$$\mathcal{M} := \bigcup_{t \in \Gamma} \Big(\{t\} \times \mathcal{A}(0, 1; 1/p(t) + \lambda(t)) \Big).$$

(a) *for each point $(t, \mu) \in \mathcal{M}$, the map*

$$\sigma_{t,\mu} : \{S^{(n)}\} \cup \{aI^{(n)} : a \in PC^{n \times n}\} \to \mathbb{C}^{2n \times 2n},$$

given by

$$\sigma_{t,\mu}(S^{(n)}) = \begin{pmatrix} E & O \\ O & -E \end{pmatrix}, \quad \sigma_{t,\mu}(aI^{(n)}) = \begin{pmatrix} a_{11}(t,\mu) & a_{12}(t,\mu) \\ a_{21}(t,\mu) & a_{22}(t,\mu) \end{pmatrix},$$

where

$$\begin{aligned} a_{11}(t,\mu) &:= a(t+0)\mu + a(t-0)(1-\mu), \\ a_{12}(t,\mu) &= a_{21}(t,\mu) := (a(t+0) - a(t-0))\sqrt{\mu(1-\mu)}, \\ a_{22}(t,\mu) &:= a(t+0)(1-\mu) + a(t-0)\mu, \end{aligned}$$

and O and E are the zero and identity $n \times n$ matrices, respectively, extends to a Banach algebra homomorphism

$$\sigma_{t,\mu} : \mathcal{U} \to \mathbb{C}^{2n \times 2n}$$

with the property that

$$\sigma_{t,\mu}(K) = \begin{pmatrix} O & O \\ O & O \end{pmatrix}$$

for every compact operator K on X_n;

(b) *an operator $A \in \mathcal{U}$ is Fredholm on X_n if and only if*

$$\det \sigma_{t,\mu}(A) \neq 0 \quad \text{for all} \quad (t,\mu) \in \mathcal{M};$$

(c) *the quotient algebra $\mathcal{U}^\pi$ is inverse closed in the Calkin algebra $\mathcal{B}^\pi$, that is, if an arbitrary coset $A^\pi \in \mathcal{U}^\pi$ is invertible in $\mathcal{B}^\pi$, then $(A^\pi)^{-1} \in \mathcal{U}^\pi$.*

Proof. The idea of the proof of this theorem based on the Allan-Douglas local principle and the two projections theorem is borrowed from [2].

Fix $t \in \Gamma$ and choose a function $\chi_t \in PC$ such that χ_t is continuous on $\Gamma \setminus \{t\}$, $\chi_t(t-0) = 0$, $\chi_t(t+0) = 1$, and $\chi_t(\Gamma \setminus \{t\}) \cap \mathcal{A}(0,1;1/p(t)+\lambda(t)) = \emptyset$. From (12) and Lemma 5.3 we deduce that the algebras $\mathcal{L}^\pi/\mathcal{J}_t$ and $\mathcal{W}_t = \mathrm{alg}(\mathcal{C}, p, q)$, where p, q and $\mathcal{C}$ are given by (10) and (11), respectively, satisfy all the conditions of the two projections theorem (Theorem 4.2).

(a) In view of Theorem 4.2(a), for every $\mu \in \mathcal{A}(0,1;1/p(t)+\lambda(t))$, the map $\sigma_\mu : \mathbb{C}^{n\times n} \cup \{p,q\} \to \mathbb{C}^{2n\times 2n}$ given by (6)–(7) extends to a Banach algebra homomorphism $\sigma_\mu : \mathcal{W}_t \to \mathbb{C}^{2n\times 2n}$. Then the map

$$\sigma_{t,\mu} = \sigma_\mu \circ \pi_t : \mathcal{U} \to \mathbb{C}^{2n\times 2n},$$

where $\pi_t : \mathcal{U} \to \mathcal{W}_t = \mathcal{U}^\pi/\mathcal{J}_t$ is acting by the rule $A \mapsto A^\pi + \mathcal{J}_t$, is a well-defined Banach algebra homomorphism and

$$\sigma_{t,\mu}(S^{(n)}) = 2\sigma_\mu p - \sigma_\mu e = \begin{pmatrix} E & O \\ O & -E \end{pmatrix}.$$

If $a \in PC^{n\times n}$, then in view of (9) and $(aI^{(n)})^\pi - (a_t I^{(n)})^\pi \in \mathcal{J}_t$ it follows that

$$\begin{aligned} \sigma_{t,\mu}(aI^{(n)}) &= \sigma_{t,\mu}(a_t I^{(n)}) = \sigma_\mu(a(t-0))\sigma_\mu(e-q) + \sigma_\mu(a(t+0))\sigma_\mu q \\ &= \begin{pmatrix} a_{11}(t,\mu) & a_{12}(t,\mu) \\ a_{21}(t,\mu) & a_{22}(t,\mu) \end{pmatrix}. \end{aligned}$$

From Proposition 5.2(a) it follows that $\pi_t(K) = K^\pi + \mathcal{J}_t = \mathcal{J}_t$ for every $K \in \mathcal{K}$ and every $t \in \Gamma$. Hence,

$$\sigma_{t,\mu}(K) = \sigma_\mu(0) = \begin{pmatrix} O & O \\ O & O \end{pmatrix}.$$

Part (a) is proved.

(b) From Proposition 5.2 it follows that the Fredholmness of $A \in \mathcal{U}$ is equivalent to the invertibility of $A^\pi \in \mathcal{L}^\pi$. By Theorem 4.1, the former is equivalent to the invertibility of $\pi_t(A) = A^\pi + \mathcal{J}_t$ in $\mathcal{L}^\pi/\mathcal{J}_t$ for every $t \in \Gamma$. By Theorem 4.2(b), this is equivalent to

$$\det \sigma_{t,\mu}(A) = \det \sigma_\mu \pi_t(A) \neq 0 \quad \text{for all} \quad (t,\mu) \in \mathcal{M}. \tag{14}$$

Part (b) is proved.

(c) Since $\mathcal{A}(0,1;1/p(t)+\lambda(t))$ does not separate the complex plane $\mathbb{C}$, it follows that the spectra of (13) in the algebras $\mathcal{L}^\pi/\mathcal{J}_t$ and $\mathcal{W}_t = \mathcal{U}^\pi/\mathcal{J}_t$ coincide, so we can apply Theorem 4.2(c). If A^π, where $A \in \mathcal{U}$, is invertible in $\mathcal{B}^\pi$, then (14) holds. Consequently, by Theorem 4.2(b), (c), $\pi_t(A) = A^\pi + \mathcal{J}_t$ is invertible in $\mathcal{W}_t = \mathcal{U}^\pi/\mathcal{J}_t$ for every $t \in \Gamma$. Applying Theorem 4.1 to $\mathcal{U}^\pi$, its central subalgebra $\mathcal{Z}^\pi$, and the ideals $\mathcal{J}_t$, we obtain that A^π is invertible in $\mathcal{U}^\pi$, that is, $\mathcal{U}^\pi$ is inverse closed in the Calkin algebra $\mathcal{B}^\pi$. □

6. Index of a Fredholm SIO

6.1. Functions on the cylinder $\Gamma \times [0,1]$ with an exotic topology

Let us consider the cylinder $\mathfrak{M} := \Gamma \times [0,1]$. Following [14, 15], we equip it with an exotic topology, where a neighborhood base is given as follows:

$$\Omega(t,0) := \big\{(t,x) \in \mathfrak{M} : |\tau - t| < \delta, \tau \prec t, x \in [0,1]\big\} \cup \big\{(t,x) \in \mathfrak{M} : x \in [0,\varepsilon)\big\},$$
$$\Omega(t,1) := \big\{(t,x) \in \mathfrak{M} : |\tau - t| < \delta, t \prec \tau, x \in [0,1]\big\} \cup \big\{(t,x) \in \mathfrak{M} : x \in (\varepsilon,1]\big\},$$
$$\Omega(t,x_0) := \big\{(t,x) \in \mathfrak{M} : x \in (x_0 - \delta_1, x_0 + \delta_2)\big\},$$

where $x_0 \neq 0, 0 < \delta_1 < x_0, 0 < \delta_2 < 1 - x_0$, and $0 < \varepsilon < 1$.

Note that $\mathcal{A}(z_1, z_2; r)$ has the following parametric representation

$$z(x) = z_1 + (z_2 - z_1)\omega(x,r), \quad 0 \le x \le 1,$$

where $\omega(x,r) = x$ for $r = 1/2$ and

$$\omega(x,r) := \frac{\sin(\theta x)\exp(i\theta x)}{\sin\theta\exp(i\theta)}, \quad \theta := \pi(1-2r), \quad r \neq 1/2.$$

Let Λ be the set of all piecewise continuous scalar functions having only finitely many jumps. For $a \in \Lambda$, put

$$U_a(t,x) := a(t+0)\omega(x, 1/p(t)+\lambda(t)) + a(t-0)\big(1-\omega(x,1/p(t)+\lambda(t))\big), \quad (t,x) \in \mathfrak{M}.$$

Let us consider the function

$$F(t,x) := \prod_{j=1}^{k} U_{a_j}(t,x), \quad (t,x) \in \mathfrak{M}, \tag{15}$$

where $a_j \in \Lambda$, $1 \le j \le k$, and $k \ge 1$. If $F(t,x) \neq 0$ for all $(t,x) \in \mathfrak{M}$, then F is continuous on $\mathfrak{M}$ and the image of this function is a continuous closed curve that does not pass through the origin and can be oriented in a natural way. Namely, at the points where the functions a_j are continuous, the orientation of the curve is defined correspondingly to the orientation of Γ. Along the complementary arcs connecting the one-sided limits at jumps the orientation is defined by the variation of x from 0 to 1. The index $\operatorname{ind}_{\mathfrak{M}} F$ of F is defined as the winding number of the above defined curve about the origin.

By $\mathfrak{F}(\mathfrak{M})$ we denote the class of functions $H : \mathfrak{M} \to \mathbb{C}$ satisfying the following two conditions:

(i) $H(t,x) \neq 0$ for all $(t,x) \in \mathfrak{M}$;

(ii) H can be represented as the uniform limit with respect to $(t,x) \in \mathfrak{M}$ of a sequence of functions F_s of the form (15).

The numbers $\operatorname{ind}_{\mathfrak{M}} F_s$ are independent of s starting from some number s_0. The number

$$\operatorname{ind}_{\mathfrak{M}} H := \lim_{s\to\infty} \operatorname{ind}_{\mathfrak{M}} F_s$$

will be called the index of $H \in \mathfrak{F}(\mathfrak{M})$. One can see that the index just defined is independent of the choice of a sequence F_s of the form (15).

6.2. Index formula

The matrix function

$$\mathfrak{A}(t,x) = \sigma_{t,\omega(x,1/p(t)+\lambda(t))}(A), \quad (t,x) \in \mathfrak{M},$$

is said to be the symbol of the operator $A \in \mathcal{U}$. We can write the symbol in the block form

$$\mathfrak{A}(t,x) = \begin{pmatrix} \mathfrak{A}_{11}(t,x) & \mathfrak{A}_{12}(t,x) \\ \mathfrak{A}_{21}(t,x) & \mathfrak{A}_{22}(t,x) \end{pmatrix}, \quad (t,x) \in \mathfrak{M},$$

where $\mathfrak{A}_{ij}(t,x)$ are $n \times n$ matrix functions.

Theorem 6.1. *If an operator $A \in \mathcal{U}$ is Fredholm on X_n, then the function*

$$Q_A(t,x) := \frac{\det \mathfrak{A}(t,x)}{\det \mathfrak{A}_{22}(t,0) \det \mathfrak{A}_{22}(t,1)}, \quad (t,x) \in \mathfrak{M},$$

belongs to $\mathfrak{F}(\mathfrak{M})$ and

$$\operatorname{Ind} A = -\operatorname{ind}_{\mathfrak{M}} Q_A.$$

Proof. The proof of this theorem is developed as in the classical situation [14, 15] (see also [22, 23] and [2]) in several steps. We do not present all details here, although we mention the main steps.

1) The index formula for the scalar Fredholm operator $aP + Q$ with $a \in \Lambda$:

$$\operatorname{Ind}(aP+Q) = -\operatorname{ind}_{\mathfrak{M}} U_a.$$

In a slightly different form (and in the non-weighted case) this formula was proved by Kokilashvili and Samko [30].

2) The index formula for $aP^{(n)} + Q^{(n)}$, where $a \in C^{n\times n}$:

$$\operatorname{Ind}(aP^{(n)} + Q^{(n)}) = -\frac{1}{2\pi}\{\operatorname{Arg}\det a(t)\}_\Gamma,$$

where the latter denotes the Cauchy index of the continuous function $\det a$. This formula can be proved by using standard homotopic arguments.

3) The index formula for $aP^{(n)} + Q^{(n)}$, where a is a function in $\Lambda^{n\times n}$, the set of $n \times n$ matrices with entries in Λ:

$$\operatorname{Ind}(aP^{(n)} + Q^{(n)}) = -\operatorname{ind}_{\mathfrak{M}} \det U_a.$$

A proof of this fact is based on the possibility of a representation of $a \in \Lambda^{n\times n}$ as the product $c_1 Y c_2$, where c_1 and c_2 are nonsingular continuous matrix functions and Y is an invertible upper-triangular matrix function in $\Lambda^{n\times n}$. A proof of this representation can be found, e.g., in [6, Ch. VIII, Lemma 2.2].

4) An index formula for the operators of the form

$$\sum_{j=1}^{k} (a_{j1}P^{(n)} + b_{j1}Q^{(n)}) \times \cdots \times (a_{jr}P^{(n)} + b_{jr}Q^{(n)}), \tag{16}$$

where $a_{jl}, b_{jl} \in \Lambda^{n\times n}$, $1 \le l \le r$, $k \ge 1$, can be proved by using the previous step an a procedure of linear dilation as in [14, Theorem 7.1] or [15, Theorem 3.1].

5) Every operator $A \in \mathcal{U}$ can be represented as a limit (in the operator topology) of operators of the form (16). So, the index formula in the general case follows from the fourth step by passing to the limits. Notice that if a sequence of operators $A_s \in \mathcal{U}$ converges to A, then

$$\det \mathfrak{A}^{(s)} \to \det \mathfrak{A}, \quad \det \mathfrak{A}_{11}^{(s)} \to \det \mathfrak{A}_{11}, \quad \det \mathfrak{A}_{22}^{(s)} \to \det \mathfrak{A}_{22}$$

uniformly on $\mathfrak{M}$, where $\mathfrak{A}$ and $\mathfrak{A}^{(s)}$ are the symbols of A and A_s (see [23, Theorem 3]), so passage to the limits is legitimate. □

References

[1] C. Bennett, R. Sharpley, *Interpolation of Operators.* Pure and Applied Mathematics, **129**. Academic Press, Boston, 1988.

[2] A. Böttcher, Yu.I. Karlovich, *Carleson Curves, Muckenhoupt Weights, and Toeplitz Operators.* Progress in Mathematics, **154**. Birkhäuser Verlag, Basel, Boston, Berlin, 1997.

[3] A. Böttcher, Yu.I. Karlovich, *Cauchy's singular integral operator and its beautiful spectrum.* In: "Systems, approximation, singular integral operators, and related topics" (Bordeaux, 2000), pp. 109–142, Operator Theory: Advances and Applications, **129**. Birkhäuser Verlag, Basel, 2001.

[4] A. Böttcher, Yu.I. Karlovich, I.M. Spitkovsky, *Convolution Operators and Factorization of Almost Periodic Matrix Functions.* Operator Theory: Advances and Applications, **131**. Birkhäuser, Basel, Boston, Berlin, 2002.

[5] A. Böttcher, B. Silbermann, *Analysis of Toeplitz Operators*, Springer-Verlag, Berlin, 1990.

[6] K.F. Clancey, I. Gohberg, *Factorization of Matrix Functions and Singular Integral Operators.* Operator Theory: Advances and Applications, **3**. Birkhäuser Verlag, Basel, 1981.

[7] I.I. Danilyuk, *Nonregular Boundary Value Problems in the Plane*, Nauka, Moscow, 1975 (in Russian).

[8] I.I. Danilyuk, V.Yu. Shelepov, *Boundedness in L_p of a singular operator with Cauchy kernel along a curve of bounded rotation.* Dokl. Akad. Nauk SSSR, **174** (1967), 514–517 (in Russian). English translation: Soviet Math. Dokl., **8** (1967), 654–657.

[9] L. Diening, *Maximal function on Musielak-Orlicz spaces and generalized Lebesgue spaces.* Mathematische Fakultät, Albert-Ludwigs-Universität Freiburg, Preprint Nr. 03–21 (2003).

[10] L. Dieinig, M. Růžička, *Calderón-Zygmund operators on generalized Lebesgue spaces $L^{p(\cdot)}$ and problems related to fluid dynamics.* J. Reine Angew. Math., **563** (2003), 197–220.

[11] E.M. Dynkin, *Methods of the theory of singular integrals (Hilbert transform and Calderón-Zygmund theory).* Itogi nauki i tehniki VINITI, Ser. Sovrem. probl. mat., **15** (1987), 197–292 (in Russian). English translation: Commutative harmonic analysis I. General survey. Classical aspects, Encycl. Math. Sci., **15** (1991), 167–259.

[12] T. Finck, S. Roch, B. Silbermann, *Two projections theorems and symbol calculus for operators with massive local spectra.* Math. Nachr., **162** (1993), 167–185.

[13] R.J. Fleming, J.E. Jamison, A. Kamińska, *Isometries of Musielak-Orlicz spaces.* In: "Function spaces" (Edwardsville, IL, 1990), pp. 139–154. Lecture Notes in Pure and Appl. Math., **136**. Dekker, New York, 1992.

[14] I. Gohberg, N. Krupnik, *On the algebra generated by one-dimensional singular integral operators with piecewise continuous coefficients.* Funct. Analiz i Ego Prilozh., **4**, no. 3 (1970), 26–36 (in Russian). English translation: Funct. Anal. Appl., **4** (1970), 193–201.

[15] I. Gohberg, N. Krupnik, *Singular integral operators with piecewise continuous coefficients and their symbols.* Izv. AN SSSR, Ser. matem., **35**, no. 4 (1971), 940–964 (in Russian). English translation: Math. USSR Izv., **5** (1971), 955–979.

[16] I. Gohberg, N. Krupnik, *One-Dimensional Linear Singular Integral Equations.* Vols. 1, 2, Operator Theory: Advances and Applications, **53, 54**. Birkhäuser Verlag, Basel, Boston, Berlin, 1992. Russian original: Shtiintsa, Kishinev, 1973.

[17] I. Gohberg, N. Krupnik, *Extension theorems for Fredholm and invertibility symbols.* Integr. Equat. and Oper. Theory, **17** (1993), 514–529.

[18] S. Janson, *On functions with conditions on the mean oscillation.* Ark. Math., **14** (1976), 189–196.

[19] A. Kamińska, B. Turett, *Type and cotype in Musielak-Orlicz spaces.* In: "Geometry of Banach spaces" (Strobl, 1989), pp. 165–180, London Math. Soc. Lecture Note Ser., **158**. Cambridge Univ. Press, Cambridge, 1990.

[20] N. Karapetiants, S. Samko, *Equations with Involutive Operators*, Birkhäuser Boston, Inc., Boston, MA, 2001.

[21] A.Yu. Karlovich, *Algebras of singular integral operators with piecewise continuous coefficients on reflexive Orlicz spaces.* Math. Nachr., **179** (1996), 187–222.

[22] A.Yu. Karlovich, *Singular integral operators with piecewise continuous coefficients in reflexive rearrangement-invariant spaces.* Integr. Equat. Oper. Theory, **32** (1998), 436–481.

[23] A.Yu. Karlovich, *The index of singular integral operators in reflexive Orlicz spaces.* Matem. Zametki, **64**, no. 3 (1998), 383–396 (in Russian). English translation: Math. Notes, **64** (1999), 330–341.

[24] A.Yu. Karlovich, *Algebras of singular integral operators with PC coefficients in rearrangement-invariant spaces with Muckenhoupt weights.* J. Operator Theory, **47** (2002), 303–323.

[25] A.Yu. Karlovich, *Fredholmness of singular integral operators with piecewise continuous coefficients on weighted Banach function spaces.* J. Integr. Equat. Appl., **15**, no. 3, (2003), 263–320.

[26] A.Yu. Karlovich, A.K. Lerner, *Commutators of singular integrals on generalized L^p spaces with variable exponent.* Publ. Mat. (to appear).

[27] B.V. Khvedelidze, *Linear discontinuous boundary problems in the theory of functions, singular integral equations and some of their applications.* Akad. Nauk Gruzin. SSR. Trudy Tbiliss. Mat. Inst. Razmadze, **23** (1956), 3–158 (in Russian).

[28] B.V. Khvedelidze, *The method of the Cauchy type integrals for discontinuous boundary value problems of the theory of holomorphic functions of one complex variable.* Itogi nauki i tehniki VINITI, Ser. Sovrem. probl. mat., **7** (1975), 5–162 (in Russian). English translation: J. Sov. Math., **7** (1977), 309–414.

[29] V. Kokilashvili, S. Samko, *Singular integrals in weighted Lebesgue spaces with variable exponent.* Georgian Math. J., **10**, no. 1, (2003) 145–156.

[30] V. Kokilashvili, S. Samko, *Singular integral equations in the Lebesgue spaces with variable exponent.* Proc. A. Razmadze Math. Inst., **131** (2003), 61–78.

[31] V. Kokilashvili, S. Samko, *Maximal and fractional operators in weighted $L^{p(x)}$ spaces.* Rev. Mat. Iberoamericana, **20**, no. 2 (2004), 493–515.

[32] O. Kováčik, J. Rákosník, *On spaces $L^{p(x)}$ and $W^{k,p(x)}$.* Czechoslovak Math. J., **41 (116)** (1991), no. 4, 592–618.

[33] N.Ya. Krupnik, *Banach Algebras with Symbol and Singular Integral Operators.* Operator Theory: Advances and Applications, **26**. Birkhäuser Verlag, Basel, 1987.

[34] A.K. Lerner, *Some remarks on the Hardy-Littlewood maximal function on variable L^p spaces.* Preprint (2004). Available at `http://www.math.biu.ac.il`

[35] G.S. Litvinchuk, I.M. Spitkovsky, *Factorization of Measurable Matrix Functions.* Operator Theory: Advances and Applications, **25**. Birkhäuser Verlag, Basel, 1987.

[36] S.G. Mikhlin, S. Prössdorf, *Singular Integral Operators.* Springer-Verlag, Berlin, 1986.

[37] J. Musielak, *Orlicz Spaces and Modular Spaces.* Lecture Notes in Mathematics, **1034**. Springer-Verlag, Berlin, 1983.

[38] J. Musielak, W. Orlicz, *On modular spaces.* Studia Math., **18** (1959), 49–65.

[39] E. Nakai, K. Yabuta, Pointwise multipliers for functions of bounded mean oscillation, J. Math. Soc. Japan, **37** (1985), 207–218.

[40] H. Nakano, *Modulared Semi-Ordered Linear Spaces.* Maruzen Co., Ltd., Tokyo, 1950.

[41] H. Nakano, *Topology of Linear Topological Spaces.* Maruzen Co., Ltd., Tokyo, 1951.

[42] W. Orlicz, *Über konjugierte Exponentenfolgen.* Studia Math., **3** (1931), 200–211.

[43] I. Spitkovsky, *Singular integral operators with PC symbols on the spaces with general weights.* J. Functional Analysis, **105** (1992), 129–143.

[44] D.A. Stegenga, *Bounded Toeplitz operators on H^1 and applications of the duality between H^1 and the functions of bounded mean oscillation.* Amer. Math. J., **98** (1976), 573–589.

Alexei Yu. Karlovich
Universade do Minho
Centro de Matemática
Escola de Ciências
Campus de Gualtar
4710-057, Braga, Portugal
e-mail: `oleksiy@math.uminho.pt`

Operator Theory:
Advances and Applications, Vol. 171, 189–224

Pseudodifferential Operators with Compound Slowly Oscillating Symbols

Yuri I. Karlovich

Abstract. Let $V(\mathbb{R})$ denote the Banach algebra of absolutely continuous functions of bounded total variation on $\mathbb{R}$. We study an algebra $\mathfrak{B}$ of pseudodifferential operators of zero order with compound slowly oscillating $V(\mathbb{R})$-valued symbols $(x, y) \mapsto a(x, y, \cdot)$ of limited smoothness with respect to $x, y \in \mathbb{R}$. Sufficient conditions for the boundedness and compactness of pseudodifferential operators with compound symbols on Lebesgue spaces $L^p(\mathbb{R})$ are obtained. A symbol calculus for the algebra $\mathfrak{B}$ is constructed on the basis of an appropriate approximation of symbols by infinitely differentiable ones and by use of the techniques of oscillatory integrals. A Fredholm criterion and an index formula for pseudodifferential operators $A \in \mathfrak{B}$ are obtained. These results are carried over to Mellin pseudodifferential operators with compound slowly oscillating $V(\mathbb{R})$-valued symbols. Finally, we construct a Fredholm theory of generalized singular integral operators on weighted Lebesgue spaces L^p with slowly oscillating Muckenhoupt weights over slowly oscillating Carleson curves.

Mathematics Subject Classification (2000). Primary 47G30, 47A53; Secondary 45E05, 47G10, 47L15.

Keywords. Pseudodifferential operator, compound symbol, oscillatory integral, Lebesgue space, boundedness, compactness, Fredholm theory, generalized singular integral operator, Muckenhoupt weight, Carleson curve, slow oscillation.

1. Introduction

Pseudodifferential operators with compound (double) symbols play an important role in the modern theory of linear PDE and singular integral operators (see, e.g., [13], [22], [39], [40], [42], [32], [1]). Treatments of pseudodifferential operators most frequently concentrate on operators with smooth symbols, but for a lot of applications we need to study pseudodifferential operators with symbols of limited

Partially supported by the CONACYT (México) Project No. 47220.

smoothness. Thus, studying pseudodifferential operators with symbols of minimal smoothness is of interest now (see, e.g., [6], [7], [24], [39], [41]).

Pseudodifferential operators of Mellin type were intensively studied along with the Fourier pseudodifferential operators (see, e.g., H.O. Cordes [8], J.E. Lewis and C. Parenti [23] and the references given there). The theory of Mellin pseudodifferential operators of higher order with smooth symbols is applied in the study of boundary value problems on manifolds with singularities (see, e.g., [35], [36] and the references therein), in the diffraction theory (see, e.g., [25], [26]), and in the theory of singular integral operators with singularities (see [2]–[4], [29]).

Let $\mathcal{B}_p = \mathcal{B}\big(L^p(\mathbb{R})\big)$ be the Banach algebra of bounded linear operators acting on the Lebesgue space $L^p(\mathbb{R})$, $1 < p < \infty$, and let $\mathcal{K} = \mathcal{K}_p$ be the closed two-sided ideal of all compact operators in $\mathcal{B}_p$.

The paper is devoted to studying the Fourier and Mellin pseudodifferential operators with compound symbols on Lebesgue spaces L^p over $\mathbb{R}$ and $\mathbb{R}_+$, respectively. More precisely, we study an algebra $\mathfrak{B}$ of (Fourier) pseudodifferential operators of zero order with compound slowly oscillating $V(\mathbb{R})$-valued symbols $(x, y) \mapsto a(x, y, \cdot)$ of limited smoothness with respect to $x, y \in \mathbb{R}$, where $V(\mathbb{R})$ is the Banach algebra of absolutely continuous functions of bounded total variation on $\mathbb{R}$. Sufficient conditions for the boundedness and compactness of pseudodifferential operators with compound symbols on the Lebesgue spaces $L^p(\mathbb{R})$ are obtained. A symbol calculus for the algebra $\mathfrak{B}$ is constructed on the basis of an appropriate approximation of symbols by infinitely differentiable ones and by use of the techniques of oscillatory integrals. As is well known (see, e.g., [5], [12], [1]), an operator $A \in \mathfrak{A}$ is said to be *Fredholm*, if its image is closed and the spaces $\ker A$ and $\ker A^*$ are finite-dimensional. In that case $\operatorname{Ind} A = \dim\ker A - \dim\ker A^*$ is referred to as the *index* of A. A Fredholm criterion and an index formula for the pseudodifferential operators $A \in \mathfrak{B}$ are obtained. These results are carried over to Mellin pseudodifferential operators with compound slowly oscillating $V(\mathbb{R})$-valued symbols $(r, \varrho) \mapsto a(r, \varrho, \cdot)$ of limited smoothness relative to $r, \varrho \in \mathbb{R}_+$. Finally, as an application we construct a Fredholm theory of generalized singular integral operators on weighted Lebesgue spaces L^p over slowly oscillating Carleson curves with slowly oscillating Muckenhoupt weights.

The paper extends the results of [16] and is organized as follows. By analogy with [16], in Section 2 we introduce new classes of compound symbols $C_b\big(\mathbb{R}^n, V(\mathbb{R})\big)$, $\mathfrak{S}_n$, and $\mathcal{E}_n^V \subset \mathcal{E}_n^C$ that consist of bounded continuous, uniformly continuous, and slowly oscillating $V(\mathbb{R})$-valued functions on $\mathbb{R}^n$, respectively. Approximations of the functions in these classes by infinitely differentiable ones are constructed by the scheme of [8, Chapter 3] with estimates caused by the considered classes of symbols. Here we also introduce a subset $\widetilde{\mathcal{E}}_n \subset \mathcal{E}_n^V$ of symbols which allows us to construct a Fredholm theory for pseudodifferential operators with compound symbols in $\widetilde{\mathcal{E}}_2$.

In Section 3 we study (Fourier) pseudodifferential operators with compound finitely differentiable (with respect to spatial variables) $V(\mathbb{R})$-valued symbols in

$C_b(\mathbb{R} \times \mathbb{R}, V(\mathbb{R}))$ making use of the techniques of oscillatory integrals (see, e.g., [22], [42], [30], [32]).

In Section 4, we establish sufficient conditions for the boundedness and compactness of (Fourier) pseudodifferential operators with compound $V(\mathbb{R})$-valued symbols of limited smoothness relative to spatial variables on the Lebesgue spaces $L^p(\mathbb{R})$, $1 < p < \infty$. The proofs are based on the results of Sections 2–3 and [16].

In Section 5 we construct the symbol calculus for pseudodifferential operators with compound slowly oscillating $V(\mathbb{R})$-valued symbols of limited smoothness with respect to spatial variables and obtain a Fredholm criterion and an index formula for such operators making use of the results of [16] and Section 4.

In Section 6 we carry the results obtained for Fourier pseudodifferential operators with compound symbols over to Mellin pseudodifferential operators with compound slowly oscillating $V(\mathbb{R})$-valued symbols of limited smoothness with respect to spatial variables $r, \varrho \in \mathbb{R}_+$.

The next two sections are devoted to an application of Mellin pseudodifferential operators to constructing a Fredholm theory for generalized singular integral operators on weighted Lebesgue spaces $L^p(\Gamma, w)$ with slowly oscillating Muckenhoupt weights w over slowly oscillating Carleson curves Γ.

Given an oriented rectifiable simple arc Γ in the plane and a function f in $L^1(\Gamma)$, the Cauchy singular integral $S_\Gamma f$,

$$(S_\Gamma f)(t) := \lim_{\varepsilon \to 0} \frac{1}{\pi i} \int_{\Gamma \setminus \Gamma(t,\varepsilon)} \frac{f(\tau)}{\tau - t} \, d\tau, \quad \Gamma(t, \varepsilon) := \Big\{ \tau \in \Gamma : |\tau - t| < \varepsilon \Big\},$$

exists for almost all $t \in \Gamma$. We consider S_Γ as an operator on the weighted L^p space $L^p(\Gamma, w)$ with the norm $\|f\| := \|fw\|_{L^p(\Gamma)}$ where $1 < p < \infty$ and $w : \Gamma \to [0, \infty]$ is a measurable function such that $w^{-1}(\{0, \infty\})$ has measure zero. After a long development, which culminated with the work by Hunt, Muckenhoupt, Wheeden [14] and David [9], it became clear that S_Γ is a well-defined and bounded operator on $L^p(\Gamma, w)$ if and only if

$$\sup_{\varepsilon > 0} \sup_{t \in \Gamma} \frac{1}{\varepsilon} \bigg(\int_{\Gamma(t,\varepsilon)} w(\tau)^p |d\tau| \bigg)^{1/p} \bigg(\int_{\Gamma(t,\varepsilon)} w(\tau)^{-q} |d\tau| \bigg)^{1/q} < \infty, \tag{1.1}$$

where $1/p + 1/q = 1$ (see also [11] and [1]). Following [3], we write A_p for the set of all pairs (Γ, w) satisfying (1.1). Using Hölder's inequality, it is easily seen that (1.1) implies that

$$\sup_{\varepsilon > 0} \sup_{t \in \Gamma} |\Gamma(t, \varepsilon)| / \varepsilon < \infty, \tag{1.2}$$

where $|\Gamma(t, \varepsilon)|$ stands for the Lebesgue (length) measure of $\Gamma(t, \varepsilon)$. Condition (1.1) is called the Muckenhoupt condition. The curves Γ satisfying (1.2) are named Carleson or Ahlfors-David curves (see [1]).

Applications of Mellin pseudodifferential operators with compound (double) symbols to studying algebras of singular integral operators with slowly oscillating coefficients on weighted Lebesgue spaces with slowly oscillating Muckenhoupt weights over slowly oscillating Carleson curves were considered in [29], [2], [3], [4] (see also [1], [32] and the references therein). Singular integral operators with some slowly oscillating shifts were investigated in [17], pseudodifferential operators with slowly oscillating shifts were studied in [31]. In all these papers the symbols of pseudodifferential operators were infinitely differentiable. Now we apply the results on Mellin pseudodifferential operators with compound $V(\mathbb{R})$-valued symbols of limited smoothness (with respect to spatial variables).

In Section 7 we introduce slowly oscillating data of generalized singular integral operators. In Section 8 we obtain a Fredholm criterion and an index formula for generalized singular integral operators with slowly oscillating coefficients on weighted Lebesgue spaces $L^p(\Gamma, w)$ with slowly oscillating Muckenhoupt weights w over slowly oscillating unbounded Carleson curves Γ with endpoints $t \in \mathbb{C}$ and ∞, making use of the results of Sections 4 to 6. Let α be an orientation-preserving diffeomorphism of $\Gamma^0 = \Gamma \setminus \{t, \infty\}$ onto itself that is slowly oscillating at the endpoints of Γ, and let V_α be the corresponding shift operator, $V_\alpha \varphi = \varphi \circ \alpha$. We call the operators of the form $V_\alpha S_\Gamma V_\alpha^{-1}$ *generalized singular integral operators*. The Fredholm result obtained in Section 8 essentially depends on all slowly oscillating data: the curve, the weight, the shift, and the coefficients.

2. Oscillating compound symbols and their approximation

Slowly oscillating functions. For $n \in \mathbb{N}$ and a point $x = (x_1, \dots, x_n) \in \mathbb{R}^n$, fix the norm $\|x\| = \max\{|x_1|, \dots, |x_n|\}$. Given a continuous function $a : \mathbb{R}^n \to \mathbb{C}$, let $C_b(\mathbb{R}^n) := C(\mathbb{R}^n) \cap L^\infty(\mathbb{R}^n)$ and let $cm_x(a) := cm_{x,1}(a)$ where

$$cm_{x,\varepsilon}(a) := \max\Big\{|a(x+y) - a(x)| : \ y \in \mathbb{R}^n, \ \ \|y\| \le \varepsilon\Big\} \tag{2.1}$$

for $\varepsilon > 0$ is a local oscillation of a at a point $x \in \mathbb{R}^n$.

According to [8, p. 122], a function $a \in C_b(\mathbb{R}^n)$ is called *slowly oscillating at* ∞ if

$$\lim_{\|x\|\to\infty} cm_x(a) = 0 \tag{2.2}$$

or, equivalently, $\lim\limits_{\|x\|\to\infty} cm_{x,\varepsilon}(a) = 0$ for every $\varepsilon > 0$.

Let $SO(\mathbb{R}^n)$ be the set of all functions in $C_b(\mathbb{R}^n)$ which are slowly oscillating at ∞. By [8, Chapter 3, Lemma 10.4], $SO(\mathbb{R}^n)$ is a C^*-algebra being the closure in $L^\infty(\mathbb{R}^n)$ of the algebra

$$SO^\infty(\mathbb{R}^n) := \Big\{a \in C_b^\infty(\mathbb{R}^n) : \ \lim_{\|x\|\to\infty} \partial_x^\alpha a(x) = 0, \ |\alpha| > 0\Big\}. \tag{2.3}$$

Here $C_b^\infty(\mathbb{R}^n)$ is the set of all infinitely differentiable functions $a : \mathbb{R}^n \to \mathbb{C}$ that are bounded with all their partial derivatives, $\partial_{x_j} = \partial/\partial_{x_j}$, $\partial_x^\alpha = \partial_{x_1}^{\alpha_1} \dots \partial_{x_n}^{\alpha_n}$ and $|\alpha| = \alpha_1 + \dots + \alpha_n$ for every multi-index $\alpha = (\alpha_1, \dots, \alpha_n)$ with non-negative integers α_j. For $a \in SO(\mathbb{R}^n)$, an approximation $a_\varepsilon \in SO^\infty(\mathbb{R}^n)$ $(\varepsilon \to 0)$ can be chosen in the form $a_\varepsilon = \psi_\varepsilon * a$ where

$$\psi_\varepsilon(x) = \varphi_\varepsilon(x_1)\varphi_\varepsilon(x_2)\cdots\varphi_\varepsilon(x_n), \quad \varphi_\varepsilon(x_j) = \varepsilon^{-1}\varphi(x_j/\varepsilon) \quad \text{for} \quad \varepsilon > 0, \tag{2.4}$$

$$\varphi \in C_0^\infty(\mathbb{R}), \quad \operatorname{supp}\varphi \subset [-1,1], \quad \varphi \geq 0, \quad \int_{\mathbb{R}} \varphi(t)dt = 1, \tag{2.5}$$

and $C_0^\infty(\mathbb{R}^n)$ denotes the set of all infinitely differentiable complex-valued functions on $\mathbb{R}^n$ with compact support.

Functions of bounded total variation. Let a be an absolutely continuous function on $\mathbb{R}$ of bounded total variation $V(a)$ where

$$V(a) := \sup\left\{\sum_{k=1}^{n} |a(x_k) - a(x_{k-1})| : -\infty < x_0 < x_1 < \cdots < x_n < +\infty, \; n \in \mathbb{N}\right\}.$$

As is known (see, e.g., [27, Chapter VIII, § 3; Chapter IX, § 4]), if $V(a) < \infty$, then the limits $a(\pm\infty) = \lim_{x\to\pm\infty} a(x)$ exist and thus a is continuous on $\overline{\mathbb{R}} = [-\infty, +\infty]$, $a' \in L^1(\mathbb{R})$, and

$$a(x) = \int_{-\infty}^{x} a'(y)dy + a(-\infty) \quad \text{for } x \in \mathbb{R}, \quad V(a) = \int_{\mathbb{R}} |a'(y)|dy.$$

The set $V(\mathbb{R})$ of all absolutely continuous functions on $\mathbb{R}$ of bounded total variation is a Banach space with the norm $\|a\|_V := \|a\|_C + V(a)$ where $\|a\|_C := \sup\{|a(x)| : x \in \mathbb{R}\}$ and $\|ab\|_V \leq \|a\|_V \|b\|_V$.

Bounded symbols. In what follows we denote by $L^\infty(\mathbb{R}^n, V(\mathbb{R}))$ the set of all functions $a : \mathbb{R}^n \times \overline{\mathbb{R}} \to \mathbb{C}$ such that $x \mapsto a(x,\cdot)$ is a bounded measurable $V(\mathbb{R})$-valued function on $\mathbb{R}^n$. Since the Banach space $V(\mathbb{R})$ is separable, we conclude according to [37, Chapter IV, Theorem 23_2] that every measurable $V(\mathbb{R})$-valued function is a limit a.e. of a sequence of simple measurable functions $\widetilde{a}_k : \mathbb{R}^n \to V(\mathbb{R})$ having only finite sets of values $b_i \in V(\mathbb{R})$, with measurable pre-images $\widetilde{a}_k^{-1}(b_i)$. This implies that the functions $x \mapsto a(x,\lambda)$ for all $\lambda \in \overline{\mathbb{R}}$ and the function $x \mapsto \|a(x,\cdot)\|_V$ are measurable on $\mathbb{R}^n$ as limits a.e. of corresponding sequences of simple measurable functions. Therefore, $a(\cdot,\lambda) \in L^\infty(\mathbb{R}^n)$ for every $\lambda \in \overline{\mathbb{R}}$, and the function $x \mapsto \|a(x,\cdot)\|_V$, where

$$\|a(x,\cdot)\|_V := \max_{\lambda\in\overline{\mathbb{R}}} |a(x,\lambda)| + \int_{\mathbb{R}} |\partial_\lambda a(x,\lambda)|d\lambda, \tag{2.6}$$

belongs to $L^\infty(\mathbb{R}^n)$. Clearly, $L^\infty(\mathbb{R}^n, V(\mathbb{R}))$ is a Banach algebra with the norm

$$\|a\|_{L^\infty(\mathbb{R}^n, V(\mathbb{R}))} := \operatorname*{ess\,sup}_{x\in\mathbb{R}^n} \|a(x,\cdot)\|_V. \tag{2.7}$$

Bounded continuous symbols. Let $C_b(\mathbb{R}^n, V(\mathbb{R}))$ stand for the set of all functions $a : \mathbb{R}^n \times \overline{\mathbb{R}} \to \mathbb{C}$ such that $x \mapsto a(x,\cdot)$ is a bounded continuous $V(\mathbb{R})$-valued function on $\mathbb{R}^n$. If $a \in C_b(\mathbb{R}^n, V(\mathbb{R}))$, the functions $x \mapsto a(x,\lambda)$ for all $\lambda \in \overline{\mathbb{R}}$ and the function $x \mapsto \|a(x,\cdot)\|_V$ given by (2.6) belong to $C_b(\mathbb{R}^n)$. Clearly, $C_b(\mathbb{R}^n, V(\mathbb{R}))$ is a Banach subalgebra of $L^\infty(\mathbb{R}^n, V(\mathbb{R}))$ with the norm

$$\|a\|_{C_b(\mathbb{R}^n,V(\mathbb{R}))} := \sup_{x\in\mathbb{R}^n} \big\|a(x,\cdot)\big\|_V. \tag{2.8}$$

Let $\mathfrak{S}_n$ be the Banach subalgebra of all functions $a(x,\lambda)$ in $C_b(\mathbb{R}^n, V(\mathbb{R}))$ such that the $V(\mathbb{R})$-valued function $x \mapsto a(x,\cdot)$ is uniformly continuous on $\mathbb{R}^n$ and

$$\lim_{|h|\to 0} \sup_{x\in\mathbb{R}^n} \big\|a(x,\cdot) - a^h(x,\cdot)\big\|_V = 0 \tag{2.9}$$

where $a^h(x,\lambda) := a(x,\lambda+h)$ for all $(x,\lambda) \in \mathbb{R}^n \times \mathbb{R}$.

By analogy with [16, Theorem 2.1] we prove the following.

Theorem 2.1. *Every function $a(x,\lambda) \in \mathfrak{S}_n$ can be approximated in the norm of $C_b(\mathbb{R}^n, V(\mathbb{R}))$ by functions $a_\varepsilon(x,\lambda) \in \mathfrak{S}_n$ $(\varepsilon \to 0)$ such that $\partial_x^\alpha \partial_\lambda^j a_\varepsilon(x,\lambda) \in \mathfrak{S}_n$ for all multi-indices $\alpha = (\alpha_1,\dots,\alpha_n)$ with $|\alpha| = 0,1,2,\dots$ and for all $j = 0,1,2,\dots$.*

Proof. Fix a function φ given by (2.5) and define functions ψ_ε and φ_ε as in (2.4). For every $\varepsilon > 0$, every multi-index α and every $k = 0,1,2,\dots$, we set

$$I_\alpha := \int_{\mathbb{R}^n} \big|\partial_x^\alpha \psi_\varepsilon(x)\big|dx < \infty, \quad I_k := \int_{\mathbb{R}} \big|\varphi_\varepsilon^{(k)}(\mu)\big|d\mu < \infty. \tag{2.10}$$

Following [8, Chapter 3, Lemma 10.4], we construct approximations a_ε in the form

$$a_\varepsilon(x,\lambda) = \iint_{\mathbb{R}^{n+1}} \psi_\varepsilon(x-y)\varphi_\varepsilon(\lambda-\mu)\, a(y,\mu)dyd\mu \tag{2.11}$$

for $(x,\lambda) \in \mathbb{R}^n \times \mathbb{R}$. Let $|\alpha|, j = 0,1,2,\dots$. Since $\psi_\varepsilon \in C_0^\infty(\mathbb{R}^n)$ and $\varphi_\varepsilon \in C_0^\infty(\mathbb{R})$, we obtain

$$\big[\partial_x^\alpha \partial_\lambda^j a_\varepsilon\big](x,\lambda) = \iint_{\mathbb{R}^{n+1}} \partial_y^\alpha \psi_\varepsilon(y)\, \varphi_\varepsilon^{(j)}(\lambda-\mu)\, a(x-y,\mu)dyd\mu. \tag{2.12}$$

Hence, according to (2.4), (2.5) and (2.10), and because $a \in C_b(\mathbb{R}^n, C(\overline{\mathbb{R}}))$, we get

$$\big\|\big[\partial_x^\alpha \partial_\lambda^j a_\varepsilon\big](x,\cdot)\big\|_C \le I_j \int_{\mathbb{R}^n} \big|\partial_y^\alpha \psi_\varepsilon(y)\big| \big\|a(x-y,\cdot)\big\|_C dy. \tag{2.13}$$

On the other hand, along with (2.12),

$$\big[\partial_x^\alpha \partial_\lambda^{j+1} a_\varepsilon\big](x,\lambda) = \iint_{\mathbb{R}^{n+1}} \partial_y^\alpha \psi_\varepsilon(y)\, \varphi_\varepsilon^{(j)}(\lambda-\mu)\, \partial_\mu a(x-y,\mu)dyd\mu. \tag{2.14}$$

From (2.14) and (2.10) it follows that

$$\begin{aligned} V\big(\big[\partial_x^\alpha \partial_\lambda^j a_\varepsilon\big](x,\cdot)\big) &\le \int_{\mathbb{R}} \big|\varphi_\varepsilon^{(j)}(\lambda-\mu)\big|d\lambda \left(\int_{\mathbb{R}^n} \big|\partial_y^\alpha \psi_\varepsilon(y)\big|dy \int_{\mathbb{R}} \big|\partial_\mu a(x-y,\mu)\big|d\mu\right) \\ &\le I_j \int_{\mathbb{R}^n} \big|\partial_y^\alpha \psi_\varepsilon(y)\big| V\big(a(x-y,\cdot)\big)dy. \end{aligned} \tag{2.15}$$

Combining (2.13) and (2.15) we conclude that

$$\begin{aligned}\big\|\big[\partial_x^\alpha\partial_\lambda^j a_\varepsilon\big](x,\cdot)\big\|_V &\le I_j\int_{\mathbb{R}^n}\big|\partial_y^\alpha\psi_\varepsilon(y)\big|\big\|a(x-y,\cdot)\big\|_V\,dy\\ &\le \max_{\|y\|\le\varepsilon}\big\|a(x-y,\cdot)\big\|_V\,I_\alpha I_j. \end{aligned}\tag{2.16}$$

As $\big[\partial_x^\alpha\partial_\lambda^j a_\varepsilon\big]^h=\partial_x^\alpha\partial_\lambda^j\big(a^h\big)_\varepsilon$, the estimate (2.16) implies that

$$\big\|\big[\partial_x^\alpha\partial_\lambda^j a_\varepsilon\big](x,\cdot)-\big[\partial_x^\alpha\partial_\lambda^j a_\varepsilon\big](z,\cdot)\big\|_V\le\max_{\|y\|\le\varepsilon}\big\|a(x-y,\cdot)-a(z-y,\cdot)\big\|_V\,I_\alpha I_j,$$

$$\big\|[\partial_x^\alpha\partial_\lambda^j a_\varepsilon](x,\cdot)-[\partial_x^\alpha\partial_\lambda^j a_\varepsilon]^h(x,\cdot)\big\|_V\le\max_{\|y\|\le\varepsilon}\big\|a(x-y,\cdot)-a^h(x-y,\cdot)\big\|_V\,I_\alpha I_j.$$

Hence, $\partial_x^\alpha\partial_\lambda^j a_\varepsilon\in C_b\big(\mathbb{R}^n,V(\mathbb{R})\big)$ for all $|\alpha|,j=0,1,2,\dots$, and, moreover,

$$\lim_{h\to 0}\sup_{\|x-z\|\le h}\big\|\big[\partial_x^\alpha\partial_\lambda^j a_\varepsilon\big](x,\cdot)-\big[\partial_x^\alpha\partial_\lambda^j a_\varepsilon\big](z,\cdot)\big\|_V=0,$$

$$\lim_{|h|\to 0}\sup_{x\in\mathbb{R}^n}\big\|[\partial_x^\alpha\partial_\lambda^j a_\varepsilon](x,\cdot)-[\partial_x^\alpha\partial_\lambda^j a_\varepsilon]^h(x,\cdot)\big\|_V=0.$$

Thus, all the $V(\mathbb{R})$-valued functions $x\mapsto\big[\partial_x^\alpha\partial_\lambda^j a_\varepsilon\big](x,\cdot)$ ($|\alpha|,j=0,1,2,\dots$) are uniformly continuous on $\mathbb{R}^n$ and satisfy (2.9), that is, belong to the algebra $\mathfrak{S}_n$.

Since according to (2.3),

$$\iint_{\mathbb{R}^{n+1}}\big|\psi_\varepsilon(y)\,\varphi_\varepsilon(\mu)\big|dyd\mu=\iint_{\mathbb{R}^{n+1}}\psi_\varepsilon(y)\,\varphi_\varepsilon(\mu)dyd\mu=1,\tag{2.17}$$

we obtain

$$\begin{aligned}\partial_\lambda^j a(x,\lambda)-\partial_\lambda^j a_\varepsilon(x,\lambda) &= \iint_{\mathbb{R}^{n+1}}\psi_\varepsilon(x-y)\,\varphi_\varepsilon(\lambda-\mu)\big[\partial_\lambda^j a(x,\lambda)-\partial_\lambda^j a(y,\mu)\big]dyd\mu\\ &= \iint_{\mathbb{R}^{n+1}}\psi_\varepsilon(y)\,\varphi_\varepsilon(\lambda-\mu)\big[\partial_\mu^j a(x,\mu)-\partial_\mu^j a(x-y,\mu)\big]dyd\mu\\ &\quad+\iint_{\mathbb{R}^{n+1}}\psi_\varepsilon(x-y)\,\varphi_\varepsilon(\mu)\big[\partial_\lambda^j a(x,\lambda)-\partial_\lambda^j a(x,\lambda-\mu)\big]dyd\mu,\end{aligned}$$

where $j=0,1$, $\operatorname{supp}\psi_\varepsilon\subset[-\varepsilon,\varepsilon]^n$ and $\operatorname{supp}\varphi_\varepsilon\subset[-\varepsilon,\varepsilon]$. Hence, taking into account (2.5), we infer from the latter equality for $j=0$ and $j=1$, respectively, that

$$\begin{aligned}\big\|a(x,\cdot)-a_\varepsilon(x,\cdot)\big\|_C &\le \int_{\mathbb{R}^n}\big|\psi_\varepsilon(y)\big|\big\|a(x,\cdot)-a(x-y,\cdot)\big\|_C\,dy\\ &\quad+\int_{\mathbb{R}}\big|\varphi_\varepsilon(\mu)\big|\big\|a(x,\cdot)-a^{-\mu}(x,\cdot)\big\|_C\,d\mu,\end{aligned}\tag{2.18}$$

$$\begin{aligned}
&V\big(a(x,\cdot)-a_\varepsilon(x,\cdot)\big)=\int_{\mathbb{R}}\big|\partial_\lambda a(x,\lambda)-\partial_\lambda a_\varepsilon(x,\lambda)\big|d\lambda\\
&\le\Big(\int_{\mathbb{R}}\big|\varphi_\varepsilon(\lambda-\mu)\big|d\lambda\Big)\Big(\int_{\mathbb{R}^n}\big|\psi_\varepsilon(y)\big|dy\int_{\mathbb{R}}\big|\partial_\mu a(x,\mu)-\partial_\mu a(x-y,\mu)\big|d\mu\Big)\\
&+\Big(\int_{\mathbb{R}^n}\big|\psi_\varepsilon(x-y)\big|dy\Big)\Big(\int_{\mathbb{R}}\big|\varphi_\varepsilon(\mu)\big|d\mu\int_{\mathbb{R}}\big|\partial_\lambda a(x,\lambda)-\partial_\lambda a(x,\lambda-\mu)\big|d\lambda\Big)\\
&=\int_{\mathbb{R}^n}\big|\psi_\varepsilon(y)\big|V\big(a(x,\cdot)-a(x-y,\cdot)\big)dy+\int_{\mathbb{R}}\big|\varphi_\varepsilon(\mu)\big|V\big(a(x,\cdot)-a^{-\mu}(x,\cdot)\big)d\mu.
\end{aligned}\tag{2.19}$$

From (2.8), (2.18) and (2.19) it follows that

$$\begin{aligned}
\big\|a-a_\varepsilon\big\|_{C_b(\mathbb{R}^n,V(\mathbb{R}))}&=\sup_{x\in\mathbb{R}^n}\big\|a(x,\cdot)-a_\varepsilon(x,\cdot)\big\|_V\\
&\le\sup_{x\in\mathbb{R}^n}\Big(\int_{\mathbb{R}^n}\big|\psi_\varepsilon(y)\big|\big\|a(x,\cdot)-a(x-y,\cdot)\big\|_V\,dy\\
&\quad+\int_{\mathbb{R}}\big|\varphi_\varepsilon(\mu)\big|\big\|a(x,\cdot)-a^{-\mu}(x,\cdot)\big\|_V\,d\mu\Big)\\
\le\sup_{x\in\mathbb{R}^n}\max_{\|y\|\le\varepsilon}\big\|a(x,\cdot)-a(x-y,\cdot)\big\|_V&+\sup_{x\in\mathbb{R}^n}\max_{|h|\le\varepsilon}\big\|a(x,\cdot)-a^{h}(x,\cdot)\big\|_V.
\end{aligned}$$

Hence $\lim\limits_{\varepsilon\to 0}\big\|a-a_\varepsilon\big\|_{C_b(\mathbb{R}^n,V(\mathbb{R}))}=0$ in view of the uniform continuity of the $V(\mathbb{R})$-valued function $x\mapsto a(x,\cdot)$ on $\mathbb{R}^n$ and according to (2.9). □

Corollary 2.2. *If $\partial_x^\alpha\partial_\lambda^j a(x,\lambda)\in\mathfrak{S}_n$ for all multi-indices $\alpha=(\alpha_1,\dots,\alpha_n)$ with $|\alpha|=0,1,2,\dots,N$ and for all $j=0,1,2,\dots,M$, then for all mentioned α and j, $\partial_x^\alpha\partial_\lambda^j a(x,\lambda)=\lim\limits_{\varepsilon\to 0}\partial_x^\alpha\partial_\lambda^j a_\varepsilon(x,\lambda)$ in the norm of $C_b\big(\mathbb{R}^n,V(\mathbb{R})\big)$.*

Proof. By virtue of Theorem 2.1, the function $a(x,\lambda)\in\mathfrak{S}_n$ can be approximated in the norm of $C_b\big(\mathbb{R}^n,V(\mathbb{R})\big)$ by the functions

$$a_\varepsilon(x,\lambda)=\big[\psi_\varepsilon(x)\varphi_\varepsilon(\lambda)\big]*a(x,\lambda)\in\mathfrak{S}_n.\tag{2.20}$$

Since $\partial_x^\alpha\partial_\lambda^j a(x,\lambda)\in\mathfrak{S}_n$ too, from (2.20) it follows that

$$\partial_x^\alpha\partial_\lambda^j a_\varepsilon(x,\lambda)=\big[\psi_\varepsilon(x)\varphi_\varepsilon(\lambda)\big]*\partial_x^\alpha\partial_\lambda^j a(x,\lambda)\in\mathfrak{S}_n.$$

Hence the function $\partial_x^\alpha\partial_\lambda^j a(x,\lambda)$ is approximated in the norm of $C_b\big(\mathbb{R}^n,V(\mathbb{R})\big)$ by the functions $\partial_x^\alpha\partial_\lambda^j a_\varepsilon(x,\lambda)$. □

Slowly oscillating symbols. By analogy with (2.1), for $a\in C_b\big(\mathbb{R}^n,V(\mathbb{R})\big)$, we define

$$\begin{aligned}
cm_x^C(a)&:=\max\Big\{\big\|a(x+y,\cdot)-a(x,\cdot)\big\|_C:\ y\in\mathbb{R}^n,\ \|y\|\le 1\Big\},\\
cm_x^V(a)&:=\max\Big\{\big\|a(x+y,\cdot)-a(x,\cdot)\big\|_V:\ y\in\mathbb{R}^n,\ \|y\|\le 1\Big\}.
\end{aligned}\tag{2.21}$$

Let $\mathcal{E}_n^C$ and $\mathcal{E}_n^V$ be the subsets of all functions $a : \mathbb{R}^n \times \overline{\mathbb{R}} \to \mathbb{C}$ in $\mathfrak{S}_n$ that slowly oscillate at ∞, that is, satisfy the conditions

$$\lim_{\|x\|\to\infty} cm_x^C(a) = 0 \quad \text{and} \quad \lim_{\|x\|\to\infty} cm_x^V(a) = 0, \tag{2.22}$$

respectively. Thus, $\mathcal{E}_n^C$ is contained in $SO\big(\mathbb{R}^n, C(\overline{\mathbb{R}})\big)$, the C^*-algebra of all bounded continuous $C(\overline{\mathbb{R}})$-valued functions on $\mathbb{R}^n$ that slowly oscillate at ∞. Analogously, $\mathcal{E}_n^V$ is contained in $SO\big(\mathbb{R}^n, V(\mathbb{R})\big)$, the Banach algebra of all bounded continuous $V(\mathbb{R})$-valued functions on $\mathbb{R}^n$ that slowly oscillate at ∞. Obviously, each function a in $SO\big(\mathbb{R}^n, C(\overline{\mathbb{R}})\big)$ or in $SO\big(\mathbb{R}^n, V(\mathbb{R})\big)$ automatically is uniformly continuous on $\mathbb{R}^n$ with values in $C(\overline{\mathbb{R}})$ or in $V(\mathbb{R})$, and for every $\lambda \in \overline{\mathbb{R}}$ the function $x \mapsto a(x, \lambda)$ belongs to $SO(\mathbb{R}^n)$. Obviously, the sets $\mathcal{E}_n^C = \mathfrak{S}_n \cap SO\big(\mathbb{R}^n, C(\overline{\mathbb{R}})\big)$ and $\mathcal{E}_n^V = \mathfrak{S}_n \cap SO\big(\mathbb{R}^n, V(\mathbb{R})\big)$ are Banach subalgebras of $C_b\big(\mathbb{R}^n, V(\mathbb{R})\big)$.

Theorem 2.3. *Let $\mathcal{E}_n \in \{\mathcal{E}_n^C, \mathcal{E}_n^V\}$. Every function $a(x,\lambda) \in \mathcal{E}_n$ can be approximated in the norm of $C_b\big(\mathbb{R}^n, V(\mathbb{R})\big)$ by functions $a_\varepsilon \in \mathcal{E}_n$ $(\varepsilon \to 0)$ such that $\partial_x^\alpha \partial_\lambda^j a_\varepsilon(x,\lambda) \in \mathcal{E}_n$ for all multi-indices $\alpha = (\alpha_1, \dots, \alpha_n)$ with $|\alpha| = 0, 1, 2, \dots$ and for all $j = 0, 1, 2, \dots$. In particular, if $|\alpha| > 0$, then*

$$\begin{aligned} &\lim_{\|x\|\to\infty} \big\|\big[\partial_x^\alpha \partial_\lambda^j a_\varepsilon\big](x,\cdot)\big\|_C = 0 \quad \text{for } a \in \mathcal{E}_n^C, \\ &\lim_{\|x\|\to\infty} \big\|\big[\partial_x^\alpha \partial_\lambda^j a_\varepsilon\big](x,\cdot)\big\|_V = 0 \quad \text{for } a \in \mathcal{E}_n^V. \end{aligned} \tag{2.23}$$

Proof. By Theorem 2.1, all the derivatives $\partial_x^\alpha \partial_\lambda^j a_\varepsilon$ $(|\alpha|, j = 0, 1, 2, \dots)$ of the functions a_ε $(\varepsilon > 0)$ given by (2.11) are in $\mathfrak{S}_n$. Moreover, in view of (2.21), (2.22) and the inequalities

$$\big\|\big[\partial_x^\alpha \partial_\lambda^j a_\varepsilon\big](x,\cdot) - \big[\partial_x^\alpha \partial_\lambda^j a_\varepsilon\big](z,\cdot)\big\|_C \le \max_{\|y\|\le\varepsilon} \big\|a(x-y,\cdot) - a(z-y,\cdot)\big\|_C I_\alpha I_j,$$

$$\big\|\big[\partial_x^\alpha \partial_\lambda^j a_\varepsilon\big](x,\cdot) - \big[\partial_x^\alpha \partial_\lambda^j a_\varepsilon\big](z,\cdot)\big\|_V \le \max_{\|y\|\le\varepsilon} \big\|a(x-y,\cdot) - a(z-y,\cdot)\big\|_V I_\alpha I_j$$

deduced, respectively, from (2.13) and (2.16), we conclude that a_ε and all its derivatives satisfy the condition

$$\begin{aligned} \lim_{\|x\|\to\infty} cm_x^C(\partial_x^\alpha \partial_\lambda^j a_\varepsilon) &= \lim_{\|x\|\to\infty} cm_x^C(a) = 0 \quad \text{if } a \in \mathcal{E}_n^C, \\ \lim_{\|x\|\to\infty} cm_x^V(\partial_x^\alpha \partial_\lambda^j a_\varepsilon) &= \lim_{\|x\|\to\infty} cm_x^V(a) = 0 \quad \text{if } a \in \mathcal{E}_n^V, \end{aligned} \tag{2.24}$$

and consequently belong to $\mathcal{E}_n$. Furthermore, if $|\alpha| > 0$, then $\displaystyle\int_{\mathbb{R}} \partial_y^\alpha \psi_\varepsilon(y)dy = 0$ because $\psi_\varepsilon(x) = \varphi_\varepsilon(x_1)\varphi_\varepsilon(x_2)\cdots\varphi_\varepsilon(x_n)$ and $\varphi_\varepsilon(x_j) = 0$ for $|x_j| > \varepsilon$. Therefore,

$$\big[\partial_x^\alpha \partial_\lambda^j a_\varepsilon\big](x,\lambda) = \iint_{\mathbb{R}^{n+1}} \partial_y^\alpha \psi_\varepsilon(y)\, \varphi_\varepsilon^{(j)}(\lambda-\mu)\big[a(x-y,\mu) - a(x,\mu)\big]dyd\mu.$$

Hence,

$$\left\|\left[\partial_x^k \partial_\lambda^j a_\varepsilon\right](x,\cdot)\right\|_C \le \max_{\|y\|\le\varepsilon} \left\|a(x-y,\cdot)-a(x,\cdot)\right\|_C I_\alpha I_j,$$
$$\left\|\left[\partial_x^k \partial_\lambda^j a_\varepsilon\right](x,\cdot)\right\|_V \le \max_{\|y\|\le\varepsilon} \left\|a(x-y,\cdot)-a(x,\cdot)\right\|_V I_\alpha I_j,$$

which together with (2.24) imply (2.23) for all $|\alpha| \in \mathbb{N}$ and $j = 0,1,2,\dots$. □

Theorem 2.3 and Corollary 2.2 immediately imply the following.

Corollary 2.4. *If $\mathcal{E}_n \in \{\mathcal{E}_n^C, \mathcal{E}_n^V\}$ and $\partial_x^\alpha \partial_\lambda^j a(x,\lambda) \in \mathcal{E}_n$ for all multi-indices $\alpha = (\alpha_1,\dots,\alpha_n)$ with $|\alpha| = 0,1,2,\dots,N$ and for all $j = 0,1,2,\dots,M$, then for all mentioned α and j satisfying the condition $|\alpha| > 0$,*

$$\begin{aligned} \lim_{\|x\|\to\infty} \left\|\left[\partial_x^\alpha \partial_\lambda^j a\right](x,\cdot)\right\|_C = 0 \quad if \ \ a \in \mathcal{E}_n^C, \\ \lim_{\|x\|\to\infty} \left\|\left[\partial_x^\alpha \partial_\lambda^j a\right](x,\cdot)\right\|_V = 0 \quad if \ \ a \in \mathcal{E}_n^V. \end{aligned} \tag{2.25}$$

Special slowly oscillating symbols. Let $a(x,\lambda) \in C_b(\mathbb{R}^n, V(\mathbb{R}))$. Taking into account the fact that $\partial_\lambda a(x,\lambda) \in C_b(\mathbb{R}^n, L^1(\mathbb{R}))$, for $M \in \mathbb{R}$ we put

$$V_M^{+\infty}(a(x,\cdot)) := \int_M^{+\infty} |\partial_\lambda a(x,\lambda)| d\lambda, \ \ V_{-\infty}^M(a(x,\cdot)) := \int_{-\infty}^M |\partial_\lambda a(x,\lambda)| d\lambda. \tag{2.26}$$

To construct a Fredholm theory for pseudodifferential operators with compound symbols, we need to introduce the following subset of slowly oscillating symbols:

$$\widetilde{\mathcal{E}}_n := \left\{a \in \mathcal{E}_n^C : \lim_{M\to-\infty} \sup_{x\in\mathbb{R}^n} V_{-\infty}^M(a(x,\cdot)) = \lim_{M\to+\infty} \sup_{x\in\mathbb{R}^n} V_M^{+\infty}(a(x,\cdot)) = 0\right\}. \tag{2.27}$$

Obviously, $\widetilde{\mathcal{E}}_n$ is a Banach subalgebra of the Banach algebra $C_b(\mathbb{R}^n, V(\mathbb{R}))$.

Theorem 2.5. *Every function $a \in \widetilde{\mathcal{E}}_n$ belongs to $\mathcal{E}_n^V$ and can be approximated in the norm of $C_b(\mathbb{R}^n, V(\mathbb{R}))$ by functions $a_\varepsilon \in \widetilde{\mathcal{E}}_n$ ($\varepsilon \to 0$) such that $\partial_x^\alpha \partial_\lambda^j a_\varepsilon(x,\lambda) \in \widetilde{\mathcal{E}}_n$ for all multi-indices $\alpha = (\alpha_1,\dots,\alpha_n)$ with $|\alpha| = 0,1,2,\dots$ and for all $j = 0,1,2,\dots$.*

Proof. By Theorem 2.3, each function $a \in \mathcal{E}_n^C$ can be approximated in $C_b(\mathbb{R}^n, V(\mathbb{R}))$ by functions $a_\varepsilon \in \mathcal{E}_n^C$ such that $\partial_x^\alpha \partial_\lambda^j a_\varepsilon(x,\lambda) \in \mathcal{E}_n^C$ for all $|\alpha|, j = 0,1,2,\dots$. Analogously to (2.15) for every $M > 0$ and every $\varepsilon > 0$ we get

$$\begin{aligned} V_M^{+\infty}\left(\left[\partial_x^\alpha \partial_\lambda^j a_\varepsilon\right](x,\cdot)\right) &\le \int_{\mathbb{R}^n} \left|\partial_y^\alpha \psi_\varepsilon(x-y)\right| dy \left(\int_{\mathbb{R}} \left|\varphi_\varepsilon^{(j)}(\mu)\right| d\mu \int_M^{+\infty} \left|\partial_\lambda a(y,\lambda-\mu)\right| d\lambda\right) \\ &\le \sup_{x\in\mathbb{R}^n} V_{M-\varepsilon}^{+\infty}(a(x,\cdot)) I_\alpha I_j, \end{aligned} \tag{2.28}$$

where I_α, I_j are given by (2.10). Similarly,

$$V_{-\infty}^{-M}\left(\left[\partial_x^\alpha \partial_\lambda^j a_\varepsilon\right](x,\cdot)\right) \le \sup_{x\in\mathbb{R}^n} V_{-\infty}^{-M+\varepsilon}(a(x,\cdot)) I_k I_j. \tag{2.29}$$

Finally, (2.28), (2.29), and (2.27) yield that $\partial_x^\alpha \partial_\lambda^j a_\varepsilon(x,\lambda) \in \widetilde{\mathcal{E}}_n$ for all $|\alpha|, j = 0, 1, 2, \ldots$.

Furthermore, from the first equality in (2.25), from (2.27) and from the estimate

$$V\big([\partial_x^\alpha \partial_\lambda^j a_\varepsilon](x,\cdot)\big) \le 2M \big\|[\partial_x^\alpha \partial_\lambda^{j+1} a_\varepsilon](x,\cdot)\big\|_C + \Big(V_{-\infty}^{-M} + V_M^{+\infty}\Big)\big([\partial_x^\alpha \partial_\lambda^j a_\varepsilon](x,\cdot)\big)$$

it follows that

$$\lim_{\|x\|\to\infty} \big\|[\partial_x^\alpha \partial_\lambda^j a_\varepsilon](x,\cdot)\big\|_V = 0 \quad \text{for all } |\alpha| = 1, 2, \ldots \text{ and all } j = 0, 1, 2, \ldots,$$

which in its turn implies that all the functions $\partial_x^\alpha \partial_\lambda^j a_\varepsilon$ ($|\alpha|, j = 0, 1, 2, \ldots$) belong to $SO\big(\mathbb{R}^n, V(\mathbb{R})\big)$. Since $\lim_{\varepsilon\to\infty} \|a - a_\varepsilon\|_{C_b(\mathbb{R}^n, V(\mathbb{R}))} = 0$ and $SO(\mathbb{R}^n, V(\mathbb{R}))$ is a Banach subalgebra of $C_b\big(\mathbb{R}^n, V(\mathbb{R})\big)$, the function $a \in \widetilde{\mathcal{E}}_n$ also belongs to $SO\big(\mathbb{R}^n, V(\mathbb{R})\big)$. Finally, $a \in \mathcal{E}_n^V$ because $\mathcal{E}_n^C = \mathfrak{S}_n \cap SO\big(\mathbb{R}^n, C(\overline{\mathbb{R}})\big)$, $\mathcal{E}_n^V = \mathfrak{S}_n \cap SO\big(\mathbb{R}^n, V(\mathbb{R})\big)$, and hence $\mathcal{E}_n^V = \mathcal{E}_n^C \cap SO\big(\mathbb{R}^n, V(\mathbb{R})\big)$. □

3. Oscillatory integrals and pseudodifferential operators

Let $\chi \in C_0^\infty(\mathbb{R}\times\mathbb{R})$ and $\chi(y,\eta) = 1$ in a neighborhood of the origin. Set $\chi_\varepsilon(y,\eta) = \chi(\varepsilon y, \varepsilon\eta)$. If for a function a defined on $\mathbb{R}\times\mathbb{R}$ the limit

$$\lim_{\varepsilon\to 0} \frac{1}{2\pi} \iint_{\mathbb{R}^2} \chi_\varepsilon(y,\eta) a(y,\eta) e^{-iy\eta}\, dy d\eta$$

exists and does not depend on the choice of the cut-off function χ, then it is called the (double) *oscillatory integral of* a and is denoted by $\mathrm{Os}\big[a(y,\eta)e^{-iy\eta}\big]$ (see, e.g., [38, Chapter 1], [42, Vol. 1, Chapter 1], or [30, Chapter 2]).

Clearly, if $a \in L^1(\mathbb{R}\times\mathbb{R})$, then, by the Lebesgue dominated convergence theorem (see, e.g., [33, Theorem I.11]), $\mathrm{Os}\big[a(y,\eta)e^{-iy\eta}\big]$ exists and

$$\mathrm{Os}\big[a(y,\eta)e^{-iy\eta}\big] = \frac{1}{2\pi} \iint_{\mathbb{R}^2} a(y,\eta) e^{-iy\eta}\, dy d\eta. \tag{3.1}$$

In what follows we will use the notation

$$\langle y\rangle^{-2} = (1+y^2)^{-1}, \quad \langle D_\eta\rangle^2 = I - \partial_\eta^2, \quad D_\eta = -i\partial_\eta \tag{3.2}$$

and will apply the following regularization of oscillatory integrals (see, e.g., [30, Theorem 2.1.3]), which is based on the relations

$$\langle y\rangle^{-2}\langle D_\eta\rangle^2 e^{-iy\eta} = e^{-iy\eta}, \quad \langle \eta\rangle^{-2}\langle D_y\rangle^2 e^{-iy\eta} = e^{-iy\eta}, \tag{3.3}$$

integrating by parts, and on the Lebesgue dominated convergence theorem.

Lemma 3.1. *If $\partial_\eta^j \partial_y^k a(y,\eta) \in C_b\big(\mathbb{R}, V(\mathbb{R})\big)$ for all $k = 0, 1, 2$ and $j = 0, 1$, then the oscillatory integral $\mathrm{Os}\big[a(y,\eta)e^{-iy\eta}\big]$ exists and*

$$\mathrm{Os}\big[a(y,\eta)e^{-iy\eta}\big] = \frac{1}{2\pi} \iint_{\mathbb{R}^2} \langle y\rangle^{-2}\langle D_\eta\rangle^2 \Big\{\langle\eta\rangle^{-2}\langle D_y\rangle^2 \big\{a(y,\eta)\big\}\Big\} e^{-iy\eta}\, dy d\eta. \tag{3.4}$$

Proof. Choosing a cut-off function $\chi \in C_0^\infty(\mathbb{R}\times\mathbb{R})$ and taking into account the relations (3.3), we deduce by integrating by parts that

$$\begin{aligned}&\iint_{\mathbb{R}^2} \chi_\varepsilon(y,\eta)a(y,\eta)e^{-iy\eta}dyd\eta\\ &= \iint_{\mathbb{R}^2} \chi_\varepsilon(y,\eta)a(y,\eta)\langle\eta\rangle^{-2}\langle D_y\rangle^2\Big\{\langle y\rangle^{-2}\langle D_\eta\rangle^2\{e^{-iy\eta}\}\Big\}dyd\eta\\ &= \iint_{\mathbb{R}^2} \langle y\rangle^{-2}\langle D_\eta\rangle^2\Big\{\langle\eta\rangle^{-2}\langle D_y\rangle^2\{\chi_\varepsilon(y,\eta)a(y,\eta)\}\Big\}\,e^{-iy\eta}dyd\eta.\end{aligned} \tag{3.5}$$

In view of the relations

$$\partial_\eta\{\langle\eta\rangle^{-2}\} = -2\eta\langle\eta\rangle^{-4},\quad \partial_\eta^2\{\langle\eta\rangle^{-2}\} = (6\eta^2-2)\langle\eta\rangle^{-6},$$

we conclude that

$$\sup_{k=1,2}\big|\partial_\eta^k\{\langle\eta\rangle^{-2}\}\big| \le 2\langle\eta\rangle^{-2}\quad\text{for all}\quad \eta\in\mathbb{R}.$$

Consequently,

$$\begin{aligned}\big|\langle D_\eta\rangle^2\{\langle\eta\rangle^{-2}f(y,\eta)\}\big| &= \big|\big[\langle\eta\rangle^{-2}+\big(2-6\eta^2\big)\langle\eta\rangle^{-6}\big]f(y,\eta)\\ &+\ 4\eta\langle\eta\rangle^{-4}\partial_\eta f(y,\eta)-\langle\eta\rangle^{-2}\partial_\eta^2 f(y,\eta)\big|\\ &\le\ \langle\eta\rangle^{-2}\big[3\big|f(y,\eta)\big|+2\big|\partial_\eta f(y,\eta)\big|+\big|\partial_\eta^2 f(y,\eta)\big|\big].\end{aligned} \tag{3.6}$$

Taking $f(y,\eta) = \langle D_y\rangle^2\{\chi_\varepsilon(y,\eta)a(y,\eta)\}$, we deduce from (3.6) that

$$\begin{aligned}&\Big|\langle y\rangle^{-2}\langle D_\eta\rangle^2\Big\{\langle\eta\rangle^{-2}\langle D_y\rangle^2\{\chi_\varepsilon(y,\eta)a(y,\eta)\}\Big\}e^{-iy\eta}\Big|\\ &\le 3\langle y\rangle^{-2}\langle\eta\rangle^{-2}\big|f(y,\eta)\big|+2\langle y\rangle^{-2}\langle\eta\rangle^{-2}\big|\partial_\eta f(y,\eta)\big|+\langle y\rangle^{-2}\big|\partial_\eta^2 f(y,\eta)\big|.\end{aligned} \tag{3.7}$$

As $\partial_\eta^2\partial_y^k\{\chi_\varepsilon(y,\eta)a(y,\eta)\} \in C_b\big(\mathbb{R},L^1(\mathbb{R})\big)$ and $\partial_\eta^j\partial_y^k\{\chi_\varepsilon(y,\eta)a(y,\eta)\} \in C_b(\mathbb{R}\times\mathbb{R})$ for all $k=0,1,2$ and $j=0,1$, each summand in (3.7) belongs to $L^1(\mathbb{R}\times\mathbb{R})$, and hence the last double integral in (3.5) exists.

Taking into account the facts that the function

$$(y,\eta)\mapsto\ \langle y\rangle^{-2}\langle D_\eta\rangle^2\Big\{\langle\eta\rangle^{-2}\langle D_y\rangle^2\{a(y,\eta)\}\Big\}\,e^{-iy\eta}$$

belongs to $L^1(\mathbb{R}\times\mathbb{R})$ and, therefore, the double integral on the right of (3.4) exists, we proceed to prove the equality (3.4).

By the definition of oscillatory integrals and by (3.5),

$$\begin{aligned}&\mathrm{Os}\big[a(y,\eta)e^{-iy\eta}\big] = \lim_{\varepsilon\to 0}\frac{1}{2\pi}\iint_{\mathbb{R}^2}\chi_\varepsilon(y,\eta)a(y,\eta)e^{-iy\eta}dyd\eta\\ &= \lim_{\varepsilon\to 0}\frac{1}{2\pi}\iint_{\mathbb{R}^2}\langle y\rangle^{-2}\langle D_\eta\rangle^2\Big\{\langle\eta\rangle^{-2}\langle D_y\rangle^2\{\chi_\varepsilon(y,\eta)a(y,\eta)\}\Big\}\,e^{-iy\eta}dyd\eta.\end{aligned} \tag{3.8}$$

Applying the Lebesgue dominated convergence theorem and taking into account the fact that for $k+j>0$ all the partial derivatives $\partial_y^k\partial_\eta^j\chi_\varepsilon(y,\eta)$ uniformly converge on compacts to zero as $\varepsilon \to 0$, we infer from (3.8) that

$$\begin{aligned}
&\mathrm{Os}\big[a(y,\eta)e^{-iy\eta}\big] \\
&= \lim_{\varepsilon\to 0}\frac{1}{2\pi}\iint_{\mathbb{R}^2}\chi_\varepsilon(y,\eta)\langle y\rangle^{-2}\langle D_\eta\rangle^2\Big\{\langle\eta\rangle^{-2}\langle D_y\rangle^2\{a(y,\eta)\}\Big\}\,e^{-iy\eta}dyd\eta \\
&= \frac{1}{2\pi}\iint_{\mathbb{R}^2}\langle y\rangle^{-2}\langle D_\eta\rangle^2\Big\{\langle\eta\rangle^{-2}\langle D_y\rangle^2\{a(y,\eta)\}\Big\}\,e^{-iy\eta}dyd\eta.
\end{aligned}$$

Thus the oscillatory integral $\mathrm{Os}\big[a(y,\eta)e^{-iy\eta}\big]$ exists, equals the double integral in (3.4) and hence does not depend on the choice of the cut-off function χ_ε. □

Lemma 3.2. *If $\partial_y^k a(x,y,\lambda) \in C_b\big(\mathbb{R}\times\mathbb{R}, V(\mathbb{R})\big)$ for $k = 0,1,2$, then the (Fourier) pseudodifferential operator A defined for every $u \in C_0^\infty(\mathbb{R})$ by the iterated integral*

$$(Au)(x) := \frac{1}{2\pi}\int_{\mathbb{R}} d\lambda \int_{\mathbb{R}} a(x,y,\lambda)e^{i(x-y)\lambda}u(y)dy, \quad x \in \mathbb{R}, \tag{3.9}$$

can be represented in the form

$$(Au)(x) = \mathrm{Os}\big[a(x,x+y,\lambda)u(x+y)e^{-iy\lambda}\big] \tag{3.10}$$

where the oscillatory integral depends on the parameter $x \in \mathbb{R}$.

Proof. Applying the equality $\langle\lambda\rangle^{-2}\langle D_y\rangle^2\{e^{-iy\lambda}\} = e^{-iy\lambda}$, we infer by integrating by parts that

$$\begin{aligned}
\int_{\mathbb{R}} a(x,y,\lambda)u(y)e^{i(x-y)\lambda}dy &= \int_{\mathbb{R}} a(x,x+y,\lambda)u(x+y)\langle\lambda\rangle^{-2}\langle D_y\rangle^2\{e^{-iy\lambda}\}dy \\
&= \langle\lambda\rangle^{-2}\int_{\mathbb{R}}\langle D_y\rangle^2\{a(x,x+y,\lambda)u(x+y)\}e^{-iy\lambda}dy,
\end{aligned}$$

where for every $x,\lambda \in \mathbb{R}$ the function

$$\langle D_y\rangle^2\{a(x,x+y,\lambda)u(x+y)\} = \big[\langle D_y\rangle^2\{a\}u - 2(\partial_y a)u' - au''\big](x,x+y,\lambda)$$

belongs to the space $L^1(\mathbb{R})$ with respect to y because $u \in C_0^\infty(\mathbb{R})$ and $\partial_y^k a(x,y,\lambda) \in C_b\big(\mathbb{R}\times\mathbb{R}, V(\mathbb{R})\big)$ for $k = 0,1,2$. Therefore, the function

$$(x,\lambda) \mapsto \int_{\mathbb{R}}\langle D_y\rangle^2\{a(x,x+y,\lambda)u(x+y)\}e^{-iy\lambda}dy$$

belongs to $C_b(\mathbb{R}\times\mathbb{R})$, and hence in (3.9) the iterated integral

$$\begin{aligned}
(Au)(x) &= \frac{1}{2\pi}\int_{\mathbb{R}} d\lambda\int_{\mathbb{R}} a(x,y,\lambda)u(y)e^{i(x-y)\lambda}dy \\
&= \frac{1}{2\pi}\int_{\mathbb{R}}\langle\lambda\rangle^{-2}d\lambda\int_{\mathbb{R}}\langle D_y\rangle^2\{a(x,x+y,\lambda)u(x+y)\}e^{-iy\lambda}dy
\end{aligned} \tag{3.11}$$

is well defined.

Since $u \in C_0^\infty(\mathbb{R})$, the function

$$(y,\lambda) \mapsto \langle\lambda\rangle^{-2}\langle D_y\rangle^2\{a(x,x+y,\lambda)u(x+y)\}\,e^{-iy\lambda}$$

belongs to $L^1(\mathbb{R}\times\mathbb{R})$ for every $x\in\mathbb{R}$. Therefore, by analogy with Lemma 3.1, for any cut-off function $\chi\in C_0^\infty(\mathbb{R}\times\mathbb{R})$ and for every $x\in\mathbb{R}$, we deduce with the help of (3.1) that

$$\begin{aligned}&\mathrm{Os}\big[a(x,x+y,\lambda)u(x+y)e^{-iy\lambda}\big]\\&=\lim_{\varepsilon\to 0}\frac{1}{2\pi}\iint_{\mathbb{R}^2}\chi_\varepsilon(y,\lambda)a(x,x+y,\lambda)u(x+y)e^{-iy\lambda}dyd\lambda\\&=\frac{1}{2\pi}\iint_{\mathbb{R}^2}\langle\lambda\rangle^{-2}\langle D_y\rangle^2\{a(x,x+y,\lambda)u(x+y)\}\,e^{-iy\lambda}dyd\lambda.\qquad(3.12)\end{aligned}$$

Thus the oscillatory integral $\mathrm{Os}\big[a(x,x+y,\lambda)u(x+y)e^{-iy\lambda}\big]$ exists and equals (3.12). Finally, applying the Fubini theorem to the latter convergent double integral, we deduce (3.10) from (3.11) and (3.12). □

Thus, under the conditions of Lemma 3.2, the pseudodifferential operator A with compound symbol $a(x,y,\lambda)$ can be defined via the oscillatory integral (3.10) depending on the parameter. In that form the operator A can be extended to functions in $C_b^2(\mathbb{R})$ whenever $\partial_\lambda^j\partial_y^k a(x,y,\lambda)\in C_b\big(\mathbb{R}\times\mathbb{R},V(\mathbb{R})\big)$ for $k=0,1,2$ and $j=0,1$. In what follows let $C_b^n(\mathbb{R})$ stand for the Banach space of n times continuously differentiable functions $f:\mathbb{R}\to\mathbb{C}$ with the norm

$$\|f\|_{C_b^n(\mathbb{R})}:=\sup_{x\in\mathbb{R}}\sum_{k=0}^{n}\Big|f^{(k)}(x)\Big|<\infty.$$

Below we also use the following simple relations:

$$\int_{\mathbb{R}}\langle y\rangle^{-2}dy=\pi,\qquad\int_{\mathbb{R}\setminus[-M,M]}\langle y\rangle^{-2}dy<2/M\quad\text{for all }\ M>0.\qquad(3.13)$$

Lemma 3.3. *If $\partial_\lambda^j\partial_y^k a(x,y,\lambda)\in C_b\big(\mathbb{R}\times\mathbb{R},V(\mathbb{R})\big)$ for $k=0,1,2$ and $j=0,1$, then the pseudodifferential operator A given by (3.10) is bounded from the space $C_b^2(\mathbb{R})$ into the space $C_b(\mathbb{R})$ and*

$$\|A\|\le\max_{k=0,1,2}\sup_{x,y\in\mathbb{R}}\Big(3\pi\big\|\partial_y^k a(x,y,\cdot)\big\|_C+2\pi\big\|\partial_\lambda\partial_y^k a(x,y,\cdot)\big\|_C+V\big(\partial_\lambda\partial_y^k a(x,y,\cdot)\big)\Big).$$

Proof. Let $u\in C_b^2(\mathbb{R})$. Obviously, $\partial_\lambda^j\partial_y^k\{a(x,x+y,\lambda)u(x+y)\}\in C_b\big(\mathbb{R}\times\mathbb{R},V(\mathbb{R})\big)$ for $k=0,1,2$ and $j=0,1$. Therefore Lemma 3.1 implies that

$$\begin{aligned}&(Au)(x)=\mathrm{Os}\big[a(x,x+y,\lambda)u(x+y)e^{-iy\lambda}\big]\\&=\frac{1}{2\pi}\iint_{\mathbb{R}^2}\langle y\rangle^{-2}\langle D_\lambda\rangle^2\Big\{\langle\lambda\rangle^{-2}\langle D_y\rangle^2\{a(x,x+y,\lambda)u(x+y)\}\Big\}\,e^{-iy\lambda}dyd\lambda\end{aligned}\qquad(3.14)$$

where the latter integral is a usual double integral of an $L^1(\mathbb{R}\times\mathbb{R})$ function for every $x\in\mathbb{R}$. From (3.14) it follows that

$$\begin{aligned}(Au)(x)=\frac{1}{2\pi}\iint_{\mathbb{R}^2}&\langle y\rangle^{-2}\langle D_\lambda\rangle^2\Big\{\langle\lambda\rangle^{-2}\Big[\langle D_y\rangle^2\{a(x,x+y,\lambda)\}u(x+y)\\&-2\partial_y a(x,x+y,\lambda)\partial_y u(x+y)-a(x,x+y,\lambda)\partial_y^2u(x+y)\Big]\Big\}\,e^{-iy\lambda}dyd\lambda.\end{aligned}$$

By the Fubini theorem, the latter double integral is represented via iterated integrals as follows:

$$\begin{aligned}
&(Au)(x)\\
&= \frac{1}{2\pi}\int_{\mathbb{R}}\langle y\rangle^{-2}u(x+y)\left(\int_{\mathbb{R}}\langle D_\lambda\rangle^2\Big\{\langle\lambda\rangle^{-2}\langle D_y\rangle^2\{a(x,x+y,\lambda)\}\Big\}\,e^{-iy\lambda}d\lambda\right)dy\\
&-\frac{1}{\pi}\int_{\mathbb{R}}\langle y\rangle^{-2}\partial_y u(x+y)\left(\int_{\mathbb{R}}\langle D_\lambda\rangle^2\{\langle\lambda\rangle^{-2}\partial_y a(x,x+y,\lambda)\}e^{-iy\lambda}d\lambda\right)dy\\
&-\frac{1}{2\pi}\int_{\mathbb{R}}\langle y\rangle^{-2}\partial_y^2 u(x+y)\left(\int_{\mathbb{R}}\langle D_\lambda\rangle^2\{\langle\lambda\rangle^{-2}a(x,x+y,\lambda)\}e^{-iy\lambda}d\lambda\right)dy.
\end{aligned}$$

Hence, $Au\in C_b(\mathbb{R})$ and, because $\int_{\mathbb{R}}\langle y\rangle^{-2}dy=\pi$, we obtain

$$\big\|Au\big\|_{C_b(\mathbb{R})}\le\big\|u\big\|_{C_b^2(\mathbb{R})}\max_{k=0,1,2}\sup_{x,y\in\mathbb{R}}\int_{\mathbb{R}}\big|\langle D_\lambda\rangle^2\{\langle\lambda\rangle^{-2}\partial_y^k a(x,y,\lambda)\}\big|d\lambda, \tag{3.15}$$

Since in view of (3.6),

$$\big|\langle D_\lambda\rangle^2\{\langle\lambda\rangle^{-2}f(x,y,\lambda)\}\big|\le\langle\lambda\rangle^{-2}\big[3\big|f(x,y,\lambda)\big|+2\big|\partial_\lambda f(x,y,\lambda)\big|+\big|\partial_\lambda^2 f(x,y,\lambda)\big|\big],$$

taking $f(x,y,\lambda)=\partial_y^k a(x,y,\lambda)$, we obtain

$$\begin{aligned}
&\int_{\mathbb{R}}\big|\langle D_\lambda\rangle^2\{\langle\lambda\rangle^{-2}\partial_y^k a(x,y,\lambda)\}\big|d\lambda\\
&\le 3\pi\big\|\partial_y^k a(x,y,\cdot)\big\|_C+2\pi\big\|\partial_\lambda\partial_y^k a(x,y,\cdot)\big\|_C+\big\|\partial_\lambda^2\partial_y^k a(x,y,\cdot)\big\|_{L^1(\mathbb{R})}.
\end{aligned}\tag{3.16}$$

Finally, (3.15) and (3.16) imply the estimate of $\|A\|$ in the lemma. □

Given $u\in C_0^\infty(\mathbb{R})$, let

$$\widehat{u}(\lambda):=(\mathcal{F}u)(\lambda):=\int_{\mathbb{R}}u(x)e^{-ix\lambda}dx\quad(\lambda\in\mathbb{R})$$

be its Fourier transform.

Lemma 3.4. *Let all the conditions of Lemma 3.3 hold. Then the pseudodifferential operator A, given by (3.10) for $u\in C_0^\infty(\mathbb{R})$, can be represented in the form*

$$\big[\sigma_A(x,D)u\big](x)=\frac{1}{2\pi}\int_{\mathbb{R}}d\lambda\int_{\mathbb{R}}\sigma_A(x,\lambda)e^{i(x-y)\lambda}u(y)dy,\quad x\in\mathbb{R}, \tag{3.17}$$

where the symbol $\sigma_A(x,\lambda)$ is given by the oscillatory integral

$$\sigma_A(x,\lambda)=\mathrm{Os}\big[a(x,x+y,\lambda+\eta)e^{-iy\eta}\big],\quad(x,\lambda)\in\mathbb{R}\times\mathbb{R}. \tag{3.18}$$

Proof. By Lemmas 3.2 and 3.3,

$$(Au)(x)=\mathrm{Os}\big[a(x,x+y,\eta)u(x+y)e^{-iy\eta}\big],$$

and the operator A is bounded from the space $C_b^2(\mathbb{R})$ into the space $C_b(\mathbb{R})$. Therefore, taking $u(x) = e^{i\lambda x} \in C_b^2(\mathbb{R})$ and following [40, Chapter 2, Theorem 3.8], we get

$$\begin{aligned} \sigma_A(x,\lambda) &:= e^{-ix\lambda}A(e^{i(\cdot)\lambda}) = \mathrm{Os}\big[e^{-ix\lambda}a(x,x+y,\eta)e^{i(x+y)\lambda}e^{-iy\eta}\big] \\ &= \mathrm{Os}\big[a(x,x+y,\eta)e^{-iy(\eta-\lambda)}\big] = \mathrm{Os}\big[a(x,x+y,\lambda+\eta)e^{-iy\eta}\big], \end{aligned} \tag{3.19}$$

which gives (3.18). Substituting the first equality of (3.19) into (3.17) we infer for $u \in C_0^\infty(\mathbb{R})$ that

$$\begin{aligned} \big[\sigma_A(x,D)u\big](x) &= \frac{1}{2\pi}\int_{\mathbb{R}} d\lambda \int_{\mathbb{R}} \sigma_A(x,\lambda)e^{i(x-y)\lambda}u(y)dy = \frac{1}{2\pi}\int_{\mathbb{R}} e^{ix\lambda}\sigma_A(x,\lambda)\widehat{u}(\lambda)d\lambda \\ &= \frac{1}{2\pi}\int_{\mathbb{R}} \big[A(e^{i(\cdot)\lambda})\big](x)\,\widehat{u}(\lambda)d\lambda = \left[A\left(\frac{1}{2\pi}\int_{\mathbb{R}}\widehat{u}(\lambda)e^{i(\cdot)\lambda}d\lambda\right)\right](x) = (Au)(x), \end{aligned}$$

which completes the proof. □

If $\partial_\lambda^j a(x,y,\lambda) \in C_b\big(\mathbb{R}\times\mathbb{R}, V(\mathbb{R})\big)$ for $j = 0,1,2$, then we will write

$$\begin{aligned} F\big[a(x,y,\lambda)\big] &:= 3\big|a(x,y,\lambda)\big| + 2\big|\partial_\lambda a(x,y,\lambda)\big| + \big|\partial_\lambda^2 a(x,y,\lambda)\big|, \\ F_C\big[a(x,y,\cdot)\big] &:= 3\big\|a(x,y,\cdot)\big\|_C + 2\big\|\partial_\lambda a(x,y,\cdot)\big\|_C + \big\|\partial_\lambda^2 a(x,y,\cdot)\big\|_C, \\ \widetilde{F}_V\big[a(x,y,\cdot)\big] &:= 3V\big(a(x,y,\cdot)\big) + 2V\big(\partial_\lambda a(x,y,\cdot)\big) + V\big(\partial_\lambda^2 a(x,y,\cdot)\big), \\ F_V\big[a(x,y,\cdot)\big] &:= 3\big\|a(x,y,\cdot)\big\|_V + 2\big\|\partial_\lambda a(x,y,\cdot)\big\|_V + \big\|\partial_\lambda^2 a(x,y,\cdot)\big\|_V. \end{aligned} \tag{3.20}$$

Clearly, from (3.20) it follows that

$$F_V\big[a(x,y,\cdot)\big] = F_C\big[a(x,y,\cdot)\big] + \widetilde{F}_V\big[a(x,y,\cdot)\big], \tag{3.21}$$

$$\sup_{\lambda\in\mathbb{R}} F\big[a(x,\lambda)\big] \le F_C\big[a(x,y,\cdot)\big], \quad \int_{\mathbb{R}} F\big[\partial_\lambda a(x,y,\lambda)\big]d\lambda = \widetilde{F}_V\big[a(x,y,\cdot)\big]. \tag{3.22}$$

Theorem 3.5. *If $\partial_\lambda^j\partial_y^k a(x,y,\lambda) \in C_b\big(\mathbb{R}\times\mathbb{R}, V(\mathbb{R})\big)$ for all $k,j = 0,1,2$, then $\sigma_A(x,\lambda)$ given by (3.18) belongs to $C_b\big(\mathbb{R}, V(\mathbb{R})\big)$. If $\partial_\lambda^j\partial_y^k a(x,y,\lambda) \in \mathfrak{S}_2$ for all $k,j = 0,1,2$, then $\sigma_A(x,\lambda)$ belongs to $\mathfrak{S}_1$. If $\partial_\lambda^j\partial_y^k a(x,y,\lambda)$ belong to $\mathcal{E}_2^C$ (respectively, to $\mathcal{E}_2^V$) for all $k,j = 0,1,2$, then $\sigma_A(x,\lambda)$ belongs to $\mathcal{E}_1^C$ (respectively, to $\mathcal{E}_1^V$).*

Proof. According to Lemma 3.1, the oscillatory integral

$$\sigma_A(x,\lambda) = \mathrm{Os}\big[a(x,x+y,\lambda+\eta)e^{-iy\eta}\big]$$

with the parameters $x,\lambda \in \mathbb{R}$ can be represented by the convergent double integral

$$\frac{1}{2\pi}\iint_{\mathbb{R}^2} \langle y\rangle^{-2}\langle D_\eta\rangle^2\Big\{\langle\eta\rangle^{-2}\langle D_y\rangle^2\big\{a(x,x+y,\lambda+\eta)\big\}\Big\}\,e^{-iy\eta}dyd\eta. \tag{3.23}$$

Setting $f(x,y,\lambda+\eta) := \langle D_y\rangle^2\{a(x,x+y,\lambda+\eta)\}$ in (3.6) and taking into account the first equality in (3.20), we obtain

$$\Big|\langle D_\eta\rangle^2\Big\{\langle\eta\rangle^{-2}\langle D_y\rangle^2\{a(x,x+y,\lambda+\eta)\}\Big\}\Big| \le \langle\eta\rangle^{-2}\Big(3\big|\langle D_y\rangle^2\{a(x,x+y,\lambda+\eta)\}\big|$$
$$+\ 2\big|\partial_\lambda\langle D_y\rangle^2\{a(x,x+y,\lambda+\eta)\}\big| + \big|\partial_\lambda^2\langle D_y\rangle^2\{a(x,x+y,\lambda+\eta)\}\big|\Big)$$
$$= \langle\eta\rangle^{-2}F\big[\langle D_y\rangle^2\{a(x,x+y,\lambda+\eta)\}\big]. \tag{3.24}$$

Therefore, for every $x \in \mathbb{R}$ we deduce from (3.23), (3.24) and the second equality in (3.20) that

$$\big\|\sigma_A(x,\cdot)\big\|_C \le \frac{1}{2\pi}\iint_{\mathbb{R}^2}\langle y\rangle^{-2}\langle\eta\rangle^{-2}F_C\big[\langle D_y\rangle^2\{a(x,x+y,\cdot)\}\big]\,dyd\eta. \tag{3.25}$$

Let E be the union of pairwise disjoint intervals (c_k,d_k), $k=1,2,\dots,n$. As

$$\sum_{k=1}^n\big|\sigma_A(x,d_k)-\sigma_A(x,c_k)\big|$$
$$\le \frac{1}{2\pi}\iint_{\mathbb{R}^2}\langle y\rangle^{-2}\sum_{k=1}^n\Big|\langle D_\eta\rangle^2\big\{a(x,x+y,d_k+\eta)-a(x,x+y,c_k+\eta)\big\}\Big|dyd\eta$$
$$\le \frac{1}{2\pi}\int_{\mathbb{R}}\langle y\rangle^{-2}dy\int_{\mathbb{R}}d\eta\int_E\Big(\big|\partial_\lambda a(x,x+y,\lambda+\eta)\big| + \big|\partial_\lambda^3 a(x,x+y,\lambda+\eta)\big|\Big)d\lambda$$
$$\le \frac{1}{2\pi}\int_{\mathbb{R}}\langle y\rangle^{-2}dy\int_E\Big(V\big(a(x,x+y,\cdot)\big) + V\big(\partial_\lambda^2 a(x,x+y,\cdot)\big)\Big)d\lambda$$
$$\le \frac{1}{2}\sup_{x,y\in\mathbb{R}}\Big(V\big(a(x,y,\cdot)\big) + V\big(\partial_\lambda^2 a(x,y,\cdot)\big)\Big)\sum_{k=1}^n(d_k-c_k),$$

we conclude that for every $x \in \mathbb{R}$ the function $\lambda \mapsto \sigma_A(x,\lambda)$ is absolutely continuous on $\mathbb{R}$.

Let us prove now that actually $\sigma_A(x,\lambda) \in C_b\big(\mathbb{R}, V(\mathbb{R})\big)$. Let

$$f(x,y,\eta,\lambda_k,\lambda_{k-1}) := a(x,x+y,\lambda_k+\eta) - a(x,x+y,\lambda_{k-1}+\eta)$$

where $-\infty < \lambda_0 < \lambda_1 < \dots < \lambda_n < +\infty$. From (3.23) and (3.6) it follows that

$$\frac{1}{2\pi}\sum_{k=1}^n\bigg|\iint_{\mathbb{R}^2}\langle y\rangle^{-2}\langle D_\eta\rangle^2\Big\{\langle\eta\rangle^{-2}\langle D_y\rangle^2\{f(x,y,\eta,\lambda_k,\lambda_{k-1})\}\Big\}\,e^{-iy\eta}dyd\eta\bigg|$$
$$\le \frac{1}{2\pi}\iint_{\mathbb{R}^2}\langle y\rangle^{-2}\langle\eta\rangle^{-2}dyd\eta\sum_{k=1}^n\Big(3\big|\langle D_y\rangle^2\{f(x,y,\eta,\lambda_k,\lambda_{k-1})\}\big|$$
$$+\ 2\big|\partial_\lambda\langle D_y\rangle^2\{f(x,y,\eta,\lambda_k,\lambda_{k-1})\}\big| + \big|\partial_\lambda^2\langle D_y\rangle^2\{f(x,y,\eta,\lambda_k,\lambda_{k-1})\}\big|\Big).$$

Therefore, passing in the latter expression to the supremum with respect to all partitions $-\infty < \lambda_0 < \lambda_1 < \dots < \lambda_n < +\infty$ $(n \in \mathbb{N})$ and applying the third equality in (3.20), we obtain

$$\begin{aligned} V\big(\sigma_A(x,\cdot)\big) \;&\le\; \frac{1}{2\pi}\iint_{\mathbb{R}^2}\langle y\rangle^{-2}\langle\eta\rangle^{-2}\Big(3V\big(\langle D_y\rangle^2\{a(x,x+y,\cdot)\}\big)\\ &+\;2V\big(\partial_\lambda\langle D_y\rangle^2\{a(x,x+y,\cdot)\}\big)+V\big(\partial_\lambda^2\langle D_y\rangle^2\{a(x,x+y,\cdot)\}\big)\Big)dyd\eta\\ &=\;\frac{1}{2\pi}\iint_{\mathbb{R}^2}\langle y\rangle^{-2}\langle\eta\rangle^{-2}\widetilde{F}_V\big[\langle D_y\rangle^2\{a(x,x+y,\cdot)\}\big]dyd\eta. \end{aligned} \tag{3.26}$$

The inequalities (3.25) and (3.26) together with the fourth equality in (3.20) and (3.21) imply that

$$\begin{aligned} &\big\|\sigma_A(x,\cdot)\big\|_V\le\frac{1}{2\pi}\iint_{\mathbb{R}^2}\langle y\rangle^{-2}\langle\eta\rangle^{-2}F_V\big[\langle D_y\rangle^2\{a(x,x+y,\cdot)\}\big]dyd\eta\\ &\le\frac{\pi}{2}\sup_{x,y\in\mathbb{R}}F_V\big[\langle D_y\rangle^2\{a(x,y,\cdot)\}\big]=\frac{\pi}{2}\sup_{x,y\in\mathbb{R}}\Big(3\big\|\langle D_y\rangle^2\{a(x,y,\cdot)\}\big\|_V\\ &+2\big\|\partial_\lambda\langle D_y\rangle^2\{a(x,y,\cdot)\}\big\|_V+\big\|\partial_\lambda^2\langle D_y\rangle^2\{a(x,y,\cdot)\}\big\|_V\Big)<\infty. \end{aligned} \tag{3.27}$$

Taking into account (3.13) and applying a representation of the form

$$\int_{\mathbb{R}}\langle y\rangle^{-2}f(x,y)dy=\int_{\mathbb{R}\setminus[-M,M]}\langle y\rangle^{-2}f(x,y)dy+\int_{[-M,M]}\langle y\rangle^{-2}f(x,y)dy,$$

we infer from the first inequality in (3.27) that

$$\begin{aligned} \big\|\sigma_A(x,\cdot)-\sigma_A(x_0,\cdot)\big\|_V\;&\le\;\frac{\pi}{2}\sup_{|y|\le M}F_V\big[\langle D_y\rangle^2\{a(x,x+y,\cdot)-a(x_0,x_0+y,\cdot)\}\big]\\ &+\;\frac{2}{M}\sup_{x,y\in\mathbb{R}}F_V\big[\langle D_y\rangle^2\{a(x,y,\cdot)\}\big]. \end{aligned} \tag{3.28}$$

Since the $V(\mathbb{R})$-valued functions $(x,y)\mapsto\partial_\lambda^j\partial_y^k a(x,y,\lambda)\in C_b\big(\mathbb{R}\times\mathbb{R},V(\mathbb{R})\big)$ are continuous on $\mathbb{R}\times\mathbb{R}$ for all $k,j=0,1,2$, we conclude from (3.28) and (3.25) that $\sigma_A(x,\lambda)\in C_b\big(\mathbb{R},V(\mathbb{R})\big)$.

Let now $\partial_\lambda^j\partial_y^k a(x,y,\lambda)\in\mathfrak{S}_2$ for all $k,j=0,1,2$. Passing to limit in (3.28) as $M\to\infty$, we obtain

$$\big\|\sigma_A(x,\cdot)-\sigma_A(x_0,\cdot)\big\|_V\le\frac{\pi}{2}\sup_{y\in\mathbb{R}}F_V\big[\langle D_y\rangle^2\{a(x,x+y,\cdot)-a(x_0,x_0+y,\cdot)\}\big]. \tag{3.29}$$

Since the $V(\mathbb{R})$-valued functions $(x,y)\mapsto\partial_\lambda^j\partial_y^k a(x,y,\cdot)$ are uniformly continuous on $\mathbb{R}\times\mathbb{R}$ for all $k,j=0,1,2$, we conclude from (3.29) that the $V(\mathbb{R})$-valued function $x\mapsto\sigma_A(x,\cdot)$ is uniformly continuous on $\mathbb{R}$.

Analogously to (3.29), the second inequality in (3.27) gives

$$\big\|\sigma_A(x,\cdot)-\sigma_A^h(x,\cdot)\big\|_V\le\frac{\pi}{2}\sup_{x,y\in\mathbb{R}}F_V\big[\langle D_y\rangle^2\{a(x,y,\cdot)-a^h(x,y,\cdot)\}\big]. \tag{3.30}$$

The estimate (3.30) and the relations

$$\lim_{|h|\to 0}\sup_{x,y\in\mathbb{R}}\big\|\partial_\lambda^j\partial_y^k a(x,y,\cdot)-\partial_\lambda^j\partial_y^k a^h(x,y,\cdot)\big\|_V=0\quad (k,j=0,1,2) \tag{3.31}$$

imply that

$$\lim_{|h|\to 0}\sup_{x\in\mathbb{R}}\big\|\sigma_A(x,\cdot)-\sigma_A^h(x,\cdot)\big\|_V=0. \tag{3.32}$$

Consequently, the function $\sigma_A(x,\lambda)$ belongs to the Banach algebra $\mathfrak{S}_1$.

Let now $\partial_\lambda^j\partial_y^k a(x,y,\lambda)\in\mathcal{E}_2^V$ for all $k,j=0,1,2$. Hence, for these k,j,

$$\lim_{|x|\to\infty}\max_{|y|\le M,\,|h|\le 1}\Big\|\partial_\lambda^j\partial_y^k a(x,x+y,\cdot)-\partial_\lambda^j\partial_y^k a(x+h,x+y+h,\cdot)\Big\|_V=0. \tag{3.33}$$

Fix $\varepsilon>0$ and choose $M>0$ such that

$$\frac{2}{M}\sup_{x,y\in\mathbb{R}}F_V\big[\langle D_y\rangle^2\{a(x,y,\cdot)\}\big]<\frac{\varepsilon}{2}. \tag{3.34}$$

By (3.33), for all sufficiently large $|x|$ we get

$$\frac{\pi}{2}\sup_{|y|\le M,\,|h|\le 1}F_V\big[\langle D_y\rangle^2\{a(x,x+y,\cdot)-a(x+h,x+h+y,\cdot)\}\big]<\frac{\varepsilon}{2}. \tag{3.35}$$

Finally, setting $x_0=x+h$ in (3.28), we deduce from (3.28), (3.34) and (3.35) that

$$\lim_{|x|\to\infty}\max_{|h|\le 1}\big\|\sigma_A(x,\cdot)-\sigma_A(x+h,\cdot)\big\|_V=0 \tag{3.36}$$

and hence $\sigma_A(x,\lambda)\in\mathcal{E}_1^V$.

Analogously, starting from (3.25) instead of (3.28), one can prove that if $\partial_\lambda^j\partial_y^k a(x,y,\lambda)\in\mathcal{E}_2^C$ for all $k,j=0,1,2$, then $\sigma_A(x,\lambda)\in\mathcal{E}_1^C$. □

Lemma 3.6. *If $\partial_\lambda^j\partial_y^k a(x,y,\lambda)\in C_b\big(\mathbb{R}\times\mathbb{R},V(\mathbb{R})\big)$ for all $k,j=0,1,2,3$, then*

$$\sigma_A(x,\lambda)=a(x,x,\lambda)+r(x,\lambda), \tag{3.37}$$

where

$$r(x,\lambda)=\frac{1}{2\pi}\int_0^1 d\theta\iint_{\mathbb{R}^2}\langle y\rangle^{-2}\langle D_\eta\rangle^2\Big\{\langle\eta\rangle^{-2}\langle D_y\rangle^2\{D_\eta\partial_x a(x,x+\theta y,\lambda+\eta)\}\Big\}e^{-iy\eta}dyd\eta. \tag{3.38}$$

Proof. From (3.18) it follows that

$$\begin{aligned}\sigma_A(x,\lambda)&=\mathrm{Os}\big[a(x,x+y,\lambda+\eta)-a(x,x,\lambda+\eta))e^{-iy\eta}\big]\\&+\mathrm{Os}\big[a(x,x,\lambda+\eta)e^{-iy\eta}\big].\end{aligned} \tag{3.39}$$

Applying the equality

$$a(x,x+y,\lambda+\eta)-a(x,x,\lambda+\eta)=y\int_0^1\big[\partial_y a\big](x,x+\theta y,\lambda+\eta)d\theta$$

to the first oscillatory integral in (3.39) and changing in it the order of integration on the basis of the Fubini theorem, we obtain

$$\begin{aligned}
&\mathrm{Os}\big[a(x,x+y,\lambda+\eta)-a(x,x,\lambda+\eta))e^{-iy\eta}\big]\\
&=\lim_{\varepsilon\to 0}\frac{1}{2\pi}\iint_{\mathbb{R}^2}\chi_\varepsilon(y,\eta)\left(\int_0^1\big[\partial_y a\big](x,x+\theta y,\lambda+\eta)d\theta\right)ye^{-iy\eta}dyd\eta\\
&=\lim_{\varepsilon\to 0}\frac{1}{2\pi}\iint_{\mathbb{R}^2}D_\eta\left(\chi_\varepsilon(y,\eta)\int_0^1\big[\partial_y a\big](x,x+\theta y,\lambda+\eta)d\theta\right)e^{-iy\eta}dyd\eta\\
&=\lim_{\varepsilon\to 0}\frac{1}{2\pi}\int_0^1 d\theta\iint_{\mathbb{R}^2}D_\eta\{\chi_\varepsilon(y,\eta)\}\big[\partial_y a\big](x,x+\theta y,\lambda+\eta)e^{-iy\eta}dyd\eta\\
&+\lim_{\varepsilon\to 0}\frac{1}{2\pi}\int_0^1 d\theta\iint_{\mathbb{R}^2}\chi_\varepsilon(y,\eta)\big[D_\eta\partial_y a\big](x,x+\theta y,\lambda+\eta)e^{-iy\eta}dyd\eta.
\end{aligned}$$

Since the iterated integral on the right of (3.38) absolutely converges, we infer by analogy with the proof of Lemma 3.1 that

$$\begin{aligned}
&\lim_{\varepsilon\to 0}\frac{1}{2\pi}\int_0^1 d\theta\iint_{\mathbb{R}^2}D_\eta\{\chi_\varepsilon(y,\eta)\}\big[\partial_y a\big](x,x+\theta y,\lambda+\eta)e^{-iy\eta}dyd\eta=0,\\
&\lim_{\varepsilon\to 0}\frac{1}{2\pi}\int_0^1 d\theta\iint_{\mathbb{R}^2}\chi_\varepsilon(y,\eta)\big[D_\eta\partial_y a\big](x,x+\theta y,\lambda+\eta)e^{-iy\eta}dyd\eta=r(x,\lambda),
\end{aligned}$$

where $r(x,\lambda)$ is given by (3.38). Hence

$$\mathrm{Os}\big[a(x,x+y,\lambda+\eta)-a(x,x,\lambda+\eta))e^{-iy\eta}\big]=r(x,\lambda). \tag{3.40}$$

Since $\partial_\lambda^j a(x,x,\lambda)\in C_b\big(\mathbb{R},V(\mathbb{R})\big)$ for $j=0,1$, we deduce from [16, Lemma 6.4] that

$$\mathrm{Os}\big[a(x,x,\lambda+\eta)e^{-iy\eta}\big]=a(x,x,\lambda). \tag{3.41}$$

Finally, (3.37) follows from (3.38), (3.40) and (3.41). □

By analogy with [16, Lemma 9.6] we obtain the following important property of the function (3.38).

Lemma 3.7. *If $\partial_\lambda^j\partial_y^k a(x,y,\lambda)\in\mathcal{E}_2^C$ for all $k,j=0,1,2,3$, then the function $r(x,\lambda)$ given by (3.38) satisfies the condition*

$$\lim_{x^2+\lambda^2\to\infty} r(x,\lambda)=0. \tag{3.42}$$

Proof. From (3.38), (3.24) and (3.20) it follows that

$$\begin{aligned}
&|r(x,\lambda)|\\
&\le \frac{1}{2\pi}\max_{\theta\in[0,1]}\int_{\mathbb{R}}\langle y\rangle^{-2}dy\int_{\mathbb{R}}\Big|\langle D_\eta\rangle^2\Big\{\langle\eta\rangle^{-2}\langle D_y\rangle^2\big\{D_\eta\big[\partial_y a\big](x,x+\theta y,\lambda+\eta)\big\}\Big\}\Big|d\eta\\
&\le \frac{1}{\pi}\max_{k=1,3}\max_{\theta\in[0,1]}\int_{\mathbb{R}}\langle y\rangle^{-2}dy\int_{\mathbb{R}}\langle\eta\rangle^{-2}F\big[\big[\partial_\eta\partial_y^k a\big](x,x+\theta y,\lambda+\eta)\big]d\eta\\
&\le \max_{k=1,3}\bigg(\bigg(\int_{\mathbb{R}\setminus[-N,N]}\langle y\rangle^{-2}dy\bigg)\sup_{x,y\in\mathbb{R}}F_C\big[\partial_\eta\partial_y^k a(x,y,\cdot)\big]\\
&\quad+\bigg(\frac{1}{\pi}\int_{-N}^{N}\langle y\rangle^{-2}dy\bigg)\sup_{|y|\le N}\int_{\mathbb{R}}\langle\eta-\lambda\rangle^{-2}F\big[\partial_\eta\partial_y^k a(x,x+y,\eta)\big]d\eta\bigg),
\end{aligned}\tag{3.43}$$

where $N>0$. Hence for sufficiently large $M>0$ and $|\lambda|>M$, we obtain

$$\begin{aligned}
|r(x,\lambda)|\ \le\ &\max_{k=1,3}\bigg(\frac{2}{N}\sup_{x,y\in\mathbb{R}}F_C\big[\partial_\eta\partial_y^k a(x,y,\cdot)\big]+\frac{1}{|\lambda|-M}\sup_{|y|\le N}\widetilde{F}_V\big[\partial_y^k a(x,x+y,\cdot)\big]\\
&+\sup_{|y|\le N}\int_{\mathbb{R}\setminus[-M,M]}F\big[\partial_\eta\partial_y^k a(x,x+y,\eta)\big]d\eta\bigg).
\end{aligned}\tag{3.44}$$

Since $\partial_\lambda^j\partial_y^k a(x,y,\lambda)\in C_b\big(\mathbb{R}\times\mathbb{R},V(\mathbb{R})\big)$ for all $k,j=0,1,2,3$, from [16, Lemma 4.2] it follows that the latter summand in (3.44) tends to zero as $M\to\infty$, uniformly with respect to $x\in[-K,K]$ for any $K>0$. Thus, choosing sufficiently large $N,M>0$ and $|\lambda|>M$, we infer that

$$\lim_{|\lambda|\to\infty}\max_{|x|\le K}|r(x,\lambda)|=0.\tag{3.45}$$

On the other hand, from the first two inequalities in (3.43) it follows that

$$\begin{aligned}
|r(x,\lambda)|&\le\max_{k=1,3}\max_{\theta\in[0,1]}\int_{\mathbb{R}}\langle y\rangle^{-2}F_C\big[\big[\partial_\lambda\partial_y^k a\big](x,x+\theta y,\cdot)\big]dy\\
&\le\max_{k=1,3}\bigg(\frac{2}{N}\sup_{x,y\in\mathbb{R}}F_C\big[\partial_\lambda\partial_y^k a(x,y,\cdot)\big]+\pi\sup_{|y|\le N}F_C\big[\partial_\lambda\partial_y^k a(x,x+y,\cdot)\big]\bigg).
\end{aligned}$$

Since $\partial_\lambda\partial_y^k a(x,y,\lambda)\in\mathcal{E}_2^C$ for $(k=0,1,2,3)$, we infer from Corollary 2.2 and the latter estimate that

$$\lim_{|x|\to\infty}\max_{\lambda\in\overline{\mathbb{R}}}|r(x,\lambda)|=0.\tag{3.46}$$

Finally, (3.45) and (3.46) imply (3.42). □

4. Boundedness and compactness of pseudodifferential operators

Let $f\in C_0^\infty(\mathbb{R})$ and its Fourier transform be given by

$$(\mathcal{F}f)(\lambda):=\widehat{f}(\lambda):=\int_{\mathbb{R}}f(x)e^{-ix\lambda}dx.\tag{4.1}$$

For $-\infty < a < b < +\infty$, we consider the *partial sum operators* $S_{(a,b)}$ given by

$$\big(S_{(a,b)}f\big)(x) = \frac{1}{2\pi}\int_a^b \widehat{f}(\lambda)e^{ix\lambda}d\lambda, \quad x \in \mathbb{R}.$$

Such operators can be represented in the form

$$S_{(a,b)} = \mathcal{F}^{-1}\chi_{(a,b)}\mathcal{F} = \frac{1}{2}\Big[e^{iax}S_{\mathbb{R}}e^{-iax}I - e^{ibx}S_{\mathbb{R}}e^{-ibx}I\Big] \tag{4.2}$$

where $\chi_{(a,b)}$ is the characteristic function of the interval (a,b) and $S_{\mathbb{R}}$ is the Cauchy singular integral operator defined for functions $\varphi \in L^p(\mathbb{R})$ by

$$(S_{\mathbb{R}}\varphi)(x) = \lim_{\varepsilon\to 0}\frac{1}{\pi i}\int_{\mathbb{R}\setminus(x-\varepsilon,x+\varepsilon)}\frac{\varphi(t)}{t-x}dt, \quad x \in \mathbb{R}. \tag{4.3}$$

Since the operator $S_{\mathbb{R}}$ is bounded on all the spaces $L^p(\mathbb{R})$ with $1 < p < \infty$ (see, e.g., [12], [1]), from (4.2) it follows that the operators $S_{(a,b)}$ also are bounded on these spaces.

According to the celebrated Carleson-Hunt theorem on almost everywhere convergence (more precisely, by its integral analog for $L^p(\mathbb{R})$, see [6] and [20]), the maximal operator S_* given by

$$(S_*f)(x) = \sup_{-\infty<a<b<+\infty}\big|(S_{(a,b)}f)(x)\big|, \quad x \in \mathbb{R},$$

is bounded on every space $L^p(\mathbb{R})$, $1 < p < \infty$ (also see [10, p. 18]). In particular, for almost every $x \in \mathbb{R}$,

$$\sup_{\lambda\in\mathbb{R}}\big|(S_{(0,\lambda)}f)(x)\big| \le (S_*f)(x) < \infty. \tag{4.4}$$

Theorem 4.1. [16, Theorem 3.1] *If $\sigma \in L^\infty\big(\mathbb{R}, V(\mathbb{R})\big)$, then the pseudodifferential operator $\sigma(x,D)$ defined for functions $u \in C_0^\infty(\mathbb{R})$ by the iterated integral*

$$\big[\sigma(x,D)u\big](x) = \frac{1}{2\pi}\int_{\mathbb{R}}d\lambda\int_{\mathbb{R}}\sigma(x,\lambda)e^{i(x-y)\lambda}u(y)dy, \quad x \in \mathbb{R}, \tag{4.5}$$

extends to a bounded linear operator on every Lebesgue space $L^p(\mathbb{R})$, $p \in (1,\infty)$, and

$$\big\|\sigma(x,D)\big\|_{\mathcal{B}(L^p(\mathbb{R}))} \le 2\,\big\|\sigma\big\|_{L^\infty(\mathbb{R},V(\mathbb{R}))}\big\|S_*\big\|_{\mathcal{B}(L^p(\mathbb{R}))}. \tag{4.6}$$

Theorem 4.2. *If $\partial_\lambda^j\partial_y^k a(x,y,\lambda) \in C_b\big(\mathbb{R}\times\mathbb{R}, V(\mathbb{R})\big)$ for all $k,j = 0,1,2$, then the pseudodifferential operator A defined for functions $u \in C_0^\infty(\mathbb{R})$ by the iterated integral*

$$(Au)(x) := \frac{1}{2\pi}\int_{\mathbb{R}}d\lambda\int_{\mathbb{R}}a(x,y,\lambda)e^{i(x-y)\lambda}u(y)dy, \quad x \in \mathbb{R}, \tag{4.7}$$

extends to a bounded linear operator on every Lebesgue space $L^p(\mathbb{R})$, $p \in (1,\infty)$, and

$$\big\|A\big\|_{\mathcal{B}(L^p(\mathbb{R}))} \le 2\sup_{x\in\mathbb{R}}\big\|\sigma_A(x,\cdot)\big\|_V\big\|S_*\big\|_{\mathcal{B}(L^p(\mathbb{R}))}, \tag{4.8}$$

where $\sigma_A(x,\lambda)$ *is given by* (3.18) *and*

$$\begin{aligned}\big\|\sigma_A(x,\cdot)\big\|_V &\le \frac{\pi}{2}\sup_{x,y\in\mathbb{R}}\Big(3\big\|\langle D_y\rangle^2\{a(x,y,\cdot)\}\big\|_V\\ &+ 2\big\|\partial_\lambda\langle D_y\rangle^2\{a(x,y,\cdot)\}\big\|_V+\big\|\partial_\lambda^2\langle D_y\rangle^2\{a(x,y,\cdot)\}\big\|_V\Big).\end{aligned}$$

Proof. By Lemma 3.4, the pseudodifferential operator A defined for functions $u\in C_0^\infty(\mathbb{R})$ by the iterated integral (4.7) can be represented in the form

$$(Au)(x)=\big[\sigma_A(x,D)u\big](x)=\frac{1}{2\pi}\int_{\mathbb{R}}d\lambda\int_{\mathbb{R}}\sigma_A(x,\lambda)e^{i(x-y)\lambda}u(y)dy,\quad x\in\mathbb{R},$$

with $\sigma_A(x,\lambda)$ given by (3.18). By Theorem 3.5, $\sigma_A(x,\lambda)\in C_b\big(\mathbb{R},V(\mathbb{R})\big)$ and it satisfies the estimate (3.27). It remains to apply Theorem 4.1. □

Theorem 4.3. [16, Theorem 4.4] *If* $\sigma(x,\lambda)\in\mathfrak{S}_1$ *and*

$$\lim_{x^2+\lambda^2\to\infty}\sigma(x,\lambda)=0, \tag{4.9}$$

then the pseudodifferential operator $\sigma(x,D)$ *is compact on every Lebesgue space* $L^p(\mathbb{R})$, $1<p<\infty$.

Theorem 4.4. *If* $\partial_\lambda^j\partial_y^k a(x,y,\lambda)\in\mathcal{E}_2^C$ *for all* $k,j=0,1,2$, *then the pseudodifferential operator*

$$r(x,D)=\sigma_A(x,D)-\widetilde{a}(x,D), \tag{4.10}$$

where $\widetilde{a}(x,\lambda)=a(x,x,\lambda)$, *is compact on every Lebesgue space* $L^p(\mathbb{R})$, $1<p<\infty$.

Proof. Clearly, due to the conditions of the theorem, $\widetilde{a}(x,\lambda)\in\mathcal{E}_1^C\subset C_b\big(\mathbb{R},V(\mathbb{R})\big)$. Theorem 3.5 implies that $\sigma_A(x,\lambda)\in\mathcal{E}_1^C$ too. Therefore, the function $r(x,\lambda)=\sigma_A(x,\lambda)-\widetilde{a}(x,\lambda)$ also belongs to the Banach algebra $\mathcal{E}_1^C$.

By Theorems 4.1 and 4.2, the pseudodifferential operators $\widetilde{a}(x,D)$, $r(x,D)$ and $\sigma_A(x,D)$ are bounded on all the spaces $L^p(\mathbb{R})$, $1<p<\infty$. For the function $a(x,y,\lambda)\in\mathcal{E}_2^C$, we construct infinitely differentiable approximations $a_\varepsilon(x,y,\lambda)\in\mathcal{E}_2^C$ by formulas (2.11). By Theorem 2.3, the derivatives $\partial_y^k\partial_\lambda^j a_\varepsilon(x,y,\lambda)$ for all $k,j=0,1,2,\dots$ also belong to the Banach algebra $\mathcal{E}_2^C\subset C_b\big(\mathbb{R},V(\mathbb{R})\big)$ and hence $\widetilde{a}_\varepsilon(x,\lambda)=a_\varepsilon(x,x,\lambda)\in\mathcal{E}_1^C$. Moreover, Theorem 2.3 and Corollary 2.2 imply that for all $k,j=0,1,2$,

$$\lim_{\varepsilon\to0}\sup_{x,y\in\mathbb{R}}\big\|\partial_y^k\partial_\lambda^j a(x,y,\cdot)-\partial_y^k\partial_\lambda^j a_\varepsilon(x,y,\cdot)\big\|_V=0, \tag{4.11}$$

which, in particular, gives

$$\lim_{\varepsilon\to0}\sup_{x\in\mathbb{R}}\big\|\widetilde{a}(x,\cdot)-\widetilde{a}_\varepsilon(x,\cdot)\big\|_V=0. \tag{4.12}$$

From (3.27) and (4.11) it follows that

$$\lim_{\varepsilon\to0}\sup_{x\in\mathbb{R}}\big\|\sigma_A(x,\cdot)-\sigma_\varepsilon(x,\cdot)\big\|_V=0 \tag{4.13}$$

where

$$\sigma_\varepsilon(x,\lambda)=\mathrm{Os}\big[a_\varepsilon(x,x+y,\lambda+\eta)e^{-iy\eta}\big],\quad (x,\lambda)\in\mathbb{R}\times\mathbb{R}. \tag{4.14}$$

Therefore, by Theorem 4.2, the pseudodifferential operators $\sigma_\varepsilon(x,D)$ are bounded on all the spaces $L^p(\mathbb{R})$, $1<p<\infty$, and

$$\lim_{\varepsilon\to 0}\big\|\sigma_\varepsilon(x,D)-\sigma_A(x,D)\big\|_{\mathcal{B}(L^p(\mathbb{R}))}=0. \tag{4.15}$$

Analogously, we deduce from (4.12) and Theorem 4.1 that the pseudodifferential operators $\widetilde{a}_\varepsilon(x,D)$ are bounded on all the spaces $L^p(\mathbb{R})$, $1<p<\infty$, and

$$\lim_{\varepsilon\to 0}\big\|\widetilde{a}_\varepsilon(x,D)-\widetilde{a}(x,D)\big\|_{\mathcal{B}(L^p(\mathbb{R}))}=0. \tag{4.16}$$

From (4.10), (4.15) and (4.16) it follows that

$$\lim_{\varepsilon\to 0}\|r_\varepsilon(x,D)-r(x,D)\|_{\mathcal{B}(L^p(\mathbb{R}))}=0$$

where $r_\varepsilon(x,D)$ is the pseudodifferential operator with the symbol $r_\varepsilon(x,\lambda):=\sigma_\varepsilon(x,\lambda)-\widetilde{a}_\varepsilon(x,\lambda)\in\mathcal{E}_1^C$ and $\sigma_\varepsilon(x,\lambda)$ is given by (4.14). Since $r_\varepsilon(x,\lambda)\in\mathcal{E}_1^C\subset\mathfrak{S}_1$, we infer from Lemma 3.7 and Theorem 4.3 that each pseudodifferential operator $r_\varepsilon(x,D)$ is compact on all the spaces $L^p(\mathbb{R})$. Therefore, the operator $r(x,D)=\lim_{\varepsilon\to 0} r_\varepsilon(x,D)$ also is compact on the spaces $L^p(\mathbb{R})$, $1<p<\infty$. □

5. Fredholm theory for pseudodifferential operators with compound symbols

Given $p\in(1,\infty)$, let $\mathcal{B}=\mathcal{B}(L^p(\mathbb{R}))$ denote the Banach algebra of all bounded linear operators acting on the Banach space $L^p(\mathbb{R})$, $\mathcal{K}=\mathcal{K}(L^p(\mathbb{R}))$ be the closed two-sided ideal of all compact operators in $\mathcal{B}$, and let $\mathcal{B}^\pi=\mathcal{B}/\mathcal{K}$ be the Calkin algebra of the cosets $A^\pi=A+\mathcal{K}$ where $A\in\mathcal{B}$.

Let $\mathfrak{A}$ denote the non-closed algebra of all pseudodifferential operators $a(x,D)$ with symbols $a\in\widetilde{\mathcal{E}}_1$, where the algebra $\widetilde{\mathcal{E}}_1$ is defined by (2.27). According to [16, Lemma 10.1], the closure $\overline{\mathfrak{A}}_p$ of $\mathfrak{A}$ in $\mathcal{B}(L^p(\mathbb{R}))$ contains all compact operators $K\in\mathcal{B}(L^p(\mathbb{R}))$.

Let $\mathfrak{B}$ be the non-closed algebra of all pseudodifferential operators A of the form (3.9) with compound symbols in the class

$$\widehat{\mathcal{E}}_2:=\big\{a(x,y,\lambda):\ \partial_\lambda^j\partial_y^k a(x,y,\lambda)\in\widetilde{\mathcal{E}}_2,\ k,j=0,1,2\big\}.$$

Since $\widetilde{\mathcal{E}}_2\subset\mathcal{E}_2^C$ and since $\widetilde{a}(x,\lambda)\in\widetilde{\mathcal{E}}_1$ whenever $a(x,y,\lambda)\in\widehat{\mathcal{E}}_2$, we immediately deduce from Theorems 4.1, 4.2 and 4.4 that the closure $\overline{\mathfrak{B}}_p$ of $\mathfrak{B}$ in $\mathcal{B}(L^p(\mathbb{R}))$ is a Banach subalgebra of $\overline{\mathfrak{A}}_p$. Moreover, the quotient algebra $\overline{\mathfrak{B}}_p^\pi=\{A^\pi : A\in\overline{\mathfrak{B}}_p\}\subset\overline{\mathfrak{A}}_p^\pi$ is commutative because the Banach algebra $\overline{\mathfrak{A}}_p^\pi=\overline{\mathfrak{A}}_p/\mathcal{K}$ is commutative (see [16, Theorem 11.11]).

The non-closed subalgebra $\mathfrak{B}$ consists of all operators of the form $A=\sum_i\prod_j A_{i,j}$ where $A_{i,j}$ are pseudodifferential operators of the form (3.9) with compound symbols $a_{i,j}\in\widehat{\mathcal{E}}_2$, i,j run through finite subsets, and products $\prod_j A_{i,j}$ are ordered. Since the operators $a(x,D)$ with symbols $a(x,\lambda)\in\mathcal{E}_1^C$ commute to within compact operators (see [16, Theorem 8.3]), we infer from Theorem 4.4 that every operator $A = \sum_i \prod_j A_{i,j} \in \mathfrak{B}$ can be represented as the sum of the pseudodifferential operator with the compound symbol $a(x, y, \lambda) = \sum_i \prod_j a_{i,j}(x, y, \lambda) \in \widehat{\mathcal{E}}_2$ and a compact operator. On the other hand, we can assign to A the pseudodifferential operator $\widetilde{a}(x, D) \in \mathfrak{A}$ with the usual symbol $\widetilde{a}(x, \lambda) := \sum_i \prod_j a_{i,j}(x, x, \lambda) \in \widehat{\mathcal{E}}_1$, and the operator $A - \widetilde{a}(x, D)$ again is compact. We call the function $\widetilde{a}(x, \lambda)$ the *symbol* of the operator $A = \sum_i \prod_j A_{i,j} \in \mathfrak{B}$. Thus, the symbol algebra generated by the symbols $\widetilde{a}(x, \lambda)$ is commutative.

As a result, the operator $A \in \mathfrak{B}$ is Fredholm on the space $L^p(\mathbb{R})$ $(1 < p < \infty)$ if and only if the corresponding operator $\widetilde{a}(x, D) \in \mathfrak{A}$ is as well, and in this case their indices coincide.

Finally, we derive from [16, Theorems 12.2 and 12.5] the following result (cf. [29], [16]).

Theorem 5.1. *The pseudodifferential operator $A \in \mathfrak{B}$ of the form (3.9) with a compound symbol $a(x, y, \lambda) \in \widehat{\mathcal{E}}_2$ is Fredholm on the Lebesgue space $L^p(\mathbb{R})$ $(1 < p < \infty)$ if and only if*

$$\inf_{x \in \mathbb{R}} |a(x, x, \pm\infty)| > 0, \quad \liminf_{x \to \pm\infty} \min_{\lambda \in \overline{\mathbb{R}}} |a(x, x, \lambda)| > 0. \tag{5.1}$$

In the case A is Fredholm

$$\operatorname{Ind} A = \lim_{r \to +\infty} \frac{1}{2\pi} \Big\{ \arg a(x, x, \lambda) \Big\}_{(x,\lambda) \in \partial \Pi_r} \tag{5.2}$$

where $\Pi_r = [-r, r] \times \overline{\mathbb{R}}$ and $\big\{ \arg a(x, x, \lambda) \big\}_{(x,x,\lambda) \in \partial \Pi_r}$ denotes the increment of $\arg a(x, x, \lambda)$ when the point (x, λ) traces the boundary $\partial \Pi_r$ of Π_r counter-clockwise.

6. Mellin pseudodifferential operators with compound symbols

Let $\mathbb{R}_+ := (0, \infty)$, $\mathbb{R}_+^n = (\mathbb{R}_+)^n$, and let $L^\infty\big(\mathbb{R}_+^n, V(\mathbb{R})\big)$ stand for the set of all functions $a : \mathbb{R}_+^n \times \overline{\mathbb{R}} \to \mathbb{C}$ such that $x \mapsto a(x, \cdot)$ is a bounded measurable $V(\mathbb{R})$-valued function on $\mathbb{R}_+^n$. The set $L^\infty\big(\mathbb{R}_+^n, V(\mathbb{R})\big)$ becomes a Banach algebra if we equip it with the norm

$$\big\|a\big\|_{L^\infty(\mathbb{R}_+^n, V(\mathbb{R}))} := \operatorname*{ess\,sup}_{x \in \mathbb{R}_+^n} \big\|a(x, \cdot)\big\|_V < \infty. \tag{6.1}$$

Theorem 6.1. *If $a \in L^\infty\big(\mathbb{R}_+, V(\mathbb{R})\big)$, then the Mellin pseudodifferential operator $OP(a)$, defined for functions $f \in C_0^\infty(\mathbb{R}_+)$ by the iterated integral*

$$\big[OP(a)f\big](r) = \frac{1}{2\pi} \int_{\mathbb{R}} d\lambda \int_{\mathbb{R}_+} a(r, \lambda) \left(\frac{r}{\varrho}\right)^{i\lambda} f(\varrho) \frac{d\varrho}{\varrho}, \quad r \in \mathbb{R}_+, \tag{6.2}$$

extends to a bounded linear operator on every Lebesgue space $L^p(\mathbb{R}_+, d\mu)$ with the measure $d\mu = d\varrho/\varrho$ $(1 < p < \infty)$, and

$$\big\|a(x, D)\big\|_{\mathcal{B}(L^p(\mathbb{R}_+, d\mu))} \le 2\big\|a\big\|_{L^\infty(\mathbb{R}_+, V(\mathbb{R}))} \big\|S_*\big\|_{\mathcal{B}(L^p(\mathbb{R}))}. \tag{6.3}$$

Proof. Let E be the isometric isomorphism

$$E : L^p(\mathbb{R}_+, d\mu) \to L^p(\mathbb{R}), \quad (Ef)(x) = f(e^x), \quad x \in \mathbb{R}. \tag{6.4}$$

Applying the transform $A \mapsto EAE^{-1}$ to the Mellin pseudodifferential operator $OP(a)$ given by (6.2) we get the equality

$$E\,[OP(a)]\,E^{-1} = b(x, D)$$

where $b(x,\lambda) = a(e^x, \lambda)$ and $b(x, D)$ is the pseudodifferential operator given for functions $\varphi \in C_0^\infty(\mathbb{R})$ by

$$[b(x, D)\varphi](x) = \frac{1}{2\pi}\int_{\mathbb{R}} d\lambda \int_{\mathbb{R}} b(x,\lambda)e^{i(x-y)\lambda}\varphi(y)dy, \quad x \in \mathbb{R}.$$

Since $b \in L^\infty\big(\mathbb{R}, V(\mathbb{R})\big)$ if and only if $a \in L^\infty\big(\mathbb{R}_+, V(\mathbb{R})\big)$ and since

$$\big\|b\big\|_{L^\infty(\mathbb{R},V(\mathbb{R}))} = \big\|a\big\|_{L^\infty(\mathbb{R}_+,V(\mathbb{R}))},$$

Theorem 6.1 immediately follows from Theorem 4.1. □

Obviously, for the function $b(r,\varrho,\lambda) = a(\log r, \log\varrho, \lambda)$, the conditions

$$\partial_\lambda^j \partial_y^k a(x, y, \lambda) \in C_b\big(\mathbb{R}\times\mathbb{R}, V(\mathbb{R})\big) \quad \text{for all } k, j = 0, 1, 2,$$

are equivalent to the conditions

$$\partial_\lambda^j (\varrho\partial_\varrho)^k b(r,\varrho,\lambda) \in C_b\big(\mathbb{R}_+\times\mathbb{R}_+, V(\mathbb{R})\big) \quad \text{for all } k, j = 0, 1, 2. \tag{6.5}$$

Thus, similarly to Theorem 6.1, applying (6.4) and Theorem 4.2, we easily obtain the following.

Theorem 6.2. *If conditions* (6.5) *are fulfilled, then the Mellin pseudodifferential operator* $B = OP(b)$ *defined for functions* $f \in C_0^\infty(\mathbb{R}_+)$ *by the iterated integral*

$$\big[OP(b)f\big](r) = \frac{1}{2\pi}\int_{\mathbb{R}} d\lambda \int_{\mathbb{R}_+} b(r,\varrho,\lambda)\left(\frac{r}{\varrho}\right)^{i\lambda} f(\varrho)\frac{d\varrho}{\varrho}, \quad r \in \mathbb{R}_+, \tag{6.6}$$

extends to a bounded linear operator on every Lebesgue space $L^p(\mathbb{R}_+, d\mu)$, $p \in (1,\infty)$, *and*

$$\big\|B\big\|_{\mathcal{B}(L^p(\mathbb{R}_+,d\mu))} \le 2\sup_{x\in\mathbb{R}}\big\|\sigma_A(x,\cdot)\big\|_V \big\|S_*\big\|_{\mathcal{B}(L^p(\mathbb{R}))}, \tag{6.7}$$

where $\sigma_A(x,\lambda)$ *is given by* (3.18) *with*

$$a(x, x+y, \lambda+\eta) = b(e^x, e^{x+y}, \lambda+\eta), \quad x, y, \lambda, \eta \in \mathbb{R}, \tag{6.8}$$

and $\sigma_A(x,\lambda)$ *satisfies the estimate*

$$\begin{aligned}\sup_{x\in\mathbb{R}}\big\|\sigma_A(x,\cdot)\big\|_V \;&\le\; \frac{\pi}{2}\sup_{r,\varrho\in\mathbb{R}_+}\Big(3\big\|\langle\widetilde{D}_\varrho\rangle^2\{b(r,\varrho,\cdot)\}\big\|_V + 2\big\|\partial_\lambda\langle\widetilde{D}_\varrho\rangle^2\{b(r,\varrho,\cdot)\}\big\|_V \\ &+\; \big\|\partial_\lambda^2\langle\widetilde{D}_\varrho\rangle^2\{b(r,\varrho,\cdot)\}\big\|_V\Big) \quad \textit{where} \quad \langle\widetilde{D}_\varrho\rangle^2 = I - (\varrho\partial_\varrho)^2.\end{aligned}$$

A function $a \in C_b(\mathbb{R}_+)$ is called slowly oscillating at 0 and at $+\infty$ if the function $x \mapsto a(e^x) \in C_b(\mathbb{R})$ is slowly oscillating at the points $\pm\infty$. Thus (see [28] and [34]), a function $a \in C_b(\mathbb{R}_+)$ is slowly oscillating at 0 if for every $\lambda \in (0,1)$,

$$\lim_{r\to 0} \max\big\{|a(x) - a(y)| : x, y \in [\lambda r, r]\big\} = 0 \tag{6.9}$$

or, equivalently,

$$\lim_{r\to 0} \max\big\{|a(x) - a(y)| : x, y \in [r/2, r]\big\} = 0. \tag{6.10}$$

Analogously, a function $a \in C_b(\mathbb{R}_+)$ is slowly oscillating at ∞ if

$$\lim_{r\to \infty} \max\big\{|a(x) - a(y)| : x, y \in [r/2, r]\big\} = 0. \tag{6.11}$$

Clearly (see, e.g., [3] or Corollary 2.4), if $a \in C_b^n(\mathbb{R}_+)$, then it slowly oscillates at 0 if and only if

$$\lim_{r\to 0} |ra'(r)| = 0, \tag{6.12}$$

which in its turn is equivalent to the relations

$$\sup_{r\in\mathbb{R}_+} |(rd_r)^j a(r)| < \infty \quad (j = 0, 1, \dots, n), \tag{6.13}$$

$$\lim_{r\to 0} |(rd_r)^j a(r)| = 0 \quad (j = 1, 2, \dots, n) \tag{6.14}$$

where $(rd_r)a(r) = ra'(r)$.

Below we need the following classes of compound symbols for Mellin pseudodifferential operators. Let

$$\begin{aligned}
\mathcal{E}_2^C(\mathbb{R}_+) &:= \big\{b(r,\varrho,\lambda) \in C_b\big(\mathbb{R}_+ \times \mathbb{R}_+, V(\mathbb{R})\big) : b(e^x, e^y, \lambda) \in \mathcal{E}_2^C\big\},\\
\widetilde{\mathcal{E}}_2(\mathbb{R}_+) &:= \big\{b(r,\varrho,\lambda) \in C_b\big(\mathbb{R}_+ \times \mathbb{R}_+, V(\mathbb{R})\big) : b(e^x, e^y, \lambda) \in \widetilde{\mathcal{E}}_2\big\},\\
\widehat{\mathcal{E}}_2(\mathbb{R}_+) &:= \big\{b(r,\varrho,\lambda) \in C_b\big(\mathbb{R}_+ \times \mathbb{R}_+, V(\mathbb{R})\big) : b(e^x, e^y, \lambda) \in \widehat{\mathcal{E}}_2\big\}
\end{aligned} \tag{6.15}$$

Theorem 4.4 implies the following.

Theorem 6.3. *If $\partial_\lambda^j (\varrho\partial_\varrho)^k b(r,\varrho,\lambda) \in \mathcal{E}_2^C(\mathbb{R}_+)$ for all $k, j = 0, 1, 2$, then the pseudodifferential operator $OP(b) - OP(\widetilde{b})$, where $OP(b)$ is given by (6.6) and $OP(\widetilde{b})$ is of the form (6.2) with $\widetilde{b}(r,\lambda) = b(r,r,\lambda)$, is compact on every Lebesgue space $L^p(\mathbb{R}_+, d\mu)$, $1 < p < \infty$.*

Obviously, we can easily rewrite Theorem 5.1 for Mellin pseudodifferential operators with compound symbols $b(r,\varrho,\lambda) \in \widehat{\mathcal{E}}_2(\mathbb{R}_+)$. Let $\mathfrak{M}$ be the Banach subalgebra of $\mathcal{B}\big(L^p(\mathbb{R}_+, d\mu)\big)$ generated by all Mellin pseudodifferential operators with compound symbols $b(r,\varrho,\lambda) \in \widehat{\mathcal{E}}_2(\mathbb{R}_+)$. Below we consider an application of such pseudodifferential operators to generalized singular integral operators on weighted Lebesgue spaces on some Carleson curves.

Remark 6.4. All the previous results remain valid if in the definitions of the algebras $SO(\mathbb{R}^n)$, $SO^\infty(\mathbb{R}^n)$, $\mathcal{E}_n^C$, $\mathcal{E}_n^V$ (that is, in (2.2), (2.3), and (2.22)), and in equations (2.23) and (2.25), we replace $\lim\limits_{\|x\|\to\infty}$, where $x = (x_1, \dots, x_n) \in \mathbb{R}^n$, with the two limits $\lim\limits_{x_1,\dots,x_n\to+\infty}$ and $\lim\limits_{x_1,\dots,x_n\to-\infty}$, assume in addition the uniform continuity of functions in $SO(\mathbb{R}^n)$ and $SO^\infty(\mathbb{R}^n)$, and modify accordingly the definitions of the algebras $\widetilde{\mathcal{E}}_n$, $\widehat{\mathcal{E}}_2$, $\mathcal{E}_2^C(\mathbb{R}_+)$, $\widetilde{\mathcal{E}}_2(\mathbb{R}_+)$, and $\widehat{\mathcal{E}}_2(\mathbb{R}_+)$. In what follows we will use these algebras defined in this new way.

7. Oscillating data

Following [3], we introduce the slowly oscillating data.

Slowly oscillating functions. Let $SO(\mathbb{R}_+)$ stand for the set of all functions in $C_b(\mathbb{R}_+)$ which are slowly oscillating at 0 and at ∞. Thus, $SO(\mathbb{R}_+)$ is a C^*-subalgebra of $L^\infty(\mathbb{R}_+)$.

Slowly oscillating curves. Let Γ be an unbounded oriented simple arc with the starting point t and the terminating point ∞ that is defined by

$$\Gamma = \left\{\tau = t + re^{i\theta(r)} : r \in \mathbb{R}_+\right\} \tag{7.1}$$

where θ is a real-valued function in $C^3(\mathbb{R}_+)$. We say that Γ is a slowly oscillating curve (at the endpoints t and ∞) if the function $r\theta'(r)$ is slowly oscillating at 0 and ∞. A slow oscillation of $r\theta'(r)$ at 0 and ∞ means in the case of $\theta \in C^3(\mathbb{R}_+)$ that

$$\sup_{r\in\mathbb{R}_+} |(rd_r)^j\theta(r)| < \infty \quad (j = 1, 2, 3), \tag{7.2}$$

$$\lim_{r\to 0}(rd_r)^j\theta(r) = 0, \quad \lim_{r\to\infty}(rd_r)^j\theta(r) = 0 \quad (j = 2, 3). \tag{7.3}$$

Condition (7.1) says that Γ may be parameterized by the distance to the starting point t. Note that $\theta(r)$ may be unbounded as $r \to 0$ and $r \to \infty$. Since

$$|d\tau| = \sqrt{1 + (r\theta'(r))^2}\, dr,$$

condition (7.2) for $j = 1$ ensures that Γ is a Carleson curve (see, e.g., [1]).

Slowly oscillating weights. Let Γ be a slowly oscillating curve as above. We call a function $w : \Gamma \to (0, +\infty)$ a slowly oscillating weight (at the endpoints t and ∞ of Γ) if

$$w\big(t + re^{i\theta(r)}\big) = e^{v(r)}, \quad r \in \mathbb{R}_+, \tag{7.4}$$

where v is a real-valued function in $C^3(\mathbb{R}_+)$ and $rv'(r)$ is slowly oscillating at the points 0 and ∞. The latter means that

$$\sup_{r\in\mathbb{R}_+} |(rD_r)^j v(r)| < \infty \quad (j = 1, 2, 3), \tag{7.5}$$

$$\lim_{r\to 0}(rD_r)^j v(r) = 0, \quad \lim_{r\to\infty}(rD_r)^j v(r) = 0 \quad (j = 2, 3). \tag{7.6}$$

One can show (see, e.g., [1, Theorem 2.36] and [18, Section 5]) that the Muckenhoupt condition (1.1) is satisfied if and only if

$$\begin{aligned} -1/p < \liminf_{r\to 0} rv'(r) \le \limsup_{r\to 0} rv'(r) < 1/q, \\ -1/p < \liminf_{r\to \infty} rv'(r) \le \limsup_{r\to \infty} rv'(r) < 1/q. \end{aligned} \tag{7.7}$$

We denote by A_p^0 the set of all pairs $(\Gamma, w) \in A_p$ in which Γ is a slowly oscillating curve and w is a slowly oscillating weight. Thus, $(\Gamma, w) \in A_p^0$ means that (7.1) to (7.7) are true with functions θ and v in $C^3(\mathbb{R}_+)$.

Slowly oscillating shifts. Let $(\Gamma, w) \in A_p^0$ and let α be an orientation-preserving diffeomorphism of Γ onto itself such that $\log|\alpha'| \in C_b(\Gamma^0)$ where $\Gamma^0 = \Gamma \setminus \{t, \infty\}$. We call α a slowly oscillating shift (at the endpoints t and ∞ of Γ) if

$$\alpha\big(t + re^{i\theta(r)}\big) = t + re^{\omega(r)} \exp\big(i\theta\big(re^{\omega(r)}\big)\big), \quad r \in \mathbb{R}_+, \tag{7.8}$$

where ω is a real-valued function in $C^3(\mathbb{R}_+)$ and the functions ω and $r\omega'(r)$ are slowly oscillating at 0 and ∞. Then ω, $r\omega'(r) \in SO(\mathbb{R}_+)$.

Since $r\omega'(r)$ is slowly oscillating at 0, the slow oscillation of ω at 0 and ∞ is equivalent to the property:

$$\lim_{r\to 0} r\omega'(r) = 0, \quad \lim_{r\to\infty} r\omega'(r) = 0. \tag{7.9}$$

Thus, in contrast to (7.5) and (7.6), slow oscillation of α means that

$$\sup_{r\in\mathbb{R}_+} |(rD_r)^j \omega(r)| < \infty \quad (j = 0, 1, 2, 3),$$

$$\lim_{r\to 0} (rD_r)^j \omega(r) = 0, \quad \lim_{r\to\infty} (rD_r)^j v(r) = 0 \quad (j = 1, 2, 3).$$

If α is a slowly oscillating shift at the point t (respectively, at ∞), then the derivative α', given by

$$\alpha'(\tau) = \frac{1 + r\omega'(r)}{1 + ir\theta'(r)} \Big(1 + ire^{\omega(r)}\theta'\big(re^{\omega(r)}\big)\Big) \exp\Big(\omega(r) + i\theta\big(re^{\omega(r)}\big) - i\theta(r)\Big) \tag{7.10}$$

for $\tau = t + re^{i\theta(r)} \in \Gamma$, is a slowly oscillating function at the point t (respectively, at ∞) (cf. [17] and [15]). The inverse assertion is false in general.

Since $\log|\alpha'| \in C_b(\Gamma^0)$, we conclude that $\inf_{r\in\mathbb{R}_+} |1 + r\omega'(r)| > 0$, whence we infer due to (7.9) that

$$\inf_{r\in\mathbb{R}_+} (1 + r\omega'(r)) > 0. \tag{7.11}$$

Slowly oscillating coefficients. Let $(\Gamma, w) \in A_p^0$. We denote by $SO^n(\Gamma)$ the set of all functions $c_\Gamma : \Gamma \to \mathbb{C}$ such that

$$c_\Gamma\big(t + re^{i\theta(r)}\big) = c(r), \ \ r \in \mathbb{R}_+,$$

where $c \in C_b^n(\mathbb{R}_+) \cap SO(\mathbb{R}_+)$, respectively. Below we assume that coefficients belong to $SO^2(\Gamma)$.

8. Algebra of generalized singular integral operators

Let $\mathcal{D}_{\Gamma,w}$ denote the smallest closed subalgebra of $\mathcal{B}(L^p(\Gamma,w))$ containing the set

$$\{c_\Gamma I : c_\Gamma \in SO^2(\Gamma)\} \cup \{S_\Gamma\} \cup \{V_\alpha S_\Gamma V_\alpha^{-1}\}, \tag{8.1}$$

where V_α is the shift operator, $V_\alpha\varphi = \varphi\circ\alpha$. The Banach algebra $\mathcal{D}_{\Gamma,w}$ plays an important role in studying singular integral operators with shifts (see, e.g., [21], [19], [17]). Here, in general, the shift operator V_α is an isometric isomorphism from the space $L^p(\Gamma,\widetilde{w})$ with the weight $\widetilde{w} = (w\circ\alpha^{-1})\big|(\alpha^{-1})'\big|^{1/p}$ onto the space $L^p(\Gamma,w)$. This makes a contribution of the properties of the shift α in the Fredholm theory of operators $D\in\mathcal{D}_{\Gamma,w}$. To get the boundedness of the operator $V_\alpha S_\Gamma V_\alpha^{-1}$ on $L^p(\Gamma,w)$, we need to assume that along with (7.7),

$$\begin{aligned} -1/p < \liminf_{r\to 0} r\widetilde{v}\,'(r) \le \limsup_{r\to 0} r\widetilde{v}\,'(r) < 1/q,\\ -1/p < \liminf_{r\to \infty} r\widetilde{v}\,'(r) \le \limsup_{r\to \infty} r\widetilde{v}\,'(r) < 1/q. \end{aligned} \tag{8.2}$$

where $\widetilde{w}(t+re^{i\theta(r)}) = e^{\widetilde{v}(r)}$ for $r\in\mathbb{R}_+$.

We will show below that the Banach algebra $\mathcal{D}_{\Gamma,w}$ can be imbedded into the Banach algebra of Mellin pseudodifferential operators $\mathfrak{M}\subset\mathcal{B}(L^p(\mathbb{R}_+,d\mu))$ as a Banach algebra $\mathcal{C}_{\Gamma,w}$.

Let $(\Gamma,w)\in A_p^0$. The map Φ given by

$$(\Phi f)(r) = e^{v(r)} r^{1/p} f(t+re^{i\theta(r)}),\quad r\in\mathbb{R}_+, \tag{8.3}$$

is an isomorphism of $L^p(\Gamma,w)$ onto $L^p(\mathbb{R}_+,d\mu)$. Consider the map

$$\Psi : \mathcal{B}(L^p(\Gamma,w)) \to \mathcal{B}(L^p(\mathbb{R}_+,d\mu)),\quad A\mapsto \Phi A\Phi^{-1}.$$

Then $\mathcal{C}_{\Gamma,w} := \Psi(\mathcal{D}_{\Gamma,w})$.

For a real-valued function $b\in C^n(\mathbb{R}_+)$, let

$$m_b(r,\varrho) = (b(r)-b(\varrho))/(\log r - \log\varrho). \tag{8.4}$$

Along with $m_b(r,\varrho)$, for a real-valued function $a\in C^n(\mathbb{R})$, we put

$$\widetilde{m}_a(x,y) = (a(x)-a(y))/(x-y). \tag{8.5}$$

Lemma 8.1. *If $a\in C^3(\mathbb{R})$ and $a'\in SO(\mathbb{R})$, then the functions $(x,y)\mapsto \partial_x^j\widetilde{m}_a(x,y)$ belong to $SO(\mathbb{R}^2)$ for all $j=0,1,2$, where according to Remark 6.4,*

$$SO(\mathbb{R}^2) := \Big\{f\in C_b(\mathbb{R}^2) : \lim_{x,y\to+\infty} cm_{(x,y)}(f) = \lim_{x,y\to-\infty} cm_{(x,y)}(f) = 0\Big\}. \tag{8.6}$$

Proof. As $a\in C^3(\mathbb{R})$ and $a'\in SO(\mathbb{R}) = \mathrm{clos}_{L^\infty(\mathbb{R})} SO^\infty(\mathbb{R})$, (2.3) implies that

$$a', a'', a''' \in C_b(\mathbb{R}),\qquad \lim_{x\to\infty} a''(x) = \lim_{x\to\infty} a'''(x) = 0. \tag{8.7}$$

Hence, by the mean value theorem, the function $\widetilde{m}_a$ is bounded on $\mathbb{R}^2$. Since a' is continuous on $\mathbb{R}$ and since for every $z\in\mathbb{R}$,

$$\lim_{x,y\to z}\big[\widetilde{m}_a(x,y) - a'(z)\big] = \lim_{x,y\to z}\frac{\int_y^x \big[a'(t)-a'(z)\big]dt}{x-y} = 0,$$

the function $(x, y) \mapsto \widetilde{m}_a(x, y)$ is continuous at the points (z, z) and hence on the whole plane $\mathbb{R}^2$. Thus, $\widetilde{m}_a \in C_b(\mathbb{R}^2)$. On the other hand, because

$$\widetilde{m}_a(x, y) = \frac{1}{x - y} \int_y^x a'(\tau) d\tau = \int_0^1 a'\big(y + t(x - y)\big) dt,$$

$$\widetilde{m}_a(x + h, y + s) - \widetilde{m}_a(x, y) = \int_0^1 \int_{y+t(x-y)}^{y+s+t(x+h-y-s)} a''(\tau)\, d\tau dt, \tag{8.8}$$

we conclude from (2.1), (8.8), and (8.7) that

$$\lim_{x,y\to+\infty} cm_{(x,y)}(\widetilde{m}_a) = \lim_{x,y\to-\infty} cm_{(x,y)}(\widetilde{m}_a) = 0,$$

whence $\widetilde{m}_a \in SO(\mathbb{R}^2)$ in view of (8.6).

Further, from the equalities

$$\partial_x \widetilde{m}_a(x, y) = \frac{a'(x) - \widetilde{m}_a(x, y)}{x - y} = \frac{\int_y^x \big[a'(x) - a'(t)\big] dt}{(x - y)^2} = \frac{\int_y^x \int_t^x a''(\tau) d\tau dt}{(x - y)^2},$$

$$\lim_{x,y\to z} \big[\partial_x \widetilde{m}_a(x, y) - a''(z)/2\big] = \lim_{x,y\to z} \frac{\int_y^x \int_t^x \big[a''(\tau) - a''(z)\big] d\tau dt}{(x - y)^2} = 0$$

it follows, respectively, that the function $(x, y) \mapsto \partial_x \widetilde{m}_a(x, y)$ is bounded on $\mathbb{R}^2$ and continuous at every point (z, z). Consequently, $\partial_x \widetilde{m}_a \in C_b(\mathbb{R}^2)$. As

$$\partial_x \widetilde{m}_a(x, y) = \int_0^1 \int_0^1 (1 - t)\, a''\big(y + [t + \tau(1 - t)](x - y)\big) d\tau dt,$$

$$\partial_x \widetilde{m}_a(x + h, y + s) - \partial_x \widetilde{m}_a(x, y) = \int_0^1 \int_0^1 \int_{y+[t+\tau(1-t)](x-y)}^{y+s+[t+\tau(1-t)](x+h-y-s)} (1 - t)\, a'''(\xi)\, d\xi d\tau dt,$$

and the points $y + [t + \tau(1 - t)](x - y)$ are in the segment with the endpoints x, y for all $t, \tau \in [0, 1]$, we conclude from (2.1), (8.8), (8.7), and (8.6) that $\partial_x \widetilde{m}_a \in SO(\mathbb{R}^2)$.

Analogously, the equalities

$$\partial_x^2 \widetilde{m}_a(x, y) = \frac{2 \int_y^x \int_t^x \big[a''(x) - a''(\tau)\big] d\tau dt}{(x - y)^3} = \frac{2 \int_y^x \int_t^x \int_\tau^x a^{(3)}(\xi) d\xi d\tau dt}{(x - y)^3},$$

$$\lim_{x,y\to z} \big[\partial_x^2 \widetilde{m}_a(x, y) - a^{(3)}(z)/3\big] = \lim_{x,y\to z} \frac{2 \int_y^x \int_t^x \int_\tau^x \big[a^{(3)}(\xi) - a^{(3)}(z)\big] d\xi d\tau dt}{(x - y)^3} = 0,$$

where $z \in \mathbb{R}$, imply that $\partial_x^2 \widetilde{m}_a \in C_b(\mathbb{R}^2)$. Finally, since

$$\begin{aligned} &\partial_x^2 \widetilde{m}_a(x+h, y+s) - \partial_x^2 \widetilde{m}_a(x, y) \\ &= 2 \int_0^1 \int_0^1 \int_0^1 (1-t)^2 (1-\tau) \big[a'''\big(\theta(x+h, y+s, t, \tau, \xi)\big) - a'''\big(\theta(x, y, t, \tau, \xi)\big)\big] d\xi d\tau dt, \end{aligned}$$

where the points $\theta(x, y, t, \tau, \xi) := y + [t + (\tau + \xi(1 - \tau))(1 - t)](x - y)$ belong to the segment with the endpoints x, y, we again infer from (2.1), (8.8), (8.7), and (8.6) that $\partial_x^2 \widetilde{m}_a \in SO(\mathbb{R}^2)$. □

Lemma 8.1 immediately gives the following.

Corollary 8.2. *If $a \in C^3(\mathbb{R}_+)$ and $ra'(r) \in SO(\mathbb{R}_+)$, then for all $j = 0, 1, 2$,*

$$(r\partial_r)^j m_a(r, \varrho) \in SO(\mathbb{R}_+^2) := \Big\{b(r, \varrho) \in C_b(\mathbb{R}_+^2) : b(e^x, e^y) \in SO(\mathbb{R}^2)\Big\}.$$

The following theorem reveals our interest in Mellin pseudodifferential operators with slowly oscillating compound symbols.

Theorem 8.3. *Let $(\Gamma, w) \in A_p^0$ and*

$$0 < \inf_{r,\varrho\in\mathbb{R}_+} \big(1/p + m_v(r, \varrho)\big) \leq \sup_{r,\varrho\in\mathbb{R}_+} \big(1/p + m_v(r, \varrho)\big) < 1, \tag{8.9}$$

$$0 < \inf_{r,\varrho\in\mathbb{R}_+} \frac{1/p + m_v(r, \varrho)}{1 + m_\omega(r, \varrho)} \leq \sup_{r,\varrho\in\mathbb{R}_+} \frac{1/p + m_v(r, \varrho)}{1 + m_\omega(r, \varrho)} < 1. \tag{8.10}$$

If $c_\Gamma \in SO^2(\Gamma)$ then $\Psi(c_\Gamma I) = cI$. For the operators $S_\Gamma \in \mathcal{B}\big(L^p(\Gamma, w)\big)$ and $V_\alpha S_\Gamma V_\alpha^{-1} \in \mathcal{B}\big(L^p(\Gamma, w)\big)$, we have

$$\Psi(S_\Gamma) = OP(\sigma), \quad \Psi(V_\alpha S_\Gamma V_\alpha^{-1}) = OP(\sigma_\alpha),$$

where for $(r, \varrho, \lambda) \in \mathbb{R}_+ \times \mathbb{R}_+ \times \mathbb{R}$,

$$\sigma(r, \varrho, \lambda) := \frac{1 + i\varrho\theta'(\varrho)}{1 + im_\theta(r, \varrho)} \coth\left(\pi\frac{\lambda + i(1/p + m_v(r, \varrho))}{1 + im_\theta(r, \varrho)}\right), \tag{8.11}$$

$$\sigma_\alpha(r, \varrho, \lambda) := \frac{1 + \varrho\omega'(\varrho) + i\varrho\gamma'(\varrho)}{1 + m_\omega(r, \varrho) + im_\gamma(r, \varrho)} \coth\left(\pi\frac{\lambda + i(1/p + m_v(r, \varrho))}{1 + m_\omega(r, \varrho) + im_\gamma(r, \varrho)}\right) \tag{8.12}$$

$\gamma(r) = \theta(re^{\omega(r)})$, and the functions σ and σ_α belong to $\widehat{\mathcal{E}}_2(\mathbb{R}_+)$.

Proof. One can easily check that (8.10) implies (8.2) for the weight $\widetilde{w}$, and hence $(\Gamma, \widetilde{w}) \in A_p^0$. Thus the operator S_Γ is bounded on the space $L^p(\Gamma, \widetilde{w})$ and therefore the operator $V_\alpha S_\Gamma V_\alpha^{-1}$ is bounded on the space $L^p(\Gamma, w)$ together with S_Γ.

The function (8.14) is calculated in [2]. A straightforward computation on the basis of (7.8) and (8.3) shows that for $r > 0$,

$$\begin{aligned}
&[\Psi(V_\alpha S_\Gamma V_\alpha^{-1}) f](r) \\
&= \frac{1}{\pi i}\int_{\mathbb{R}_+} \frac{e^{v(r)-v(\varrho)}\,(r/\varrho)^{1/p}\,(1 + \varrho\,\omega'(\varrho) + i\varrho\,\gamma'(\varrho))\,e^{\omega(\varrho)+i\gamma(\varrho)}}{\varrho\,e^{\omega(\varrho)+i\gamma(\varrho)} - r\,e^{\omega(r)+i\gamma(r)}} f(\varrho)\,d\varrho \\
&= \frac{1}{\pi i}\int_{\mathbb{R}_+} \frac{(r/\varrho)^{1/p+m_v(r,\varrho)}\,(1 + \varrho\,\omega'(\varrho) + i\varrho\,\gamma'(\varrho))}{1 - (r/\varrho)^{1+m_\omega(r,\varrho)+im_\gamma(r,\varrho)}} f(\varrho)\,\frac{d\varrho}{\varrho} \qquad =: I_0.
\end{aligned}$$

Further, taking into account (8.10) and using the Mellin transform identity

$$\frac{1}{\pi i}\,\frac{\delta\,x^\mu}{1 - x^\delta} = \frac{1}{2\pi}\int_{\mathbb{R}} \coth\left[\pi(\lambda + i\mu)/\delta\right] x^{i\lambda}\,d\lambda \quad (x > 0,\ \mathrm{Re}\,\delta \geq 1,\ 0 < \mu < 1)$$

(see, e.g., [2]), we infer for $x = (r/\varrho)^{1+m_\omega(r,\varrho)}$ and

$$\mu = (1/p + m_v(r, \varrho))/(1 + m_\omega(r, \varrho)), \quad \delta = 1 + im_\gamma(r, \varrho)/(1 + m_\omega(r, \varrho)),$$

that

$$\begin{aligned}
I_0 &= \frac{1}{\pi i}\int_{\mathbb{R}_+}\int_{\mathbb{R}}\coth\left[\pi\left(\lambda+i\frac{1/p+m_v(r,\varrho)}{1+m_\omega(r,\varrho)}\right)\left(1+i\frac{m_\gamma(r,\varrho)}{1+m_\omega(r,\varrho)}\right)^{-1}\right]\\
&\times\ (1+\varrho\,\omega'(\varrho)+i\varrho\,\gamma'(\varrho))\left(1+i\frac{m_\gamma(r,\varrho)}{1+m_\omega(r,\varrho)}\right)^{-1}\left(\frac{r}{\varrho}\right)^{i\lambda(1+m_\omega(r,\varrho))}d\lambda\,f(\varrho)\frac{d\varrho}{\varrho}\\
&= \frac{1}{\pi i}\int_{\mathbb{R}}d\lambda\int_{\mathbb{R}_+}\sigma_\alpha(r,\varrho,\lambda)\left(\frac{r}{\varrho}\right)^{i\lambda}f(\varrho)\frac{d\varrho}{\varrho},
\end{aligned}$$

which gives (8.15).

It is easily seen with the help of Corollary 8.2 that the functions σ, σ_α belong to the class $\widehat{\mathcal{E}}_2(\mathbb{R}_+)$ given by (6.15). □

Finally, we deduce from Theorems 5.1, 6.3 and 8.3 the following Fredholm result.

Theorem 8.4. *Let $1 < p < \infty$, $(\Gamma, w) \in A_p^0$, α is a slowly oscillating shift on Γ, and let (8.9)–(8.10) hold. Then the operator*

$$B=\sum_i\prod_j\left((c_{i,j})_\Gamma I+(d_{i,j})_\Gamma S_\Gamma+(g_{i,j})_\Gamma V_\alpha S_\Gamma V_\alpha^{-1}\right)\in\mathcal{D}_{\Gamma,w}, \tag{8.13}$$

with coefficients $(c_{i,j})_\Gamma$, $(d_{i,j})_\Gamma$, $(g_{i,j})_\Gamma \in SO^2(\Gamma)$ and i, j running through finite subsets, is Fredholm on the space $L^p(\Gamma, w)$ if and only if

$$\inf_{r\in\mathbb{R}_+}|b(r,r,\pm\infty)|>0,\quad \liminf_{r\to 0}\min_{\lambda\in\overline{\mathbb{R}}}|b(r,r,\lambda)|>0,\quad \liminf_{r\to+\infty}\min_{\lambda\in\overline{\mathbb{R}}}|b(r,r,\lambda)|>0,$$

where

$$b(r,r,\lambda):=\sum_i\prod_j\Big(c_{i,j}(r)+d_{i,j}(r)\sigma(r,r,\lambda)+g_{i,j}(r)\sigma_\alpha(r,r,\lambda)\Big),$$

$$\sigma(r,r,\lambda):=\coth\left(\pi\frac{\lambda+i(1/p+rv'(r))}{1+ir\theta'(r)}\right), \tag{8.14}$$

$$\sigma_\alpha(r,r,\lambda):=\coth\left(\pi\frac{\lambda+i(1/p+rv'(r))}{1+r\omega'(r)+ir\gamma'(r)}\right), \tag{8.15}$$

and $\gamma(r) = \theta(re^{\omega(r)})$. In the case B is Fredholm

$$\operatorname{Ind}B=\lim_{m\to+\infty}\frac{1}{2\pi}\Big\{\arg b(r,r,\lambda)\Big\}_{(r,\lambda)\in\partial\Pi_m} \tag{8.16}$$

where $\Pi_m = [1/m, m] \times \overline{\mathbb{R}}$ and $\{\arg b(r,r,\lambda)\}_{(r,r,\lambda)\in\partial\Pi_m}$ denotes the increment of $\arg b(r,r,\lambda)$ when the point (r,λ) traces the boundary $\partial\Pi_m$ of Π_m counter-clockwise.

Note that Theorems 8.3 and 8.4 remain true without the property of slow oscillation of $\omega(r)$. In the same manner we can study the algebra of generalized singular integral operators containing several operators $V_{\alpha_n}S_\Gamma V_{\alpha_n}^{-1}$, $n = 1, 2, \dots, n_0$.

Acknowledgment

The author is grateful to the referee for useful comments and suggestions.

References

[1] A. Böttcher and Yu.I. Karlovich, *Carleson Curves, Muckenhoupt Weights, and Toeplitz Operators.* Progress in Mathematics **154**, Birkhäuser, Verlag, Basel, Boston, Berlin, 1997.

[2] A. Böttcher, Yu.I. Karlovich, and V.S. Rabinovich, *Mellin pseudodifferential operators with slowly varying symbols and singular integral on Carleson curves with Muckenhoupt weights.* Manuscripta Math. **95** (1998), 363–376.

[3] A. Böttcher, Yu.I. Karlovich, and V.S. Rabinovich, *The method of limit operators for one-dimensional singular integrals with slowly oscillating data.* J. Operator Theory **43** (2000), 171–198.

[4] A. Böttcher, Yu.I. Karlovich, and V.S. Rabinovich, *Singular integral operators with complex conjugation from the viewpoint of pseudodifferential operators.* Operator Theory: Advances and Applications **121** (2001), 36–59.

[5] A. Böttcher and B. Silbermann, *Analysis of Toeplitz Operators.* Akademie-Verlag, Berlin, 1989 and Springer-Verlag, Berlin, Heidelberg, New York, 1990.

[6] R.R. Coifman and Y. Meyer, *Au delà des opérateurs pseudodifférentiels.* Astérisque **57** (1978), 1–184.

[7] H.O. Cordes, *On compactness of commutators of multiplications and convolutions, and boundedness of pseudodifferential operators.* J. Funct. Anal. **18** (1975), 115–131.

[8] H.O. Cordes, *Elliptic Pseudo-Differential Operators – An Abstract Theory.* Lecture Notes in Math. **756**, Springer, Berlin, 1979.

[9] D. David, *Opérateurs integraux singuliers sur certaines courbes du plan complexe.* Ann. Sci. École Norm. Sup. **17** (1984), 157–189.

[10] J. Duoandikoetxea, *Fourier Analysis.* American Mathematical Society, Providence, RI, 2000.

[11] E.M. Dynkin and B.P. Osilenker, *Weighted norm estimates for singular integrals and their applications.* J. Soviet Math. **30** (1985), 2094–2154.

[12] I. Gohberg and N. Krupnik, *One-Dimensional Linear Singular Integral Equations.* Vols. **1** and **2**, Birkhäuser, Basel, 1992; Russian original, Shtiintsa, Kishinev, 1973.

[13] L. Hörmander, *The Analysis of Linear Partial Differential Operators.* Vols. **1**–**4**, Springer, Berlin, 1983–1985.

[14] R. Hunt, B. Muckenhoupt, and R. Wheeden, *Weighted norm inequalities for the conjugate function and Hilbert transform.* Trans. Amer. Math. Soc. **176** (1973), 227–251.

[15] A.Yu. Karlovich, Yu.I. Karlovich, and A.B. Lebre, *Invertibility of functional operators with slowly oscillating non-Carleman shifts.* Operator Theory: Advances and Applications **142** (2003), 147–174.

[16] Yu.I. Karlovich, *An algebra of pseudodifferential operators with slowly oscillating symbols.* Proc. London Math. Soc., to appear.

[17] Yu.I. Karlovich and A.B. Lebre, *Algebra of singular integral operators with a Carleman backward slowly oscillating shift.* Integral Equations and Operator Theory **41** (2001), 288–323.

[18] Yu.I. Karlovich and E. Ramírez de Arellano, *Singular integral operators with fixed singularities on weighted Lebesgue spaces.* Integral Equations and Operator Theory **48** (2004), 331–363.

[19] Yu.I. Karlovich and B. Silbermann, *Fredholmness of singular integral operators with discrete subexponential groups of shifts on Lebesgue spaces.* Math. Nachr. **272** (2004), 55–94.

[20] C.E. Kenig and P.A. Tomas, *Maximal operators defined by Fourier multipliers.* Studia Math. **68** (1980), 79–83.

[21] V.G. Kravchenko and G.S. Litvinchuk, *Introduction to the Theory of Singular Integral Operators with Shift.* Mathematics and its Applications. Kluwer Academic Publishers, v. **289**, Dordrecht, Boston,London, 1994.

[22] H. Kumano-go, *Pseudodifferential Operators.* MIT Press, Cambridge, MA, 1974.

[23] J.E. Lewis and C. Parenti, *Pseudodifferential operators of Mellin type.* Comm. Part. Diff. Equ. **8** (1983), 477–544.

[24] J. Marschall, *Weighted L^p-estimates for pseudo-differential operators with nonregular symbols.* Z. Anal. Anwendungen. **10** (1991), 493–501.

[25] R. Melrous, *Transformation of boundary value problems.* Singularities in Boundary Value Problems. Proc. NATO Adv. Stud. Inst., Dordrecht, 1981, 133–168.

[26] R. Melrous and J. Sjöstrand, *Singularities of boundary value problems, I.* Comm. Pure Appl. Math. **31** (1978), 593–617.

[27] I.P. Natanson, *Theory of Functions of Real Variable.* Vol. **1**, Frederick Ungar Publ. Co., New York, 1964.

[28] S.C. Power, *Fredholm Toeplitz operators and slow oscillation.* Canad. J. Math. **XXXII** (1980), 1058–1071.

[29] V.S. Rabinovich, *Algebras of singular integral operators on compound contours with nodes that are logarithmic whirl points.* Russ. Acad. Sci. Izv. Math. **60** (1996), 1261–1292.

[30] V.S. Rabinovich, *An introductory course on pseudodifferential operators,* Textos de Matemática-1, Centro de Matemática Aplicada, Instituto Superior Técnico, Lisboa, 1998.

[31] V.S. Rabinovich, *Pseudodifferential operators on R^n with variable shifts.* Zeitschrift für Analysis und ihre Adwendungen, Journal for Analysis and its Applications **22** (2003), No. 2, 315–338.

[32] V. Rabinovich, S. Roch, and B. Silbermann, *Limit Operators and Their Applications in Operator Theory.* Birkhäuser, Basel, 2004.

[33] M. Reed and B. Simon, *Methods of Modern Mathematical Physics. 1. Functional Analysis.* Academic Press, New York, 1972.

[34] D. Sarason, *Toeplitz operators with piecewise quasicontinuous symbols.* Indiana Univ. Math. J. **26** (1977), 817–838.

[35] B.W. Schulze, *Pseudo-Differential Operators on Manifolds with Singularities.* North-Holland, Amsterdam, 1991.

[36] B.W. Schulze, *Pseudo-Differential Boundary Value Problems, Conical Singularities and Asymptotics.* Akademie-Verlag, Berlin, 1994.

[37] L. Schwartz, *Analyse Mathématique.* Vol. **1**, Hermann, 1967.

[38] M.A. Shubin, *Pseudodifferential Operators and Spectral Theory.* Springer, Berlin, 1987; Russian original, Nauka, Moscow, 1978.

[39] E.M. Stein, *Harmonic Analysis: Real-Variable Methods, Orthogonality, and Oscillatory Integrals.* Princeton Univ. Press, Princeton, NJ, 1993.

[40] M.E. Taylor, *Pseudodifferential Operators.* Princeton Univ. Press, Princeton, NJ, 1981.

[41] M.E. Taylor, *Tools for PDE. Pseudodifferential Operators, Paradifferential Operators, and Layer Potentials.* American Mathematical Society, Providence, RI, 2000.

[42] F. Treves, *Introduction to Pseudodifferential and Fourier Integral Operators.* Vols. **1** and **2**, Plenum Press, New York, 1982.

Yuri I. Karlovich
Facultad de Ciencias
Universidad Autónoma del Estado de Morelos
Av. Universidad 1001, Col. Chamilpa
C.P. 62209 Cuernavaca
Morelos, México
e-mail: karlovich@buzon.uaem.mx

Operator Theory:
Advances and Applications, Vol. 171, 225–244

Extension of Operator Lipschitz and Commutator Bounded Functions

Edward Kissin, Victor S. Shulman and Lyudmila B. Turowska

1. Introduction and preliminaries

Let $(B(H), \|\cdot\|)$ be the algebra of all bounded operators on an infinite-dimensional Hilbert space H. Let $B(H)_{sa}$ be the set of all selfadjoint operators in $B(H)$. Throughout the paper we denote by α a compact subset of $\mathbb{R}$ and by $B(H)_{sa}(\alpha)$ the set of all operators in $B(H)_{sa}$ with spectrum in α:

$$B(H)_{sa}(\alpha) = \{A = A^* \in B(H) : \ \mathrm{Sp}(A) \subseteq \alpha\}.$$

We will use similar notations $\mathcal{A}_{sa}$, $\mathcal{A}_{sa}(\alpha)$ for a Banach $*$-algebra $\mathcal{A}$. Each bounded Borel function g on α defines, via the spectral theorem, a map $A \to g(A)$ from $B(H)_{sa}(\alpha)$ into $B(H)$. Various smoothness conditions when imposed on this map define the corresponding classes of operator-smooth functions.

Definition 1.1.

(i) *A function g on α in $\mathbb{R}$ is operator Lipschitzian if there is $D > 0$ such that*

$$\|g(A) - g(B)\| \leq D\|A - B\| \textit{ for } A, B \in B(H)_{sa}(\alpha).$$

(ii) *A function g on an open subset Γ of $\mathbb{R}$ is operator Lipschitzian if it is operator Lipschitzian on each compact subset of Γ.*

For infinite-dimensional H, the classes of operator Lipschitz functions defined above do not depend on the dimension of H. For finite-dimensional spaces, the classes defined in (i) coincide with the class of all functions Lipschitzian on α in the usual sense, that is, such that $|g(x) - g(y)| \leq D\,|x - y|$ for some $D > 0$ and all $x, y \in \alpha$.

The study of operator Lipschitz functions was initiated by Daletskii and Krein [DK] and motivated by various problems in the scattering theory. Following their paper, there have been significant articles by Birman and Solomyak [BS1], [BS2], Johnson and Williams [JW], Farforovskaya [F], Peller [Pe] and others (see the bibliography in [KS3]) which consider "scalar" smoothness properties of operator Lipschitz functions. In [KS2] the first two authors investigated the action of operator Lipschitz functions on the domains of $*$-derivations of C^*-algebras. It was shown that operator Lipschitz functions are exactly the functions which act on the domains of all *weakly closed* $*$-derivations of C^*-algebras.

In this paper we study the following extension problems for operator Lipschitz functions.

Question 1. Let g be an operator Lipschitz function on intervals $[a, b]$ and $[c, d]$ with $b < c$. Is it operator Lipschitzian on $[a, b] \cup [c, d]$?

Question 2. Let a function g be operator Lipschitzian on an interval $[a, b]$. Can it be extended to an operator Lipschitz function on a larger interval $[a, c]$?

Question 3. Let a continuous function g on $[a, b]$ be operator Lipschitzian on an infinite set of closed intervals covering $[a, b)$. Under which conditions is it operator Lipschitzian on $[a, b]$?

We show that Question 1 always has a positive answer as long as $b \neq c$. This implies that a function on an open subset of $\mathbb{R}$ is operator Lipschitzian if and only if it is operator Lipschitzian in a neighborhood of each point. For $b = c$, the answer, in general, is negative and we find some conditions when it is positive.

We also show that Question 2 has a positive answer if $\mathrm{supp}(g) \subset [a, b)$. However, we do not know the full answer to the more subtle case when $\mathrm{supp}(g) = [a, b]$ nor the full answer to Question 3. We prove that under some conditions on the behavior of g near the point b these questions have positive answers.

As an important consequence of these results, we construct in Section 5 a large variety of operator Lipschitz functions which are not continuously differentiable. The problem of the existence of such functions was posed by Williams in [W] and the first example was given in [KS3]. The existence of such functions allows us to distinguish between the classes of operator Lipschitz and operator differentiable functions (see [KS2]).

In this paper we study the extension of operator Lipschitz functions in a wider framework. Apart from the standard operator norm, we consider also other unitarily invariant norms on $B(H)$ and the classes of operator Lipschitz functions with respect to these norms. This approach gives rise to a rich variety of functional spaces.

Recall (see [GK]) that a two-sided ideal J of $B(H)$ is *symmetrically normed* (s.n.) if it is a Banach space with respect to a norm $\|\cdot\|_J$ and

$$\|AXB\|_J \leq \|A\|\|X\|_J\|B\| \text{ for } A, B \in B(H) \text{ and } X \in J.$$

It is a $*$-ideal and, by the Calkin theorem, it lies in the ideal $C(H)$ of all compact operators on H. An important class of s.n. ideals is constituted by Schatten ideals C_p, $1 \leq p < \infty$. We will write $\|\cdot\|_p$ instead of $\|\cdot\|_{C_p}$. We denote $C_\infty = C(H)$ and $C_b = B(H)$.

Definition 1.2. *Let J be an s.n. ideal and let Γ be an open subset of $\mathbb{R}$, or one of the finite or infinite intervals $(a, d]$, $[a, d)$.*

(i) *A function g on a compact α in $\mathbb{R}$ is J-Lipschitzian, if there is $D > 0$ such that*

$$g(A) - g(B) \in J \text{ and } \|g(A) - g(B)\|_J \leq D\|A - B\|_J \text{ for } A, B \in J_{sa}(\alpha). \quad (1.1)$$

Denote by $D(g, \alpha)$ the minimal value of D for which the above inequality holds.

(ii) *A function g on Γ is J-Lipschitzian if it is J-Lipschitzian on each compact set in Γ.*

It should be noted (see [KS3]) that the class of operator Lipschitz functions coincides with the classes of C_∞-Lipschitz functions and of C_1-Lipschitz functions.

Davies [D] and Farforovskaya [F] studied the smoothness properties of C_p-Lipschitz functions. For arbitrary symmetrically normed ideals J, the spaces of J-Lipschitz functions, their hierarchy and properties were investigated in [KS2] and [KS3]. De Pagter, Sukochev and Witvliet studied in [PSW] properties of Lipschitz functions on non-commutative L_p-spaces.

Let $\mathcal{A}, \mathcal{B}$ be C^*-algebras and let a Banach space $\mathcal{X}$ be a left $\mathcal{A}$- and right $\mathcal{B}$-module. It is a Banach $(\mathcal{A}, \mathcal{B})$-bimodule if the actions commute and are bounded: there is $M = M(\mathcal{A}, \mathcal{B}, \mathcal{X}) > 0$ such that

$$\|AX\|_\mathcal{X} \leq M\|A\|\|X\|_\mathcal{X} \text{ and } \|XB\|_\mathcal{X} \leq M\|B\|\|X\|_\mathcal{X}, \text{ for } A \in \mathcal{A},\ B \in \mathcal{B},\ X \in \mathcal{X}. \quad (1.2)$$

We write '$\mathcal{A}$-bimodule' instead of $(\mathcal{A}, \mathcal{A})$-bimodule and $[A, X]$ instead of the commutator $AX - XA$ for $A \in \mathcal{A}$ and $X \in \mathcal{X}$. Note that all symmetrically normed ideals are Banach $B(H)$-bimodules and Banach $\mathcal{A}$-modules for each C^*-subalgebra $\mathcal{A}$ of $B(H)$.

The following notion of commutator $\mathcal{X}$-bounded functions has no "scalar" analogues.

Definition 1.3. *Let $\mathcal{X}$ be a Banach $\mathcal{A}$-bimodule and let Γ be an open set in $\mathbb{R}$, or one of the finite or infinite intervals $(a, d]$, $[a, d)$.*

(i) *A continuous function g on α is commutator $(\mathcal{A}, \mathcal{X})$-bounded, if there is $K > 0$ such that*

$$\|[g(A), X]\|_\mathcal{X} \leq K\|[A, X]\|_\mathcal{X}, \text{ for } X \in \mathcal{X} \text{ and } A \in \mathcal{A}_{sa}(\alpha). \quad (1.3)$$

Denote by $K(g, \alpha)$ the minimal constant K for which (1.3) holds.

(ii) *A continuous function on Γ is commutator $(\mathcal{A}, \mathcal{X})$-bounded, if it is commutator $(\mathcal{A}, \mathcal{X})$-bounded on each compact in Γ.*

The proof of the following result can be found in [D, F, KS1].

Theorem 1.4. *Let J be a symmetrically normed ideal and α be a compact subset of $\mathbb{R}$. The space of J-Lipschitz (respectively, operator Lipschitz) functions on α coincides with the space of all commutator $(B(H), J)$-bounded (respectively, $(B(H), B(H))$-bounded) functions on α.*

We extend now the above result to C^*-algebras.

Theorem 1.5. *Let $\mathcal{A}$ be an infinite-dimensional C^*-algebra of operators on H and let $\alpha = [-a, a]$.*

(i) *If J is a separable s.n. ideal, then the space of commutator $(\mathcal{A}, J)$-bounded functions on α coincides with the space of J-Lipschitz functions.*
(ii) *The spaces of commutator $(\mathcal{A}, B(H))$-bounded functions on α coincides with the space of operator Lipschitz functions.*

Proof. Let $\mathcal{B}$ be the weak closure of $\mathcal{A}$ and $B \in \mathcal{B}_{sa}(\alpha)$. By Kaplansky's density theorem, there are A_n in $\mathcal{A}_{sa}$ such that $\|A_n\| \leq \|B\|$ and A_n converge to B in the strong operator topology ($A_n \overset{sot}{\to} B$). Hence $A_n \in \mathcal{A}_{sa}(\alpha)$. Let g be a commutator $(\mathcal{A}, J)$-bounded function on α. Since g is uniformly approximated by polynomials on α and since $P(A_n) \overset{sot}{\to} P(B)$ for each polynomial P, we have $g(A_n) \overset{sot}{\to} g(B)$. Hence

$$[g(A_n), X] \overset{sot}{\to} [g(B), X] \text{ for } X \in B(H). \tag{1.4}$$

Let J be a separable s.n. ideal. By Theorem III.6.3 of [GK], $\|[A_n, X] - [B, X]\|_J \to 0$ for $X \in J$. Hence

$$\overline{\lim}\|[g(A_n), X]\|_J \leq K\overline{\lim}\|[A_n, X]\|_J = K\|[B, X]\|_J.$$

Then $[g(B), X] \in J$ and it follows from the above inequality, from (1.4) and from Theorem III.5.1 of [GK] that $\|[g(B), X]\|_J \leq K\|[B, X]\|_J$. Thus g is a commutator $(\mathcal{B}, J)$-bounded function on α.

Since $\mathcal{B}$ is an infinite-dimensional W^*-algebra, there is a self-adjoint operator B with $\mathrm{Sp}(B) = \alpha$. It follows from Theorem 3.4 of [KS1] that

$$\|[g(A), X]\|_J \leq K\|[A, X]\|_J \text{ for every } A \in B(H)_{sa}(\alpha).$$

Hence g is a commutator $(B(H), J)$-bounded function on α.

Conversely, each commutator $(B(H), J)$-bounded function on α is continuous (see [KS3]), so it is commutator $(\mathcal{A}, J)$-bounded. From this and from Theorem 1.4 we have that the spaces of commutator $(\mathcal{A}, J)$-bounded and of J-Lipschitz functions on α coincide. This proves part (i).
All commutator $(\mathcal{A}, B(H))$-bounded functions on α are commutator $(\mathcal{A}, C(H))$-bounded. By (i), they are all commutator $(B(H), C(H))$-bounded on α. It was proved in [KS3] that the spaces of commutator $(B(H), C(H))$-bounded functions on α, of commutator $(B(H), B(H))$-bounded functions and of operator Lipschitz functions coincide. This completes the proof of (ii). □

The possibility to reduce the study of J-Lipschitz functions to the study of commutator bounded functions is very useful, since it linearizes the problems and enables one to avoid the complicated techniques of double operator integrals.

2. Commutator $(\mathcal{A}, \mathcal{X})$-bounded functions

Let $\mathcal{A}, \mathcal{B}$ be C^*-algebras and let $\mathcal{X}$ be a Banach $(\mathcal{A}, \mathcal{B})$-bimodule. If $\mathrm{Sp}(A) \cap \mathrm{Sp}(B) = \emptyset$, for $A \in \mathcal{A}$ and $B \in \mathcal{B}$, then, by Rosenblum's Theorem (see [RR]), there is $C > 0$ such that

$$\|AX - XB\|_{\mathcal{X}} \geq C\|X\|_{\mathcal{X}} \text{ for all } X \in \mathcal{X}. \tag{2.1}$$

The constant C can be estimated via the distance between $\mathrm{Sp}(A)$ and $\mathrm{Sp}(B)$. The next two results are known (see [BR] and the references there). We include their proofs for the convenience of the reader.

Lemma 2.1. *Let T be a bounded operator on a Banach space such that*

$$\mathrm{Sp}(T) \subset (\infty, -\delta) \cup (\delta, \infty) \text{ and } \|\exp(isT)\| \leq R, \text{ for all } s \in \mathbb{R},$$

where R, δ are some positive numbers. Then $\|T^{-1}\| \leq C_0 R\delta^{-1}$ for some universal constant C_0.

Proof. Let f be a smooth odd function in $L^2(\mathbb{R})$ that coincides with $1/t$ on $(-\infty, -1) \cup (1, \infty)$. Denote by $\widehat{f}$ its Fourier transform. Since $f' \in L^2(\mathbb{R})$, we have $s\widehat{f}(s) \in L^2(\mathbb{R})$ whence $|s|\widehat{f}(s) \in L^2(\mathbb{R})$. Hence $(1 + |s|)\widehat{f}(s) \in L^2(\mathbb{R})$. Since $(1 + |s|)^{-1} \in L^2(\mathbb{R})$, $\widehat{f} \in L^1(\mathbb{R})$.

Set $C_0 = \int_{-\infty}^{\infty} |\widehat{f}(s)|ds$ and let $f_\delta(t) = \frac{1}{\delta} f(\frac{t}{\delta})$. Then $f_\delta(t) = 1/t$ for $|t| \geq \delta$ whence $f_\delta(T) = T^{-1}$. Since $\widehat{f}(\delta s)$ is the Fourier transform of f_δ,

$$f_\delta(T) = \int_{-\infty}^{\infty} \widehat{f}(\delta s) \exp(-isT)ds.$$

Therefore $\|T^{-1}\| \leq \int_{-\infty}^{\infty} |\widehat{f}(\delta s)|Rds \leq C_0 R\delta^{-1}$. □

For disjoint subsets α, β of $\mathbb{R}$, let $\delta(\alpha, \beta) = \inf_{t\in\alpha, s\in\beta} |t - s|$ be the distance between them.

Lemma 2.2. *Let $\mathcal{A}$ and $\mathcal{B}$ be C^*-algebras and let $\mathcal{X}$ be a Banach $(\mathcal{A}, \mathcal{B})$-bimodule. If $\delta = \delta(\alpha, \beta) > 0$, for disjoint subsets α and β of $\mathbb{R}$, then*

$$\|AX - XB\|_{\mathcal{X}} \geq C_0^{-1}M^{-2}\delta\|X\|_{\mathcal{X}}, \text{ for } X \in \mathcal{X},\ A \in \mathcal{A}_{sa}(\alpha),\ B \in \mathcal{B}_{sa}(\beta),$$

where M is the constant in (1.2) and $C_0 > 0$ is the constant in Lemma 2.1.

Proof. For $A \in \mathcal{A}_{sa}(\alpha)$ and $B \in \mathcal{B}_{sa}(\beta)$, the operators L_A: $X \in \mathcal{X} \to AX$, R_B: $X \in \mathcal{X} \to XB$ act on $\mathcal{X}$ and commute. Set $T = L_A - R_B$. To prove the lemma, it suffices to show that $\|T^{-1}\| \leq C_0 M^2\delta^{-1}$. It is easy to see that

$$\begin{aligned}\|\exp(isT)\| &= \|\exp(isL_A)\exp(-isR_B)\| = \|L_{\exp(isA)}R_{\exp(-isB)}\| \\ &\leq M^2\|\exp(isA)\|\|\exp(-isB)\| = M^2.\end{aligned}$$

Since $\mathrm{Sp}(T) \subset \{t - s : t \in \alpha,\ s \in \beta\} \subset (-\infty, -\delta) \cup (\delta, \infty)$, it only remains to apply Lemma 2.1. □

In the rest of this paper $\mathcal{A}$ will be a W^*-algebra on a Hilbert space H and $\mathcal{X}$ be a Banach $\mathcal{A}$-bimodule. Let g be a continuous function on α and $A \in \mathcal{A}_{sa}(\alpha)$. Let $E(t)$ be the spectral resolution of the identity for A continuous from the right in the strong operator topology and let $P(\gamma)$, for $\gamma \subseteq \alpha$, be the corresponding spectral measure of A. Let $E(t-0) = s\text{-}\lim_{s\to t-0} E(s)$ in the strong operator topology. Then $E(t) = P((-\infty, t])$ and $E(t-0) = P((-\infty, t))$ belong to $\mathcal{A}$. Set $H_\gamma = P(\gamma)H$. Then $\mathrm{Sp}(A|_{H_\gamma}) \subseteq \gamma$ and

$$P(\gamma)g(A) = g(A)P(\gamma) = P(\gamma)g(P(\gamma)A) = g(A|_{H_\gamma}) \oplus \mathbf{0}|_{H_\gamma^\perp}. \tag{2.2}$$

Corollary 2.3. *Let β and γ be disjoint closed subsets of α and let g be a continuous function on α. For each $A \in \mathcal{A}_{sa}(\alpha)$ and $X \in \mathcal{X}$,*

$$\|P(\gamma)g(A)XP(\beta)\|_{\mathcal{X}} \le C_0\delta(\gamma,\beta)^{-1}M^5 \sup_{t\in\gamma}|g(t)|\|[A,X]\|_{\mathcal{X}},$$
$$\|P(\beta)Xg(A)P(\gamma)\|_{\mathcal{X}} \le C_0\delta(\gamma,\beta)^{-1}M^5 \sup_{t\in\gamma}|g(t)|\|[A,X]\|_{\mathcal{X}}.$$

Proof. From (2.2) we have $\|P(\gamma)g(A)\| = \|g(A|_{H_\gamma})\| \le \sup_{t\in\gamma}|g(t)|$. Hence, by (1.2),

$$\begin{aligned}\|P(\gamma)g(A)XP(\beta)\|_{\mathcal{X}} &\le M\|P(\gamma)g(A)\|\|P(\gamma)XP(\beta)\|_{\mathcal{X}}\\ &\le M\sup_{t\in\gamma}|g(t)|\|P(\gamma)XP(\beta)\|_{\mathcal{X}}.\end{aligned}$$

Set $T = A|_{H_\gamma}, S = A|_{H_\beta}$. Then $\mathrm{Sp}(T) \subseteq \gamma$, $\mathrm{Sp}\ (S) \subseteq \beta$. Since $\mathcal{Y} = P(\gamma)\mathcal{X}P(\beta)$ is a Banach left $P(\gamma)\mathcal{A}P(\gamma)$-module and right $P(\beta)\mathcal{A}P(\beta)$-module with constant

$$M(P(\gamma)\mathcal{A}P(\gamma), P(\beta)\mathcal{A}P(\beta), \mathcal{Y}) \le M(\mathcal{A},\mathcal{A},\mathcal{X}) = M,$$

we obtain from (1.2) and Lemma 2.2 that

$$\begin{aligned}M^2\|[A,X]\|_{\mathcal{X}} &\ge \|P(\gamma)[A,X]P(\beta)\|_{\mathcal{X}} = \|TP(\gamma)XP(\beta) - P(\gamma)XP(\beta)S\|_{\mathcal{Y}}\\ &\ge C_0^{-1}M^{-2}\delta(\gamma,\beta)\|P(\gamma)XP(\beta)\|_{\mathcal{Y}} = C_0^{-1}M^{-2}\delta(\gamma,\beta)\|P(\gamma)XP(\beta)\|_{\mathcal{X}}.\end{aligned}$$

Combining the above two inequalities, we obtain the first inequality of the corollary. The proof of the second inequality is similar. □

For $\beta = [a,d]$, $P(\beta) = E(d) - E(a-0)$. Set

$$\widetilde{P}(\beta) = E(d) - E(a) = P((a,d]) \text{ and } \widetilde{H}_\beta = \widetilde{P}(\beta)H.$$

Then $\mathrm{Sp}(A|_{\widetilde{H}_\beta}) \subseteq \beta$.

Remark 2.4. *Let β and γ be closed intervals in α. The inequalities in Corollary* 2.3 *hold if $P(\beta)$ is replaced by $\widetilde{P}(\beta)$ or (and) $P(\gamma)$ by $\widetilde{P}(\gamma)$.*

Now we turn our attention to commutator $(\mathcal{A},\mathcal{X})$-bounded functions. The set of all such functions on α is an algebra with respect to the usual operations. Supplied with the norm

$$|f|_\alpha = \|f\|_\infty + K(f,\alpha)$$

it becomes a differential Banach algebra in the sense of [BK, KS]. The result below shows that this algebra is quite rich.

Proposition 2.5. *Let $g \in L^1(\mathbb{R})$ with Fourier transform $\widehat{g}$ satisfy*

$$\int_{\mathbb{R}} |s\widehat{g}(s)| < \infty.$$

For all $\mathcal{A}$ and $\mathcal{X}$, g is commutator $(\mathcal{A}, \mathcal{X})$-bounded on each $\alpha \subset \mathbb{R}$ and $K(g, \alpha) \leq \frac{M^2}{2\pi} \int_{\mathbb{R}} |s\widehat{g}(s)|$.

Proof. Let $X \in \mathcal{X}$ and $A \in \mathcal{A}$. Repeating the argument of Lemma 2 [Po], we have

$$[e^A, X] = \int_0^1 e^{tA}[A, X]e^{(1-t)A} dt.$$

The function g acts on $B(H)_{sa}$ by the formula

$$g(B) = \int_{\mathbb{R}} e^{-isB}\widehat{g}(s)ds \quad \text{for } B \in B(H)_{sa}.$$

Hence it follows from (1.2) that

$$\begin{aligned}
\|[g(A), X]\|_{\mathcal{X}} &= \left\| \int_{\mathbb{R}} \widehat{g}(s)[e^{-isA}, X]ds \right\|_{\mathcal{X}} \\
&\leq \int_{\mathbb{R}} |\widehat{g}(s)| \left(\int_0^1 M^2 \|e^{-istA}\| \|[sA, X]\|_{\mathcal{X}} \|e^{-is(1-t)A}\| dt \right) ds \\
&= M^2 \|[A, X]\|_{\mathcal{X}} \int_{\mathbb{R}} |s\widehat{g}(s)| ds
\end{aligned}$$

which proves the lemma. □

Let g be a continuous function on α and $A \in \mathcal{A}_{sa}(\alpha)$. Let γ be a subset of α. By (2.2), $0 \in \mathrm{Sp}(P(\gamma)A) \subseteq \gamma \cup \{0\}$. If $0 \notin \alpha$, $\mathrm{Sp}(P(\gamma)A)$ does not lie in α, so g does not act on $P(\gamma)A$.

Lemma 2.6. *Let g be a commutator $(\mathcal{A}, \mathcal{X})$-bounded function on a compact $\beta \subseteq \alpha$. For any subsets γ, δ of β and all $A \in \mathcal{A}_{sa}(\alpha)$ and $X \in \mathcal{X}$,*

$$\|P(\gamma)[g(A), X]P(\gamma)\|_{\mathcal{X}} \leq K(g, \beta)M^2 \|[A, X]\|_{\mathcal{X}}, \tag{2.3}$$

$$\|P(\gamma)[g(A), X]P(\delta)\|_{\mathcal{X}} \leq K(g, \beta)M^4 \|[A, X]\|_{\mathcal{X}}. \tag{2.4}$$

Proof. Choose $\lambda \in \gamma$ and set $B = P(\gamma)A + \lambda(\mathbf{1} - P(\gamma))$. Then $\mathrm{Sp}(B) \subseteq \gamma \subseteq \beta$. By (2.2), $P(\gamma)g(A) = P(\gamma)g(B)$. Since $P(\gamma)A = P(\gamma)B$, we obtain (2.3) by applying (1.3):

$$\begin{aligned}
\|P(\gamma)[g(A), X]P(\gamma)\|_{\mathcal{X}} &= \|[g(B), P(\gamma)XP(\gamma)]\|_{\mathcal{X}} \leq K(g, \beta)\|[B, P(\gamma)XP(\gamma)]\|_{\mathcal{X}} \\
&= K(g, \beta)\|P(\gamma)[A, X]P(\gamma)\|_{\mathcal{X}} \leq K(g, \beta)M^2 \|[A, X]\|_{\mathcal{X}}.
\end{aligned}$$

From (1.2) and (2.3) we obtain (2.4)

$$\begin{aligned}\|P(\gamma)[g(A),X]P(\delta)\|_{\mathcal{X}} &= \|P(\gamma)P(\beta)[g(A),X]P(\beta)P(\delta)\|_{\mathcal{X}}\\ &\le M^2\|P(\beta)[g(A),X]P(\beta)\|_{\mathcal{X}} \le K(g,\beta)M^4\|[A,X]\|_{\mathcal{X}}.\ \square\end{aligned}$$

Proposition 2.7. *Let a continuous function g be commutator $(\mathcal{A},\mathcal{X})$-bounded on compact subsets α and β of $\mathbb{R}$. If $\overline{\alpha\setminus\beta}\cap\overline{\beta\setminus\alpha}=\emptyset$ then g is commutator $(\mathcal{A},\mathcal{X})$-bounded on $\alpha\cup\beta$.*

Proof. Let $A\in\mathcal{A}_{sa}(\alpha\cup\beta)$. Set $\gamma_1=\alpha\setminus\beta$, $\gamma_2=\alpha\cap\beta$, $\gamma_3=\beta\setminus\alpha$ and $P_i=P(\gamma_i)$. Then $\alpha\cup\beta=\gamma_1\cup\gamma_2\cup\gamma_3$ and

$$\|[g(A),X]\|_{\mathcal{X}} \le \sum_{i,j}\|P_i[g(A),X]P_j\|_{\mathcal{X}}, \text{ for } X\in\mathcal{X}.$$

To prove the proposition, it suffices to show that, for all $i,j\in\{1,2,3\}$,

$$\|P_i[g(A),X]P_j\|_{\mathcal{X}} \le C_{ij}\|[A,X]\|_{\mathcal{X}},$$

where C_{ij} do not depend on A and X.

If $\gamma_i\cup\gamma_j\subseteq\alpha$, it follows from (2.4) that

$$\|P_i[g(A),X]P_j\|_{\mathcal{X}} \le M^4K(g,\alpha)\|A,X]\|_{\mathcal{X}}.$$

If $\gamma_i\cup\gamma_j\subseteq\beta$ then $\|P_i[g(A),X]P_j\|_{\mathcal{X}}\le M^4K(g,\beta)\|A,X]\|_{\mathcal{X}}$.

Since $\overline{\gamma_1}$ and $\overline{\gamma_3}$ are disjoint, we have from (1.2) and Corollary 2.3,

$$\begin{aligned}&\|P_1[g(A),X]P_3\|_{\mathcal{X}} \le M^2\|P(\overline{\gamma_1})[g(A),X]P(\overline{\gamma_3})\|_{\mathcal{X}}\\ &\le M^2\|P(\overline{\gamma_1})g(A)XP(\overline{\gamma_3})\|_{\mathcal{X}} + M^2\|P(\overline{\gamma_1})Xg(A)P(\overline{\gamma_3})\|_{\mathcal{X}}\\ &\le 2C_0\delta(\overline{\gamma_1},\overline{\gamma_3})^{-1}M^7r\|[A,X]\|_{\mathcal{X}},\end{aligned}$$

where $r=\sup_{t\in\alpha\cup\beta}|g(t)|$. We get the same inequality, if we exchange P_1 and P_3. □

Note that the proof of Proposition 2.7 gives the following estimate for $K(g,\alpha\cup\beta)$:

$$K(g,\alpha\cup\beta) \le 7M^4\max(K(g,\alpha),K(g,\beta)) + 4C_0\delta(\overline{\gamma_1},\overline{\gamma_3})^{-1}M^7\sup_{t\in\alpha\cup\beta}|g(t)|. \quad (2.5)$$

Corollary 2.8. *If a function is commutator $(\mathcal{A},\mathcal{X})$-bounded on open sets Γ_i, $i\in I$, for some index set I, then it is commutator $(\mathcal{A},\mathcal{X})$-bounded on their union Γ.*

Proof. Without loss of generality, assume that all Γ_i are open intervals. Since any compact subset α of Γ is contained in the union of a finite number of Γ_i, it suffices to consider finite sets I. A simple induction reduces the problem to the case of two intervals.

If $\alpha\subset\Gamma_1\cup\Gamma_2$, it can be represented as the union of compact sets β in Γ_1 and γ in Γ_2 such that $\overline{\gamma\setminus\beta}\cap\overline{\beta\setminus\gamma}=\emptyset$. By Proposition 2.7, g is commutator $(\mathcal{A},\mathcal{X})$-bounded on $\Gamma_1\cup\Gamma_2$. □

We obtain immediately that *locally* commutator $(\mathcal{A},\mathcal{X})$-bounded functions are commutator $(\mathcal{A},\mathcal{X})$-bounded.

Corollary 2.9. *A continuous function g is commutator $(\mathcal{A},\mathcal{X})$-bounded on an open set Γ in $\mathbb{R}$ if and only if it is commutator $(\mathcal{A},\mathcal{X})$-bounded on a neighborhood of each point of Γ.*

If g is a continuous function on $\alpha = [a,d]$ (a can be $-\infty$ and d can be ∞), we denote by $\tilde{g}$ its extension to $\mathbb{R}$ that equals 0 outside α.

Corollary 2.10. *Let g be a commutator $(\mathcal{A},\mathcal{X})$-bounded function on α.*

(i) *If* $\operatorname{supp}(g) \subset [a,d)$, *then $\tilde{g}$ is commutator $(\mathcal{A},\mathcal{X})$-bounded on $[a,\infty)$.*
(ii) *If* $\operatorname{supp}(g) \subset (a,d]$, *then $\tilde{g}$ is commutator $(\mathcal{A},\mathcal{X})$-bounded on $(-\infty,d]$.*

Proof. If $\operatorname{supp}(g) \subset [a,d)$, there is $c < d$ such that $\operatorname{supp}(g) \subseteq [a,c]$. For each $r > d$, $\tilde{g}(t) \equiv 0$ on $[c,r]$ and, therefore, is commutator $(\mathcal{A},\mathcal{X})$-bounded on $[c,r]$. By Proposition 2.7, it is commutator $(\mathcal{A},\mathcal{X})$-bounded on $[a,r]$. This proves part (i). The proof of (ii) is identical. □

Making use of Theorem 1.4, we have

Corollary 2.11. *Let $\mathcal{A}$ be a W^*-subalgebra of $B(H)$ and J be a separable symmetrically normed ideal of $B(H)$. Proposition* 2.7, *Corollaries* 2.8, 2.9 *and* 2.10 *hold if the words "commutator $(\mathcal{A},\mathcal{X})$-bounded" are replaced by "J-Lipschitzian" or "operator Lipschitzian".*

It follows from Proposition 2.7 that a commutator bounded function on two intervals, which are either disjoint or intersect by an interval, is commutator bounded on their union. The case when the intersection consists of one point is more subtle and will be considered in the next section.

3. Extension of $\mathbf{C}_p$-Lipschitz functions from a closed interval

In Corollary 2.10 we proved that if g is a commutator $(\mathcal{A},\mathcal{X})$-bounded function on $[d,c]$ and $\operatorname{supp}(g)$ does not contain d, then its extension $\widetilde{g}$ over d is commutator $(\mathcal{A},\mathcal{X})$-bounded on $(-\infty,c]$. In this section we consider the case when $d \in \operatorname{supp}(g)$ and find some sufficient conditions for $\widetilde{g}$ to be commutator bounded on $(-\infty,c]$.

We say that a function g on $[a,c]$ is *square-summable* at $d \in [a,c)$ *from the right* if there are $\lambda_n \searrow d$ such that

$$G = \sum_{n=1}^{\infty}\left(\frac{g_n}{\lambda_n - d}\right)^2 < \infty \text{ where } g_n = \sup\{|g(t)-g(d)| : t \in [d,\lambda_{n-1}]\}. \quad (3.1)$$

Similarly, we define the square-summability of g at d *from the left.*

If g is square-summable at d from the right then $\frac{g_n}{\lambda_n - d} \to 0$, so g is differentiable at point d from the right and $g'(d) = 0$. Conversely, if $g'(d) = g''(d) = 0$ then g is square-summable at d.

The proof of the next lemma is evident.

Lemma 3.1. *Let g and h be functions on $[a,c]$ and let h be square-summable from the right at $d \in [a,c)$. Then g is square-summable from the right at d, if there are $\varepsilon, M > 0$ such that*

$$|g(t) - g(d)| \le M\,|h(t) - h(d)| \ \text{ for } t \in (d, d+\varepsilon).$$

The condition of the square-summability can be written in a simpler way. Denote by g^e the monotone envelope of g for $t \ge d$, that is, $g^e(t) = \sup_{d\le s\le t} |g(s) - g(d)|$.

Proposition 3.2. *The following conditions are equivalent:*

(i) *g is square-summable at $t = d$ from the right;*
(ii) $\sum_{n=1}^{\infty} 2^n g^e(d + 2^{-\frac{n}{2}})^2 < \infty$;
(iii) $\sum_{n=1}^{\infty} g^e(d + n^{-\frac{1}{2}})^2 < \infty$;
(iv) $\int_0^1 g^e(d+t)^2 \frac{dt}{t^3} < \infty$.

Proof. The function $f(x) = g^e(d + x^{-1/2})^2$, for $x \in [\delta, \infty)$ with $\delta = (c-d)^{-2}$, is non-increasing and non-negative. The conditions (i)–(iv) can be rewritten for f as follows:

$$\text{(i}'\text{) there are } x_n \nearrow \infty \text{ such that } \sum_{n=1}^{\infty} x_n f(x_{n-1}) < \infty;$$

$$\text{(ii}'\text{)} \sum_{n=1}^{\infty} 2^n f(2^n) < \infty; \quad \text{(iii}'\text{)} \sum_{n=1}^{\infty} f(n) < \infty; \quad \text{(iv}'\text{)} \int_0^1 f(x)dx < \infty.$$

The equivalence of (ii′), (iii′) and (iv′) is well known, (ii′) $\longrightarrow$ (i′) is clear. Suppose that (i′) holds. Then (iii′) also holds, since

$$\sum_n f(n) = \sum_k \left(\sum_{x_k < n \le x_{k+1}} f(n) \right) \le \sum_k (x_{k+1} - x_k + 1) f(x_k)$$
$$\le \sum_k x_{k+1} f(x_k) < \infty. \qquad \square$$

Our next aim is to prove that any commutator $(\mathcal{A}, \mathcal{X})$-bounded function on $[d,c]$, square summable from the right at d, can be extended to a commutator $(\mathcal{A}, \mathcal{X})$-bounded function on $(-\infty, c]$. To do this we need to restrict the class of bimodules.

Let $\mathcal{A}$ be a W^*-algebra. We say that a left Banach $\mathcal{A}$-module $\mathcal{X}$ is *subsquare* if, for all projections $P \in \mathcal{A}$,

$$\|X\|_{\mathcal{X}}^2 \le \|PX\|_{\mathcal{X}}^2 + \|(\mathbf{1} - P)X\|_{\mathcal{X}}^2, \ \text{ for } X \in \mathcal{X}.$$

Let $\mathcal{X}$ be subsquare, let $X_n \in \mathcal{X}$ and let $P_n \in \mathcal{A}$ be mutually orthogonal projections. If $\sum_n \|P_n X_n\|_{\mathcal{X}}^2 < \infty$ then $\sum_n P_n X_n$ converges in $\mathcal{X}$ and

$$\|\sum_n P_n X_n\|_{\mathcal{X}}^2 \le \sum_n \|P_n X_n\|_{\mathcal{X}}^2. \tag{3.2}$$

Similarly, we define right subsquare Banach $\mathcal{A}$-modules. A Banach $\mathcal{A}$-bimodule is subsquare if it is subsquare as a left and right module.

All Schatten ideals C_p, $2 \le p < \infty$, are subsquare. Indeed, for $X \in C_p$, $\|X\|_p^2 = \|X^*X\|_{\frac{p}{2}}$. Since $\frac{p}{2} \ge 1$,

$$\begin{aligned}\|X\|_p^2 &= \|X^*(P + (\mathbf{1} - P))X\|_{\frac{p}{2}} \le \|X^*PX\|_{\frac{p}{2}} + \|X^*(\mathbf{1} - P)X\|_{\frac{p}{2}} \\ &= \|PX\|_p^2 + \|(\mathbf{1} - P)X\|_p^2.\end{aligned}$$

In the same way it follows that $C_\infty = C(H)$ and $C_b = B(H)$ are also subsquare.

Theorem 3.3. *Suppose that a Banach $\mathcal{A}$-bimodule $\mathcal{X}$ is subsquare. Let g be a commutator $(\mathcal{A}, \mathcal{X})$-bounded function on $\alpha = [d, c]$ and $g(d) = 0$. If g is square-summable at $t = d$ from the right, then its extension $\widetilde{g}$ is commutator $(\mathcal{A}, \mathcal{X})$-bounded on $(-\infty, c]$.*

Proof. Since g is square-summable at d from the right, there are $\lambda_n \searrow d$ such that $\lambda_0 = c$ and (3.1) holds. Set $\alpha_n = [\lambda_n, \lambda_{n-1}]$. For some $p \le d$, set $\gamma = [p, c]$ and $\beta = [p, d]$. For $A = A^* \in \mathcal{A}$ with $\mathrm{Sp}(A) \subseteq \gamma$, let $E(t)$ be its spectral resolution of the identity. Then $E(c) = \mathbf{1}$ and $E(p - 0) = \mathbf{0}$. Set $Q = E(d)$ and $P = \mathbf{1} - Q$. Then $Q = E(d) - E(p - 0) = P(\beta)$ and $P = E(c) - E(d) = P((d, c])$.

Since $\mathrm{supp}(\widetilde{g}) \subseteq [d, c]$ and $\mathrm{Sp}(QA|_{QH}) \subseteq [p, d]$, we have from (2.2) that $Q\tilde{g}(A) = \tilde{g}(A)Q = Q\tilde{g}(QA) = \mathbf{0}$. Hence, for $X \in \mathcal{X}$,

$$\begin{aligned}\|[\tilde{g}(A), X]\|_{\mathcal{X}} &\le \|P[\tilde{g}(A), X]P\|_{\mathcal{X}} + \|P[\tilde{g}(A), X]Q\|_{\mathcal{X}} \\ &\quad + \|Q[\tilde{g}(A), X]P\|_{\mathcal{X}} + \|Q[\tilde{g}(A), X]Q\|_{\mathcal{X}} \\ &= \|P[\tilde{g}(A), X]P\|_{\mathcal{X}} + \|P\tilde{g}(A)XQ]\|_{\mathcal{X}} + \|QX\tilde{g}(A)P]\|_{\mathcal{X}}.\end{aligned}$$

Since $\tilde{g}$ is commutator $(\mathcal{A}, \mathcal{X})$-bounded on α with $K(\tilde{g}, \alpha) = K(g, \alpha)$, we have from (2.3)

$$\|P[\tilde{g}(A), X]P\|_{\mathcal{X}} \le K(g, \alpha)M^2\|[A, X]\|_{\mathcal{X}}.$$

For $\widetilde{P}(\alpha_n) = E(\lambda_{n-1}) - E(\lambda_n)$, we obtain from Corollary 2.3 and Remark 2.4 that

$$\begin{aligned}&\max\{\|\widetilde{P}(\alpha_n)\tilde{g}(A)XP(\beta)]\|_{\mathcal{X}}, \|P(\beta)X\tilde{g}(A)\widetilde{P}(\alpha_n)]\|_{\mathcal{X}}\} \\ &\le C_0^{-1}M^5\delta(\alpha_n, \beta)^{-1}\sup_{t\in\alpha_n}|g(t)|\|[A, X]\|_{\mathcal{X}} \le C_0^{-1}M^5(\lambda_n - d)^{-1}g_n\|[A, X]\|_{\mathcal{X}}.\end{aligned}$$

Since $E(t)$ is strongly continuous from the right, $s\text{-}\lim_{n\to\infty} E(\lambda_n) = E(d) = Q = P(\beta)$ and

$$P = \mathbf{1} - Q = E(c) - s\text{-}\lim_{n\to\infty} E(\lambda_n) = s\text{-}\lim_{n\to\infty} \sum_{k=1}^{n} \widetilde{P}(\alpha_n).$$

By (3.1) and (3.2),

$$\begin{aligned}\|P\tilde{g}(A)XQ]\|_{\mathcal{X}}^2 &= \|\sum_{n=1}^{\infty} \widetilde{P}(\alpha_n)\tilde{g}(A)XP(\beta)]\|_{\mathcal{X}}^2 \\ &\le \sum_{n=1}^{\infty} \|\widetilde{P}(\alpha_n)\tilde{g}(A)XP(\beta)]\|_{\mathcal{X}}^2 \le GC_0^{-2}M^{10}\|[A, X]\|_{\mathcal{X}}^2.\end{aligned}$$

Similarly, $\|QX\tilde{g}(A)P]\|_{\mathcal{X}}^2 \le GC_0^{-2}M^{10}\|[A,X]\|_{\mathcal{X}}^2$. Therefore

$$\|[\tilde{g}(A),X]\|_{\mathcal{X}} \le K\|[A,X]\|_{\mathcal{X}}$$

where $K = K(g,\alpha)M^2 + 2G^{\frac{1}{2}}C_0^{-1}M^5$. □

Corollary 3.4. *Let $p \in [1,\infty] \cup b$, let g be a C_p-Lipschitz function on $[d,c]$ and let $g(d) = 0$. If g is square-summable at $t = d$ from the right, then its extension $\widetilde{g}$ is C_p-Lipschitzian on $(-\infty, c]$.*

Proof. For $p \in [2,\infty] \cup b$, the ideal C_p is subsquare, so the result follows from Theorems 1.4 and 3.3.

For $1 \le p < 2$, g is C_p-Lipschitzian if and only if it is C_q-Lipschitzian for $q = \frac{p}{p-1}$, if $p \neq 1$, and for $q = \infty$, if $p = 1$ (see [KS3]). Combining this with the case $p \ge 2$, we complete the proof. □

Let f and h be C_p-Lipschitz functions on $[a,d]$ and $[d,c]$, respectively, and let $f(d) = h(d) = \lambda$. Suppose that they have, respectively, left and right derivatives r, s at d. Consider the function

$$g(t) = \begin{cases} f(t) \text{ for } t \in [a,d], \\ h(t) \text{ for } t \in [d,c]. \end{cases}$$

Recall that any function square-summable from the left (right) at point d has zero left (right) derivative in d.

Corollary 3.5. *Let $r = f'(d)$ and $s = h'(d)$. Let the function $f(t) - r(t-d)$ be square-summable at point d from the left and let the function $h(t) - s(t-d)$ be square-summable at d from the right.*

(i) *If $1 < p < \infty$, then g is C_p-Lipschitzian on $[a,c]$.*
(ii) *If $p \in \{1,\infty,b\}$ and $r = s$, then g is C_p-Lipschitzian on $[a,c]$.*

Proof. Let $1 < p < \infty$. It is known (see [D]) that the function

$$\xi(t) = \begin{cases} r(t-d) + \lambda \text{ for } t \in [a,d], \\ s(t-d) + \lambda \text{ for } t \in [d,c], \end{cases}$$

is C_p-Lipschitzian for all r, s. Set $\theta = g - \xi$, $\theta_1 = \theta|_{[a,d]}$ and $\theta_2 = \theta|_{[d,c]}$.

The function θ_1 on $[a,d]$ is square-summable at point d from the left and $\theta_1(d) = 0$. It is C_p-Lipschitzian, since it is a linear combination of C_p-Lipschitz functions f and $r(t-d) + \lambda$. By Corollary 3.4, its extension $\widetilde{\theta_1}$ to $[a,\infty)$ is a C_p-Lipschitz function. Similarly, the extension $\widetilde{\theta_2}$ to $(-\infty,c]$ of the function θ_2 on $[d,c]$ is C_p-Lipschitzian. Hence the function

$$\theta = \widetilde{\theta_1}|_{[a,c]} + \widetilde{\theta_2}|_{[a,c]}$$

is C_p-Lipschitzian on $[a,c]$. Thus g is C_p-Lipschitzian on $[a,c]$. Part (i) is proved.

The classes of C_p-Lipschitz functions, for $p \in \{1,\infty,b\}$, coincide and all the functions are differentiable at each point (see [JW]). This is why we set $r = s$ in (ii). In this case $\xi = r(t-d) + \lambda$ is C_p-Lipschitzian on $[a,c]$. By the above argument, g is C_p-Lipschitzian on $[a,c]$. □

It was proved in [KS3] that the function

$$g(t) = t^2 \sin(1/t), \text{ for } t \neq 0, \text{ and } g(0) = 0,$$

is operator Lipschitzian on $[0,1]$, although it is not continuously differentiable at $t = 0$. By Proposition 3.2(iv), $h(t) = t^2$ is square-summable at $t = 0$ from the right. Since $|g(t)| \leq h(t)$, it follows from Lemma 3.1 that g is square-summable at $t = 0$ from the right. Hence, by Theorem 3.3, the extension of g to $(-\infty, 1]$ is operator Lipschitzian.

4. Extension of locally C_p-Lipschitz functions to closed intervals

It was proved in Section 2 that a function is J-Lipschitzian on (a,c) if and only if it is locally J-Lipschitzian. In this section we find some conditions for locally C_p-Lipschitz functions to be C_p-Lipschitzian on $[a,c]$. As before, we first consider this problem for commutator $(\mathcal{A}, C_p)$-bounded functions.

To begin with, note that if g is commutator $(\mathcal{A}, \mathcal{X})$-bounded on $[a,d]$, it is commutator $(\mathcal{A}, \mathcal{X})$-bounded on (a,d) and

$$K(g,\alpha) \leq K(g,[a,d]), \text{ for each compact } \alpha \text{ in } [a,d].$$

To prove the converse statement, we need the following result.

Lemma 4.1. *Let g be a function on α and let λ be a non-isolated point in α. If (1.3) holds for all $A \in \mathcal{A}_{sa}(\alpha)$ with* $\mathrm{Ker}(A - \lambda\mathbf{1}) = \{0\}$, *then it holds for all $A \in \mathcal{A}_{sa}(\alpha)$.*

Proof. Let P be the projection on $L = \mathrm{Ker}(A - \lambda\mathbf{1}) \neq \{0\}$. Set $T_n = A + \lambda_n P$, where $\lambda \neq \lambda_n \in \alpha$ and $\lambda_n \to \lambda$. Then $T_n \in \mathcal{A}_{sa}$, $\mathrm{Ker}(T_n - \lambda\mathbf{1}) = \{0\}$ and $\mathrm{Sp}(T_n) \subseteq \alpha$. Hence $\|[g(T_n), X]\|_{\mathcal{X}} \leq K\|[T_n, X]\|_{\mathcal{X}}$. Since $T_n \to A$ and

$$g(T_n) = g(A|_{H\ominus L}) \oplus g(\lambda_n)P \to g(A|_{H\ominus L}) \oplus g(\lambda)P = g(A),$$

we have $\|[g(A), X]\|_{\mathcal{X}} \leq K\|[A, X]\|_{\mathcal{X}}$. □

Let $\mathcal{A}$ be a W^*-algebra. A Banach $\mathcal{A}$-bimodule $\mathcal{X}$ is *approximate* if there is $C > 0$ such that, for each sequence of projections P_n in $\mathcal{A}$ converging to $\mathbf{1}$ in the strong operator topology ($P_n \overset{sot}{\to} \mathbf{1}$),

$$||X||_{\mathcal{X}} \leq C \sup_n ||P_n X P_n||_{\mathcal{X}} \text{ for all } X \in \mathcal{X}.$$

In particular, $B(H)$ and all separable and all dual to separable symmetrically normed ideals of $B(H)$ are approximate Banach $B(H)$-bimodules (see [GK]).

Proposition 4.2. *Let $\mathcal{X}$ be an approximate Banach $\mathcal{A}$-bimodule. A commutator $(\mathcal{A}, \mathcal{X})$-bounded function g on (a,d) is commutator $(\mathcal{A}, \mathcal{X})$-bounded on $[a,d]$ if and only if*

$$K = \sup\{K(g,\alpha) : \ \alpha \subset (a,d)\} < \infty.$$

Proof. We only need to prove the "if" part. Let $A = A^*$ and $\mathrm{Sp}(A) \subseteq [a,d]$. By Lemma 4.1, we only have to prove (1.3) in the case when $\mathrm{Ker}(A - a\mathbf{1}) =$

$\mathrm{Ker}(A-d\mathbf{1})=\{0\}$. Let $E(t)$ be the spectral resolution of the identity for A. For $d_n\nearrow d$ and $a_n\searrow a$,

$$E(d_n)\overset{sot}{\to}E(d-0)=\mathbf{1}-P_{\mathrm{Ker}(A-d\mathbf{1})}=\mathbf{1}\text{ and }E(a_n)\overset{sot}{\to}E(a)=P_{\mathrm{Ker}(A-a\mathbf{1})}=\mathbf{0}.$$

Set $P_n=E(d_n)-E(a_n)$. Then $\mathrm{Sp}(A|_{P_nH})\subseteq[a_n,d_n]$ and $P_n\overset{sot}{\to}\mathbf{1}$. Choose some λ which lies in all $[a_n,d_n]$ and set $A_n=P_nA+\lambda(\mathbf{1}-P_n)$. Then $\mathrm{Sp}(A_n)\subseteq[a_n,d_n]$ and, by (2.2), $P_ng(A_n)=P_ng(P_nA_n)=P_ng(P_nA)$. Since $\mathcal{X}$ is approximate, we have from (1.2) and (2.2)

$$\begin{aligned}\|[g(A),X]\|_{\mathcal{X}}&\le C\sup\|P_n[g(A),X]P_n\|_{\mathcal{X}}=C\sup\|[P_ng(P_nA),P_nXP_n]\|_{\mathcal{X}}\\&=C\sup\|[g(A_n),P_nXP_n]\|_{\mathcal{X}}\le CK(g,d_n-a_n)\sup\|[A_n,P_nXP_n]\|_{\mathcal{X}}\\&\le CK\sup\|P_n[A,X]P_n\|_{\mathcal{X}}=CKM^2\|[A,X]\|_{\mathcal{X}}.\end{aligned}$$

Thus g is commutator $(\mathcal{A},\mathcal{X})$-bounded on $[a,d]$. □

Corollary 4.3. *Let J be an approximate s.n. ideal. A J-Lipschitz function g on (a,d) is J-Lipschitzian on $[a,d]$ if and only if* $\sup\{D(g,\alpha):\alpha\subset(a,d)\}<\infty$.

Proof. By Proposition 3.4 of [KS3], g is J-Lipschitzian on a compact α if and only if it is commutator $(B(H),J)$-bounded. Moreover, $D(g,\alpha)=K(g,\alpha)$. Hence the result follows from Proposition 4.2. □

Let $\{P_n\}$ be a set of mutually orthogonal projections in $B(H)$. Then (see [GK]), for $A\in C_p$, $p\in[1,\infty)$,

$$\|A\|_p\ge\|\sum_nP_nAP_n\|_p,\tag{4.1}$$

$$\|\sum_nP_nA_nP_n\|_p^p=\sum_n\|P_nA_nP_n\|_p^p.\tag{4.2}$$

We need the following extension of these formulae.

Lemma 4.4. *Let $\{P_n\}$ be a set of mutually orthogonal projections in $B(H)$ and let $\{Q_n\}$ be another set of mutually orthogonal projections. Let $p\in[1,\infty)$. For each $A\in C_p$,*

$$\|A\|_p^p\ge\|\sum_nP_nAQ_n\|_p^p=\sum_n\|P_nAQ_n\|_p^p.\tag{4.3}$$

Proof. Set $X=\sum_nP_nAQ_n$. Since

$$\|Y^*Y\|_{p/2}=\|Y\|_p^2=\|Y^*\|_p^2=\|YY^*\|_{p/2}\text{ for }Y\in C_p,\tag{4.4}$$

we obtain the proof of the equality in (4.3):

$$\begin{aligned}\|\sum_nP_nAQ_n\|_p^p&\overset{(4.4)}{=}\|X^*X\|_{p/2}^{p/2}=\|\sum_nQ_nA^*P_nAQ_n\|_{p/2}^{p/2}\\&\overset{(4.2)}{=}\sum_n\|Q_nA^*P_nAQ_n\|_{p/2}^{p/2}\overset{(4.4)}{=}\sum_n\|P_nAQ_n\|_p^p.\end{aligned}$$

Let $2 \le p$. Since $\|Q_nYQ_n\|_p \le \|Q_n\|\|Y\|_p\|Q_n\| = \|Y\|_p$, for $Y \in C_p$, we have from the above formula

$$\begin{aligned}\|\sum_n P_nAQ_n\|_p^p &= \sum_n \|Q_nA^*P_nAQ_n\|_{p/2}^{p/2} \le \sum_n \|A^*P_nA\|_{p/2}^{p/2} \\ &\overset{(4.4)}{=} \sum_n \|P_nAA^*P_n\|_{p/2}^{p/2} \overset{(4.2)}{=} \|\sum_n P_nAA^*P_n\|_{p/2}^{p/2} \\ &\overset{(4.1)}{\le} \|AA^*\|_{p/2}^{p/2} \overset{(4.4)}{=} \|A\|_p^p.\end{aligned}$$

This completes the proof of (4.3) for $2 \le p$.

For $1 < p < 2$, set $q = \frac{p}{p-1}$ and consider the bounded functional F_X on C_q defined by the formula

$$F_X(Y) = \operatorname{Tr}(XY) = \operatorname{Tr}\left(\sum_n P_nAQ_nY\right) = \operatorname{Tr}\left(A\sum_n Q_nYP_n\right).$$

Since $q > 2$, it follows from (4.3) that $\|\sum_n Q_nYP_n\|_q \le \|Y\|_q$. Hence

$$|F_X(Y)| = \left|\operatorname{Tr}\left(A\sum_n Q_nYP_n\right)\right| \le \|A\|_p\|\sum_n Q_nYP_n\|_q \le \|A\|_p\|Y\|_q.$$

Since $\|X\|_p = \|F_X\|$ (see [GK]), we have $\|X\|_p \le \|A\|_p$. Thus (4.3) is proved for $1 < p < 2$.

For $p = 1$, we have $q = \infty$ and the proof is the same as above. □

We shall now prove the main result of this section.

Theorem 4.5. *Let $\mathcal{A}$ be a W^*-algebra and let $p \in [2, \infty] \cup b$. Let g be a continuous function on $[a, c]$. Let $\{\lambda_n\}_{n=0}^{\infty}$ in $[a, c]$ be such that $\lambda_0 = c$ and $\lambda_n \searrow a$ and let*

$$S = \sum_{n=1}^{\infty}\left(\frac{g_n}{\lambda_n - \lambda_{n+1}}\right)^2 < \infty, \text{ where } g_n = \sup_{t\in[a,\lambda_{n-1}]} |g(t) - g(a)|. \qquad (4.5)$$

Set $\alpha_n = [\lambda_n, \lambda_{n-2}]$. The function g is commutator $(\mathcal{A}, C_p)$-bounded on $[a, c]$ if and only if

1) *g is commutator $(\mathcal{A}, C_p)$-bounded on all α_n;*
2) *$\sup_n\{K(g, \alpha_n)\} = K < \infty$.*

Proof. The "only if" part is evident. Without loss of generality we assume that $g(a) = 0$. Let $A = A^*$, $\operatorname{Sp}(A) \subseteq [a, c]$ and let $E(\lambda)$ be its spectral resolution of identity. By Lemma 4.1, we only need to consider the case when $\operatorname{Ker}(A - a\mathbf{1}) = \{0\}$, so that $s\text{-}\lim_{n\to\infty} E(\lambda_n) = 0$. Consider the projections $P_n = E(\lambda_{n-1}) - E(\lambda_n) = P((\lambda_n, \lambda_{n-1}])$. Then

$$E(\lambda_{n-1}) = \sum_{j=n}^{\infty} P_j \quad \text{and} \quad \mathbf{1} = \sum_{j=1}^{\infty} P_j.$$

Therefore

$$\begin{aligned}
\|[g(A),X]\|_p &= \|\sum_{n=1}^{\infty}\sum_{j=1}^{\infty} P_n[g(A),X]P_j\|_p \\
&\le \|\sum_{n=1}^{\infty} P_n[g(A),X]P_n\|_p + \|\sum_{n=1}^{\infty} P_n[g(A),X]P_{n+1}\|_p + \|\sum_{n=1}^{\infty} P_{n+1}[g(A),X]P_n\|_p \\
&\quad + \|\sum_{n=1}^{\infty} P_n[g(A),X]E(\lambda_{n+1})\|_p + \|\sum_{n=1}^{\infty} E(\lambda_{n+1})[g(A),X]P_n\|_p. \qquad (4.6)
\end{aligned}$$

We will evaluate all terms in (4.6).

In our case $M=1$ in (1.2). Since all P_n commute with $g(A)$, since $(\lambda_n,\lambda_{n-1}]\subset \alpha_{n+1}$ and $(\lambda_{n+1},\lambda_n]\subset\alpha_{n+1}$, we have from Lemma 2.6

$$\|P_i[g(A),X]P_j\|_p \le K(g,\alpha_{n+1})\|[A,X]\|_p \le K\|[A,X]\|_p, \qquad (4.7)$$

$$\begin{aligned}
\|P_i[g(A),X]P_j\|_p &= \|P_i[g(A),P_iXP_j]P_j\|_p \\
&\le K(g,\alpha_{n+1})\|[A,P_iXP_j]\|_p \le K\|[A,P_iXP_j]\|_p, \qquad (4.8)
\end{aligned}$$

where i and j are either n or $n+1$. Hence, if p is ∞ or b, then

$$\|\sum_{n=1}^{\infty} P_n[g(A),X]P_n\|_p = \sup_n \|P_n[g(A),X]P_n\|_p \overset{(4.7)}{\le} K\|[A,X]\|_p,$$

$$\|\sum_{n=1}^{\infty} P_n[g(A),X]P_{n+1}\|_p = \sup_n \|P_n[g(A),X]P_{n+1}\|_p \overset{(4.7)}{\le} K\|[A,X]\|_p,$$

$$\|\sum_{n=1}^{\infty} P_{n+1}[g(A),X]P_n\|_p = \sup_n \|P_{n+1}[g(A),X]P_n\|_p \overset{(4.7)}{\le} K\|[A,X]\|_p.$$

If $p\in[2,\infty)$ then

$$\begin{aligned}
&\|\sum_{n=1}^{\infty} P_n[g(A),X]P_n\|_p^p \overset{(4.3)}{=} \sum_{n=1}^{\infty}\|P_n[g(A),X]P_n\|_p^p \overset{(4.8)}{\le} K^p\sum_{n=1}^{\infty}\|[A,P_nXP_n]\|_p^p \\
&= K^p\sum_{n=1}^{\infty}\|P_n[A,X]P_n]\|_p^p \overset{(4.3)}{\le} K^p\|[A,X]]\|_p^p.
\end{aligned}$$

Similarly,

$$\|\sum_{n=1}^{\infty} P_n[g(A),X]P_{n+1}\|_p^p \le K^p\|[A,X]]\|_p^p$$

and

$$\|\sum_{n=1}^{\infty} P_{n+1}[g(A),X]P_n\|_p^p \le K^p\|[A,X]]\|_p^p.$$

Therefore, for all $p \in [2,\infty] \cup b$,

$$\|\sum_{n=1}^{\infty} P_n[g(A),X]P_n\|_p + \|\sum_{n=1}^{\infty} P_n[g(A),X]P_{n+1}\|_p + \|\sum_{n=1}^{\infty} P_{n+1}[g(A),X]P_n\|_p$$
$$\leq 3K\|[A,X]\|_p. \tag{4.9}$$

Set $\gamma_n = [\lambda_n, \lambda_{n-1}]$. Using Corollary 2.3 and Remark 2.4, we have

$$\|P_n[g(A),X]E(\lambda_{n+1})\|_p \leq \|P_n g(A) X E(\lambda_{n+1})\|_p + \|P_n X g(A) E(\lambda_{n+1})\|_p$$

$$\leq C_0\delta(\gamma_n,[a,\lambda_{n+1}])^{-1}\|[A,X]\|_p\left(\sup_{t\in\gamma_n}|g(t)| + \sup_{t\in[a,\lambda_{n+1}]}|g(t)|\right) \leq \frac{2C_0 g_n}{\lambda_n - \lambda_{n+1}}\|[A,X]\|_p.$$

Since all C_p are subsquare (see Section 3), it follows from the above formula

$$\|\sum_{n=1}^{\infty} P_n[g(A),X]E(\lambda_{n+1})\|_p \overset{(3.2)}{\leq} \left(\sum_{n=1}^{\infty}\|P_n[g(A),X]E(\lambda_{n+1})\|_p^2\right)^{1/2} \leq 2C_0 S\|[A,X]\|_p.$$

Similarly,

$$\sum_{n=1}^{\infty}\|E(\lambda_{n+1})[g(A),X]P_n\|_p \leq 2C_0 S\|[A,X]\|_p.$$

Substituting this and (4.9) in (4.6), we obtain that

$$\|[g(A),X]\|_p \leq (3K + 4C_0 S)\|[A,X]\|_p,$$

so g is commutator $(\mathcal{A}, C_p)$-bounded on $[a,c]$, for $p \in [2,\infty] \cup b$. □

Although conditions (3.1) and (4.5) look similar, comparing them, one can see that $G < S$. Thus condition (4.5) is stronger than condition (3.1) of square-summability of functions. We do not know whether condition (4.5) in Theorem 4.5 can be replaced by more natural and weaker condition (3.1).

Corollary 4.6. *Let $p \in [1,\infty] \cup b$ and let g be a continuous function on $[a,c]$. Let $\{\lambda_n\}_{n=0}^{\infty}$ in $[a,c]$ be such that $\lambda_0 = c$ and $\lambda_n \searrow a$ and let*

$$S = \sum_{n=1}^{\infty}\left(\frac{g_n}{\lambda_n - \lambda_{n+1}}\right)^2 < \infty, \text{ where } g_n = \sup_{t\in[a,\lambda_{n-1}]}|g(t) - g(a)|.$$

Set $\alpha_n = [\lambda_n, \lambda_{n-2}]$. The function g is C_p-Lipschitzian on $[a,c]$ if and only if
1) *g is C_p-Lipschitzian on all α_n;*
2) *$\sup_n\{D(g,\alpha_n)\} = D < \infty$ (see Definition 1.1).*

Proof. It follows from Proposition 3.4 of [KS3] that g is C_p-Lipschitzian on a compact α if and only if it is commutator $(B(H), C_p)$-bounded. Moreover, $D(g,\alpha) = K(g,\alpha)$. Hence, for $p \in [2,\infty] \cup b$, the result follows from Theorem 4.5.

For $1 \leq p < 2$, g is C_p-Lipschitzian if and only if it is C_q-Lipschitzian for $q = \frac{p}{p-1}$, if $p \neq 1$, and for $q = \infty$, if $p = 1$ (see [KS3]). Moreover, the corresponding constants $D(g,\alpha)$ are the same in both cases. Combining this with the case $p \geq 2$, we complete the proof. □

5. Operator Lipschitz functions with discontinuous derivative

In this section we consider only the case $\mathcal{X} = B(H)$ and show that Theorem 4.5 allows us to construct a large variety of operator Lipschitz functions with discontinuous derivatives. We will call commutator $(B(H), B(H))$-bounded functions just commutator bounded.

Let φ be an infinitely differentiable, non-negative function on $\mathbb{R}$ such that

$$\operatorname{supp}(\varphi) = [-1, 1], \ \max_{t\in\mathbb{R}}(\varphi(t)) = 1, \ \varphi'(-\frac{1}{2}) = 1, \ \varphi'(\frac{1}{2}) = -1.$$

Let $\{\sigma_n\}$ be such that $0 < \sigma_{n+1} < \sigma_n < \frac{1}{4}$ and $\sum_{n=1}^{\infty} \sigma_n^2 < \infty$. Set

$$d_n = \frac{3}{2^{n+1}}, \ a_n = \frac{\sigma_n}{2^n} \text{ and } \varphi_n(t) = a_n\varphi\left(\frac{t-d_n}{a_n}\right).$$

Set $\operatorname{supp}(\varphi_n) = \Delta_n$. Then

$$\sup_{t\in\mathbb{R}} |\varphi_n(t)| = a_n, \quad \Delta_n = [d_n - a_n, d_n + a_n] \subset [2^{-n}, 2^{-n+1}]$$

and $\Delta_n \cap \Delta_k = \emptyset$, if $n \neq k$. The function

$$g(t) = \sum_{n=1}^{\infty} \varphi_n(t) \tag{5.1}$$

is infinitely differentiable on $\mathbb{R} - \{0\}$ and $\operatorname{supp}(g) = \{0\} \cup (\cup_{n=1}^{\infty}\Delta_n) \subseteq [0, 1]$. It is differentiable but not continuously differentiable at $t = 0$. Indeed,

$$\frac{g(t) - g(0)}{t} = \frac{g(t)}{t} = 0, \ \text{if } t \notin \cup_{n=1}^{\infty}\Delta_n,$$

$$\frac{g(t) - g(0)}{t} = \frac{g(t)}{t} \leq \frac{a_n}{d_n - a_n}, \ \text{if } t \in \Delta_n.$$

Since $\frac{a_n}{d_n - a_n} \to 0$, as $n \to \infty$, we have that g is differentiable at $t = 0$ and $g'(0) = 0$.

On the other hand, the points $d_n \pm \frac{a_n}{2}$ lie in Δ_n, tend to 0 and

$$g'(d_n \pm \frac{a_n}{2}) = \varphi_n'(d_n \pm \frac{a_n}{2}) = \varphi'(\pm\frac{1}{2}) = \mp 1.$$

Theorem 5.1. *The function g in (5.1) is commutator bounded on $\mathbb{R}$.*

Proof. Since φ is infinitely differentiable and has compact support, it is commutator bounded on $\mathbb{R}$ (see Proposition 2.5), that is, there exists $K > 0$ such that $\|[\varphi(A), X]\| \leq K\|[A, X]\|$, for all $X \in B(H)$ and $A = A^*$. Therefore

$$\begin{aligned}\|[\varphi_n(A), X]\| &= \|[a_n\varphi\left(\frac{1}{a_n}(A - d_n\mathbf{1})\right), X]\| \\ &\leq K\|[\frac{1}{a_n}(A - d_n\mathbf{1}), a_nX]\| = K\|[A, X]\|. \end{aligned} \tag{5.2}$$

Hence φ_n are commutator bounded on $\mathbb{R}$ and $K(\varphi_n, \gamma) \leq K$ for each compact subset γ of $\mathbb{R}$.

For some $c > 1$, set $\lambda_0 = c$, $\lambda_n = 2^{-n+1}$, for $n \geq 1$, and $\alpha_n = [\lambda_n, \lambda_{n-2}]$, for $n \geq 2$. Since $\operatorname{supp}(\varphi_n) = [d_n - a_n, d_n + a_n] \subset [2^{-n}, 2^{-n+1}]$, we have $g|_{\alpha_2} = \varphi_1|_{\alpha_2}$ and $g|_{\alpha_n} = (\varphi_n + \varphi_{n-1})|_{\alpha_n}$, for $n \geq 3$. Therefore g is commutator bounded on each α_n, $K(g, \alpha_2) = K(\varphi_1, \alpha_2) \leq K$ and

$$K(g, \alpha_n) = K(\varphi_n + \varphi_{n-1}, \alpha_n) \leq K(\varphi_n, \alpha_n) + K(\varphi_{n-1}, \alpha_n) \leq 2K.$$

Finally, let us show that condition (4.5) holds. We have $g_1 = g_2 = \frac{\sigma_1}{2}$ and

$$g_n = \sup_{t \in [0, \lambda_{n-1}]} |g(t)| = \sup_{k \geq n-1} \left(\sup_{t \in \mathbb{R}} |\varphi_k(t)| \right) = \sup_{k \geq n-1} a_k = \frac{\sigma_{n-1}}{2^{n-1}},$$

for $n > 1$. Therefore

$$\sum_{n=1}^{\infty} \left(\frac{g_n}{\lambda_n - \lambda_{n+1}} \right)^2 = \sigma_1 + \sum_{n=2}^{\infty} (2\sigma_{n-1})^2 < \infty.$$

Thus, by Theorem 4.5, g is commutator bounded on $[0, c]$. Since c is arbitrary, g is commutator bounded on $[0, \infty)$. Since g is square-summable at $t = 0$ from the right and $g|_{(-\infty,0]} \equiv 0$, we have from Theorem 3.3 that g is commutator bounded on $\mathbb{R}$. □

Combining Theorems 1.4 and 5.1 yields

Corollary 5.2. *The function g in (5.1) is operator Lipschitzian on $\mathbb{R}$.*

References

[BK] B. Blackadar and J. Cuntz, *Differentiable Banach algebra norms and smooth subalgebras of C^*-algebras*, J. Operator Theory, **26** (1991), 255–282.

[BR] R. Bhatia and P. Rosenthal, *How and why to solve the equation $AX - XB = Y$*, Bull. London Math. Soc., **29** (1997), no. 1, 1–21.

[BS1] M.S. Birman and M.Z. Solomyak 1, *Stieltjes double-integral operators. II*, (Russian) Prob. Mat. Fiz. **2** (1967), 26–60.

[BS2] M.S. Birman and M.Z. Solomyak, *Stieltjes double-integral operators. III*, (Russian) Prob. Mat. Fiz. **6** (1973), 28–54.

[DK] J.L. Daletskii and S.G. Krein, *Integration and differentiation of functions of hermitian operators and applications to the theory of perturbations*, A.M.S. Translations (2) **47** (1965), 1–30.

[D] E.B. Davies, *Lipschitz continuity of functions of operators in the Schatten classes*, J. London Math. Soc., **37** (1988), 148–157.

[F] Yu.B. Farforovskaya, *Example of a Lipschitz function of selfadjoint operators that gives a non-nuclear increment under a nuclear perturbation*, J. Soviet Math., **4** (1975), 426–433.

[GK] I.C. Gohberg and M.G. Kreĭn, *Introduction to the theory of linear non-selfadjoint operators in Hilbert space*, Izdat. "Nauka", Moscow 1965 448 pp.

[JW] B.E. Johnson and J.P. Williams, *The range of a normal derivation*, Pacific J. Math., **58** (1975), 105–122.

[KS] E. Kissin and V.S. Shulman, *Differential properties of some dense subalgebras of C^*-algebras*, Proc. Edinburgh Math. Soc., Proc. Edinburgh Math. Soc., **37** (1994), 399–422.

[KS1] E. Kissin and V.S. Shulman, *On the range inclusion of normal derivations: variations on a theme by Johnson, Williams and Fong*, Proc. London Math. Soc., **83** (2001), 176–198.

[KS2] E. Kissin and V.S. Shulman, *Classes of operator-smooth functions. II. Operator-differentiable functions*, Integral Equations Operator Theory **49** (2004), 165–210.

[KS3] E. Kissin and V.S. Shulman, *Classes of operator-smooth functions. I. Operator-Lipschitz functions*, Proc. Edinb. Math. Soc., **48** (2005), 151–173.

[PSW] B. de Pagter, F.A. Sukochev, H. Witvliet, *Double operator integrals*, J. Funct. Anal., **192** (2002), 52–111.

[Pe] V.V. Peller, *Hankel operators in the theory of perturbations of unitary and self-adjoint operators*, Funktsional. Anal. i Prilozhen., **19** (1985), 37–51, 96.

[Po] R. Powers, *A remark on the domain of an unbounded derivation of a C^*-algebra*, J. Functional Analysis **18** (1975), 85–95.

[RR] H. Radjavi and P. Rosenthal, *Invariant Subspaces*, Ergebnisse der Mathematik und ihrer Grenzgebiete, Band 77, Springer-Verlag, New York-Heidelberg, 1973, xi+219 pp.

[W] J.P. Williams, *Derivation ranges: open problems*, in "Topics in modern operator theory (Timişoara/Herculane, 1980)", pp. 319–328, Operator Theory: Adv. Appl., 2, Birkhäuser, Basel-Boston, Mass., 1981.

Edward Kissin
Department of Computing,
Communications Technology and Mathematics
London Metropolitan University
Holloway Road , London N7 8DB, Great Britain
e-mail: e.kissin@londonmet.ac.uk

Victor S. Shulman
Department of Mathematics
Vologda State Technical University
Vologda, Russia
e-mail: shulman_v@yahoo.com

Lyudmila B. Turowska
Department of Mathematics
Chalmers University of Technology
Gothenburg, Sweden
e-mail: turowska@math.chalmers.se

Operator Theory:
Advances and Applications, Vol. 171, 245–257

On the Kernel of Some One-dimensional Singular Integral Operators with Shift

Viktor G. Kravchenko and Rui C. Marreiros

Abstract. An estimate for the dimension of the kernel of the singular integral operator with shift $\left(I + \sum_{j=1}^{n} a_j(t)U^j\right) P_+ + P_- : L_2(\mathbb{R}) \to L_2(\mathbb{R})$ is obtained, where $P_\pm$ are the Cauchy projectors, $(U\varphi)(t) = \varphi(t+h)$, $h \in \mathbb{R}^+$, is the shift operator and $a_j(t)$ are continuous functions on the one point compactification of $\mathbb{R}$. The roots of the polynomial $1 + \sum_{j=1}^{n} a_j(\infty)\eta^j$ are assumed to belong all simultaneously either to the interior of the unit circle or to its exterior.

Mathematics Subject Classification (2000). Primary 47G10, Secondary 45P05.

Keywords. Singular integral operators, shift operators.

1. Introduction

We consider the singular integral operator with shift, on the real line $\mathbb{R}$,

$$T = AP_+ + P_- : L_2(\mathbb{R}) \to L_2(\mathbb{R}), \tag{1}$$

where

$$A = I + \sum_{j=1}^{n} a_j(t)U^j, \tag{2}$$

I is the identity operator, $a_j \in C(\overset{o}{\mathbb{R}})$, $j = \overline{1,n}$ are continuous functions, $\overset{o}{\mathbb{R}} = \mathbb{R} \cup \{\infty\}$ is the one point compactification of $\mathbb{R}$,

$$(U\varphi)(t) = \varphi(t+h), \quad h \in \mathbb{R}^+,$$

This research was supported by Fundação para a Ciência e Tecnologia (Portugal) through Centro de Matemática e Aplicações of Instituto Superior Técnico.

is an isometric shift operator, and

$$P_{\pm} = \frac{1}{2}(I \pm S)$$

are mutually complementary projection operators, where

$$(S\varphi)(t) = (\pi i)^{-1} \int_{\mathbb{R}} \varphi(\tau)(\tau - t)^{-1} d\tau$$

is the operator of singular integration.

We note that for the singular integral operator with shift in $L_2^n(\mathbb{R})$

$$T(A_1, A_2) = A_1 P_+ + A_2 P_-,$$

where

$$A_1 = a_1 I + b_1 U, \quad A_2 = a_2 I + b_2 U$$

and $a_1, a_2, b_1, b_2 \in C^{n\times n}(\overset{o}{\mathbb{R}})$, the Fredholmness conditions and the index formulas are known [6]. The Fredholm criterion can be formulated as follows: the operator $T(A_1, A_2)$ is Fredholm in $L_2^n(\mathbb{R})$ if and only if the functional operators A_1 and A_2 are continuously invertible in $L_2^n(\mathbb{R})$. The spectral properties of the operator $T(A_1, A_2)$ have been less studied (see [7], [8]), even for the case of a so-called Carleman shift, i.e., a diffeomorphism of a curve onto itself which, after a finite number of iterations, will coincide with the identity transform. For the case of a non-Carleman shift (such as $t+h$ on $\mathbb{R}$), the only works known to the authors are [1] and [9].

In [1] an estimate for $\dim\ker T$ was obtained, with

$$T = (I - cU)P_+ + P_- : L_2(\mathbb{T}) \to L_2(\mathbb{T}),$$

where $(U\varphi)(t) = |\alpha'(t)|^{\frac{1}{2}} \varphi(\alpha(t))$ and α a non-Carleman shift on the unit circle $\mathbb{T}$. In particular, it was shown that:

For any continuous function $c \in C(\mathbb{T})$ such that

$$|c(\tau_j)| < 1, \quad j = \overline{1, s},$$

where $\{\tau_1, \ldots, \tau_s\}$ are the fixed points of the shift α, there exists a polynomial

$$r(t) = \prod_{k=1}^{m} (t - \lambda_k), \quad |\lambda_k| > 1, \quad k = \overline{1, m}$$

such that the condition

$$\left| r(t)c(t)r^{-1}(\alpha(t)) \right| < 1, \quad t \in \mathbb{T},$$

holds.

Furthermore, using some auxiliary results, it was shown that

$$\dim\ker T \leq m.$$

In [9] an analogous estimate for $\dim\ker T$ was obtained for the operator

$$T = (I - cU)P_+ + P_- : L_2^n(\mathbb{T}) \to L_2^n(\mathbb{T}),$$

with matrix coefficients, satisfying one of the conditions:

1. $\sigma[c(\tau_j)] \subset \mathbb{T}_+$, $j = \overline{1, s}$,
2. $\sigma[c(\tau_j)] \subset \mathbb{T}_-$, $j = \overline{1, s}$, and $\det c(t) \neq 0, \forall t \in \mathbb{T}$,

where $\mathbb{T}_+ (\mathbb{T}_-)$ denotes the interior (exterior, respectively) of the unit disk and $\sigma(g)$ is the spectrum of a matrix $g \in \mathbb{C}^{n \times n}$.

In the present paper, an estimate for the dimension of the kernel of the operator (1) is obtained under conditions corresponding to 1 and 2 in the matrix case above. Then we consider two subtypes of the operator (1) and examples which show that our estimate, in a certain sense, is sharp.

2. The main proposition

From now on, the Cauchy index of a continuous function $f \in C(\Gamma)$, on Γ, for $\Gamma = \mathbb{T}, \overset{o}{\mathbb{R}}$, will be denoted by $\operatorname{ind} f$, i.e.,

$$\operatorname{ind} f = \frac{1}{2\pi} \{\arg f(t)\}_{t \in \Gamma}.$$

We define the polynomial

$$A_\infty(\eta) = 1 + \sum_{j=1}^{n} a_j(\infty)\eta^j, \quad \eta \in \mathbb{T}. \tag{3}$$

The invertibility of the operator (2) implies (see, for instance, p. 142–145 in [6])

$$A_\infty(\eta) \neq 0, \quad \eta \in \mathbb{T}. \tag{4}$$

Therefore for the index $k(A)$ of the polynomial $A_\infty(\eta)$ we have

$$0 \leq k(A) \leq n.$$

Consider now the matrix operator (see [6], [10])

$$\widetilde{T} = \widetilde{A} P_+ + P_- : L_2^n(\mathbb{R}) \to L_2^n(\mathbb{R}),$$

with

$$\widetilde{A} = I + aU, \tag{5}$$

where

$$a(t) = \begin{pmatrix} a_1(t) & a_2(t) & \cdots & a_{n-1}(t) & a_n(t) \\ & & -E_{n-1} & & O_{(n-1)\times 1} \end{pmatrix}$$

and E_n is the unit matrix.

Our main result in this section is the following:

Proposition 2.1. *The operator T is a Fredholm operator in $L_2(\mathbb{R})$ if and only if the operator $\widetilde{T}$ is a Fredholm operator in $L_2^n(\mathbb{R})$. In this case,* $\dim\ker T = \dim\ker \widetilde{T}$ *and* $\dim\operatorname{coker} T = \dim\operatorname{coker} \widetilde{T}$.

Proof. As we know, the Fredholmness of a bounded linear operator T is preserved under multiplication by invertible operators, as are the numbers $\dim\ker T$ and $\dim\operatorname{coker} T$.

We multiply $\widetilde{T}$ on the right by the invertible operator

$$N = \begin{pmatrix} 0 & 0 & \cdots & 0 & I \\ 0 & 0 & \cdots & I & UP_+ \\ \cdots & & & & \\ 0 & I & \cdots & U^{n-3}P_+ & U^{n-2}P_+ \\ I & UP_+ & \cdots & U^{n-2}P_+ & U^{n-1}P_+ \end{pmatrix}.$$

Using $UP_+ = P_+U$, we obtain

$$\widetilde{T}N = \begin{pmatrix} D_1 & D_2 & \cdots & D_{n-1} & T \\ 0 & 0 & \cdots & I & 0 \\ \cdots & & & & \\ 0 & I & \cdots & 0 & 0 \\ I & 0 & \cdots & 0 & 0 \end{pmatrix},$$

where $D_1 = a_nUP_+, D_2 = (a_{n-1}U + a_nU^2)P_+, \ldots, D_{n-1} = (a_2U + a_3U^2 + \ldots + a_nU^{n-1})P_+$.

The operator $\widetilde{T}N$ is Fredholm if and only if the operator T is Fredholm. Moreover, the defect numbers of $\widetilde{T}N$ and T coincide. □

We define

$$D_\infty(\eta) = E_n + a(\infty)\eta, \quad \eta \in \mathbb{T}.$$

If the necessary condition for the invertibility of the matrix operator (5) (see p. 118–120 in [6])

$$\det D_\infty(\eta) \neq 0, \quad \eta \in \mathbb{T},$$

is satisfied, then for the index $k(\widetilde{A})$ of the polynomial $\det D_\infty(\eta)$ we have

$$0 \leq k(\widetilde{A}) \leq n.$$

We note that

$$\det D_\infty(\eta) = A_\infty(\eta). \tag{6}$$

Then

$$k(\widetilde{A}) = k(A) \equiv \operatorname{ind} A_\infty.$$

Furthermore, denoting by $\lambda_i(a,\infty)$, the eigenvalues of the matrix $a(\infty)$, and by η_i the roots of the polynomial $A_\infty(\eta)$, we have $\lambda_i^{-1}(a,\infty) = -\eta_i$, $i = \overline{1,n}$, taking into account equality (6). Therefore we note that $\operatorname{ind} A_\infty$ coincides with the number of the roots of the polynomial $A_\infty(\eta)$ that are situated inside the unit disk, or, equivalently, with the eigenvalues of the matrix $a(\infty)$ that are situated outside the unit disk.

In general, the necessary and sufficient conditions of invertibility for the operator (2), and so the Fredholmness conditions for the operator (1), can not be expressed in an explicit form. Instead the conditions are expressed by a particular choice of a solution of the homogeneous functional equation associated with the operator A. In the two extreme cases, the invertibility conditions for the operator (2) and also Fredholmness conditions for the operator (1), can be written in a simple form:

Proposition 2.2. [6] *Let T be the operator defined by (1) with coefficients $a_j \in C(\overset{o}{\mathbb{R}})$, $j = \overline{1,n}$, $A_\infty(\eta)$ be defined by (3) and let (4) be fulfilled. The following assertions hold:*

1. *If* $\operatorname{ind} A_\infty = 0$, *then the operator T is Fredholm and*
$$\operatorname{Ind} T = 0.$$
2. *If* $\operatorname{ind} A_\infty = n$ *and* $a_n(t) \neq 0, \forall t \in \overset{o}{\mathbb{R}}$, *then the operator T is Fredholm and*
$$\operatorname{Ind} T = \operatorname{ind} a_n^{-1}.$$

Remark. If $n = 1$, then the conditions 1 and 2 of Proposition 2.2 are not only sufficient but also necessary for the Fredholmness of the operator T. Both cases mentioned in the proposition are considered in this paper.

We will need the following result.

Proposition 2.3. [9] *For any continuous matrix function $d \in C^{n\times n}(\overset{o}{\mathbb{R}})$ such that*
$$\sigma[d(\infty)] \subset \mathbb{T}_+,$$
there exist an induced matrix norm $\|\cdot\|_0$ and a rational matrix r satisfying the conditions
$$\max_{t\in\overset{o}{\mathbb{R}}} \left\|r(t)d(t)r^{-1}(t+h)\right\|_0 < 1 \tag{7}$$
and
$$P_+ r^{\pm 1} P_+ = r^{\pm 1} P_+. \tag{8}$$

Let R_d be the set of all rational matrices r satisfying the conditions (7), (8),
$$l_1(r) = \sum_{i=1}^{n} \max_{j=\overline{1,n}} l_{i,j},$$
where $l_{i,j}$ is the degree of the element $r_{i,j}(t) = p_{i,j}\left(\frac{t-i}{t+i}\right)$ ($p_{i,j}$ is a polynomial) of the rational matrix r, and
$$l(d) = \min_{r\in R_d} \{l_1(r)\}. \tag{9}$$

Proposition 2.4. *If* $\operatorname{ind} A_\infty = 0$ *and*
$$a(t) = \begin{pmatrix} a_1(t) & a_2(t) & \cdots & a_{n-1}(t) & a_n(t) \\ & & -E_{n-1} & & O_{(n-1)\times 1} \end{pmatrix},$$
then the estimate
$$\dim\ker T \leq l(a), \tag{10}$$
holds.

Proof. Since $\operatorname{ind} A_\infty = 0$, i.e., $\sigma[a(\infty)] \subset \mathbb{T}_+$, we can apply the Proposition 2.3 to the matrix function a so that there exists a rational matrix r satisfying the conditions (7) and (8). For the matrix a, let $l(a)$ be defined by (9). Then, from Propositions 2.2 and 2.3 in [9], we obtain the estimate (10). $\square$

It is known that (see, for instance, [11]; see also [2], [3], [4]) if $a \in C^{n\times n}(\overset{o}{\mathbb{R}})$ satisfies

$$\det a(t) \neq 0, \quad \forall t \in \overset{o}{\mathbb{R}},$$

then the continuous matrix function a admits the factorization in $L_2^{n\times n}(\mathbb{R})$

$$a = a_- \Lambda a_+, \tag{11}$$

where

$$(t-i)^{-1} a_-^{\pm 1} \in \left[\widehat{L}_2^-(\mathbb{R})\right]^{n\times n}, \quad (t+i)^{-1} a_+^{\pm 1} \in \left[\widehat{L}_2^+(\mathbb{R})\right]^{n\times n},$$
$$\Lambda = \operatorname{diag}\left\{\left(\frac{t-i}{t+i}\right)^{\varkappa_j}\right\},$$

$\varkappa_j$ are the partial indices of the factorization with $\varkappa_1 \geq \varkappa_2 \geq \cdots \geq \varkappa_n$, $\widehat{L}_2^{\pm}$ are the spaces of the Fourier transforms of the functions of $L_2^{\pm}$, respectively, and, as usual, $L_2^+ = P_+ L_2$, $L_2^- = P_- L_2 \oplus \mathbb{C}$.

Proposition 2.5. *If the conditions* $\operatorname{ind} A_\infty = n$, $a_n(t) \neq 0, \forall t \in \overset{o}{\mathbb{R}}$, (11) *with*

$$(t-i)^{-1} a_-^{\pm 1}, \ (t+i)^{-1} a_+^{\pm 1} \in L_\infty^{n\times n}(\mathbb{R}),$$

are fulfilled, then the estimate

$$\dim\ker T \leq l(a^{-1}) + \sum_{\varkappa_j < 0} |\varkappa_j|, \tag{12}$$

holds.

Proof. From (11), according to Proposition 2.4 in [9], the estimate

$$\dim\ker \widetilde{T} \leq \dim\ker(I - a^{-1}U^{-1}P_+) + \sum_{\varkappa_j<0} |\varkappa_j|, \tag{13}$$

holds. Furthermore, as $\operatorname{ind} A_\infty = n$, i.e., $\sigma[a(\infty)] \subset \mathbb{T}_-$, we can see that a^{-1} satisfies the conditions of Proposition 2.3, i.e., for a^{-1} there exists a rational matrix r satisfying the conditions (7) and (8). For this matrix a^{-1}, let $l(a^{-1})$ be defined by (9). Then Proposition 2.4 yields

$$\dim\ker(I - a^{-1}U^{-1}P_+) \leq l(a^{-1}).$$

Together with (13) we get (12). $\square$

If $n = 2$ the estimate (12) can be written in a simpler form.

Corollary 2.1. *Let $n = 2$. Under the conditions of Proposition 2.5, the estimate*

$$\dim\ker T \leq l(a^{-1}) - \min\{0, \operatorname{ind} a_2\}, \tag{14}$$

holds.

Proof. If $n = 2$ the matrix a has the form

$$a = \begin{pmatrix} a_1 & a_2 \\ -1 & 0 \end{pmatrix}. \tag{15}$$

We compute

$$\begin{pmatrix} 0 & 1 \\ 1 & 0 \end{pmatrix} a^T = \begin{pmatrix} a_2 & 0 \\ a_1 & -1 \end{pmatrix}.$$

The partial indices $\varkappa_{1,2}$ of the factorization

$$a^T = b_+ \Lambda b_-,$$

are equal to $\operatorname{ind} a_2$ and 0 if $\operatorname{ind} a_2 \leq 1$ (see p. 147–148 in [11]). Obviously the partial indices of the factorization

$$a = a_- \Lambda a_+,$$

are the same. Therefore negative partial indices are possible only if $\operatorname{ind} a_2 < 0$ which implies (14). □

3. Some particular cases and examples

Having the estimate (14) in mind, we will consider two subtypes of the operator (1) with $n = 2$. The first one is an operator T, such that $\dim\ker T = -\operatorname{ind} a_2$; the number $l(a^{-1})$ will not play any role in this case. For the second subtype we will consider examples of operators T with $\dim\ker T = l(a^{-1})$ and $\dim\ker T = l(a^{-1}) - \operatorname{ind} a_2$.

$D_+(D_-)$ will denote the upper (lower, respectively) half plane $(D_+(D_-) = \{z \in \mathbb{C} : \operatorname{Im} z > 0 (< 0)\})$.

We consider the singular integral operator with shift

$$T_1 = \left[I + a_1 U + a_2 U^2\right] P_+ + P_- : L_2(\mathbb{R}) \to L_2(\mathbb{R}), \tag{16}$$

where $a_{1,2} \in C(\overset{o}{\mathbb{R}})$ have analytic continuation into D_-.

As we noted in Section 1, the operator T_1 is Fredholm in $L_2(\mathbb{R})$ if and only if the operator

$$A_1 = I + a_1 U + a_2 U^2$$

is continuously invertible in $L_2(\mathbb{R})$. Under the conditions on $a_{1,2}$, the operator A_1 satisfies the equalities

$$P_- A_1 P_- = A_1 P_-, \quad P_+ A_1 P_- = 0.$$

From this, it follows that

$$T_1^{(-1)} = (P_+ + A_1 P_-) A_1^{-1}$$

is a right inverse operator to T_1. Therefore, if the operator T_1 is Fredholm, then

$$\operatorname{Ind} T_1 = \dim\ker T_1.$$

Let

$$A_{1,\infty}(\eta) = 1 + a_1(\infty)\eta + a_2(\infty)\eta^2, \quad \eta \in \mathbb{T}, \tag{17}$$

and

$$A_{1,\infty}(\eta) \neq 0, \quad \eta \in \mathbb{T}. \tag{18}$$

Taking into account Proposition 2.2, we can write

Proposition 3.1. *Let T_1 be the operator defined by (16) with coefficients $a_{1,2} \in C(\overset{o}{\mathbb{R}})$ having analytic continuation into D_-, $A_{1,\infty}(\eta)$ be defined by (17) and let (18) be fulfilled. The following assertions hold:*

1. *If $\operatorname{ind} A_{1,\infty} = 0$, then the operator T_1 is invertible.*
2. *If $\operatorname{ind} A_{1,\infty} = 2$ and $a_2(t) \neq 0, \forall t \in \overset{o}{\mathbb{R}}$, then the operator T_1 is right invertible and*

$$\dim\ker T_1 = \operatorname{ind} a_2^{-1}.$$

Let us consider again the case $\operatorname{ind} A_{1,\infty} = 2$. From the Proof of the Corollary 2.1, since $\operatorname{ind} a_2 \leq 0$, the partial indices of the matrix (15) are $\varkappa_1 = \operatorname{ind} a_2$ and $\varkappa_2 = 0$.

It is not difficult to construct an example of an operator T_1, satisfying the conditions of assertion 2 of Proposition 3.1, such that $l(a^{-1}) = 0$, and we would have equality in estimate (14).

Now we consider the operator

$$T_2 = \left[I + a_1U + a_2U^2\right] P_+ + P_- : L_2(\mathbb{R}) \to L_2(\mathbb{R}), \tag{19}$$

where $a_{1,2} \in C(\overset{o}{\mathbb{R}})$, $a_1(t) = a_-(t) + a_+(t)$, $a_2(t) = a_-(t)a_+(t+h)$, $a_\pm$ have analytic continuation into $D_\pm$, respectively.

Let

$$A_2 = I + (a_-(t) + a_+(t))U + a_-(t)a_+(t+h)U^2$$

and

$$A_{2,\infty}(\eta) = 1 + (\nu_1 + \nu_2)\eta + \nu_1\nu_2\eta^2, \quad \eta \in \mathbb{T}, \tag{20}$$

where $\nu_1 \equiv a_-(\infty)$, $\nu_2 \equiv a_+(\infty)$. The invertibility of the operator A_2 implies

$$A_{2,\infty}(\eta) \neq 0, \quad \eta \in \mathbb{T}. \tag{21}$$

Then $|\eta_{1,2}| \neq 1$, where $\eta_{1,2}$ are the roots of the polynomial $A_{2,\infty}(\eta)$. We note that

$$\eta_{1,2} = -\frac{1}{\nu_{1,2}}. \tag{22}$$

The operator T_2 can be written as the product

$$T_2 = T_{2,1} \cdot T_{2,2}, \tag{23}$$

where

$$T_{2,1} = A_{2,1}P_+ + P_-, \quad T_{2,2} = A_{2,2}P_+ + P_-,$$

and

$$A_{2,1} = I + a_- U, \quad A_{2,2} = I + a_+ U.$$

Taking into account the remark after Proposition 2.2, and the equalities (23), (22), under conditions on the coefficients $a_\pm$ similar to those on a_1 for the operator T with $n = 1$, depending on the value of ind$A_{2,\infty}$, it follows that the operator T_2 is Fredholm if and only if $\operatorname{ind} A_{2,\infty} = k$, $0 \le k \le 2$.

Making use of equality (23) we analyze the operator T_2 in more detail. The operator $T_{2,1}$ is invertible when $|\nu_1| < 1$, and is right invertible when $|\nu_1| > 1$. The operator $T_{2,2}$ is invertible when $|\nu_2| < 1$, and is left invertible when $|\nu_2| > 1$. We have

$$\operatorname{Ind} T_2 = \operatorname{Ind} T_{2,1} + \operatorname{Ind} T_{2,2} = \dim\ker T_{2,1} - \dim\operatorname{coker} T_{2,2}.$$

Denoting

$$\operatorname{ind} a_- \equiv -\gamma_1, \quad \operatorname{ind} a_+ \equiv \gamma_2, \quad \gamma_{1,2} \ge 0,$$

we can write

$$\begin{aligned}
&\operatorname{Ind} T_2 = 0, \text{ when } |\nu_{1,2}| < 1, \ (\operatorname{ind} A_{2,\infty} = 0),\\
&\operatorname{Ind} T_2 = -\gamma_2, \text{ when } |\nu_1| < 1, \ |\nu_2| > 1, \ (\operatorname{ind} A_{2,\infty} = 1),\\
&\operatorname{Ind} T_2 = \gamma_1, \text{ when } |\nu_1| > 1, \ |\nu_2| < 1, \ (\operatorname{ind} A_{2,\infty} = 1),\\
&\operatorname{Ind} T_2 = \gamma_1 - \gamma_2, \text{ when } |\nu_{1,2}| > 1, \ (\operatorname{ind} A_{2,\infty} = 2).
\end{aligned}$$

Moreover, we can show that the following equalities hold:

$$\dim\ker T_2 = 0, \text{ when } |\nu_{1,2}| < 1 \text{ or } |\nu_1| < 1, \ |\nu_2| > 1. \tag{24}$$

$$\dim\ker T_2 = \gamma_1, \text{ when } |\nu_1| > 1, \ |\nu_2| < 1 \text{ or } |\nu_{1,2}| > 1. \tag{25}$$

The equality (24) is obvious. To show the equality (25), we take into account that:

- The operator $T_{2,1}$ is right invertible and $T_{2,1}^{(-1)} = (P_+ + A_{2,1}P_-)A_{2,1}^{-1}$ is a right inverse operator to $T_{2,1}$, and
- Since the operator $T_{2,2}$ is left invertible, $\dim\ker T_{2,2} = 0$.

We consider the projector onto the kernel of the operator $T_{2,1}$,

$$\Pi = I - T_{2,1}^{(-1)} T_{2,1} = -(I - A_{2,1})P_+ A_{2,1}^{-1} P_-.$$

We verify that

$$\Pi T_{2,2} = \Pi.$$

Therefore Π is also a projector onto the kernel of the operator T_2. Thus

$$\dim\ker T_2 = \dim\ker T_{2,1} = \gamma_1.$$

We collect the results on the operator T_2 in the following proposition:

Proposition 3.2. *Let T_2 be the operator defined by* (19) *with coefficients*

$$a_{1,2} \in \overset{o}{C}(\dot{\mathbb{R}}), \quad a_1(t) = a_-(t) + a_+(t), \quad a_2(t) = a_-(t)a_+(t+h), \quad a_\pm$$

having analytic continuation into $D_\pm$, respectively, $A_{2,\infty}(\eta)$ be defined by (20) *and let* (21) *be fulfilled.*

The operator T_2 is Fredholm if and only if one of the following conditions is fulfilled:

1. $\operatorname{ind} A_{2,\infty} = 0$. *Moreover, the operator T_2 is invertible.*
2. $\operatorname{ind} A_{2,\infty} = 1$ *with*

   (a) $|a_-(\infty)| < 1$, $|a_+(\infty)| > 1$ *and* $a_+(t) \neq 0, \forall t \in \overset{o}{\mathbb{R}}$. *Moreover, the operator T_2 is left invertible and*

$$\dim \operatorname{coker} T_2 = \operatorname{ind} a_+.$$

   (b) $|a_-(\infty)| > 1$, $|a_+(\infty)| < 1$ *and* $a_-(t) \neq 0, \forall t \in \overset{o}{\mathbb{R}}$. *Moreover, the operator T_2 is right invertible and*

$$\dim \ker T_2 = -\operatorname{ind} a_-.$$

3. $\operatorname{ind} A_{2,\infty} = 2$ *and* $a_\pm(t) \neq 0, \forall t \in \overset{o}{\mathbb{R}}$. *Moreover,*

$$\dim \ker T_2 = -\operatorname{ind} a_- \quad \text{and} \quad \dim \operatorname{coker} T_2 = \operatorname{ind} a_+.$$

Let us consider two particular operators of the type of T_2.

Example 1. Let

$$a_-(t) = 2\frac{t+i}{t-i}, \quad a_+(t) = 2\frac{t-i}{t+i}, \quad (U\varphi)(t) = \varphi(t+1).$$

In this case

$$\dim \ker T_2 = 1.$$

The partial indices $\varkappa_{1,2}$ of the factorization of the matrix

$$a(t) = \begin{pmatrix} a_-(t) + a_+(t) & a_-(t)a_+(t+1) \\ -1 & 0 \end{pmatrix},$$

are $\varkappa_1 = \varkappa_2 = 0$ (see p. 147–148 in [11]). Now we apply the estimate (14) to the operator T_2:

$$\dim \ker T_2 \leq l(a^{-1}).$$

We have

$$a^{-1}(t) = \begin{pmatrix} 0 & -1 \\ \frac{1}{4}\frac{t-i}{t+i}\frac{t+1+i}{t+1-i} & \frac{1}{2}\left(1 + \left(\frac{t-i}{t+i}\right)^2\right)\frac{t+1+i}{t+1-i} \end{pmatrix}.$$

It is easily seen that

$$\rho[a^{-1}(\infty)] < 1, \text{ but, } \max_{t \in \overset{o}{\mathbb{R}}} \rho[a^{-1}(t)] > 1,$$

where $\rho(g)$ denotes the spectral radius of the matrix g. According to the proof of the Proposition 2.5, let

$$r(t) = \begin{pmatrix} \frac{t+2+3i}{t+i} & 0 \\ 0 & 1 \end{pmatrix}.$$

Then we have

$$\max_{t\in\overset{o}{\mathbb{R}}} \rho[r(t)a^{-1}(t)r^{-1}(t+1)] =$$

$$= \max_{t\in\overset{o}{\mathbb{R}}} \rho \begin{pmatrix} 0 & -\dfrac{t+2+3i}{t+i} \\ \dfrac{1}{4}\dfrac{t-i}{t+i}\dfrac{t+1+i}{t+1-i}\dfrac{t+1+i}{t+3+3i} & \dfrac{1}{2}\left(1+\left(\dfrac{t-i}{t+i}\right)^2\right)\dfrac{t+1+i}{t+1-i} \end{pmatrix} < 1.$$

Thus $l(a^{-1}) = 1$. We get

$$1 = \dim\ker T_2 \leq 1.$$

Example 2. Now let

$$a_-(t) = 2\left(\frac{t+i}{t-i}\right)^2, \quad a_+(t) = 2\frac{t-i}{t+i}, \quad (U\varphi)(t) = \varphi(t+1).$$

In this case

$$\dim\ker T_2 = 2.$$

The partial indices $\varkappa_{1,2}$ of the factorization of the matrix

$$a(t) = \begin{pmatrix} a_-(t) + a_+(t) & a_-(t)a_+(t+1) \\ -1 & 0 \end{pmatrix},$$

are $\varkappa_1 = -1$ and $\varkappa_2 = 0$. Applying estimate (14) to the operator T_2 yields

$$\dim\ker T_2 \leq 1 + l(a^{-1}).$$

We have

$$a^{-1}(t) = \begin{pmatrix} 0 & -1 \\ \dfrac{1}{4}\left(\dfrac{t-i}{t+i}\right)^2 \dfrac{t+1+i}{t+1-i} & \dfrac{1}{2}\left(1+\left(\dfrac{t-i}{t+i}\right)^3\right)\dfrac{t+1+i}{t+1-i} \end{pmatrix}.$$

It is easy to check that

$$\rho[a^{-1}(\infty)] < 1, \text{ but, } \max_{t\in\overset{o}{\mathbb{R}}} \rho[a^{-1}(t)] > 1.$$

As in the previous example, let

$$r(t) = \begin{pmatrix} \dfrac{t+2+3i}{t+i} & 0 \\ 0 & 1 \end{pmatrix}.$$

Then we have

$$\max_{t\in\overset{o}{\mathbb{R}}} \rho[r(t)a^{-1}(t)r^{-1}(t+1)]$$

$$= \max_{t\in\overset{o}{\mathbb{R}}} \rho \begin{pmatrix} 0 & -\dfrac{t+2+3i}{t+i} \\ \dfrac{1}{4}\left(\dfrac{t-i}{t+i}\right)^2 \dfrac{t+1+i}{t+1-i}\dfrac{t+1+i}{t+3+3i} & \dfrac{1}{2}\left(1+\left(\dfrac{t-i}{t+i}\right)^3\right)\dfrac{t+1+i}{t+1-i} \end{pmatrix} < 1.$$

Thus $l(a^{-1}) = 1$. We obtain

$$2 = \dim\ker T_2 \leq 2.$$

References

[1] Baturev, A.A., Kravchenko, V.G. and Litvinchuk, G.S., *Approximate methods for singular integral equations with a non-Carleman shift*, J. Integral Equations Appl., **8** (1996), 1–17.

[2] Clancey, K. and Gohberg, I., *Factorization of Matrix Functions and Singular Integral Operators*, Operator Theory: Advances and Applications, vol. 3, Birkhäuser Verlag, Basel, 1981.

[3] Duduchava, R.B., *Convolution integral equations with discontinuous presymbols, singular integral equations with fixed singularities and their applications to problems in mechanics*, Trudy Tbilisskogo Mat. Inst. Akad. Nauk Gruz. SSR, Tbilisi, 1979 (in Russian).

[4] Gohberg, I. and Krupnik, N., *One-Dimensional Linear Singular Integral Equations, vol. I and II*, Operator Theory: Advances and Applications, vol. 53 and 54, Birkhäuser Verlag, Basel, 1992.

[5] Horn, R.A. and Johnson C.R., *Matrix Analysis*, Cambridge University Press, Cambridge, 1996.

[6] Kravchenko, V.G. and Litvinchuk, G.S., *Introduction to the Theory of Singular Integral Operators with Shift*, Mathematics and its Applications, vol. 289, Kluwer Academic Publishers, Dordrecht, 1994.

[7] Kravchenko, V.G., Lebre, A.B. and Rodriguez, J.S., *Factorization of singular integral operators with a Carleman shift and spectral problems*, J. Integral Equations Appl., **13** (2001), 339–383.

[8] Kravchenko, V.G., Lebre, A.B. and Rodriguez, J.S., *Factorization of singular integral operators with a Carleman shift via factorization of matrix functions*, Operator Theory: Advances and Applications, vol. 142, p. 189–211, Birkhäuser Verlag, Basel, 2003.

[9] Kravchenko, V.G. and Marreiros, R.C., *An estimate for the dimension of the kernel of a singular operator with a non-Carleman shift, Factorization, Singular Operators and Related Problems*, p. 197–204, Kluwer Academic Publishers, Dordrecht, 2003.

[10] Krupnik, N., *Banach Algebras with Symbol and Singular Integral Operators*, Operator Theory: Advances and Applications, vol. 26, Birkhäuser Verlag, Basel, 1987.

[11] Litvinchuk, G.S. and Spitkovskii, I.M., *Factorization of Measurable Matrix Functions*, Operator Theory: Advances and Applications, vol. 25, Birkhäuser Verlag, Basel, 1987.

Viktor G. Kravchenko
Departamento de Matemática
Faculdade de Ciências e Tecnologia
Universidade do Algarve
8005-139 Faro, Portugal
e-mail: vkravch@ualg.pt

Rui C. Marreiros
Departamento de Matemática
Faculdade de Ciências e Tecnologia
Universidade do Algarve
8005-139 Faro, Portugal
e-mail: rmarrei@ualg.pt

Operator Theory:
Advances and Applications, Vol. 171, 259–280

The Fredholm Property of Pseudodifferential Operators with Non-smooth Symbols on Modulation Spaces

Vladimir S. Rabinovich and Steffen Roch

Abstract. The aim of the paper is to study the Fredholm property of pseudodifferential operators in the Sjöstrand class OPS_w where we consider these operators as acting on the modulation spaces $M^{2,p}(\mathbb{R}^N)$. These spaces are introduced by means of a time-frequency partition of unity. The symbol class S_w does not involve any assumptions on the smoothness of its elements.

In terms of their limit operators, we will derive necessary and sufficient conditions for operators in OPS_w to be Fredholm. In particular, it will be shown that the Fredholm property and, thus, the essential spectra of operators in this class are independent of the modulation space parameter $p \in (1, \infty)$.

Mathematics Subject Classification (2000). Primary 47G30; Secondary 35S05, 47L80.

Keywords. Pseudodifferential operators, modulation spaces, Fredholm property, limit operators.

1. Introduction

This paper is devoted to the study of the Fredholm property of pseudodifferential operators in the Sjöstrand class OPS_w. The class S_w of Sjöstrand symbols and the corresponding class OPS_w of pseudodifferential operators were introduced in [8, 9]. This class contains the Hörmander class $OPS^0_{0,0}$ and other interesting classes of pseudodifferential operators. One feature of the class S_w is that no assumptions on the smoothness of its elements are made.

Sjöstrand [8, 9] considers operators in OPS_w as acting on $L^2(\mathbb{R}^N)$. He proves the boundedness of these operators and shows that OPS_w is an inverse closed Banach subalgebra of the algebra $L(L^2(\mathbb{R}^N))$ of all bounded linear operators on $L^2(\mathbb{R}^N)$.

Supported by CONACYT project 43432.

Applications in time-frequency analysis had lead to an increasing interest in pseudodifferential operators in classes similar to OPS_w but acting on several kinds of modulation spaces (see, for instance [1, 3, 2, 11]). These spaces are defined by means of a so-called time-frequency partition of unity (i.e., a partition of unity on the phase space).

Whereas the main emphasis in [1, 3, 2, 11] is on boundedness results, we are going to examine the Fredholm property of pseudodifferential operators in OPS_w on modulation spaces which seems to have not been considered earlier. Our approach is based on the limit operators method. An introduction to this method as well as several applications of limit operators to other quite general operator classes can be found in the monograph [6] (see also the references therein). For several of these operator classes (including OPS_w and the Hörmander class $OPS^0_{0,0}$), the limit operators approach seems to be the only available method to treat the Fredholm property.

The present paper is organized as follows. In Section 2 we recall some auxiliary material from [5] and [6]. In particular, we introduce the Wiener algebra $\mathcal{W}(\mathbb{Z}^N, X)$ of band-dominated operators with operator-valued coefficients acting on the spaces $l^p(\mathbb{Z}^N, X)$ where X is a Banach space. For operators belonging to the so-called rich subalgebra $\mathcal{W}^\$(\mathbb{Z}^N, X)$ of $\mathcal{W}(\mathbb{Z}^N, X)$ we formulate necessary and sufficient conditions for their Fredholmness. It will turn out that the Fredholm property and, thus, the essential spectrum of an operator $A \in \mathcal{W}^\$(\mathbb{Z}^N, X)$ are independent of $p \in (1, \infty)$.

Section 3 is devoted to modulation spaces and their discretizations. Given a time-frequency partition of unity by pseudodifferential operators

$$\sum_{\alpha \in \mathbb{Z}^{2N}} \Phi_\alpha^* \Phi_\alpha = I,$$

the modulation space $M^{2,p}(\mathbb{R}^N)$ is defined as the space of all distributions $u \in S'(\mathbb{R}^N)$ with

$$\|u\|_{M^{2,p}(\mathbb{R}^N)} := \left(\sum_{\alpha \in \mathbb{Z}^{2N}} \|\Phi_\alpha u\|^p_{L^2(\mathbb{R}^N)} \right)^{1/p} < \infty$$

if $p \in [1, \infty)$ and with

$$\|u\|_{M^{2,\infty}(\mathbb{R}^N)} := \sup_{\alpha \in \mathbb{Z}^{2N}} \|\Phi_\alpha u\|_{L^2(\mathbb{R}^N)} < \infty$$

in case $p = \infty$. In Section 4, we introduce the continuous analogue $\mathcal{W}(\mathbb{R}^N)$ of the discrete Wiener algebra $\mathcal{W}(\mathbb{Z}^N, X)$ by imposing conditions on the decay of the operators $\Phi_\alpha A \Phi^*_{\alpha-\gamma}$. More precisely, an operator A belongs to $\mathcal{W}(\mathbb{R}^N)$ if

$$\|A\|_{\mathcal{W}(\mathbb{R}^N)} := \sum_{\gamma \in \mathbb{Z}^{2N}} \sup_{\alpha \in \mathbb{Z}^{2N}} \|\Phi_\alpha A \Phi^*_{\alpha-\gamma}\|_{L(L^2(\mathbb{R}^N))} < \infty.$$

We prove that the operators in $\mathcal{W}(\mathbb{R}^N)$ act boundedly on $M^{2,p}(\mathbb{R}^N)$ for every $p \in [1, \infty]$ and that $\mathcal{W}(\mathbb{R}^N)$ is an inverse closed subalgebra of $L(M^{2,p}(\mathbb{R}^N))$.

Via discretization, the results recalled in Section 2 apply to yield necessary and sufficient conditions for the Fredholmness on $M^{2,p}(\mathbb{R}^N)$ of operators in the so-called rich subalgebra $\mathcal{W}^\$(\mathbb{R}^N)$ of $\mathcal{W}(\mathbb{R}^N)$. Moreover, the essential spectrum of $A \in W^\$(\mathbb{R}^N)$ proves to be independent of $p \in (1, \infty)$.

In the concluding fifth section, we apply the description of operators in OPS_w derived in [1] to conclude that $OPS_w \subset W^\$(\mathbb{R}^N)$. Thus, the results of the previous sections specify to give Fredholm criteria for pseudodifferential operators in OPS_w acting on modulation spaces $M^{2,p}(\mathbb{R}^N)$ in terms of limit operators. One consequence is the independence of the essential spectrum of an operator $A \in OPS_w$ of the modulation space parameter p.

Notice that a criterion for the Fredholmness of pseudodifferential operators in $OPS^0_{0,0}$ acting on $L^2(\mathbb{R}^N)$ was obtained in [5] by similar techniques (see also Chapter 4 in [6]).

2. Operators in the discrete Wiener algebra

2.1. Band-dominated operators and $\mathcal{P}$-Fredholmness

Given a complex Banach space X, let $L(X)$ and $K(X)$ stand for the Banach algebra of all bounded linear operators on X and for its closed ideal of all compact operators, respectively. For each positive integer N, each real number $p \geq 1$, and each complex Banach space X, let $l^p(\mathbb{Z}^N, X)$ denote the Banach space of all functions $f : \mathbb{Z}^N \to X$ with

$$\|f\|_{l^p(\mathbb{Z}^N,X)} := \left(\sum_{x \in \mathbb{Z}^N} \|f(x)\|_X^p \right)^{1/p} < \infty.$$

Further, let $l^\infty(\mathbb{Z}^N, X)$ refer to the Banach space of all bounded functions $f : \mathbb{Z}^N \to X$ with norm

$$\|f\|_{l^\infty(\mathbb{Z}^N,X)} := \sup_{x \in \mathbb{Z}^N} \|f(x)\|_X,$$

and write $c_0(\mathbb{Z}^N, X)$ for the closed subspace of $l^\infty(\mathbb{Z}^N, X)$ which consists of all functions f with

$$\lim_{x \to \infty} \|f(x)\|_X = 0.$$

For $1 \leq p < \infty$, the Banach dual space of $l^p(\mathbb{Z}^N, X)$ can be identified in a standard way with $l^q(\mathbb{Z}^N, X^*)$ where $1/p + 1/q = 1$, and the dual of $c_0(\mathbb{Z}^N, X)$ is isomorphic to $l^1(\mathbb{Z}^N, X^*)$. Moreover, if X is a reflexive Banach space, then the spaces $l^p(\mathbb{Z}^N, X)$ are reflexive for $1 < p < \infty$. If $X = H$ is a Hilbert space with inner product $\langle ., . \rangle_H$, then $l^2(\mathbb{Z}^N, H)$ becomes a Hilbert space on defining an inner product by

$$\langle f, g \rangle := \sum_{x \in \mathbb{Z}^N} \langle f(x), g(x) \rangle_H.$$

In what follows, we agree upon using the notation $E(X)$ to refer to one of the spaces $l^p(\mathbb{Z}^N, X)$ with $1 < p < \infty$ or $c_0(\mathbb{Z}^N, X)$, whereas we will write $E^\infty(X)$ if one of the spaces $E(X)$, $l^1(\mathbb{Z}^N, X)$ or $l^\infty(\mathbb{Z}^N, X)$ is taken into consideration.

For $n \in \mathbb{N}$, we denote the operator of multiplication by the characteristic function of the discrete cube $I_n := \{x \in \mathbb{Z}^N : |x|_\infty := \max_{1\le j\le N} |x_j| \le n\}$ by P_n. This operator acts boundedly on each of the spaces $E^\infty(X)$. We let $\mathcal{P}$ refer to the set of all operators P_n with $n \in \mathbb{N}$ and set $Q_n := I - P_n$. Following the terminology introduced in [6], an operator $K \in L(E^\infty(X))$ is called *$\mathcal{P}$-compact* if

$$\lim_{n\to\infty} \|KQ_n\|_{E^\infty(X)} = \lim_{n\to\infty} \|Q_nK\|_{E^\infty(X)} = 0.$$

We write $K(E^\infty(X), \mathcal{P})$ for the set of all $\mathcal{P}$-compact operators and $L(E^\infty(X), \mathcal{P})$ for the set of all operators $A \in L(E^\infty(X))$ for which both AK and KA are $\mathcal{P}$-compact whenever K is $\mathcal{P}$-compact. Then $L(E^\infty(X), \mathcal{P})$ is a closed subalgebra of $L(E^\infty(X))$ which contains $K(E^\infty(X), \mathcal{P})$ as a closed ideal.

Definition 2.1. *An operator $A \in L(E^\infty(X), \mathcal{P})$ is called a $\mathcal{P}$-*Fredholm operator *if the coset $A + K(E^\infty(X), \mathcal{P})$ is invertible in the quotient algebra*

$$L(E^\infty(X), \mathcal{P})/K(E^\infty(X), \mathcal{P}),$$

i.e., if there exist operators $B, C \in L(E^\infty(X), \mathcal{P})$ and $K, L \in K(E^\infty(X), \mathcal{P})$ such that $BA = I + K$ and $AC = I + L$.

This definition is equivalent to the following one.

Definition 2.2. *An operator $A \in L(E^\infty(X), \mathcal{P})$ is $\mathcal{P}$-Fredholm if and only if there exist an $m \in \mathbb{N}$ and operators $L_m, R_m \in L(E^\infty(X), \mathcal{P})$ such that*

$$L_mAQ_m = Q_mAR_m = Q_m.$$

$\mathcal{P}$-Fredholmness is often referred to as *local invertibility at infinity*. If X has finite dimension, then these definitions become equivalent to the usual definition of Fredholmness, which says that an operator is Fredholm if both its kernel and its cokernel have finite dimension.

For $k \in \mathbb{Z}^N$, let $\hat{V}_k$ stand for the operator of shift by k,

$$(\hat{V}_ku)(x) = f(x - k), \quad x \in \mathbb{Z}^N.$$

Clearly, $\hat{V}_k \in L(E^\infty(X))$ and $\|\hat{V}_k\|_{L(E^\infty(X))} = 1$.

Definition 2.3. *A* band operator *on $E^\infty(X)$ is a finite sum of the form $\sum_\alpha a_\alpha \hat{V}_\alpha$ where $\alpha \in \mathbb{Z}^N$ and $a_\alpha \in l^\infty(\mathbb{Z}^N, L(X))$. An operator is* band-dominated *if it is the uniform limit in $L(E^\infty(X))$ of a sequence of band operators.*

In case $X = \mathbb{C}$ and $N = 1$, and with respect to the standard basis of $E^\infty(X)$, band operators are given by matrices with finite band width, which justifies this notion. Observe also that the class of band operators is independent of the concrete space $E^\infty(X)$ whereas the class of band-dominated operators, which we denote by $\mathcal{A}(E^\infty(X))$, depends heavily on $E^\infty(X)$. It is easy to see that $\mathcal{A}(E^\infty(X))$ is a closed subalgebra both of $L(E^\infty(X))$ and of $L(E^\infty(X), \mathcal{P})$.

Definition 2.4. *Let $A \in L(E^\infty(X))$, and let $h : \mathbb{N} \to \mathbb{Z}^N$ be a sequence which tends to infinity. An operator $A_h \in L(E^\infty(X))$ is called a* limit operator of A *with respect to the sequence h if*

$$\begin{aligned}\lim_{n\to\infty} &\|P_k(\hat{V}_{-h(n)}A\hat{V}_{h(n)} - A_h)\|_{L(E^\infty(X))} \\ &= \lim_{n\to\infty} \|(\hat{V}_{-h(n)}A\hat{V}_{h(n)} - A_h)P_k\|_{L(E^\infty(X))} = 0 \end{aligned} \tag{1}$$

for every $k \in \mathbb{N}$. The set of all limit operators of A will be denoted by $\sigma_{op}(A)$ and is called the operator spectrum *of A. Let further $\mathcal{H}$ stand for the set of all sequences $h : \mathbb{N} \to \mathbb{Z}^N$ which tend to infinity, and let $\mathcal{A}^\$(E^\infty(X))$ refer to the set of all operators $A \in \mathcal{A}(E^\infty(X))$ enjoying the following property: Every sequence $h \in \mathcal{H}$ possesses a subsequence g for which the limit operator A_g exists. We refer to the operators in $\mathcal{A}^\$(E^\infty(X))$ as* rich *band-dominated operators.*

Obviously, richness is a compactness condition with respect to the convergence defined by (1).

The following is our main result on $\mathcal{P}$-Fredholmness of rich band-dominated operators. For its proof see [6], Theorem 2.2.1.

Theorem 2.5. *An operator $A \in \mathcal{A}^\$(E^\infty(X))$ is $\mathcal{P}$-Fredholm if and only if each of its limit operators is invertible and if the norms of their inverses are uniformly bounded, i.e.,*

$$\sup\left\{\|(A_h)^{-1}\|_{L(E^\infty(X))} : A_h \in \sigma_{op}(A)\right\} < \infty.$$

2.2. The discrete Wiener algebra

The statement of Theorem 2.5 has a more satisfactory form for band-dominated operators which belongs to the discrete Wiener algebra, in which case the uniform boundedness of the inverses of the limit operators follows from their invertibility.

Let $\mathcal{W}(\mathbb{Z}^N, X)$ denote the set of all band-dominated operators of the form

$$A = \sum_{\alpha\in\mathbb{Z}^N} a_\alpha \hat{V}_\alpha$$

where the coefficients $a_\alpha \in l^\infty(\mathbb{Z}^N, L(X))$ are subject to the condition

$$\|A\|_{\mathcal{W}(\mathbb{Z}^N, X)} := \sum_{\alpha\in\mathbb{Z}^N} \|a_\alpha\|_{l^\infty(\mathbb{Z}^N, L(X))} < \infty. \tag{2}$$

Provided with usual operations and with the norm (2), the set $\mathcal{W}(\mathbb{Z}^N, X)$ becomes a Banach algebra, the so-called *discrete Wiener algebra*. The estimate

$$\|A\|_{L(E^\infty(X))} \le \|A\|_{\mathcal{W}(\mathbb{Z}^N, X)}$$

shows that $\mathcal{W}(\mathbb{Z}^N, X)$ is a non-closed subalgebra of $\mathcal{A}(E^\infty(X))$.

One of the remarkable properties of the discrete Wiener algebra is its inverse closedness.

Proposition 2.6. *The Wiener algebra $\mathcal{W}(\mathbb{Z}^N, X)$ is inverse closed in every algebra $L(E^\infty(X))$.*

Otherwise stated: If an operator $A \in \mathcal{W}(\mathbb{Z}^N, X)$ acts on $E^\infty(X)$ and is invertible there, then $A^{-1} \in \mathcal{W}(\mathbb{Z}^N, X)$ again. A proof is in [6], Theorem 2.5.2. An immediate consequence of the inverse closedness is the independence of the spectrum of an operator $A \in \mathcal{W}(\mathbb{Z}^N, X)$, thought of as acting on one of the spaces $E^\infty(X)$, on the concrete choice of that space.

Set $\mathcal{W}^\$(\mathbb{Z}^N, X) := \mathcal{W}(\mathbb{Z}^N, X) \cap \mathcal{A}^\$(E^\infty(X))$, and let $A \in \mathcal{W}^\$(\mathbb{Z}^N, X)$. We consider this operator on one of the spaces $E^\infty(X)$ and determine its limit operators with respect to this space. It turns out that the operator spectrum of A does not depend on the choice of that space and that all limit operators of A belong to the Wiener algebra $\mathcal{W}(\mathbb{Z}^N, X)$ again. The following is Theorem 2.5.7 in [6].

Theorem 2.7. *Let X be a reflexive Banach space. The following assertions are equivalent for an operator $A \in \mathcal{W}^\$(\mathbb{Z}^N, X)$:*

(a) *there is a space $E(X)$ on which A is $\mathcal{P}$-Fredholm;*
(b) *there is a space $E(X)$ such that all limit operators of A are invertible on that space;*
(c) *all limit operators of A are invertible on $l^\infty(\mathbb{Z}^N, X)$;*
(d) *all limit operators of A are invertible on $l^\infty(\mathbb{Z}^N, X)$ and the norms of their inverses are uniformly bounded;*
(e) *all limit operators of A are invertible on all spaces $E^\infty(X)$ and the $L(E^\infty(X))$-norms of their inverses are uniformly bounded;*
(f) *A is $\mathcal{P}$-Fredholm operator on each of the spaces $E(X)$.*

Let $A \in L(E^\infty(X), \mathcal{P}))$. We say that the complex number λ belongs to the $\mathcal{P}$-spectrum of A if the operator $A - \lambda I$ is not $\mathcal{P}$-Fredholm on $E^\infty(X)$. We denote the $\mathcal{P}$-spectrum of A by $\sigma_\mathcal{P}(A|E^\infty(X))$ or $\sigma_\mathcal{P}(A)$ for short. The common spectrum of A will be denoted by $\sigma(A|E^\infty(X))$ or simply by $\sigma(A)$.

Theorem 2.8. *Let X be a reflexive Banach space and $A \in \mathcal{W}^\$(\mathbb{Z}^N, X)$. Then the $\mathcal{P}$-spectrum of A, considered as an operator on $E(X)$, is equal to*

$$\sigma_\mathcal{P}(A|E(X)) = \bigcup_{A_h \in \sigma_{op}(A)} \sigma(A_h|E(X)). \tag{3}$$

Moreover, neither the operator spectrum of A, nor the $\mathcal{P}$-spectrum of A, nor the spectra of the limit operators of A on the right-hand side of (3) *depend on the choice of $E(X)$.*

If the space X has finite dimension, then the $\mathcal{P}$-spectrum of A is the common essential spectrum of that operator; that is, the spectrum of the coset $A + K(E^\infty(X))$ in the Calkin algebra $L(E^\infty(X))/K(E^\infty(X))$. In this setting, the rich Wiener algebra coincides with the full Wiener algebra. Hence, Theorem 2.8 has the following corollary.

Theorem 2.9. *Let X be a finite-dimensional space. Then the essential spectrum of $A \in \mathcal{W}(\mathbb{Z}^N, X)$ does not depend on the choice of $E(X)$, and is given by* (3).

3. Operators on modulation spaces

In the following two sections we define the modulation spaces and consider the continuous counterparts of the band-dominated operators and the Wiener algebra. The discrete and the continuous world are linked by a certain discretization operation which we are going to introduce first.

3.1. Time-frequency discretization

Recall that a function $a \in C^\infty(\mathbb{R}^N \times \mathbb{R}^N)$ belongs to the Hörmander class $S^0_{0,\,0}$ if, for all $r,\, t \in \mathbb{N}$,

$$|a|_{r,\,t} := \sum_{|\alpha|\le r,\,|\beta|\le t} \sup_{(x,\,\xi)\in\mathbb{R}^N\times\mathbb{R}^N} |\partial_x^\beta \partial_\xi^\alpha a(x,\,\xi)| < \infty. \tag{4}$$

Let $a \in S^0_{0,\,0}$. The associated pseudodifferential operator $Op(a)$ (also written as $a(x,\,D)$) is defined at $u \in S(\mathbb{R}^N)$ by

$$(Op(a)u)(x) := (2\pi)^{-N} \int_{\mathbb{R}^N}\int_{\mathbb{R}^N} a(x,\,\xi) e^{i\langle x-y,\,\xi\rangle} u(y)\, dy\, d\xi. \tag{5}$$

The function a is called the symbol of $Op(a)$, and the class of all pseudodifferential operators with symbols in $S^0_{0,\,0}$ is denoted by $OPS^0_{0,\,0}$. Standard references on pseudodifferential operators are [12, 7, 10], to mention only a few.

It is well known that $OPS^0_{0,\,0}$ forms an algebra with respect to the usual sum and composition of operators. Further, the operators $Op(a) \in OPS^0_{0,\,0}$ are bounded both on the Schwartz space $S(\mathbb{R}^N)$ and on the Lebesgue space $L^2(\mathbb{R}^N)$, and

$$\|Op(a)\|_{L(L^2(\mathbb{R}^N))} \le C|a|_{2k,\,2l} \quad \text{if } 2k > N,\ 2l > N. \tag{6}$$

The latter fact is known as the Calderón-Vaillancourt theorem.

Let $A : S(\mathbb{R}^N) \to S(\mathbb{R}^N)$ be a bounded linear operator. An operator A^t is called the *formal adjoint* of A if

$$\langle Au,\, v\rangle = \langle u,\, A^t v\rangle \quad \text{for all } u,\, v \in S(\mathbb{R}^N). \tag{7}$$

If $A \in OPS^0_{0,\,0}$, then its formal adjoint A^t is again a pseudodifferential operator in $OPS^0_{0,\,0}$. Furthermore, if $A \in OPS^0_{0,\,0}$ acts on $L^2(\mathbb{R}^N)$, then its Hilbert space adjoint A^* also belongs to $OPS^0_{0,\,0}$. Hence, (7) can be used to define the action of $A \in OPS^0_{0,\,0}$ on the space of tempered distributions $S'(\mathbb{R}^N)$.

Our next goal is to introduce the time-frequency discretization (which is called bi-discretization in [6]). For $\gamma = (\gamma_1,\, \gamma_2) \in \mathbb{Z}^N \times \mathbb{Z}^N$, set $U_\gamma := V_{\gamma_1} E_{\gamma_2}$, where

$$(V_\alpha u)(x) := u(x-\alpha) \quad \text{and} \quad (E_\beta u)(x) := e^{i\langle \beta,\, x\rangle} u(x)$$

for $\alpha,\, \beta \in \mathbb{Z}^N$. The operators U_γ are unitary on $L^2(\mathbb{R}^N)$, and $U_\gamma^* = E_{-\gamma_2} V_{-\gamma_1} = U_\gamma^{-1}$. Note that these operators, together with the scalar unitary operators $e^{ir} I$ with $r \in \mathbb{Z}$ form a noncommutative group, the so-called discrete Heisenberg group. In particular, one has

$$U_\alpha^* = e^{i\langle \alpha_2,\, \alpha_1\rangle} U_{-\alpha}, \qquad U_\alpha U_\beta = e^{i\langle \alpha_2,\, \beta_1\rangle} U_{\alpha+\beta} \tag{8}$$

and

$$U_\alpha^* U_\beta = e^{i\langle\alpha_2,\,\alpha_1-\beta_1\rangle} U_{\beta-\alpha} = e^{i\langle\beta_2,\,\alpha_1-\beta_1\rangle} U_{\alpha-\beta}^*$$

where $\alpha = (\alpha_1,\, \alpha_2)$, $\beta = (\beta_1,\, \beta_2) \in \mathbb{Z}^N \times \mathbb{Z}^N$.

Let $f \in C_0^\infty(\mathbb{R}^N)$ be a non-negative function such that $f(x) = f(-x)$ for all x and such that $f(x) = 1$ if $|x_i| \le 2/3$ for all $i = 1, \ldots, N$ and $f(x) = 0$ if $|x_i| \ge 3/4$ for at least one i. Define a non-negative function φ on $\mathbb{R}^N$ by

$$\varphi^2(x) := \frac{f(x)}{\sum_{\beta\in\mathbb{Z}^N} f(x-\beta)}$$

and set $\varphi_\alpha(x) := \varphi(x-\alpha)$ for $\alpha \in \mathbb{Z}^N$. The family $(\varphi_\alpha)_{\alpha\in\mathbb{Z}^N}$ forms a partition of unity on $\mathbb{R}^N$ in sense that

$$\sum_{\alpha\in\mathbb{Z}^N} \varphi_\alpha^2(x) = 1 \qquad \text{for each } x \in \mathbb{R}^N.$$

For $\gamma = (\alpha,\, \beta) \in \mathbb{Z}^N \times \mathbb{Z}^N$, define ϕ_γ on $\mathbb{R}^N \times \mathbb{R}^N$ by

$$\phi_\gamma(x,\, \xi) := \varphi_\alpha(x)\varphi_\beta(\xi),$$

and write Φ_γ for the pseudodifferential operator $Op(\phi_\gamma)$. It is evident that

$$\Phi_\gamma u = \varphi_\alpha \varphi_\beta(D) u = \varphi_\alpha Op(\varphi_\beta) u$$

at $u \in S'(\mathbb{R}^N)$, and the formal adjoint of the operator Φ_γ acts as

$$\Phi_\gamma^* u = \varphi_\beta(D)\varphi_\alpha u = Op(\varphi_\beta)\varphi_\alpha u$$

at $u \in S'(\mathbb{R}^N)$.

The operators Φ_γ induce a partition of unity on the phase space $\mathbb{R}^N \times \mathbb{R}^N$ in the sense that

$$\sum_{\gamma\in\mathbb{Z}^{2N}} \Phi_\gamma^*\Phi_\gamma u = \sum_{\gamma\in\mathbb{Z}^{2N}} \Phi_\gamma\Phi_\gamma^* u = u \qquad \text{for each } u \in S'(\mathbb{R}^N) \tag{9}$$

where the series converge in $S'(\mathbb{R}^N)$. With these notations, we define the operator G of *time-frequency discretization* by

$$(Gu)_\gamma := \Phi_0 U_\gamma^* u \qquad \text{where } \gamma \in \mathbb{Z}^{2N} \text{ and } u \in S'(\mathbb{R}^N),$$

that is, we consider Gu as a vector-valued function on $\mathbb{Z}^{2N}$ with values in $S'(\mathbb{R}^N)$.

Now we are in a position to define the announced modulation spaces $M^{2,\,p}(\mathbb{R}^N)$ which will provide the framework for a localization of functions in the time-frequency domain. The modulation spaces under consideration were introduced in [4] where they are used to study the Fredholm property of pseudodifferential operators in $OPS^0_{0,\,0}$. Similar (but different) modulation spaces are considered in [3] (see also Chapter 11 of [2]).

Definition 3.1. *For $p \in [1, \infty)$, let $M^{2,p}(\mathbb{R}^N)$ denote the space of all distributions $u \in S'(\mathbb{R}^N)$ such that $(Gu)_\gamma \in L^2(\mathbb{R}^N)$ for every $\gamma \in \mathbb{Z}^{2N}$ and*

$$\|u\|_{M^{2,p}(\mathbb{R}^N)} := \left(\sum_{\gamma \in \mathbb{Z}^{2N}} \|(Gu)_\gamma\|^p_{L^2(\mathbb{R}^N)} \right)^{1/p} < \infty, \tag{10}$$

and let $L^{2,\infty}(\mathbb{R}^N)$ stand for the space of all distributions $u \in S'(\mathbb{R}^N)$ with $(Gu)_\gamma \in L^2(\mathbb{R}^N)$ for every $\gamma \in \mathbb{Z}^{2N}$ and

$$\|u\|_{M^{2,\infty}(\mathbb{R}^N)} := \sup_{\gamma \in \mathbb{Z}^{2N}} \|(Gu)_\gamma\|_{L^2(\mathbb{R}^N)} < \infty. \tag{11}$$

Since U_γ is a unitary operator on $L^2(\mathbb{R}^N)$, one can replace $(Gu)_\gamma = \Phi_0 U_\gamma^* u$ by $\Phi_\gamma u = U_\gamma \Phi_0 U_\gamma^* u$ in the definitions (10) and (11) of the norms.

The following proposition is taken from [4]. It summarizes basic properties of modulation spaces.

Proposition 3.2.

(a) *Under the norms (10) and (11), $M^{2,p}(\mathbb{R}^N)$ is a Banach space for each $p \in [1, \infty]$, and $M^{2,2}(\mathbb{R}^N)$ coincides with $L^2(\mathbb{R}^N)$.*

(b) *For $p \in [1, \infty)$, every linear continuous functional on $M^{2,p}(\mathbb{R}^N)$ is of the form*

$$v \mapsto \int_{\mathbb{R}^N} u(x)\overline{v(x)}\, dx, \tag{12}$$

with some distribution $u \in M^{2,q}(\mathbb{R}^N)$ where $1/p + 1/q = 1$. Hence, the Banach dual $M^{2,p}(\mathbb{R}^N)^*$ can be identified with $M^{2,q}(\mathbb{R}^N)$, and $M^{2,p}(\mathbb{R}^N)$ is reflexive for $p \in (1, \infty)$.

(c) *The Schwartz space $S(\mathbb{R}^N)$ is contained in $M^{2,p}(\mathbb{R}^N)$ for each $p \in [1, \infty]$, and it is dense in $M^{2,p}(\mathbb{R}^N)$ for each $p \in [1, \infty)$.*

(d) *$M^{2,p}(\mathbb{R}^N)$ is contained in $S'(\mathbb{R}^N)$ in the sense that $u \in M^{2,p}(\mathbb{R}^N)$ defines a linear functional on $S(\mathbb{R}^N)$ acting at φ via*

$$u(\varphi) := \int_{\mathbb{R}^N} u(x)\varphi(x)\, dx.$$

Moreover, if $u_n \to 0$ in $M^{2,q}(\mathbb{R}^N)$, then $u_n(\varphi) \to 0$ for each function $\varphi \in S(\mathbb{R}^N)$.

Notice that the operators $U_\gamma = V_\beta E_\alpha$ are bijective isometries on each of the spaces $M^{2,p}(\mathbb{R}^N)$ with $p \in [1, \infty]$ and that $U_\gamma^{-1} = E_{-\alpha} V_{-\beta}$.

Proposition 3.3. *The operator $G : M^{2,p}(\mathbb{R}^N) \to l^p(\mathbb{Z}^{2N}, L^2(\mathbb{R}^N))$ is an isometry, and the operator G_l^{-1} defined at $f \in l^p(\mathbb{Z}^{2N}, L^2(\mathbb{R}^N))$ by*

$$G_l^{-1} f := \sum_{\gamma \in \mathbb{Z}^{2N}} U_\gamma \Phi_0^* f(\gamma) \tag{13}$$

is a left inverse for G.

Proof. That G is an isometry is evident, and the equality $G_l^{-1}G = I$ follows from

$$G_l^{-1}Gu = \sum_{\gamma\in\mathbb{Z}^{2N}} U_\gamma \Phi_0^* \Phi_0 U_\gamma^* u = \sum_{\gamma\in\mathbb{Z}^{2N}} \Phi_\gamma^* \Phi_\gamma u = u,$$

which holds for every $u \in M^{2,p}(\mathbb{R}^N)$ due to (9) and Proposition 3.2 (d). □

Thus, the operator $Q := GG_l^{-1} : l^p(\mathbb{Z}^{2N}, L^2(\mathbb{R}^N)) \to l^p(\mathbb{Z}^{2N}, L^2(\mathbb{R}^N))$ is a projection for all $p \in [1, \infty]$. We denote its range by $\mathcal{R}_p(Q)$. Then

$$G : M^{2,p}(\mathbb{R}^N) \to \mathcal{R}_p(Q)$$

becomes an isometric bijection, and an operator $A \in L(M^{2,p}(\mathbb{R}^N))$ is similar to the operator

$$A_G := GAG_l^{-1}|_{\mathcal{R}_p(Q)} : \quad \mathcal{R}_p(Q) \to \mathcal{R}_p(Q).$$

We extend A_G to an operator $\Gamma(A)$ acting on all of $l^p(\mathbb{Z}^{2N}, L^2(\mathbb{R}^N))$ by setting

$$\Gamma(A) := A_G Q + I - Q = GAG_l^{-1} + I - Q$$

and call $\Gamma(A)$ the *time-frequency discretization* of A. Clearly,

$$G_l^{-1}\Gamma(A)G = G_l^{-1}(GAG_l^{-1} + I - GG_l^{-1})G = A.$$

Proposition 3.4. $Q \in \mathcal{W}(\mathbb{Z}^{2N}, L^2(\mathbb{R}^N))$.

Proof. The definitions of G and G_l^{-1} imply that Q acts at $f \in l^p(\mathbb{Z}^{2N}, L^2(\mathbb{R}^N))$ by

$$(Qf)(\delta) = \sum_{\gamma\in\mathbb{Z}^{2N}} \Phi_0 U_\delta^* U_{\delta-\gamma} \Phi_0^* f(\delta-\gamma) = \sum_{\gamma\in\mathbb{Z}^{2N}} R_\gamma(\delta)(\hat{V}_\gamma f)(\delta)$$

where $R_\gamma(\delta) := \Phi_0 U_\delta^* U_{\delta-\gamma}\Phi_0^*$ and where $\hat{V}_\gamma$ denotes again the discrete shift operator $(\hat{V}_\gamma f)(\delta) := f(\delta-\gamma)$ on $l^p(\mathbb{Z}^{2N}, L^2(\mathbb{R}^N))$. Choose $2k > N$. In [6], Proposition 4.3.2, it is shown that then

$$\begin{aligned} \|R_\gamma(\delta)\|_{L(L^2(\mathbb{R}^N))} &= \|\Phi_0 U_\delta^* U_{\delta-\gamma}\Phi_0^*\|_{L(L^2(\mathbb{R}^N))} \\ &= \|U_\delta \Phi_0 U_\delta^* U_{\delta-\gamma}\Phi_0^* U_{\delta-\gamma}^*\|_{L(L^2(\mathbb{R}^N))} \\ &= \|\Phi_\delta \Phi_{\delta-\gamma}^*\|_{L(L^2(\mathbb{R}^N))} \\ &\leq C(1+|\alpha|)^{-2k}(1+|\beta|)^{-2k} \end{aligned} \tag{14}$$

with a constant C independent of $\gamma = (\alpha, \beta)$. Consequently,

$$\sum_{\gamma\in\mathbb{Z}^{2N}} \|R_\gamma(\delta)\|_{L(L^2(\mathbb{R}^N))} \leq C \sum_{(\alpha,\beta)\in\mathbb{Z}^N\times\mathbb{Z}^N} (1+|\alpha|)^{-2k}(1+|\beta|)^{-2k} < \infty$$

showing that $\|Q\|_{\mathcal{W}(\mathbb{Z}^{2N}, L^2(\mathbb{R}^N))} < \infty$. □

3.2. Fredholmness and time-frequency discretization

Our next goal is to point out the relation between the Fredholmness of an operator acting on a modulation space $M^{2,p}(\mathbb{R}^N)$ and the $\mathcal{P}$-Fredholmness of its time-frequency discretization. Beginning with this subsection, we assume $p \in (1, \infty)$ unless otherwise stated.

Proposition 3.5.

(a) *For every $n \in \mathbb{N}$, the operators P_nQ and QP_n are compact on $l^p(\mathbb{Z}^{2N}, L^2(\mathbb{R}^N))$.*

(b) *The projection Q belongs to $L(l^p(\mathbb{Z}^{2N}, L^2(\mathbb{R}^N)), \mathcal{P})$.*

(c) *For $A \in L(M^{2,p}(\mathbb{R}^N))$, one has $\Gamma(A) \in L(l^p(\mathbb{Z}^{2N}, L^2(\mathbb{R}^N)), \mathcal{P})$.*

(d) *If $K \in L(l^p(\mathbb{Z}^{2N}, L^2(\mathbb{R}^N)))$ is a $\mathcal{P}$-compact operator of the form $K = QKQ$, then $G_l^{-1}KG$ is compact on $M^{2,p}(\mathbb{R}^N)$.*

(e) *The operator $A \in L(M^{2,p}(\mathbb{R}^N))$ is invertible if and only if the operator $\Gamma(A) \in L(l^p(\mathbb{Z}^{2N}, L^2(\mathbb{R}^N)))$ is invertible.*

(f) *The operator $A \in L(M^{2,p}(\mathbb{R}^N))$ is Fredholm if and only if the operator $\Gamma(A) \in L(l^p(\mathbb{Z}^{2N}, L^2(\mathbb{R}^N)))$ is $\mathcal{P}$-Fredholm.*

This proposition is proved in [5] for $p = 2$, see also Proposition 4.2.2 in [6]. The proof for general $p \in (1, \infty)$ runs similarly.

Definition 3.6. *Let $A \in L(M^{2,p}(\mathbb{R}^N))$, and let $h : \mathbb{N} \to \mathbb{Z}^{2N}$ be a sequence tending to infinity. We say that the operator $A_h \in L(M^{2,p}(\mathbb{R}^N))$ is a* limit operator of A with respect to the sequence h *if*

$$U_{h(m)}^{-1} A U_{h(m)} \to A_h \qquad \text{and} \qquad U_{h(m)}^{-1} A^* U_{h(m)} \to A_h^*$$

strongly as $m \to \infty$. The set $\sigma_{op}(A)$ of all limit operators of A is called the operator spectrum *of A.*

The following proposition describes the relation between the time-frequency discretization of the limit operators of A and the limit operators of the time-frequency discretization of A. Its proof for $p = 2$ is in [5] and Proposition 4.2.5 in [6]. The case of general $p \in (1, \infty)$ can be treated analogously.

Proposition 3.7. *Let $A \in L(M^{2,p}(\mathbb{R}^N))$, and let $h : \mathbb{N} \to \mathbb{Z}^{2N}$ be a sequence tending to infinity such that the limit operator A_h of A with respect to h exists. Then there is a subsequence g of h such that the limit operator $\Gamma(A)_g$ of $\Gamma(A)$ with respect to g exists, and there is an isometric isomorphism T_g mapping $l^p(\mathbb{Z}^{2N}, L^2(\mathbb{R}^N))$ onto itself such that*

$$\Gamma(A)_g = T_g^{-1} \Gamma(A_h) T_g.$$

We still need the counterparts of the notions of band and band-dominated operators for operators on modulation spaces.

Definition 3.8. *An operator $A \in L(S'(\mathbb{R}^N))$ is called a* band operator *if there exists an $R > 0$ such that $\Phi_\alpha A \Phi_\beta^* = 0$ for all subscripts α, $\beta \in \mathbb{Z}^{2N}$ with*

$$|\alpha - \beta| := \max_{1 \le i \le 2N} |\alpha_i - \beta_i| > R.$$

An operator $A \in L(M^{2,p}(\mathbb{R}^N))$ is called band-dominated *if it is the limit of a sequence of band operators converging to A in the norm of $L(M^{2,p}(\mathbb{R}^N))$.*

It is easy to check that the class of all band-dominated operators on $M^{2,p}(\mathbb{R}^N)$ is a closed subalgebra of $L(M^{2,p}(\mathbb{R}^N))$. We denote this algebra by $\mathcal{A}(M^{2,p}(\mathbb{R}^N))$. Further we call $A \in \mathcal{A}(M^{2,p}(\mathbb{R}^N))$ a *rich operator* if every sequence $h : \mathbb{N} \to \mathbb{Z}^{2N}$ which tends to infinity possesses a subsequence g for which the limit operator A_g exists. The set of all rich operators forms a closed subalgebra of $\mathcal{A}(M^{2,p}(\mathbb{R}^N))$ which we denote by $\mathcal{A}^\$(M^{2,p}(\mathbb{R}^N))$.

Proposition 3.9.

(a) *If $A \in \mathcal{A}(M^{2,p}(\mathbb{R}^N))$, then $\Gamma(A) \in \mathcal{A}(l^p(\mathbb{Z}^{2N}, L^2(\mathbb{R}^N))$.*

(b) *If $A \in \mathcal{A}^\$(M^{2,p}(\mathbb{R}^N))$, then $\Gamma(A) \in \mathcal{A}^\$(l^p(\mathbb{Z}^{2N}, L^2(\mathbb{R}^N))$.*

Proof. We prove assertion (a) only. The second statement follows from (a) and Proposition 3.7. First let A be a band operator on $M^{2,p}(\mathbb{R}^N)$. Then, for $u \in l^p(\mathbb{Z}^{2N}, L^2(\mathbb{R}^N))$,

$$\begin{aligned} (A_G u)(\delta) &= \sum_{\theta \in \mathbb{Z}^{2N}} \Phi_0 U_\delta^* A U_\theta \Phi_0^* u(\theta) = \sum_{\gamma \in \mathbb{Z}^{2N}} \Phi_0 U_\delta^* A U_{\delta-\gamma} \Phi_0^* u(\delta - \gamma) \\ &= \sum_{\gamma \in \mathbb{Z}^{2N}} A_\gamma(\delta)(\hat{V}_\gamma u)(\delta) \end{aligned} \tag{15}$$

where $A_\gamma(\delta) := \Phi_0 U_\delta^* A U_{\delta-\gamma} \Phi_0^*$. Since A is a band operator, all series in (15) have a finite number of non-vanishing items only. Indeed,

$$\|A_\gamma(\delta)\|_{L(L^2(\mathbb{R}^N))} = \|\Phi_\delta A \Phi_{\delta-\gamma}^*\|_{L(L^2(\mathbb{R}^N))} = 0$$

if $|\gamma| > R$ with $R > 0$ being large enough. Hence, A_G is a band operator. That the operator A_G is band-dominated whenever A is follows by an evident approximation argument (take into account that $G : M^{2,p}(\mathbb{R}^N) \to \mathcal{R}_p(Q)$ and $G_l^{-1} : \mathcal{R}_p(Q) \to M^{2,p}(\mathbb{R}^N)$ are isometries). Finally, since the projection Q belongs to the discrete Wiener algebra due to Proposition 3.4 (and is, thus, band-dominated), the operator $\Gamma(A) = A_G Q + (I - Q)$ is band-dominated for each band-dominated operator A. □

Combining Propositions 3.5, 3.7 and Theorem 2.5 we arrive at the following Fredholm criterion for rich band-dominated operators on modulation spaces.

Theorem 3.10. *An operator $A \in \mathcal{A}^\$(M^{2,p}(\mathbb{R}^N))$ is Fredholm if and only if all limit operators A_h of A are invertible and if the norms of their inverses are uniformly bounded, i.e.,*

$$\sup_{A_h \in \sigma_{op}(A)} \|A_h^{-1}\|_{L(M^{2,p}(\mathbb{R}^N))} < \infty.$$

4. The Wiener algebra on $\mathbb{R}^N$

We define the continuous analogue of the discrete Wiener algebra by imposing conditions on the decay of the norms $\|\Phi_\alpha A \Phi^*_{\alpha-\gamma}\|_{L(L^2(\mathbb{R}^N))}$ as γ tends to infinity.

Definition 4.1. *A linear operator $A : S'(\mathbb{R}^N) \to S'(\mathbb{R}^N)$ belongs to the* Wiener algebra $\mathcal{W}(\mathbb{R}^N)$ *if*

$$\|A\|_{\mathcal{W}(\mathbb{R}^N)} := \sum_{\gamma \in \mathbb{Z}^{2N}} \sup_{\alpha \in \mathbb{Z}^{2N}} \|\Phi_\alpha A \Phi^*_{\alpha-\gamma}\|_{L(L^2(\mathbb{R}^N))} < \infty. \tag{16}$$

The Wiener algebra $\mathcal{W}(\mathbb{R}^N)$ contains many interesting operators. Indeed, we will see in the next section that $\mathcal{W}(\mathbb{R}^N)$ contains the pseudodifferential operators with non-smooth symbols in the Sjöstrand class OPS_w and, thus, also the Hörmander class $OPS^0_{0,0}$. Here are some basic properties of $\mathcal{W}(\mathbb{R}^N)$.

Proposition 4.2.

(a) $\mathcal{W}(\mathbb{R}^N) \subset L(M^{2,p}(\mathbb{R}^N))$, *and*

$$\|A\|_{L(M^{2,p}(\mathbb{R}^N))} \le \|A\|_{\mathcal{W}(\mathbb{R}^N)}$$

for each $p \in [1, \infty]$ and $A \in \mathcal{W}(\mathbb{R}^N)$.

(b) *Provided with the norm* (16), *the set $\mathcal{W}(\mathbb{R}^N)$ becomes a unital Banach algebra.*

(c) *The Banach dual operator A^* of an operator $A \in \mathcal{W}(\mathbb{R}^N)$ considered as acting on $M^{2,p}(\mathbb{R}^N)$ belongs $\mathcal{W}(\mathbb{R}^N)$, too.*

Proof. (a) First let $p \in [1, \infty)$. Then

$$\begin{aligned}
\|Au\|^p_{M^{2,p}(\mathbb{R}^N)} &= \sum_{\gamma \in \mathbb{Z}^{2N}} \|\Phi_\gamma A u\|^p_{L^2(\mathbb{R}^N)} \\
&= \sum_{\gamma \in \mathbb{Z}^{2N}} \Big\| \Phi_\gamma A \sum_{\delta \in \mathbb{Z}^{2N}} \Phi^*_\delta \Phi_\delta u \Big\|^p_{L^2(\mathbb{R}^N)} \\
&\le \sum_{\gamma \in \mathbb{Z}^{2N}} \Big(\sum_{\alpha \in \mathbb{Z}^{2N}} \|\Phi_\gamma A \Phi^*_{\gamma-\alpha}\|_{L(L^2(\mathbb{R}^N))} \|\Phi_{\gamma-\alpha} u\|_{L^2(\mathbb{R}^N)} \Big)^p \\
&\le \sum_{\gamma \in \mathbb{Z}^{2N}} \Big(\sum_{\alpha \in \mathbb{Z}^{2N}} k_A(\alpha) \|\Phi_{\gamma-\alpha} u\|_{L^2(\mathbb{R}^N)} \Big)^p \\
&\le \sum_{\gamma \in \mathbb{Z}^{2N}} \Big(\sum_{\alpha \in \mathbb{Z}^{2N}} k_A(\gamma - \alpha) \|\Phi_\alpha u\|_{L^2(\mathbb{R}^N)} \Big)^p
\end{aligned}$$

where $k_A(\alpha) := \sup_{\gamma \in \mathbb{Z}^N} \|\Phi_\gamma A \Phi^*_{\gamma-\alpha}\|_{L(L^2(\mathbb{R}^N))}$. Since k_A is a sequence in $l^1(\mathbb{Z}^{2N})$, Corollary 4.1.14 in [6] implies that

$$\|Au\|_{M^{2,p}(\mathbb{R}^N)} \le \sum_{\gamma \in \mathbb{Z}^{2N}} k_A(\gamma) \Big(\sum_{\alpha \in \mathbb{Z}^{2N}} \|\Phi_\alpha u\|^p_{L^2(\mathbb{R}^N)} \Big)^{1/p} = \|A\|_{\mathcal{W}(\mathbb{R}^N)} \|u\|_{M^{2,p}(\mathbb{R}^N)}.$$

In the same way, one gets the estimate

$$\|Au\|_{M^{2,\infty}(\mathbb{R}^N)} \leq \sum_{\gamma\in\mathbb{Z}^{2N}} k_A(\gamma) \sup_{\alpha\in\mathbb{Z}^{2N}} \|\Phi_\alpha u\|_{L^2(\mathbb{R}^N)} = \|A\|_{\mathcal{W}(\mathbb{R}^N)} \|u\|_{M^{2,\infty}(\mathbb{R}^N)}.$$

(b) It is easy to verify that

$$\|AB\|_{\mathcal{W}(\mathbb{R}^N)} \leq \|A\|_{\mathcal{W}(\mathbb{R}^N)} \|B\|_{\mathcal{W}(\mathbb{R}^N)},$$

and estimate (14) shows that the identity operator I belongs to $\mathcal{W}(\mathbb{R}^N)$. Hence, $\mathcal{W}(\mathbb{R}^N)$ is a unital algebra, and its completeness with respect to the norm (16) follows straightforwardly.

(c) Let A^* be the Banach adjoint operator of A acting on $M^{2,p}(\mathbb{R}^N)$, that is

$$\int_{\mathbb{R}^N} Au\bar{v}\,dx = \int_{\mathbb{R}^N} u\overline{A^*v}\,dx, \tag{17}$$

where $u \in M^{2,p}(\mathbb{R}^N)$ and $v \in M^{2,q}(\mathbb{R}^N)$ with $1/p + 1/q = 1$. The operator A is bounded on $L^2(\mathbb{R}^N)$ since $A \in \mathcal{W}(\mathbb{R}^N)$ (Proposition 4.3.4 in [6]). Since (17) holds for arbitrary $u, v \in S(\mathbb{R}^N)$, this identity states that A^* is the adjoint operator to A considered as acting on $L^2(\mathbb{R}^N)$. Hence,

$$\|\Phi_\alpha A^*\Phi^*_{\alpha-\gamma}\|_{L(L^2(\mathbb{R}^N))} = \|\Phi_{\alpha-\gamma} A\Phi^*_\alpha\|_{L(L^2(\mathbb{R}^N))},$$

which implies that

$$\sum_{\gamma\in\mathbb{Z}^{2N}} \sup_{\alpha\in\mathbb{Z}^{2N}} \|\Phi_\alpha A^*\Phi^*_{\alpha-\gamma}\|_{L(L^2(\mathbb{R}^N))} = \sum_{\gamma\in\mathbb{Z}^{2N}} \sup_{\alpha\in\mathbb{Z}^{2N}} \|\Phi_{\alpha-\gamma} A\Phi^*_\alpha\|_{L(L^2(\mathbb{R}^N))}.$$

Replacing on the right-hand side α by $\alpha+\gamma$ and then γ by $-\gamma$ yields $A^* \in \mathcal{W}(\mathbb{R}^N)$ and $\|A^*\|_{\mathcal{W}(\mathbb{R}^N)} = \|A\|_{\mathcal{W}(\mathbb{R}^N)}$. □

Proposition 4.3.

(a) *If $A \in \mathcal{W}(\mathbb{R}^N)$, then the operators GAG_l^{-1} and $\Gamma(A)$ belong to the discrete Wiener algebra $\mathcal{W}(\mathbb{Z}^{2N}, L^2(\mathbb{R}^N))$.*

(b) *Conversely, if $B \in \mathcal{W}(\mathbb{Z}^{2N}, L^2(\mathbb{R}^N))$, then $G_l^{-1}AG$ lies in $\mathcal{W}(\mathbb{R}^N)$.*

The proof runs as that of Proposition 3.9; compare also [5] and Proposition 4.3.5 in [6].

Proposition 4.4. *$\mathcal{W}(\mathbb{R}^N)$ is inverse closed on each of the spaces $M^{2,p}(\mathbb{R}^N)$ with $p \in [1, \infty]$, i.e., if $A \in \mathcal{W}(\mathbb{R}^N)$ is invertible in $L(M^{2,p}(\mathbb{R}^N))$, then $A^{-1} \in \mathcal{W}(\mathbb{R}^N)$.*

Proof. Let the operator $A \in \mathcal{W}(\mathbb{R}^N)$ be invertible on $M^{2,p}(\mathbb{R}^N)$. Then $\Gamma(A)$ belongs to $\mathcal{W}(\mathbb{Z}^{2N}, L^2(\mathbb{R}^N))$ by Proposition 4.3, and $\Gamma(A)$ is invertible on $l^p(\mathbb{Z}^{2N}, L^2(\mathbb{R}^N))$ by Proposition 3.5 (e). From Proposition 2.6 we infer that $\Gamma(A)^{-1}$ lies in the discrete Wiener algebra $\mathcal{W}(\mathbb{Z}^{2N}, L^2(\mathbb{R}^N))$, and since

$$G_l^{-1}\Gamma(A)^{-1}GA = G_l^{-1}\Gamma(A)^{-1}GAG_l^{-1}G = G_l^{-1}\Gamma(A)^{-1}\Gamma(A)QG = I,$$

one has $G_l^{-1}\Gamma(A)^{-1}G = A^{-1} \in \mathcal{W}(\mathbb{R}^N)$ due to Proposition 4.3 (b). □

We fix a $p \in [1, \infty]$ and define the *rich Wiener algebra* by

$$\mathcal{W}^{\$}(\mathbb{R}^N) := \mathcal{W}(\mathbb{R}^N) \cap \mathcal{A}^{\$}(M^{2,p}(\mathbb{R}^N)).$$

Thus, an operator A belongs to $\mathcal{W}^{\$}(\mathbb{R}^N)$ if every sequence $h : \mathbb{N} \to \mathbb{Z}^{2N}$ possesses a subsequence g for which the limit operator A_g of A with respect to strong convergence on $M^{2,p}(\mathbb{R}^N)$ exists. It is easy to see that the limit operators A_g belong to $\mathcal{W}(\mathbb{R}^N)$ again. Thus, the definition of $\mathcal{W}^{\$}(\mathbb{R}^N)$ does not depend on the concrete choice of the parameter $p \in [1, \infty]$.

Theorem 4.5. *The following conditions are equivalent for $A \in \mathcal{W}^{\$}(\mathbb{R}^N)$:*

(a) *A is a Fredholm operator on $M^{2,p}(\mathbb{R}^N)$ for a certain $p \in (1, \infty)$;*

(b) *A is a Fredholm operator on $M^{2,p}(\mathbb{R}^N)$ for each $p \in (1, \infty)$;*

(c) *there exists a $p \in [1, \infty]$ for which all limit operators of A are invertible on $M^{2,p}(\mathbb{R}^N)$;*

(d) *all limit operators of A are invertible on every space $M^{2,p}(\mathbb{R}^N)$ with $p \in [1, \infty]$;*

(e) *all limit operators of A are uniformly invertible on each of the spaces $M^{2,p}(\mathbb{R}^N)$ with $p \in [1, \infty]$.*

This is an immediate consequence of Theorem 2.7 and Proposition 3.5 (e) and (f). The preceding theorem has the following corollary for the essential spectrum of an operator A in the rich Wiener algebra when considered on $M^{2,p}(\mathbb{R}^N)$, i.e., for the spectrum of the coset $A + K(M^{2,p}(\mathbb{R}^N))$ in the corresponding Calkin algebra.

Theorem 4.6. *Let $A \in \mathcal{W}^{\$}(\mathbb{R}^N)$. Then the essential spectrum $\sigma_{ess} A$ of A considered on $M^{2,p}(\mathbb{R}^N)$ is equal to*

$$\sigma_{ess}(A|M^{2,p}(\mathbb{R}^N)) = \bigcup_{A_h \in \sigma_{op}(A)} \sigma(A_h|M^{2,p}(\mathbb{R}^N)).$$

Moreover, the operator spectrum of A, the essential spectrum of A, and the common spectra of the limit operators of A are independent of $p \in (1, \infty)$.

5. Fredholm properties of pseudodifferential operators in the Sjöstrand class

We start with recalling the definition of the class of symbols of pseudodifferential operators introduced by Sjöstrand [8] in 1994; see also [9]. We introduce this class for $\mathbb{R}^n$ with arbitrary $n \in \mathbb{N}$. Later, we let $n = 2N$.

Let $\chi \in S(\mathbb{R}^n)$ be a function with $\int_{\mathbb{R}^n} \chi(x)\, dx = 1$. A function $a : \mathbb{R}^n \to \mathbb{C}$ belongs to the Sjöstrand class $S_w(\mathbb{R}^n)$ if

$$\|a\|_{S_w(\mathbb{R}^n)} := \int_{\mathbb{R}^n} \sup_{k \in \mathbb{Z}^n} \left| \int_{\mathbb{R}^n} e^{i\langle x,\, \xi\rangle} a(x)\chi(x-k)\, dx \right| d\xi < \infty. \tag{18}$$

Provided with the norm (18), $S_w(\mathbb{R}^n)$ becomes a Banach space. Notice that a change of the function χ gives rise to an equivalent norm on $S_w(\mathbb{R}^n)$ and leads, thus, to the same class of symbols.

There is another description of the Sjöstrand class $S_w(\mathbb{R}^n)$. In 1997, Boulkhemair [1] introduced the class $\mathcal{B}(\mathbb{R}^n)$ of all functions $a : \mathbb{R}^n \to \mathbb{C}$ which have the property

$$\|a\|_{\mathcal{B}(\mathbb{R}^n)} := \int_{\mathbb{R}^n} \sup_{x\in\mathbb{Z}^n} \left| \int_{\mathbb{R}^n} e^{-i\langle x,\, \xi\rangle} \hat{a}(\xi)\chi(\xi-\eta)d\xi \right| d\eta < \infty \tag{19}$$

where $\hat{a}$ refers to the Fourier transform of a in the sense of distributions. The norm (19) can be also written as

$$\|a\|_{\mathcal{B}(\mathbb{R}^n)} = \int_{\mathbb{R}^n} \|\chi(D-\eta)\, a\|_{L^\infty(\mathbb{R}^n)} d\eta$$

and is further equivalent to the norm

$$\|a\|_{\mathcal{B}(\mathbb{R}^n)} := \sum_{l\in\mathbb{Z}^n} \|\chi(D-l)\, a\|_{L^\infty(\mathbb{R}^n)}. \tag{20}$$

Moreover, Boulkhemair proved that the Sjöstrand class $S_w(\mathbb{R}^n)$ and his class $\mathcal{B}(\mathbb{R}^n)$ coincide. As a consequence of this fact, he derived the following very convenient constructive characterization of $S_w(\mathbb{R}^n)$.

Proposition 5.1 ([1]). *A distribution $a \in S'(\mathbb{R}^n)$ belongs to $S_w(\mathbb{R}^n)$ if and only if there exist a compact subset Q of $\mathbb{R}^n$ and a sequence of functions $(a_k)_{k\in\mathbb{Z}^n}$ in $L^\infty(\mathbb{R}^n)$ with* supp $(\hat{a}_k) \subseteq Q$ *and*

$$\sum_{k\in\mathbb{Z}^n} \|a_k\|_{L^\infty(\mathbb{R}^n)} < \infty,$$

such that

$$a(x) = \sum_{k\in\mathbb{Z}^n} e^{i\langle x,\, k\rangle} a_k(x)$$

almost everywhere.

Now let $n = 2N$ and $a \in S_w(\mathbb{R}^{2N})$. As usual, we write the independent variable in $\mathbb{R}^{2N}$ as $(x,\, \xi) \in \mathbb{R}^N \times \mathbb{R}^N$. Then the pseudodifferential operator with symbol a is defined by

$$(Op(a)u)(x) := (2\pi)^{-N} \int_{\mathbb{R}^N} \int_{\mathbb{R}^N} e^{i\langle x-y,\, \xi\rangle} a(x,\, \xi)u(y)\, dy\, d\xi$$

where $u \in S(\mathbb{R}^N)$. Let $OPS_w = OPS_w(\mathbb{R}^{2N})$ stand for the class of all pseudodifferential operators with symbols in $S_w(\mathbb{R}^{2N})$. It has been shown in [8] that the operators in OPS_w are bounded on $L^2(\mathbb{R}^N)$ and that OPS_w is an inverse closed subalgebra of $L(L^2(\mathbb{R}^N))$, i.e., if $A \in OPS_w$ is invertible on $L^2(\mathbb{R}^N)$, then $A^{-1} \in OPS_w$ again.

The Sjöstrand class S_w contains several interesting classes of pseudodifferential operators. For instance, the Hörmander class $S^0_{0,0}$ is contained in $S_w(\mathbb{R}^n)$ which can be checked as follows. Let a be in $C_b^\infty(\mathbb{R}^n)$, i.e., let

$$|a|_m := \sum_{|\alpha|\le m} \sup_{r\in\mathbb{R}^n} |\partial^\alpha a(x)| < \infty$$

for all $m \in \mathbb{N}$ (note that $S^0_{0,0} = C_b^\infty(\mathbb{R}^N \times \mathbb{R}^N)$). Then $\chi(D)a = k_0 * a$ where $k_0 \in S(\mathbb{R}^n)$ is given by

$$k_0(x) = (2\pi)^{-n} \int_{\mathbb{R}^n} e^{-i\langle x,\, \xi\rangle} \chi(\xi)\, d\xi.$$

Consequently, for $m \in \mathbb{N}$ and all multi-indices l,

$$\begin{aligned} (\chi(D-l)a)(x) &= \int_{\mathbb{R}^n} e^{-i\langle l,\, x-y\rangle} k_0(x-y)a(y)\, dy \\ &= \langle l\rangle^{-2m} \int_{\mathbb{R}^n} e^{-i\langle l,\, x-y\rangle} \langle D_y\rangle^{2m} (k_0(x-y)a(y))\, dy \end{aligned}$$

with the standard notations

$$\langle l\rangle := (1+|l|_2^2)^{1/2} \quad \text{and} \quad \langle D_y\rangle^2 := I - \Delta_y.$$

The latter estimate implies

$$\|\chi(D-l)a\|_{L^\infty(\mathbb{R}^n)} \le C_m \langle l\rangle^{-2m} |a|_{2m}$$

since $\partial_x^\alpha k_0 \in S(\mathbb{R}^n)$ for all multi-indices α. □

Similar classes of pseudodifferential operators have been considered in [2], see also [6].

To prove the inclusion of OPS_w into the Wiener algebra in Proposition 5.3 below we need the following estimates.

Proposition 5.2. *Let Q be a compact subset of $\mathbb{R}^n$, and let $f \in S'(\mathbb{R}^n)$ be a distribution with* $\operatorname{supp} \hat{f} \subset Q$. *Then $f \in C^\infty$, and for every multi-index α,*

$$\|\partial^\alpha f\|_{L^\infty(\mathbb{R}^n)} \le C_\alpha \|f\|_{L^\infty(\mathbb{R}^n)}$$

where the constant C_α depends on α only.

Proof. Let $\phi \in C_0^\infty(\mathbb{R}^n)$ be such that $\hat{f}\phi = \hat{f}$. Since $\hat{f} \in \mathcal{E}'(\mathbb{R}^n)$, the compactly supported distributions, one has

$$f(x) = (2\pi)^{-n} \hat{f}(\phi e_{-x})$$

where $e_{-x}(\xi) := e^{-i\langle x,\, \xi\rangle}$. Consequently,

$$(\partial^\alpha f)(x) = (2\pi)^{-n} \hat{f}(\psi_{\alpha,\, x}) = (2\pi)^{-n} \int_{\mathbb{R}^n} f(y) e_y(\psi_{\alpha,\, x})\, dy$$

where $\psi_{\alpha,\, x} \in C_0^\infty(\mathbb{R}^n)$ is given by

$$\psi_{\alpha,\, x}(\xi) = (-i\xi)^\alpha \phi(\xi) e^{-i\langle x,\, \xi\rangle}.$$

The linear functional e_y is continuous on $C_0^\infty(\mathbb{R}^n)$. Hence,

$$(2\pi)^{-n} e_y(\psi_{\alpha,\,x}) = (2\pi)^{-n} \int_{\mathbb{R}^n} (-i\xi)^\alpha \phi(\xi) e^{-i\langle x-y,\,\xi\rangle} d\xi =: h_\alpha(x-y).$$

Integrating by parts one finds $h_\alpha \in L^1(\mathbb{R}^n)$. Thus,

$$(\partial^\alpha f)(x) = \int_{\mathbb{R}^n} h_\alpha(x-y) f(y)\, dy,$$

whence

$$\|\partial^\alpha f\|_{L^\infty(\mathbb{R}^n)} \le \|h_\alpha\|_{L^1(\mathbb{R}^n)} \|f\|_{L^\infty(\mathbb{R}^n)}$$

for every multi-index α. □

Proposition 5.3. $OPS_w(\mathbb{R}^{2N}) \subset \mathcal{W}(\mathbb{R}^N)$.

Proof. Let $a \in S_w(\mathbb{R}^N \times \mathbb{R}^N)$. By Proposition 5.1, a can be represented as

$$a(x,\,\xi) = \sum_{(\alpha,\,\beta)\in\mathbb{Z}^N\times\mathbb{Z}^N} e^{i\langle x,\,\alpha\rangle + i\langle \xi,\,\beta\rangle} a_{\alpha\beta}(x,\,\xi) \tag{21}$$

where $\operatorname{supp} \hat{a}_{\alpha\beta}$ is contained in a compact subset Q of $\mathbb{R}^{2N}$ and

$$\sum_{(\alpha,\,\beta)\in\mathbb{Z}^N\times\mathbb{Z}^N} \|a_{\alpha\beta}\|_{L^\infty(\mathbb{R}^{2N})} < \infty.$$

Then

$$Op(a) = \sum_{(\alpha,\,\beta)\in\mathbb{Z}^N\times\mathbb{Z}^N} E_\alpha Op(a_{\alpha\beta}) V_\beta \tag{22}$$

and

$$\begin{aligned}
&\|\Phi_{(\gamma_1,\,\gamma_2)} Op(a) \Phi^*_{(\delta_1,\,\delta_2)}\| \\
&\quad = \|\Phi_0 U^*_{(\gamma_1,\,\gamma_2)} Op(a) U_{(\delta_1,\,\delta_2)} \Phi_0^*\| \\
&\quad \le \left\| \sum_{(\alpha,\,\beta)\in\mathbb{Z}^N\times\mathbb{Z}^N} e^{i\langle\alpha,\,\gamma_2\rangle} \Phi_0 E_{\alpha-\gamma_1} V_{-\gamma_2} Op(a_{\alpha\beta}) V_{\beta+\delta_2} E_{\delta_1} \Phi_0^* \right\| \\
&\quad \le \sum_{(\alpha,\,\beta)\in\mathbb{Z}^N\times\mathbb{Z}^N} \|\Phi_0 E_{\alpha-\gamma_1} V_{-\gamma_2} Op(a_{\alpha\beta}) V_{\beta+\delta_2} E_{\delta_1} \Phi_0^*\|.
\end{aligned}$$

By Proposition 5.2,

$$\|\partial_x^\gamma \partial_\xi^\delta a_{\alpha\beta}\|_{L^\infty(\mathbb{R}^{2N})} \le C_{\gamma\delta} \|a_{\alpha\beta}\|_{L^\infty(\mathbb{R}^{2N})}. \tag{23}$$

Hence (see, for instance, [6], Proposition 4.1.16),

$$\begin{aligned}
&\|\Phi_0 E_{\alpha-\gamma_1} V_{-\gamma_2} Op(a_{\alpha\beta}) V_{\beta+\delta_2} E_{\delta_1} \Phi_0^*\| \\
&\qquad \le C\, |a_{\alpha\beta}|_{2k_1,\,2k_2} (1 + |\alpha - \gamma_1 + \delta_1|)^{-2k_1} (1 + |\beta + \delta_2 - \gamma_2|)^{-2k_2},
\end{aligned}$$

where $2k_1 > N$ and $2k_2 > N$, and where the constant C is independent of $a_{\alpha\beta}$. From (23) one concludes that

$$|a_{\alpha\beta}|_{2k_1,\,2k_2} \le C\, \|a_{\alpha\beta}\|_{L^\infty(\mathbb{R}^N)}$$

with a constant C independent of $a_{\alpha\beta}$ again. So one finally has

$$\begin{aligned}&\|\Phi_{(\gamma_1,\,\gamma_2)} Op(a)\Phi^*_{(\delta_1,\,\delta_2)}\| \\ &\quad\le C \sum_{(\alpha,\,\beta)\in\mathbb{Z}^N\times\mathbb{Z}^N} \|a_{\alpha\beta}\|_{L^\infty(\mathbb{R}^N)}(1+|\alpha-\gamma_1-\delta_1|)^{-2k_1}(1+|\beta+\delta_2-\gamma_2|)^{-2k_2} \\ &\quad=: h(\gamma_1-\delta_1,\,\gamma_2-\delta_2)\end{aligned}$$

with a sequence $h \in l^1(\mathbb{Z}^N \times \mathbb{Z}^N)$. Consequently, $Op(a) \in \mathcal{W}(\mathbb{R}^N)$. □

The following corollary follows immediately from the preceding proposition in combination with Proposition 4.2 (*a*).

Corollary 5.4. *Let $a \in S_w(\mathbb{R}^{2N})$ be represented as in (21), and let $p \in [1,\,\infty]$. Then*

$$\|Op(a)\|_{L(M^{2,\,p}(\mathbb{R}^N))} \le C \sum_{(\alpha,\,\beta)\in\mathbb{Z}^N\times\mathbb{Z}^N} \|a_{\alpha\beta}\|_{L^\infty(\mathbb{R}^N)}$$

with a constant C independent of $a_{\alpha\beta}$.

We say that the symbol a belongs to the class $\mathcal{R}(\mathbb{R}^{2N})$ if there are integers $k_1,\,k_2$ with $2k_1 > N$ and $2k_2 > N$ such that a can be represented as

$$a(y) = \sum_{\gamma\in\mathbb{Z}^{2N}} e^{i\langle\gamma,\,y\rangle} a_\gamma(y)$$

where $y = (x,\,\xi) \in \mathbb{R}^N \times \mathbb{R}^N$, and where the functions $a_\gamma \in S^0_{0,\,0}$ satisfy

$$\sum_{\gamma\in\mathbb{Z}^{2N}} |a_\gamma|_{2k_1,\,2k_2} < \infty. \tag{24}$$

Proposition 5.5. *The classes $\mathcal{R}(\mathbb{R}^{2N})$ and $S_w(\mathbb{R}^{2N})$ coincide.*

Proof. Let $a \in \mathcal{R}(\mathbb{R}^{2N})$. Then

$$\begin{aligned}&\sum_{l\in\mathbb{Z}^{2N}} \|\chi(D-l)a\|_{L^\infty(\mathbb{R}^{2N})} \\ &\qquad\le \sum_{l\in\mathbb{Z}^{2N}}\sum_{\gamma\in\mathbb{Z}^{2N}} \|\chi(D-l-\gamma)a_\gamma\|_{L^\infty(\mathbb{R}^{2N})} \\ &\qquad\le C\sum_{\gamma\in\mathbb{Z}^{2N}} |a_\gamma|_{2k_1,\,2k_2} \sum_{l\in\mathbb{Z}^{2N}} (1+|l_1|)^{-2k_1}(1+|l_2|)^{-2k_2} < \infty,\end{aligned}$$

whence the inclusion $\mathcal{R}(\mathbb{R}^{2N}) \subset S_w(\mathbb{R}^{2N})$. The reverse inclusion follows from Proposition 5.1. □

The following observation will be needed to prove the richness of the operators in $OPS_w(\mathbb{R}^{2N})$.

Lemma 5.6. *Let $(A_j)_{j\in\mathbb{N}}$ be a sequence of bounded linear operators on a Hilbert space H with*

$$\sum_{j\in\mathbb{N}} \|A_j\| < \infty, \tag{25}$$

and let $A := \sum_{j\in\mathbb{N}} A_j$. Furthermore, let $(U_m)_{m\in\mathbb{N}}$ be a sequence of unitary operators on H such that the sequences $(U_m^* A_j U_m)_{m\in\mathbb{N}}$ converge strongly as $m \to \infty$ to certain operators $\tilde{A}_j$ for every j. Then the sequence $(U_m^* A U_m)_{m\in\mathbb{N}}$ converges strongly to $\tilde{A} := \sum_{j\in\mathbb{N}} \tilde{A}_j$.

Proof. Let $u \in H$ and $\varepsilon > 0$. By condition (25), there is an $n_0 \in \mathbb{N}$ such that

$$\sum_{j>n_0} \|A_j u\| < \frac{\varepsilon}{3}, \tag{26}$$

and due to strong convergence, there is an $m_0 \in \mathbb{N}$ such that, for $m > m_0$,

$$\max_{1\le j\le n_0} \|(\tilde{A}_j - U_m^* A_j U_m)u\| < \frac{\varepsilon}{3n_0}.$$

Hence, given arbitrary $u \in H$ and $\varepsilon > 0$, one finds an $m_0 \in \mathbb{N}$ such that

$$\|(\tilde{A} - U_m^* A U_m)u\| \le \sum_{j=1}^{n_0} \|(\tilde{A}_j - U_m^* A_j U_m)u\| + 2\sum_{j>n_0} \|A_j u\| < \varepsilon$$

for $m \ge m_0$. □

Proposition 5.7. *$OPS_w(\mathbb{R}^{2N}) \subset \mathcal{W}^\$(\mathbb{R}^N)$.*

Proof. Let $A := Op(a) \in OPS_w(\mathbb{R}^N \times \mathbb{R}^N)$. By Proposition 5.5, the operator A can be written as

$$A = \sum_{(\alpha,\,\beta)\in\mathbb{Z}^N\times\mathbb{Z}^N} E_\alpha Op(a_{\alpha\beta}) V_\beta$$

where

$$\sum_{(\alpha,\,\beta)\in\mathbb{Z}^N\times\mathbb{Z}^N} \|Op(a_{\alpha\beta})\| < \infty.$$

Let $h : m \mapsto h_m := (h'_m,\, h''_m) \in \mathbb{Z}^N \times \mathbb{Z}^N$ be a sequence which tends to infinity. Then, evidently,

$$U_{h_m}^* A U_{h_m} = \sum_{(\alpha,\,\beta)\in\mathbb{Z}^N\times\mathbb{Z}^N} (U_{h_m}^* E_\alpha U_{h_m})\,(U_{h_m}^* Op(a_{\alpha\beta}) U_{h_m})\,(U_{h_m}^* V_\beta U_{h_m}).$$

Since $U_\gamma = V_\beta E_\alpha$ and $U_\gamma^* = E_{-\alpha} V_{-\beta}$, one has

$$U_{h_m}^* E_\alpha U_{h_m} = e^{-i\langle\alpha,\, h''_m\rangle} E_\alpha \quad \text{and} \quad U_{h_m}^* V_\beta U_{h_m} = e^{i\langle\beta,\, h'_m\rangle} V_\beta.$$

In [6], Lemma 4.2.4, it is verified that there is a subsequence g of h such that the functions

$$\varphi_m : \alpha \mapsto e^{-i\langle\alpha,\, g''_m\rangle} \quad \text{and} \quad \gamma_m : \beta \mapsto e^{-i\langle\beta,\, g'_m\rangle}$$

converge uniformly with respect to $\alpha,\, \beta \in \mathbb{Z}^N$ to certain limit functions φ and γ as $m \to \infty$. Clearly, $|\varphi(\alpha)| = |\gamma(\beta)| = 1$ for each $\alpha,\, \beta \in \mathbb{Z}^N$. It is also easy to see that

$$U_{g_m}^* Op(a_{\alpha\beta}) U_{g_m} = Op(a_{\alpha\beta}^{g_m})$$

where

$$a_{\alpha\beta}^{g_m}(x,\, \xi) := a_{\alpha\beta}(x + g'_m,\, \xi + g'_m).$$

According to the Arzéla-Ascoli Theorem, one further finds a subsequence k of g such that the functions $a^{k_m}_{\alpha\beta}$ converge to a limit function $a^k_{\alpha\beta}$ in the topology of $C^\infty(\mathbb{R}^{2N})$. This implies (compare [6], Theorem 4.3.15) that $a^k_{\alpha\beta} \in S^0_{0,\,0}$ and that

$$U^*_{k_m} Op(a_{\alpha\beta}) U_{k_m} \to Op(a^K_{\alpha\beta}) \quad \text{strongly as } m \to \infty.$$

Applying the standard Cantor diagonal process, we finally obtain that every sequence h has a subsequence l such that

$$U^*_{l_m}(E_\alpha Op(a_{\alpha\beta}) V_\beta) U_{l_m} \to \varphi(\alpha)\gamma(\beta) E_\alpha Op(a^l_{\alpha\beta}) V_\beta$$

strongly as $m \to \infty$. Hence, the strong convergence

$$U^*_{l_m} A U_{l_m} \to A_l := \sum_{(\alpha,\,\beta)\in\mathbb{Z}^N\times\mathbb{Z}^N} \varphi(\alpha)\gamma(\beta) E_\alpha Op(a^l_{\alpha\beta}) V_\beta \tag{27}$$

as $m \to \infty$ follows from Lemma 5.6, and the strong convergence of the adjoint sequences

$$U^*_{l_m} A^* U_{l_m} \to A^*_l := \sum_{(\alpha,\,\beta)\in\mathbb{Z}^N\times\mathbb{Z}^N} \bar{\varphi}(\alpha)\bar{\gamma}(\beta) V_{-\beta} \left[Op(a^l_{\alpha\beta})\right]^* E_{-\alpha}$$

can be checked in the same way. □

Now Theorem 17 implies the following final results on the Fredholmness of pseudodifferential operators in the Sjöstrand class acting on modulation spaces.

Theorem 5.8. *The following conditions are equivalent for $A \in OPS_w$:*

(a) *A is a Fredholm operator on $M^{2,\,p}(\mathbb{R}^N)$ for a certain $p \in (1,\,\infty)$;*

(b) *A is a Fredholm operator on $M^{2,\,p}(\mathbb{R}^N)$ for each $p \in (1,\,\infty)$;*

(c) *there exists a $p \in [1,\,\infty]$ for which all limit operators of A are invertible on $M^{2,\,p}(\mathbb{R}^N)$;*

(d) *all limit operators of A are invertible on every space $M^{2,\,p}(\mathbb{R}^N)$ with $p \in [1,\,\infty]$;*

(e) *all limit operators are uniformly invertible on each of the spaces $M^{2,\,p}(\mathbb{R}^N)$ with $p \in [1,\,\infty]$.*

Corollary 5.9. *Let $A \in OPS_w$. Then the essential spectrum $\sigma_{\mathrm{ess}}(A)$ of A considered as an operator on $M^{2,\,p}(\mathbb{R}^N)$ does not depend on $p \in (1,\infty)$, and*

$$\sigma_{\mathrm{ess}}(A|M^{2,\,p}(\mathbb{R}^N)) = \bigcup_{A_h\in\sigma_{op}(A)} \sigma(A_h|L^2(\mathbb{R}^N)).$$

References

[1] A. Boulkhemair, *Remark on a Wiener type pseudodifferential algebra and Fourier integral operators*, Math. Res. Letter **4** (1997), 53–67.

[2] K. Gröchenig, *Foundation of Time-Frequency Analysis*, Birkhäuser Verlag, Boston, Basel, Berlin 2000.

[3] K. Gröchenig, C. Heil, *Modulation spaces and pseudodifferential operators*, Integral Equations Oper. Theory **34** (1999), 4, 439–457.

[4] V.S. Rabinovich, *On the Fredholmness of pseudodifferential operators on $\mathbb{R}^n$ in the scale of $L_{2,p}$-spaces*, Sib. Mat. Zh. **29** (1988), 4, 149–161 (Russian, Engl. transl. Sib. Math. J. **29**(1988), 4, 635–646).

[5] V.S. Rabinovich, S. Roch, *Wiener algebra of operators and applications to pseudodifferential operators*, J. Anal. Appl. **23** (2004), 3, 437–482.

[6] V.S. Rabinovich, S. Roch, B. Silbermann, *Limit Operators and their Applications in Operator Theory*, Operator Theory: Advances and Applications **150**, Birkhäuser Verlag, Basel, Boston, Berlin 2004.

[7] M. Shubin, *Pseudodifferential operators and Spectral Theory*, Springer-Verlag, Berlin, Heidelberg, New York, Tokio 2001.

[8] J. Sjöstrand, *An algebra of pseudodifferential operators*, Math. Res. Lett. **1**(1994), 185–192.

[9] J. Sjöstrand, *Wiener type algebras of pseudodifferential operators*, Sém. Eq. aux Dér. Part., Ecole Polytechnique, 1994/1995, Exposé n. IV.

[10] E.M. Stein, *Harmonic Analysis, Real-Variable Methods, Orthogonality, and Oscillatory Integrals*, Princeton Univ. Press, Princeton, New Jersey, 1993.

[11] K. Tachizawa, *The boundedness of pseudodifferential operators on modulation spaces*, Math. Nachr. **168** (1994), 263–277.

[12] M.E. Taylor, *Pseudodifferential Operators*, Princeton Univ. Press, Princeton, New Jersey, 1981.

Vladimir S. Rabinovich
Instituto Politecnico Nacional
ESIME Zacatenco
Avenida IPN
Mexico, D. F. 07738, Mexico
e-mail: `vladimir.rabinovich@gmail.com`

Steffen Roch
Department of Mathematics
Technical University of Darmstadt
Schlossgartenstrasse 7
D-64289 Darmstadt, Germany
e-mail: `roch@mathematik.tu-darmstadt.de`

Operator Theory:
Advances and Applications, Vol. 171, 281–322

On Indefinite Cases of Operator Identities Which Arise in Interpolation Theory

James Rovnyak and Lev A. Sakhnovich

Abstract. Operator identities involving nonnegative selfadjoint operators play a fundamental role in interpolation theory and its applications. The theory is generalized here to selfadjoint operators whose negative spectra consist of a finite number of eigenvalues of finite total multiplicity. It is shown that such identities are closely associated with generalized Nevanlinna functions by means of the Kreĭn-Langer integral representation. The Potapov fundamental matrix inequality is generalized to this situation, and it is used to formulate and solve an operator interpolation problem analogous to the definite case.

Mathematics Subject Classification (2000). Primary 47A57; Secondary 47A56, 30E05, 47B50, 46C20.

Keywords. Interpolation, operator identity, indefinite, negative squares, generalized Nevanlinna function, Kreĭn-Langer representation, Potapov, fundamental matrix inequality.

1. Introduction

One approach to interpolation theory and spectral problems for canonical differential equations is based on operator identities of the form

$$\begin{cases} AS - SA^* = i\left[\Phi_1\Phi_2^* + \Phi_2\Phi_1^*\right], \\ A, S \in \mathfrak{L}(\mathfrak{H}), \quad \Phi_1, \Phi_2 \in \mathfrak{L}(\mathfrak{G}, \mathfrak{H}), \end{cases} \tag{1.1}$$

where $\mathfrak{H}$ and $\mathfrak{G}$ are Hilbert spaces, $\dim \mathfrak{G} < \infty$, and $S = S^*$. A well-known matrix example is $A = \operatorname{diag}\{z_1, z_2, \dots, z_n\}$,

$$S = \left[\frac{w_\mu - \bar{w}_\nu}{z_\mu - \bar{z}_\nu}\right]_{\mu,\nu=1}^{n}, \quad \Phi_1 = -i\begin{bmatrix} w_1 \\ w_2 \\ \vdots \\ w_n \end{bmatrix}, \quad \Phi_2 = \begin{bmatrix} 1 \\ 1 \\ \vdots \\ 1 \end{bmatrix},$$

J. Rovnyak was supported by the National Science Foundation Grant DMS-0100437.

where $z_1, \dots, z_n$ are points in the upper half-plane and $w_1, \dots, w_n$ are any complex numbers. In the definite case, that is, when

$$S \geq 0, \tag{1.2}$$

a systematic treatment of identities (1.1) and their applications is given in [12, 13]. A large class of operator identities (1.1) satisfying (1.2) is obtained by first choosing operators $A \in \mathfrak{L}(\mathfrak{H})$ and $\Phi_2 \in \mathfrak{L}(\mathfrak{G}, \mathfrak{H})$ for some Hilbert space $\mathfrak{H}$ and $\mathfrak{G} = \mathbf{C}^m$, and an $m \times m$ matrix-valued Nevanlinna function $v(z)$. The Nevanlinna representation of $v(z)$ has the form

$$v(z) = \alpha + \beta z + \int_{-\infty}^{\infty} \left[\frac{1}{t - z} - \frac{t}{1 + t^2} \right] d\tau(t), \tag{1.3}$$

depending on data

$$\boldsymbol{\tau} = \{\tau(t), \alpha, \beta\}, \tag{1.4}$$

where α and β are selfadjoint matrices, $\beta \geq 0$, and $\tau(t)$ is a nondecreasing matrix-valued function such that the integral $\int_{-\infty}^{\infty} d\tau(t)/(1 + t^2)$ is convergent. We define operators

$$S_v = \int_{-\infty}^{\infty} (I - At)^{-1} \Phi_2 \, [d\tau(t)] \, \Phi_2^*(I - A^*t)^{-1} + FF^*, \tag{1.5}$$

$$\Phi_{1,v} = -i \int_{-\infty}^{\infty} \left[A(I - At)^{-1} + \frac{tI}{t^2 + 1} \right] \Phi_2 \, [d\tau(t)] + i\big(\Phi_2 \alpha + F\beta^{1/2}\big), \tag{1.6}$$

where $F = A^{-1}\Phi_2 \, \beta^{1/2}$ if A is invertible and $F = 0$ otherwise. Conditions for convergence of the integrals are given in [12, p. 2], and whenever the integrals converge the operators $S = S_v$, $\Phi_1 = \Phi_{1,v}$, A, and Φ_2 satisfy (1.1) and (1.2). Conversely, given an operator identity (1.1) satisfying (1.2), the abstract interpolation problem is to determine all representations of S and Φ_1 in the form $S = S_v$ and $\Phi_1 = \Phi_{1,v}$. Solutions are obtained in [12] with the aid of an operator analog of Potapov's fundamental matrix inequality. These results are applied in [12, 13] to concrete interpolation problems and spectral problems for canonical differential systems. The theory is simplest in the nondegenerate case, that is, when S is invertible.

In this paper, in place of (1.2) we assume that the negative spectrum of S consists of eigenvalues of finite total multiplicity. We replace (1.3) and (1.4) by the Kreĭn-Langer representation (2.1) of a generalized Nevanlinna function and corresponding data (2.2) and extend the definite theory. Such a generalization was initiated by A.L. Sakhnovich [10] in the scalar case. Other special cases are treated by the authors [9, 7]. We now take up the general case.

We state the Kreĭn-Langer representation of a generalized Nevanlinna function in Section 2. Our main results are formulated in Sections 3–5, with proofs deferred to Section 6. Section 3 is devoted to the construction of operators S_v and $\Phi_{1,v}$ generalizing (1.5) and (1.6). The constructions of S_v and $\Phi_{1,v}$ differ depending whether $0 \notin \sigma(A)$ or $0 \in \sigma(A)$, and these are referred to as Case 1 and

Case 2 throughout the paper. The abstract interpolation problem is formulated in Definition 3.6.

Section 4 derives an indefinite generalization of Potapov's fundamental matrix inequality. The generalization takes the form of a condition on the number of negative squares of a two-variable kernel. The results of Section 4 are used in Section 5 to characterize solutions of the abstract interpolation problem in the nondegenerate case.

Notation and preliminaries. Let $\mathbf{C}, \mathbf{C}_{\pm}$ be the complex plane and open upper and lower half-planes. Throughout, we use a finite-dimensional Hilbert space $\mathfrak{G}$, which we take to be $\mathfrak{G} = \mathbf{C}^m$ for a fixed positive integer m. Operators on $\mathfrak{G}$ are represented as $m \times m$ matrices. The **generalized Nevanlinna class** $\mathbf{N}_{\varkappa}$, $\varkappa = 0, 1, 2, \ldots$, is the set of $m \times m$ matrix-valued functions $v(z)$ which are meromorphic on $\mathbf{C}_+ \cup \mathbf{C}_-$ such that $v(z) = v(\bar{z})^*$ and the kernel $[v(z) - v(\zeta)^*]/(z - \bar{\zeta})$ has $\varkappa = \varkappa_v$ negative squares (for example, see [1, 2, 5]). We also write $\varkappa_K$ for the number of negative squares of any Hermitian kernel $K(z, \zeta)$. If S is a selfadjoint operator on a Hilbert space, $\varkappa_S$ denotes the dimension of the eigenspace for $(0, \infty)$.

We assume familiarity with Stieltjes integrals $\int_\Delta f(t)\,[d\tau(t)]\,g(t)$, where $\tau(t)$ is an $m \times m$ matrix-valued function and $f(t)$ and $g(t)$ are matrix-valued functions of orders $p \times m$ and $m \times q$. In our applications, Δ is either an interval or a finite union of intervals and $\tau(t)$ is nondecreasing on each interval in Δ. By $L^2(d\tau)$ we mean a completion of the space of continuous $\mathfrak{G}$-valued functions $g(t)$ on Δ such that $\|g\|^2 = \int_\Delta g(t)^*\,[d\tau(t)]\,g(t) < \infty$. We also use integrals of the form

$$\int_\Delta G(t)^*\,[d\tau(t)]\,F(t), \tag{1.7}$$

where $F(t)$ and $G(t)$ are continuous functions with values in $\mathfrak{L}(\mathfrak{H}, \mathfrak{G})$ and $\mathfrak{L}(\mathfrak{K}, \mathfrak{G})$ for some Hilbert spaces $\mathfrak{H}$ and $\mathfrak{K}$. If $F(t)h$ and $G(t)k$ belong to $L^2(d\tau)$ for all vectors h in $\mathfrak{H}$ and k in $\mathfrak{K}$, we define (1.7) as the unique operator in $\mathfrak{L}(\mathfrak{H}, \mathfrak{K})$ such that

$$\left\langle \left(\int_\Delta G(t)^*\,[d\tau(t)]\,F(t) \right) h, k \right\rangle_{L^2(d\tau)} = \int_\Delta [G(t)k]^* [d\tau(t)] F(t)h\,,$$

for all $h \in \mathfrak{H}$ and $k \in \mathfrak{K}$. Integrals of the type (1.7) also appear in the form

$$\int_\Delta \sum_{j=1}^r G_j(t)^* [d\tau(t)] F_j(t) = \int_\Delta \begin{bmatrix} G_1(t)^* & \cdots & G_r(t)^* \end{bmatrix} [d\tau(t)] \begin{bmatrix} F_1(t) \\ \vdots \\ F_r(t) \end{bmatrix}.$$

In practice, to prove convergence of such an integral we show that

$$\sum_{j=1}^r G_j(t)^* [d\tau(t)] F_j(t) = \sum_{k=1}^s \widetilde{G}_k(t)^* [d\tau(t)] \widetilde{F}_k(t),$$

where the integrals $\int_\Delta \widetilde{G}_j(t)^*\,[d\tau(t)]\,\widetilde{F}_j(t)$, $j = 1, \ldots, s$, exist separately.

2. Kreĭn-Langer integral representation

The Kreĭn-Langer integral representation of a generalized Nevanlinna function is given in [5] in the scalar case. The matrix case of the representation is due to Daho and Langer [2], and we use this case in a form given in [8].

Theorem 2.1. *Let $v(z)$ be an $m \times m$ matrix-valued meromorphic function such that $v(\bar{z})^* = v(z)$ on $\mathbf{C}_+ \cup \mathbf{C}_-$. A necessary and sufficient condition that $v(z)$ belong to some class $\mathbf{N}_\varkappa$, $\varkappa \geq 0$, is that it can be written in the form*

$$v(z) = \int_{-\infty}^{\infty} \left[\frac{1}{t-z} - \sum_{j=0}^{r} S_j(t,z) \right] d\tau(t) \tag{2.1}$$
$$+ R_0(z) - \sum_{j=1}^{r} R_j \left(\frac{1}{z-\alpha_j} \right)$$
$$- \sum_{k=1}^{s} \left[M_k \left(\frac{1}{z-\beta_k} \right) + M_k \left(\frac{1}{\bar{z}-\beta_k} \right)^* \right],$$

where $\alpha_1, \ldots, \alpha_r \in (-\infty, \infty)$ and $\beta_1, \ldots, \beta_s \in \mathbf{C}_+$ are distinct numbers, and

(1°) *the real line is a union of sets $\Delta_0, \Delta_1, \ldots, \Delta_r$ such that $\Delta_1, \ldots, \Delta_r$ are bounded open intervals containing $\alpha_1, \ldots, \alpha_r$ and having disjoint closures, Δ_0 is their complement, and*

$$\frac{1}{t-z} - S_j(t,z) = \frac{1}{t-z} \left(\frac{t-\alpha_j}{z-\alpha_j} \right)^{2\rho_j} \quad \text{on } \Delta_j, \quad j = 1, \ldots, r,$$
$$\frac{1}{t-z} - S_0(t,z) = \frac{1+tz}{t-z} \, \frac{(1+z^2)^{\rho_0}}{(1+t^2)^{\rho_0+1}} \quad \text{on } \Delta_0,$$

for some positive integers $\rho_1, \ldots, \rho_r$ and some nonnegative integer ρ_0, and $S_j(t,z) = 0$ off Δ_j for each $j = 0, 1, \ldots, r$;

(2°) *$\tau(t)$ is an $m \times m$ matrix-valued function which is nondecreasing on each of the $r+1$ open intervals of the real line determined by the points $\alpha_1, \ldots, \alpha_r$ such that the integral*

$$\int_{-\infty}^{\infty} \frac{(t-\alpha_1)^{2\rho_1} \cdots (t-\alpha_r)^{2\rho_r}}{(1+t^2)^{\rho_1+\cdots+\rho_r}} \, \frac{d\tau(t)}{(1+t^2)^{\rho_0+1}}$$

converges;

(3°) *for each $j = 0, 1, \ldots, r$, $R_j(z)$ is a polynomial of degree at most $2\rho_j + 1$, having selfadjoint $m \times m$ matrix coefficients, such that if a term of maximum degree $C_j z^{2\rho_j+1}$ is present then $C_j \geq 0$, and $R_1(0) = \cdots = R_r(0) = 0$;*

(4°) *for each $k = 1, \ldots, s$, $M_k(z)$ is a polynomial $\not\equiv 0$ with $m \times m$ matrix coefficients such that $M_k(0) = 0$.*

The sets $\Delta_0, \Delta_1, \ldots, \Delta_r$ in (2.1) can be chosen arbitrarily subject to the conditions in (1°).

Definition 2.2. By **Kreĭn-Langer data** we mean a collection of quantities

$$\boldsymbol{\tau} = \{\tau(t);\, \alpha_1,\dots,\alpha_r;\, \beta_1,\dots,\beta_s;\, \rho_0,\dots,\rho_r;\, \Delta_0,\dots,\Delta_r; \\ R_0(z),\dots,R_r(z);\, M_1(z),\dots,M_s(z)\} \tag{2.2}$$

having the properties listed in Theorem 2.1. Given data $\boldsymbol{\tau}$, we write $v_{\boldsymbol{\tau}}(z)$ for the associated function (2.1).

The identities

$$\frac{1}{t-z}\left(\frac{t-\alpha}{z-\alpha}\right)^{2p} = \frac{1}{t-z} + \sum_{j=0}^{2p-1} \frac{(t-\alpha)^j}{(z-\alpha)^{j+1}} \tag{2.3}$$

$$\frac{(1+z^2)^p}{(1+t^2)^{p+1}}\,\frac{1+tz}{t-z} = \frac{1}{t-z} - (t+z)\sum_{j=0}^{p-1} \frac{(1+z^2)^j}{(1+t^2)^{j+1}} \\ - t\,\frac{(1+z^2)^p}{(1+t^2)^{p+1}} \tag{2.4}$$

show that the convergence terms in (2.1) are given by

$$S_j(t,z) = -\sum_{p=0}^{2\rho_j-1} \frac{(t-\alpha_j)^p}{(z-\alpha_j)^{p+1}}\,\chi_{\Delta_j}(t), \qquad j=1,\dots,r, \tag{2.5}$$

$$S_0(t,z) = \left\{ (t+z)\sum_{p=0}^{\rho_0-1} \frac{(1+z^2)^p}{(1+t^2)^{p+1}} + t\,\frac{(1+z^2)^{\rho_0}}{(1+t^2)^{\rho_0+1}} \right\} \chi_{\Delta_0}(t). \tag{2.6}$$

3. Interpolation problem for operator identities

Throughout this section we understand that $\mathfrak{H}$ is some Hilbert space, and, as usual, $\mathfrak{G} = \mathbf{C}^m$. Our first task is to construct operators S_v and $\Phi_{1,v}$ corresponding to a given generalized Nevanlinna function $v(z)$. These operators appear in the statement of the abstract interpolation problem in Definition 3.6.

Assumptions 3.1. *Let $A \in \mathfrak{L}(\mathfrak{H})$ and $\Phi_2 \in \mathfrak{L}(\mathfrak{G},\mathfrak{H})$ be given operators, and let $v(z)$ be a generalized Nevanlinna function which is represented in the form* (2.1) *for some Kreĭn-Langer data* (2.2). *Assume that $\sigma(A)$ is a finite set that contains no point $1/\beta_k, 1/\bar\beta_k$, $k=1,\dots,s$, and no real point except perhaps* 0.
Case 1: $0 \notin \sigma(A)$. *There are no additional assumptions in this case.*
Case 2: $0 \in \sigma(A)$. *Here we assume further that* (2.1) *can be chosen such that $\rho_0 = 0$, $R_0(z)$ is constant, and*

$$\int_{\Delta_0} \langle d\tau(t)\,\Phi_2^*(I-A^*t)^{-1}h, \Phi_2^*(I-A^*t)^{-1}h\rangle < \infty, \qquad h \in \mathfrak{H}. \tag{3.1}$$

By [8, Theorem 4.1], the condition that $v(iy)/y \to 0$ as $y \to \infty$ is necessary and sufficient that a representation (2.1) can be chosen such that $\rho_0 = 0$ and $R_0(z)$ is constant.

Under the Assumptions 3.1, in both Case 1 and Case 2, we shall define operators

$$S_v = \int_{-\infty}^{\infty} \Big\{ (I-At)^{-1}\Phi_2[d\tau(t)]\Phi_2^*(I-A^*t)^{-1} - \sum_{j=0}^{r} d\tau_j(t;A,\Phi_2) \Big\} \tag{3.2}$$
$$+ \sum_{j=0}^{r} \mathfrak{R}_j + \sum_{k=1}^{s} \big[\mathfrak{M}_{1k} + \mathfrak{M}_{2k}\big],$$

$$\Phi_{1,v} = -i \int_{-\infty}^{\infty} \Big\{ A(I-At)^{-1} - \sum_{j=0}^{r} \mathfrak{S}_j(t;A) \Big\} \Phi_2\,[d\tau(t)] \tag{3.3}$$
$$- i\Big(\sum_{j=0}^{r} \widehat{\mathfrak{R}}_j + \sum_{k=1}^{s} \big[\widehat{\mathfrak{M}}_{1k} + \widehat{\mathfrak{M}}_{2k}\big] \Big),$$

which generalize (1.5) and (1.6) to the indefinite case. The terms in (3.2) and (3.3) are associated with the parts in the Kreĭn-Langer representation (2.1) of $v(z)$. The definitions differ slightly in Case 1 and Case 2 of the Assumptions 3.1, that is, according as $0 \notin \sigma(A)$ or $0 \in \sigma(A)$.

Definition of S_v and $\Phi_{1,v}$ in Case 1. In Case 1, $0 \notin \sigma(A)$ and so $\sigma(A)$ contains no real point. We first define the convergence terms $d\tau_j(t;A,\Phi_2)$ and $\mathfrak{S}_j(t;A)$ in the integral parts of (3.2) and (3.3), $j = 0,\dots,r$. These terms are defined to be zero off Δ_j, $j = 0,\dots,r$. For $j = 1,\dots,r$, we expand

$$(I-tA)^{-1}\Phi_2[d\tau(t)]\Phi_2^*(I-tA^*)^{-1} \qquad \text{and} \qquad A(I-tA)^{-1} \tag{3.4}$$

in powers of $t-\alpha_j$ using the series

$$(I-tA)^{-1} = \sum_{p=0}^{\infty} A_p(\alpha_j)(t-\alpha_j)^p, \quad A_p(\alpha_j) = A^p(I-\alpha_j A)^{-p-1}, \tag{3.5}$$

and we define $d\tau_j(t;A,\Phi_2)$ and $\mathfrak{S}_j(t;A)$ on Δ_j to be the expressions that remain after discarding all terms that are $\mathcal{O}\big((t-\alpha_j)^{2\rho_j}\big)$ as $t \to \alpha_j$.

To define $d\tau_0(t;A,\Phi_2)$ and $\mathfrak{S}_0(t;A)$ on Δ_0, we expand (3.4) in a neighborhood of infinity using

$$\begin{aligned}(I-tA)^{-1} &= (I+tA)\Big(I + A^2 - (1+t^2)A^2\Big)^{-1} \\ &= -\frac{(A^{-2}+tA^{-1})}{1+t^2}\left(I - \frac{I+A^{-2}}{1+t^2}\right)^{-1} \\ &= -\sum_{p=0}^{\infty} \frac{(A^{-2}+tA^{-1})(I+A^{-2})^p}{(1+t^2)^{p+1}},\end{aligned} \tag{3.6}$$

and collect into terms $1/(1+t^2)^\ell$ and $t/(1+t^2)^\ell$, $\ell \geq 1$. After discarding all terms that are $\mathcal{O}\big(1/(1+t^2)^{\rho_0+1}\big)$ as $|t| \to \infty$, the expressions that remain are defined to be $d\tau_0(t;A,\Phi_2)$ and $\mathfrak{S}_0(t;A)$ on Δ_0. These definitions assure that the integrals in (3.2) and (3.3) converge weakly.

In the discrete parts of (3.2) and (3.3), we define

$$\begin{cases} \mathfrak{R}_0 = \operatorname*{Res}_{\lambda=0}\Big[(A-\lambda I)^{-1}\Phi_2 R_0(\lambda^{-1})\Phi_2^*(A^*-\lambda I)^{-1}\Big], \\ \widehat{\mathfrak{R}}_0 = -\operatorname*{Res}_{\lambda=0}\Big[A(A-\lambda I)^{-1}\Phi_2 R_0(\lambda^{-1})\lambda^{-1}\Big]. \end{cases} \tag{3.7}$$

For $j=1,\dots,r$, set

$$\begin{cases} \mathfrak{R}_j = \operatorname*{Res}_{\lambda=\alpha_j}\Big[(I-\lambda A)^{-1}\Phi_2 R_j\Big(\frac{1}{\lambda-\alpha_j}\Big)\Phi_2^*(I-\lambda A^*)^{-1}\Big], \\ \widehat{\mathfrak{R}}_j = \operatorname*{Res}_{\lambda=\alpha_j}\Big[A(I-\lambda A)^{-1}\Phi_2 R_j\Big(\frac{1}{\lambda-\alpha_j}\Big)\Big]. \end{cases} \tag{3.8}$$

For $k=1,\dots,s$, set

$$\begin{cases} \mathfrak{M}_{1k} = \operatorname*{Res}_{\lambda=\beta_k}\Big[(I-\lambda A)^{-1}\Phi_2 M_k\Big(\frac{1}{\lambda-\beta_k}\Big)\Phi_2^*(I-\lambda A^*)^{-1}\Big], \\ \widehat{\mathfrak{M}}_{1k} = \operatorname*{Res}_{\lambda=\beta_k}\Big[A(I-\lambda A)^{-1}\Phi_2 M_k\Big(\frac{1}{\lambda-\beta_k}\Big)\Big], \end{cases} \tag{3.9}$$

and

$$\begin{cases} \mathfrak{M}_{2k} = \operatorname*{Res}_{\lambda=\bar\beta_k}\Big[(I-\lambda A)^{-1}\Phi_2 M_k\Big(\frac{1}{\bar\lambda-\beta_k}\Big)^*\Phi_2^*(I-\lambda A^*)^{-1}\Big], \\ \widehat{\mathfrak{M}}_{2k} = \operatorname*{Res}_{\lambda=\bar\beta_k}\Big[A(I-\lambda A)^{-1}\Phi_2 M_k\Big(\frac{1}{\bar\lambda-\beta_k}\Big)^*\Big]. \end{cases} \tag{3.10}$$

Definition of S_v and $\Phi_{1,v}$ in Case 2. Now $0\in\sigma(A)$, $\rho_0=0$, $R_0(z)=C_0$ is constant, and (3.1) holds. In this case we define

$$\begin{cases} d\tau_0(t;A,\Phi_2)=0, \qquad \mathfrak{S}_0(t;A) = -\dfrac{tI}{1+t^2}, \\ \mathfrak{R}_0 = 0, \qquad \widehat{\mathfrak{R}}_0 = -\Phi_2 C_0. \end{cases} \tag{3.11}$$

All other terms are defined as in Case 1. The integral in (3.2) converges weakly by construction. The proof that the integral in (3.3) converges weakly in Case 2 is similar to an argument in [12, p. 2].

Theorem 3.2. *Under the Assumptions* 3.1, *on* Δ_j, $j=1,\dots,r$,

$$\begin{aligned} d\tau_j(t;A,\Phi_2) &= \sum_{\ell=0}^{2\rho_j-1}(t-\alpha_j)^\ell \sum_{\substack{p+q=\ell\\ p,q\ge 0}} A_p(\alpha_j)\,\Phi_2[d\tau(t)]\Phi_2^*\,A_q(\alpha_j)^* \\ &= -\operatorname*{Res}_{\lambda=\alpha_j}\Big[(I-\lambda A)^{-1}S_j(t,\lambda)\,\Phi_2[d\tau(t)]\,\Phi_2^*(I-\lambda A^*)^{-1}\Big], \end{aligned}$$

$$\begin{aligned}\mathfrak{S}_j(t;A) &= \sum_{p=0}^{2\rho_j-1} (t-\alpha_j)^p A_p(\alpha_j) A \\ &= -\operatorname*{Res}_{\lambda=\alpha_j}\Big[A\,(I-\lambda A)^{-1}\,S_j(t,\lambda)\Big].\end{aligned}$$

If $0 \notin \sigma(A)$, *then also on* Δ_0,

$$\begin{aligned}d\tau_0(t;A,\Phi_2) &= \sum_{\ell=0}^{\rho_0-1} \frac{1}{(1+t^2)^{\ell+1}} \sum_{p=0}^{\ell} \binom{\ell}{p} \sum_{\substack{j+k=2p+2\\ j,k\geq 1}} A^{-j}\Phi_2[d\tau(t)]\,\Phi_2^* A^{*-k} \\ &\quad + \sum_{\ell=0}^{\rho_0-1} \frac{t}{(1+t^2)^{\ell+2}} \sum_{p=1}^{\ell+1} \binom{\ell+1}{p} \sum_{\substack{j+k=2p+1\\ j,k\geq 1}} A^{-j}\Phi_2[d\tau(t)]\,\Phi_2^* A^{*-k} \\ &= \operatorname*{Res}_{\lambda=0}\Big[(A-\lambda I)^{-1} S_0(t,\lambda^{-1})\,\Phi_2[d\tau(t)]\,\Phi_2^*(A^*-\lambda I)^{-1}\Big], \\ \mathfrak{S}_0(t;A) &= -S_0(t,A^{-1}) \\ &= -\operatorname*{Res}_{\lambda=0}\Big[\,\lambda^{-1}\,A\,(A-\lambda I)^{-1} S_0(t,\lambda^{-1})\,\Big].\end{aligned}$$

In the case $0 \notin \sigma(A)$, the identity

$$\mathfrak{S}_j(t;A) = -S_j(t,A^{-1}) \tag{3.12}$$

holds for all $j=0,\dots,r$.

Theorem 3.3. *Under the Assumptions* 3.1, *if*

$$R_0(z) = \sum_{p=0}^{2\rho_0+1} R_{0p}\,z^p\,, \quad R_j(z) = \sum_{p=1}^{2\rho_j+1} R_{jp}\,z^p, \quad M_k(z) = \sum_{p=1}^{\sigma_k} M_{kp}\,z^p,$$

are the polynomials in the representation (2.1) *of* $v(z)$, *then*

$$\begin{aligned}\mathfrak{R}_j &= \sum_{p=1}^{2\rho_j+1} \sum_{\substack{\mu+\nu=p+1\\ \mu,\nu\geq 1}} A_{\mu-1}(\alpha_j)\Phi_2 R_{jp}\Phi_2^* A_{\nu-1}(\alpha_j)^*\,, \\ \widehat{\mathfrak{R}}_j &= \sum_{p=1}^{2\rho_j+1} A\,A_{p-1}(\alpha_j)\Phi_2 R_{jp}\,, \\ \mathfrak{M}_{1k} &= \sum_{p=1}^{\sigma_k} \sum_{\substack{\mu+\nu=p+1\\ \mu,\nu\geq 1}} A_{\mu-1}(\beta_k)\Phi_2 M_{kp}\Phi_2^* A_{\nu-1}(\bar\beta_k)^*\,, \\ \widehat{\mathfrak{M}}_{1k} &= \sum_{p=1}^{\sigma_k} A\,A_{p-1}(\beta_k)\Phi_2 M_{kp}\,,\end{aligned}$$

$$\mathfrak{M}_{2k} = \sum_{p=1}^{\sigma_k} \sum_{\substack{\mu+\nu=p+1 \\ \mu,\nu\geq 1}} A_{\mu-1}(\bar{\beta}_k)\Phi_2 M_{kp}^* \Phi_2^* A_{\nu-1}(\beta_k)^* \,,$$

$$\widehat{\mathfrak{M}}_{2k} = \sum_{p=1}^{\sigma_k} A\, A_{p-1}(\bar{\beta}_k)\Phi_2 M_{kp}^* \,,$$

$j = 1, \ldots, r$ and $k = 1, \ldots, s$. If $0 \notin \sigma(A)$, then also

$$\mathfrak{R}_0 = \sum_{p=1}^{2\rho_0+1} \sum_{\substack{\mu+\nu=p+1 \\ \mu,\nu\geq 1}} A^{-\mu}\Phi_2 R_{0p} \Phi_2^* A^{*-\nu} \,,$$

$$\widehat{\mathfrak{R}}_0 = - \sum_{p=0}^{2\rho_0+1} A^{-p}\Phi_2 R_{0p} \,.$$

When $\varkappa = 0$, the Kreĭn-Langer representation (2.1) reduces to the Nevanlinna representation (1.3); the Nevanlinna representation is unique, and the definitions of S_v and $\Phi_{1,v}$ given above reduce to the known forms (1.5) and (1.6). In contrast, the Kreĭn-Langer representation (2.1) in general is not unique. We show that the definitions of S_v and $\Phi_{1,v}$ do not depend on the choice of representation (2.1) for $v(z)$.

Theorem 3.4. *So long as the Assumptions* 3.1 *are met, the definitions of S_v and $\Phi_{1,v}$ do not depend on the choice of Kreĭn-Langer data* (2.2) *in the representation* (2.1).

We obtain a large class of examples of the operator identity (1.1).

Theorem 3.5. *Under the Assumptions* 3.1, *in both Case* 1 *and Case* 2, *the operator S_v is selfadjoint, $\varkappa_{S_v} < \infty$, and the operators $S = S_v$ and $\Phi_1 = \Phi_{1,v}$ together with the given operators A and Φ_2 satisfy* (1.1).

Definition 3.6. The **abstract interpolation problem** for a given operator identity (1.1) is to find all generalized Nevanlinna functions $v(z)$ such that

$$S = S_v \qquad \text{and} \qquad \Phi_1 = \Phi_{1,v} \,. \tag{3.13}$$

The motivating example is classical Pick-Nevanlinna interpolation, where we choose $\mathfrak{H} = \mathbf{C}^m \oplus \cdots \oplus \mathbf{C}^m$ with n summands.

Theorem 3.7. *Let $z_1, \ldots, z_n$ be distinct points in $\mathbf{C}_+$, and set*

$$A = \begin{bmatrix} z_1 I_m & 0 & \cdots & 0 \\ 0 & z_2 I_m & \cdots & 0 \\ & & \cdots & \\ 0 & 0 & \cdots & z_n I_m \end{bmatrix}, \qquad \Phi_2 = \begin{bmatrix} I_m \\ I_m \\ \vdots \\ I_m \end{bmatrix}. \tag{3.14}$$

Let $v(z)$ be a generalized Nevanlinna function. Set $w(z) = -v(1/\bar{z})^*$, and assume that the poles of $w(z)$ are disjoint from $z_1, \ldots, z_n$. Then the operators (3.2) *and* (3.3) *are given by*

$$S_v = \left[\frac{w(z_\mu) - w(z_\nu)^*}{z_\mu - \bar{z}_\nu} \right]_{\mu,\nu=1}^n, \qquad \Phi_{1,v} = -i \begin{bmatrix} w(z_1) \\ w(z_2) \\ \vdots \\ w(z_n) \end{bmatrix}. \tag{3.15}$$

For the same A and Φ_2, the operator identity (1.1) is satisfied with

$$S = \left[\frac{w_\mu - w_\nu^*}{z_\mu - \bar{z}_\nu} \right]_{\mu,\nu=1}^n, \qquad \Phi_1 = -i \begin{bmatrix} w_1 \\ w_2 \\ \vdots \\ w_n \end{bmatrix},$$

where $w_1, \ldots, w_n$ are given matrices. Solutions of the abstract interpolation problem in this case correspond to solutions of the classical Pick-Nevanlinna interpolation problem $w(z_\mu) = w_\mu$, $\mu = 1, \ldots, n$.

4. Generalization of the fundamental matrix inequality

In this section we introduce and study two kernels, $L_v(z,\zeta)$ and $L_{v,T}(z,\zeta)$ defined by (4.2) and (4.4)–(4.5) below, which are associated with any given operator identity (1.1) and generalized Nevanlinna function $v(z)$. These kernels are related to linear fractional transformations and the fundamental matrix inequality. The fundamental matrix inequality is used to solve classical interpolation problems by Kovalishina and Potapov [4] and Katsnelson [3], for example. In the definite case, the fundamental matrix inequality is adapted to the abstract interpolation problem in [12, Theorem 1.2.1]; it asserts that the kernel $L_v(z,\zeta)$ defined by (4.2) is nonnegative on the diagonal $z = \zeta$ for any solution $v(z)$. Our generalization, Theorem 4.5, asserts that in the indefinite case the two-variable kernel $L_v(z,\zeta)$ has a finite number of negative squares for any solution $v(z)$ of an abstract interpolation problem.

If A, S, Φ_1, Φ_2 are operators which satisfy (1.1), we shall also write

$$AS - SA^* = i\,\Pi J \Pi^*, \quad \Pi = \begin{bmatrix} \Phi_1 & \Phi_2 \end{bmatrix}, \quad J = \begin{bmatrix} 0 & I \\ I & 0 \end{bmatrix}. \tag{4.1}$$

For any generalized Nevanlinna function $v(z)$, we define a kernel

$$L_v(z,\zeta) = \begin{bmatrix} S & B_v(z) \\ B_v(\zeta)^* & C_v(z,\zeta) \end{bmatrix}, \tag{4.2}$$

where

$$\begin{cases} B_v(z) = (I - zA)^{-1}\big[\Phi_1 - i\Phi_2 v(z)\big]\,, \\ C_v(z,\zeta) = \dfrac{v(z) - v(\zeta)^*}{z - \bar\zeta}\,. \end{cases} \tag{4.3}$$

We also use the transformed kernel

$$L_{v,T}(z,\zeta) = \begin{bmatrix} S & -iB_{v,T}(z) \\ iB_{v,T}(\zeta)^* & C_{v,T}(z,\zeta) \end{bmatrix} \tag{4.4}$$

defined by

$$L_{v,T}(z,\zeta) = \begin{bmatrix} I & 0 \\ L_0(\bar\zeta) & L_2(\bar\zeta) \end{bmatrix} \begin{bmatrix} S & B_v(z) \\ B_v(\zeta)^* & C_v(z,\zeta) \end{bmatrix} \begin{bmatrix} I & L_0(\bar z)^* \\ 0 & L_2(\bar z)^* \end{bmatrix}, \tag{4.5}$$

where

$$L_0(z) = iA(I - zA)^{-1} \qquad \text{and} \qquad L_2(z) = (I - zA)^{-1}\Phi_2, \tag{4.6}$$

as in [12, p. 6]. By direct calculation,

$$B_{v,T}(z) - B_{v,T}(\zeta)^* = (z - \bar\zeta)C_{v,T}(z,\zeta),$$

and hence

$$B_{v,T}(z) = B_{v,T}(\bar z)^*, \qquad C_{v,T}(z,\zeta) = \frac{B_{v,T}(z) - B_{v,T}(\zeta)^*}{z - \bar\zeta}, \tag{4.7}$$

at all points z and ζ in $\mathbf{C}_+ \cup \mathbf{C}_-$ where the functions are defined.

The nondegenerate case (S invertible) is assumed in Theorems 4.1, 4.2, 4.4 and Definition 4.3.

Theorem 4.1. *Let A, S, Φ_1, Φ_2 be operators which satisfy* (1.1) *such that S is invertible. Define*

$$\mathfrak{A}(z) = I - iz\Pi^*(I - zA^*)^{-1}S^{-1}\Pi J \tag{4.8}$$

on the set $\Omega_{\mathfrak{A}}$ of all $z \in \mathbf{C}$ such that the inverse exists. For all $\bar z, \bar\zeta \in \Omega_{\mathfrak{A}}$,

$$\frac{J - \mathfrak{A}(\bar\zeta)J\mathfrak{A}(\bar z)^*}{i(\bar\zeta - z)} = \Pi^*(I - \zeta A^*)^{-1}S^{-1}(I - zA)^{-1}\Pi. \tag{4.9}$$

If $z, \bar z \in \Omega_{\mathfrak{A}}$, $\mathfrak{A}(z)$ is invertible and $\mathfrak{A}(z)^{-1} = J\mathfrak{A}(\bar z)^* J$.

In particular, $\mathfrak{A}(z)$ has invertible values except at isolated points of its domain.

Theorem 4.2. *Let A, S, Φ_1, Φ_2 be operators which satisfy* (1.1) *such that S is invertible. Given any generalized Nevanlinna function $v(z)$, set*

$$\begin{bmatrix} P(z) \\ Q(z) \end{bmatrix} = \mathfrak{A}(z)^{-1} \begin{bmatrix} -iv(z) \\ I \end{bmatrix}. \tag{4.10}$$

Then $P(\bar z)^*Q(z) + Q(\bar z)^*P(z) = 0$ and

$$L_v(z,\zeta) = \begin{bmatrix} I & 0 \\ B_v(\zeta)^*S^{-1} & I \end{bmatrix} \begin{bmatrix} S & 0 \\ 0 & D_v(z,\zeta) \end{bmatrix} \begin{bmatrix} I & S^{-1}B_v(z) \\ 0 & I \end{bmatrix}, \tag{4.11}$$

where

$$D_v(z,\zeta) = \frac{v(z) - v(\zeta)^*}{z - \bar{\zeta}} - B_v(\zeta)^* S^{-1} B_v(z) \tag{4.12}$$
$$= i\, \frac{P(\zeta)^* Q(z) + Q(\zeta)^* P(z)}{z - \bar{\zeta}}$$

at all points where the functions are defined.

The block entries $a(z), b(z), c(z), d(z)$ of $\mathfrak{A}(z)$ are used as coefficients of a class of linear fractional transformations.

Definition 4.3. Let A, S, Φ_1, Φ_2 be operators which satisfy (1.1) such that $\sigma(A)$ is a finite set, S is invertible, and $\varkappa_S < \infty$. Write

$$\mathfrak{A}(z) = \begin{bmatrix} a(z) & b(z) \\ c(z) & d(z) \end{bmatrix}, \tag{4.13}$$

where $a(z), b(z), c(z), d(z)$ are $m \times m$ matrix-valued functions. By $\mathbf{N}(\mathfrak{A})$ we mean the set of functions

$$v(z) = i\,[a(z)P(z) + b(z)Q(z)]\,[c(z)P(z) + d(z)Q(z)]^{-1}, \tag{4.14}$$

where $P(z)$ and $Q(z)$ are $m \times m$ matrix-valued functions which are analytic on $\mathbf{C}_+ \cup \mathbf{C}_-$ except at isolated points, such that

(i) $P(\bar{z})^* Q(z) + Q(\bar{z})^* P(z) \equiv 0$;
(ii) $c(z)P(z) + d(z)Q(z)$ is invertible except at isolated points;
(iii) the kernel

$$D_{P,Q}(z,\zeta) = i\, \frac{P(\zeta)^* Q(z) + Q(\zeta)^* P(z)}{z - \bar{\zeta}}$$

has a finite number $\varkappa_{P,Q}$ of negative squares.

Theorem 4.4. *Let A, S, Φ_1, Φ_2 be operators which satisfy (1.1) such that $\sigma(A)$ is a finite set, S is invertible, and $\varkappa_S < \infty$. If $v(z) \in \mathbf{N}(\mathfrak{A})$ and has the representation (4.14), then $v(z) = v(\bar{z})^*$ at all points where the functions are defined, and $v(z) \in \mathbf{N}_{\varkappa_v}$ where*

$$\varkappa_v \le \varkappa_{P,Q} + \varkappa_S = \varkappa_{L_v}. \tag{4.15}$$

In particular, $\varkappa_{L_v} = \varkappa_{P,Q} + \varkappa_S < \infty$. Moreover,

$$\frac{v(z) - v(\zeta)^*}{z - \bar{\zeta}} = K(\zeta)^{*-1} D_{P,Q}(z,\zeta) K(z)^{-1} + B_v(\zeta)^* S^{-1} B_v(z), \tag{4.16}$$

and

$$\frac{B_{v,T}(z) - B_{v,T}(\zeta)^*}{z - \bar{\zeta}} = L_2(\bar{\zeta}) K(\zeta)^{*-1} D_{P,Q}(z,\zeta) K(z)^{-1} L_2(\bar{z})^* + B_{v,T}(\zeta)^* S^{-1} B_{v,T}(z), \tag{4.17}$$

where $K(z) = [c(z)P(z) + d(z)Q(z)]^{-1}$ except at isolated points.

A consequence of Theorem 4.4 is that the kernel $L_v(z,\zeta)$ has a finite number of negative squares in a particular case in which S is invertible. Theorems 4.5, 4.6, and 4.7 below do not presume that S is invertible. Theorem 4.5 describes another case in which $L_v(z,\zeta)$ has a finite number of negative squares, but its proof does not give a simple description of the exact value of $\varkappa_{L_v}$ as in Theorem 4.4. Theorem 4.5 can be viewed as a generalization of the fundamental matrix inequality to the indefinite setting.

Theorem 4.5. *Let A, S, Φ_1, Φ_2 be operators which satisfy (1.1), where $S = S_v$ and $\Phi_1 = \Phi_{1,v}$ are defined by (3.2) and (3.3) for some generalized Schur function $v(z)$. Then in both Case 1 and Case 2 of the Assumptions 3.1, the kernel $L_v(z,\zeta)$ defined by (4.2) has a finite number of negative squares.*

The next two results establish companions to Theorem 4.5 that provide additional necessary conditions on solutions of the abstract interpolation problem particular to Case 1 and Case 2.

Theorem 4.6. *In Theorem 4.5, Case 1, the function $B_v(z)$ in (4.2) is analytic on $\mathbf{C}_+ \cup \mathbf{C}_-$ except perhaps for poles at the poles $\beta_k, \bar{\beta}_k$, $k = 1,\dots,s$, of $v(z)$. Hence $B_v(z)$ is analytic at every point λ such that $1/\lambda \in \sigma(A)$.*

Theorem 4.7. *In Theorem 4.5, Case 2, the functions $B_v(z)$ and $B_{v,T}(z)$ in (4.2) and (4.4) satisfy*

$$\|B_v(z)\| = \mathcal{O}(1) \tag{4.18}$$

and

$$\|B_{v,T}(z)\| = \mathcal{O}\left(\frac{1}{|z|}\right) \tag{4.19}$$

as $|z| \to \infty$ in any set $D_\delta = \{z : 0 < |\arg z| < \pi - \delta\}$ where $0 < \delta < \pi$.

5. Interpolation theorems

We now begin with an operator identity $AS - SA^* = i\left[\Phi_1\Phi_2^* + \Phi_2\Phi_1^*\right]$ such that $\sigma(A)$ is a finite set, S is invertible, and $\varkappa_S < \infty$. The abstract interpolation problem (3.13) is to characterize all generalized Nevanlinna functions $v(z)$ such that $S = S_v$ and $\Phi_1 = \Phi_{1,v}$. We shall see that such a function $v(z)$ belongs to the class $\mathbf{N}(\mathfrak{A})$ introduced in Definition 4.3. We give necessary and sufficient conditions on a function $v(z)$ in $\mathbf{N}(\mathfrak{A})$ that it is a solution of the abstract interpolation problem. They assert, roughly, that the necessary conditions on the functions

$$B_v(z) = (I - zA)^{-1}\left[\Phi_1 - i\Phi_2 v(z)\right]$$

and

$$\begin{aligned} B_{v,T}(z) &= \left[SA^* + iB_v(z)\Phi_2^*\right](I - zA^*)^{-1} \\ &= (I - zA)^{-1}\left[AS - i\Phi_2 B_v(\bar{z})^*\right], \end{aligned}$$

defined as in Section 4 are sufficient in the nondegenerate case, that is, when S is invertible.

Theorem 5.1 (Interpolation in Case 1). *Let A, S, Φ_1, Φ_2 be operators which satisfy* (1.1) *such that*

- *S is invertible, and $\varkappa_S < \infty$;*
- *$\sigma(A)$ is a finite set and $\sigma(A) \cap \sigma(A^*) = \emptyset$.*

(1) *Suppose that $v(z) \in \mathbf{N}(\mathfrak{A})$, and that*

(i) *$v(z)$ has at most a removable singularity at every nonreal number λ such that $1/\lambda \in \sigma(A)$;*

(ii) *$B_v(z)$ has at most a removable singularity at every nonreal number λ such that $1/\lambda \in \sigma(A)$.*

Then $v(z)$ is a generalized Nevanlinna function, the conditions of Assumptions 3.1, *Case* 1, *are met, and $S = S_v$ and $\Phi_1 = \Phi_{1,v}$.*

(2) *Conversely, if $S = S_v$ and $\Phi_1 = \Phi_{1,v}$ for some generalized Nevanlinna function $v(z)$ having a representation* (2.1) *which satisfies Assumptions* 3.1, *Case* 1, *then $v(z)$ belongs to $\mathbf{N}(\mathfrak{A})$ and satisfies conditions* (i) *and* (ii) *in* (1).

We note a sufficient condition that the technical conditions (i) and (ii) in Theorem 5.1(1) are satisfied.

Theorem 5.2. *Conditions* (i) *and* (ii) *in Theorem* 5.1(1) *hold if $v(z)$ has a representation* (4.14) *such that every point λ satisfying $1/\lambda \in \sigma(A)$ belongs to the domain of holomorphy of $P(z)$ and $Q(z)$ and $c(\lambda)P(\lambda) + d(\lambda)Q(\lambda)$ is invertible.*

Theorem 5.3 (Interpolation in Case 2). *Let A, S, Φ_1, Φ_2 be operators which satisfy* (1.1) *such that*

- *S is invertible, and $\varkappa_S < \infty$;*
- *$\sigma(A) = \{0\}$, and $\|(I - iyA)^{-1}f\| \neq \mathcal{O}(1)$ as $|y| \to \infty$ for every $f \neq 0$ in $\mathfrak{H}$.*

(1) *Let $v(z)$ belong to $\mathbf{N}(\mathfrak{A})$, and suppose that*

(i) *$v(iy)/y \to 0$ as $|y| \to \infty$;*

(ii) *for all h in $\mathfrak{H}$ and g in $\mathbf{C}^m$, $\langle B_v(iy)g, h\rangle = \mathcal{O}(1)$ as $|y| \to \infty$;*

(iii) *for all h and k in $\mathfrak{H}$, $\langle B_{v,T}(iy)h, k\rangle = \mathcal{O}(1/|y|)$ as $|y| \to \infty$.*

Then $v(z)$ is a generalized Nevanlinna function, the conditions of Assumptions 3.1, *Case* 2, *are met, and $S = S_v$ and $\Phi_1 = \Phi_{1,v}$.*

(2) *Conversely, if $S = S_v$ and $\Phi_1 = \Phi_{1,v}$ for some generalized Nevanlinna function $v(z)$ having a representation* (2.1) *which satisfies Assumptions* 3.1, *Case* 2, *then $v(z)$ belongs to $\mathbf{N}(\mathfrak{A})$ and satisfies conditions* (i)–(iii) *in* (1).

It can be shown that the conditions on A required in Theorem 5.3 are met, for example, for

$$A = \begin{bmatrix} 0 & 0 & \cdots & 0 & 0 \\ 1 & 0 & \cdots & 0 & 0 \\ 0 & 1 & \cdots & 0 & 0 \\ & & \cdots & & \\ 0 & 0 & \cdots & 1 & 0 \end{bmatrix} \qquad \text{on} \qquad \mathbf{C}^n,$$

and for

$$(Af)(x) = i \int_0^x f(t)\, dt \qquad \text{on} \qquad L^2_m(0, \ell), \tag{5.1}$$

for any positive integers m and n.

The hypotheses in Theorem 5.3 can be weakened when $\ker A = \{0\}$.

Theorem 5.4. *Theorem* 5.3 *remains true if the hypothesis on A is changed to read:*

- $\sigma(A) = \{0\}$, $\ker A = \{0\}$, *and* $y\,\|(I - iyA)^{-1}f\| \neq \mathcal{O}(1)$ *as* $|y| \to \infty$ *for every* $f \neq 0$ *in* $\mathfrak{H}$.

Example 5.5. In Theorem 5.3(1), the conditions (i)–(iii) are satisfied in an important concrete situation. Let $\mathfrak{H} = L^2_m(0, \ell)$ and $\mathfrak{G} = \mathbf{C}^m$ for some positive integer m and positive number ℓ. Let A be given by (5.1), and assume that S has the form (see [11])

$$(Sf)(x) = \frac{d}{dx} \int_0^\ell s(x - t) f(t)\, dt,$$

where $s(x)$ is a matrix-valued function such that $s(x) = -s(-x)^*$ on $(-\ell, \ell)$ and $s(x)g \in L^2_m(-\ell, \ell)$ for every $g \in \mathfrak{G}$. A particular case of such an operator is an integral operator of the form

$$(Sf)(x) = f(x) + \int_0^\ell k(x - t) f(t)\, dt,$$

where $k(x) = k(-x)^*$ is a bounded continuous matrix-valued function on $(-\ell, \ell)$. The operator identity (1.1) is satisfied with natural choices of operators Φ_1 and Φ_2. If the integro-differential operator S is bounded, invertible, and $\varkappa_S < \infty$, then the conditions (i)–(iii) in part (1) of Theorem 5.3 are satisfied for every function $v(z)$ in $\mathbf{N}(\mathfrak{A})$ given by (4.14) such that the kernel $D_{P,Q}(z, \zeta)$ is nonnegative. This generalizes a result of A. L. Sakhnovich [10]. The method of proof is interesting and applicable in other examples. Details will appear elsewhere.

6. Proofs of the theorems

We state some elementary lemmas that will be used in what follows. The proofs of the lemmas are straightforward, and details are omitted.

Lemma 6.1. *Define* $\binom{p}{k}$ *for* $p, k \geq 0$ *by* $(1 + x)^p = \sum_{k=0}^\infty \binom{p}{k} x^k$. *If* $p, k \geq 1$,

$$\binom{p}{k} = \binom{p-1}{k} + \binom{p-1}{k-1}. \tag{6.1}$$

For $q, s, u \geq 0$,

$$\sum_{\substack{p+r=q \\ p\geq 0,\ r\geq 0}} \binom{p}{s}\binom{r}{u} = \binom{q+1}{s+u+1}. \tag{6.2}$$

The residue formulas in Lemmas 6.2, 6.3, and 6.4 are deduced from elementary expansions, such as

$$(A-\lambda I)^{-1}=\sum_{\mu=0}^{\infty}A^{-\mu-1}\lambda^{\mu},\quad (I-\lambda A)^{-1}=\sum_{p=0}^{\infty}A_p(\lambda_0)(\lambda-\lambda_0)^p,$$

where $A_p(\lambda)=A^p(I-\lambda A)^{-p-1}$. We assume here that A and C are bounded operators on appropriate spaces for which the expressions are meaningful, λ_0 and z are complex numbers, and p is a nonnegative integer.

Lemma 6.2. *If $0\notin\sigma(A)$, then*

$$\operatorname*{Res}_{\lambda=0}\frac{(A-\lambda I)^{-1}C(A^*-\lambda I)^{-1}}{\lambda^p}=\begin{cases}\displaystyle\sum_{\substack{\mu+\nu=p+1\\ \mu,\nu\geq 1}}A^{-\mu}CA^{*-\nu}, & p\geq 1,\\ 0, & p=0,\end{cases}$$

$$\operatorname*{Res}_{\lambda=0}\frac{\lambda^{-1}A(A-\lambda I)^{-1}C}{\lambda^p}=A^{-p}C,$$

$$\operatorname*{Res}_{\lambda=0}\frac{(A-\lambda I)^{-1}}{1-\lambda z}\frac{C}{\lambda^p}=\begin{cases}\displaystyle\sum_{\substack{j+k=p+1\\ j,k\geq 1}}A^{-j}C\,z^{k-1}, & p\geq 1,\\ 0, & p=0.\end{cases}$$

Lemma 6.3. (1) *If $I-\lambda_0A$ and $I-\lambda_0A^*$ are invertible, then*

$$\operatorname*{Res}_{\lambda=\lambda_0}\frac{(I-\lambda A)^{-1}C(I-\lambda A^*)^{-1}}{(\lambda-\lambda_0)^{p+1}}=\sum_{\substack{\mu+\nu=p\\ \mu,\nu\geq 0}}A_\mu(\lambda_0)CA_\nu(\bar\lambda_0)^*.$$

(2) *If $I-\lambda_0A$ is invertible and $z\neq\lambda_0$, then*

$$\operatorname*{Res}_{\lambda=\lambda_0}\frac{(I-\lambda A)^{-1}C}{(\lambda-\lambda_0)^{p+1}}=A_p(\lambda_0)C,$$

$$\operatorname*{Res}_{\lambda=\lambda_0}\frac{(I-\lambda A)^{-1}}{z-\lambda}\frac{C}{(\lambda-\lambda_0)^{p+1}}=\sum_{\mu+\nu=p}\frac{A_\nu(\lambda_0)\,C}{(z-\lambda_0)^{\mu+1}}.$$

Lemma 6.4. (1) *If $P(z)$ is a polynomial with $P(0)=0$ and $z\neq\lambda_0$, then*

$$\operatorname*{Res}_{\lambda=\lambda_0}\frac{1}{z-\lambda}P\left(\frac{1}{\lambda-\lambda_0}\right)=P\left(\frac{1}{z-\lambda_0}\right).$$

(2) *If $P(z)$ is a polynomial, then*

$$\operatorname*{Res}_{\lambda=0}\frac{1}{1-\lambda z}\frac{1}{\lambda}P\left(\frac{1}{\lambda}\right)=P(z).$$

Proof of Theorem 3.2. In each case we prove the first of the two formulas; the residue versions then follow from Lemmas 6.2 and 6.3. Writing $dT = \Phi_2[d\tau(t)]\Phi_2^*$ and using (3.5), we obtain

$$
\begin{aligned}
(I-tA)^{-1}dT(I-tA^*)^{-1} &= \sum_{p=0}^{\infty}(t-\alpha_j)^p A_p(\alpha_j)\, dT \sum_{q=0}^{\infty}(t-\alpha_j)^q A_q(\alpha_j)^* \\
&\sim \sum_{\ell=0}^{2\rho_j-1}(t-\alpha_j)^\ell \sum_{\substack{p+q=\ell\\ p,q\ge 0}} A_p(\alpha_j)\, dT\, A_q(\alpha_j)^*\,,
\end{aligned}
$$

where "$\sim$" indicates that we have dropped terms that are $\mathcal{O}\big((t-\alpha_j)^{2\rho_j}\big)$ as $t\to\alpha_j$. This yields the formula for $d\tau_j(t;A,\Phi_2)$ on Δ_j, $j=1,\dots,r$. The formula for $\mathfrak{S}_j(t;A)$, $j=1,\dots,r$, is immediate from the definition.

Now assume that $0\notin\sigma(A)$. To derive the formula for $d\tau_0(t;A,\Phi_2)$, we use (3.6) in the form

$$
(I-tA)^{-1} = -\sum_{p=0}^{\infty}\frac{A^{-2}B^p + tA^{-1}B^p}{(1+t^2)^{p+1}}\,, \qquad B = I + A^{-2}.
$$

Let "$\sim$" now indicate that we are dropping terms that are $\mathcal{O}\big(1/(1+t^2)^{\rho_0+1}\big)$ as $|t|\to\infty$. Then on Δ_0,

$$
\begin{aligned}
(I-tA)^{-1}dT(I-tA^*)^{-1} &\sim \sum_{p=0}^{\rho_0-1}\frac{A^{-2}B^p + tA^{-1}B^p}{(1+t^2)^{p+1}}\, dT \sum_{q=0}^{\rho_0-1}\frac{A^{-2}B^q + tA^{-1}B^q}{(1+t^2)^{q+1}} \\
&\sim \sum_{\ell=0}^{\rho_0-1}\frac{t}{(1+t^2)^{\ell+2}}\,P_\ell + \sum_{\ell=0}^{\rho_0-1}\frac{1}{(1+t^2)^{\ell+1}}\,Q_\ell\,.
\end{aligned}
$$

For $\ell\ge 0$, by Lemma 6.1,

$$
\begin{aligned}
P_\ell &= \sum_{p+q=\ell}\Big(A^{-2}B^p dT B^{*q} A^{*-1} + A^{-1}B^p dT B^{*q} A^{*-2}\Big) \\
&= \sum_{p+q=\ell}\sum_{\mu,\nu=0}^{\ell}\binom{p}{\mu}\binom{q}{\nu} A^{-2\mu-2} dT A^{*-2\nu-1} \\
&\quad + \sum_{p+q=\ell}\sum_{\mu,\nu=0}^{\ell}\binom{p}{\mu}\binom{q}{\nu} A^{-2\mu-1} dT A^{*-2\nu-2} \\
&= \sum_{p+q=\ell}\binom{\ell+1}{\mu+\nu+1} A^{-2\mu-2} dT A^{*-2\nu-1} \\
&\quad + \sum_{p+q=\ell}\binom{\ell+1}{\mu+\nu+1} A^{-2\mu-1} dT A^{*-2\nu-2}
\end{aligned}
$$

$$= \sum_{p=1}^{\ell+1} \binom{\ell+1}{p} \sum_{\substack{j+k=2p+1 \\ j,k\geq 1}} A^{-j} dT A^{*-k}.$$

For $\ell \geq 1$, by (6.1) and Lemma 6.1,

$$\begin{aligned}
Q_\ell = &\sum_{p+q=\ell-1} A^{-2} B^p dT B^{*q} A^{*-2} + \sum_{p+q=\ell} A^{-1} B^p dT B^{*q} A^{*-1} \\
&- \sum_{p+q=\ell-1} A^{-1} B^p dT B^{*q} A^{*-1} \\
= &\sum_{p+q=\ell-1} \sum_{\mu,\nu=0}^{\ell} \binom{p}{\mu}\binom{q}{\nu} A^{-2\mu-2} dT A^{*-2\nu-2} \\
&+ \sum_{p+q=\ell} \sum_{\mu,\nu=0}^{\ell} \binom{p}{\mu}\binom{q}{\nu} A^{-2\mu-1} dT A^{*-2\nu-1} \\
&- \sum_{p+q=\ell-1} \sum_{\mu,\nu=0}^{\ell} \binom{p}{\mu}\binom{q}{\nu} A^{-2\mu-1} dT A^{*-2\nu-1} \\
= &\sum_{\mu,\nu=0}^{\ell} \binom{\ell}{\mu+\nu+1} A^{-2\mu-2} dT A^{*-2\nu-2} \\
&+ \sum_{\mu,\nu=0}^{\ell} \binom{\ell+1}{\mu+\nu+1} A^{-2\mu-1} dT A^{*-2\nu-1} \\
&- \sum_{\mu,\nu=0}^{\ell} \binom{\ell}{\mu+\nu+1} A^{-2\mu-1} dT A^{*-2\nu-1} \\
= &\sum_{\mu,\nu=0}^{\ell} \binom{\ell}{\mu+\nu+1} A^{-2\mu-2} dT A^{*-2\nu-2} \\
&+ \sum_{\mu,\nu=0}^{\ell} \binom{\ell}{\mu+\nu} A^{-2\mu-1} dT A^{*-2\nu-1} \\
= &\sum_{p=0}^{\ell} \binom{\ell}{p} \sum_{\substack{j+k=2p+2 \\ j,k\geq 1}} A^{-j} dT A^{*-k}.
\end{aligned}$$

The last expression agrees with $Q_0 = A^{-1} dT A^{*-1}$ when $\ell = 0$, and so we obtain the formula for $d\tau_0(t; A, \Phi_2)$. Finally, by (3.6),

$$A(I - At)^{-1} = -t \sum_{p=0}^{\infty} \frac{(I + A^{-2})^p}{(1+t^2)^{p+1}} - A^{-1} \sum_{p=0}^{\infty} \frac{(I + A^{-2})^p}{(1+t^2)^{p+1}}$$

$$\begin{aligned}&\sim -t\sum_{p=0}^{\rho_0}\frac{(I+A^{-2})^p}{(1+t^2)^{p+1}} - A^{-1}\sum_{p=0}^{\rho_0-1}\frac{(I+A^{-2})^p}{(1+t^2)^{p+1}}\\ &= -S_0(t,A^{-1}),\end{aligned}$$

on Δ_0, which gives the formula for $\mathfrak{S}_0(t;A)$. □

Proof of Theorem 3.3. Calculate the residues using Lemmas 6.2 and 6.3. □

Proof of Theorem 3.4. Suppose that we have two representations $v(z) = v_{\boldsymbol{\tau}}(z) = v_{\widetilde{\boldsymbol{\tau}}}(z)$. Write $S_{\boldsymbol{\tau}}, S_{\widetilde{\boldsymbol{\tau}}}, \Phi_{1,\boldsymbol{\tau}}, \Phi_{1,\widetilde{\boldsymbol{\tau}}}$ for the operators (3.2) and (3.3) in the two representations. We show that

$$S_{\boldsymbol{\tau}} = S_{\widetilde{\boldsymbol{\tau}}} \qquad \text{and} \qquad \Phi_{1,\boldsymbol{\tau}} = \Phi_{1,\widetilde{\boldsymbol{\tau}}}. \tag{6.3}$$

The parts of $\boldsymbol{\tau}$ and $\widetilde{\boldsymbol{\tau}}$ coming from the nonreal poles of $v(z)$ are the same, and so we can assume that there are no nonreal poles. Then by (2.2),

$$\begin{aligned}\boldsymbol{\tau} &= \{\tau(t);\, \alpha_1,\dots,\alpha_r;\, -;\, \rho_0,\dots,\rho_r;\, \Delta_0,\dots,\Delta_r;\, R_0(z),\dots,R_r(z);\, -\},\\ \widetilde{\boldsymbol{\tau}} &= \{\widetilde{\tau}(t);\, \widetilde{\alpha}_1,\dots,\widetilde{\alpha}_{\widetilde{r}};\, -;\, \widetilde{\rho}_0,\dots,\widetilde{\rho}_{\widetilde{r}};\, \widetilde{\Delta}_0,\dots,\widetilde{\Delta}_{\widetilde{r}};\, \widetilde{R}_0(z),\dots,\widetilde{R}_{\widetilde{r}}(z);\, -\}.\end{aligned}$$

By [8, Corollary 3.3], we can assume that $\widetilde{\tau}(t) = \tau(t)$ in the open intervals determined by the union of the points $\alpha_1,\dots,\alpha_r$ and $\widetilde{\alpha}_1,\dots,\widetilde{\alpha}_{\widetilde{r}}$. We check (6.3) in three special cases.

Special Case A: $\widetilde{\boldsymbol{\tau}}$ is obtained from $\boldsymbol{\tau}$ by replacing one of the intervals $\Delta_1,\dots,\Delta_r$ by a smaller interval.

For example, suppose $\alpha_1 \in \widetilde{\Delta}_1 \subseteq \Delta_1$. Write $\widetilde{\Delta}_0 = \Delta_0 \cup E$, where $\widetilde{\Delta}_1 = \Delta_1 \setminus E$. Ignoring terms in (2.1) that do not change, we may take

$$\begin{aligned}v_{\boldsymbol{\tau}}(z) &= \int_{\Delta_0}\left[\frac{1}{t-z} - S_0(t,z)\right] d\tau(t) + \int_{\Delta_1}\left[\frac{1}{t-z} - S_1(t,z)\right] d\tau(t)\\ &\quad + R_0(z) - R_1\left(\frac{1}{z-\alpha_1}\right),\\ v_{\widetilde{\boldsymbol{\tau}}}(z) &= \int_{\widetilde{\Delta}_0}\left[\frac{1}{t-z} - S_0(t,z)\right] d\tau(t) + \int_{\widetilde{\Delta}_1}\left[\frac{1}{t-z} - S_1(t,z)\right] d\tau(t)\\ &\quad + \widetilde{R}_0(z) - \widetilde{R}_1\left(\frac{1}{z-\alpha_1}\right),\end{aligned}$$

where because $v_{\boldsymbol{\tau}}(z) = v_{\widetilde{\boldsymbol{\tau}}}(z)$,

$$\begin{aligned}\widetilde{R}_0(z) &= R_0(z) + \int_E S_0(t,z)\, d\tau(t),\\ \widetilde{R}_1\left(\frac{1}{z-\alpha_1}\right) &= R_1\left(\frac{1}{z-\alpha_1}\right) + \int_E S_1(t,z)\, d\tau(t).\end{aligned}$$

Theorems 3.2 and 3.3 allow us to explicitly calculate the operators in (6.3) and verify the equalities. We omit the routine calculations.

Special Case B: $\widetilde{\boldsymbol{\tau}}$ is obtained from $\boldsymbol{\tau}$ by adding a new point α_{r+1}.

By Special Case A, it can be presumed that the new point α_{r+1} lies in Δ_0. Choose any order ρ_{r+1} and any open interval $\widetilde{\Delta}_1$ which contains α_{r+1} and is contained in the interior of Δ_0. Take

$$\begin{aligned} v_{\boldsymbol{\tau}}(z) &= \int_{\Delta_0} \left[\frac{1}{t-z} - S_0(t,z) \right] d\tau(t) + R_0(z), \\ v_{\widetilde{\boldsymbol{\tau}}}(z) &= \int_{\widetilde{\Delta}_0} \left[\frac{1}{t-z} - S_0(t,z) \right] d\tau(t) + \int_{\Delta_{r+1}} \left[\frac{1}{t-z} - S_{r+1}(t,z) \right] d\tau(t) \\ &\qquad + \widetilde{R}_0(z) + \widetilde{R}_{r+1}\left(\frac{1}{z-\alpha_{r+1}} \right), \end{aligned}$$

where $\widetilde{\Delta}_0 = \Delta_0 \setminus \widetilde{\Delta}_1$ and

$$\begin{aligned} \widetilde{R}_0(z) &= R_0(z) - \int_{\Delta_{r+1}} S_0(t,z)\, d\tau(t), \\ \widetilde{R}_{r+1}\left(\frac{1}{z-\alpha_{r+1}} \right) &= - \int_{\Delta_{r+1}} S_{r+1}(t,z)\, d\tau(t). \end{aligned}$$

The identities (6.3) are again verified using Theorems 3.2 and 3.3.

From the first two special cases, it may be presumed that

$$\widetilde{\boldsymbol{\tau}} = \{\tau(t);\, \alpha_1, \dots, \alpha_r; -;\, \widetilde{\rho}_0, \dots, \widetilde{\rho}_r;\, \Delta_0, \dots, \Delta_r;\, \widetilde{R}_0(z), \dots, \widetilde{R}_{\widetilde{r}}(z); -\}.$$

To complete the proof, it remains to bring the orders ρ_j and $\widetilde{\rho}_j$, $j = 0, \dots, r$, to the same values; it then follows that $R_0(z) = \widetilde{R}_0(z), \dots, R_r(z) = \widetilde{R}_r(z)$ and $\boldsymbol{\tau} = \widetilde{\boldsymbol{\tau}}$. Thus the proof is completed with one more special case.

Special Case C: $\widetilde{\boldsymbol{\tau}}$ is obtained from $\boldsymbol{\tau}$ by replacing one of the integers $\rho_0, \dots, \rho_r$ by a larger value.

For example, suppose that $\widetilde{\rho}_1 > \rho_1$ and

$$\begin{aligned} v_{\boldsymbol{\tau}}(z) &= \int_{\Delta_1} \left[\frac{1}{t-z} - S_1(t,z) \right] d\tau(t), \\ v_{\widetilde{\boldsymbol{\tau}}}(z) &= \int_{\Delta_1} \left[\frac{1}{t-z} - \widetilde{S}_1(t,z) \right] d\tau(t) - \widetilde{R}_1\left(\frac{1}{z-\alpha_1} \right), \end{aligned}$$

where $\widetilde{S}_1(t,z)$ is given by (2.5) with ρ_1 replaced by $\widetilde{\rho}_1$ and

$$\widetilde{R}_1\left(\frac{1}{z-\alpha_1} \right) = - \int_{\Delta_0} \left[\widetilde{S}_1(t,z) - S_1(t,z) \right] d\tau(t).$$

We verify (6.3) as before using Theorems 3.2 and 3.3. It remains to treat the possibility $\widetilde{\rho}_0 > \rho_0$ in Case 1 of Assumptions 3.1 (necessarily $\rho_0 = 0$ in Case 2). This is handled similarly. □

Proof of Theorem 3.5. We assume Case 1 in the proof. Case 2 is handled with minor modifications. The selfadjointness of $S = S_v$ follows from the explicit formulas in Theorems 3.2 and 3.3. The main problem is to verify (1.1) when S and Φ_1 are corresponding parts of (3.2) and (3.3). Suppose first that

$$S = \int_{\Delta_j} \left\{ (I - At)^{-1} \Phi_2 [d\tau(t)] \Phi_2^* \, (I - A^* t)^{-1} - d\tau_j(t; A, \Phi_2) \right\}, \tag{6.4}$$

$$\Phi_1 = -i \int_{\Delta_j} \left\{ A(I - At)^{-1} - \mathfrak{S}_j(t; A) \right\} \Phi_2 \, [d\tau(t)] \,, \tag{6.5}$$

$j = 0, \dots, r$. Writing $dT = \Phi_2[d\tau(t)]\Phi_2^*$, we obtain

$$\begin{aligned} AS - SA^* &= \int_{\Delta_j} \Big\{ \Big[A(I - At)^{-1} dT - dT(I - A^*t)^{-1} A^* \Big] \\ &\qquad - \Big[A \, d\tau_j(t; A, \Phi_2) - d\tau_j(t; A, \Phi_2) \, A^* \Big] \Big\}, \\ i \left[\Phi_1 \Phi_2^* + \Phi_2 \Phi_1^* \right] &= \int_{\Delta_j} \Big\{ \Big[A(I - At)^{-1} dT - dT(I - A^*t)^{-1} A^* \Big] \\ &\qquad - \Big[\mathfrak{S}_j(t, A) dT - dT \mathfrak{S}_j^*(t, A) \Big] \Big\}. \end{aligned}$$

We show that for all $j = 0, 1, \dots, r$,

$$A \, d\tau_j(t; A, \Phi_2) - d\tau_j(t; A, \Phi_2) \, A^* = \mathfrak{S}_j(t, A) dT - dT \mathfrak{S}_j^*(t, A). \tag{6.6}$$

First assume $j = 1, \dots, r$ and $\alpha_j \neq 0$. Set $B = I + \alpha_j A(I - \alpha_j A)^{-1}$. By Theorem 3.2 and the operator identity $\sum_{j+k=n} \left(L^{j+1} X R^k - L^j X R^{k+1} \right) = L^{n+1} X - X R^{n+1}$, which holds for all $n \geq 0$,

$$\begin{aligned} & A \, d\tau_j(t; A, \Phi_2) - d\tau_j(t; A, \Phi_2) \, A^* \\ &= \sum_{\ell=0}^{2\rho_j - 1} (t - \alpha_j)^\ell \sum_{\substack{p+q=\ell \\ p,q \geq 0}} \frac{(B - I)^{p+1}}{\alpha_j^{p+1}} \, dT \, \frac{(B^* - I)^q}{\alpha_j^q} \, [(B^* - I) + I] \\ &\qquad - \sum_{\ell=0}^{2\rho_j - 1} (t - \alpha_j)^\ell \sum_{\substack{p+q=\ell \\ p,q \geq 0}} [(B - I) + I] \, \frac{(B - I)^p}{\alpha_j^p} \, dT \, \frac{(B^* - I)^{q+1}}{\alpha_j^{q+1}} \\ &= \sum_{\ell=0}^{2\rho_j - 1} \frac{(t - \alpha_j)^\ell}{\alpha^{\ell+1}} \sum_{\substack{p+q=\ell \\ p,q \geq 0}} \Big[(B - I)^{p+1} \, dT \, (B^* - I)^q \\ &\qquad\qquad - (B - I)^p \, dT \, (B^* - I)^{q+1} \Big] \\ &= \sum_{\ell=0}^{2\rho_j - 1} \frac{(t - \alpha_j)^\ell}{\alpha^{\ell+1}} \Big[(B - I)^{\ell+1} dT - dT (B^* - I)^{\ell+1} \Big] \\ &= \mathfrak{S}_j(t, A) dT - dT \mathfrak{S}_j^*(t, A). \end{aligned}$$

The case $\alpha_j = 0$ follows by continuity. Thus (6.6) holds for $j = 1, \dots, r$. A similar argument verifies (6.6) for $j = 0$. Hence (1.1) holds when S and Φ_1 are defined by (6.4) and (6.5).

The discrete parts in (3.2) and (3.3) come in two types by Theorem 3.3. One type is

$$S = \sum_{\substack{\mu+\nu=p+1\\ \mu,\nu\geq 1}} A^{-\mu}\Phi_2 X\Phi_2^* A^{*-\nu}, \qquad \Phi_1 = i\, A^{-p}\Phi_2 X,$$

where $p \geq 0$ and $X = X^*$. If $p = 0$, then $S = 0$ and $\Phi_1 = i\Phi_2 X$ and both sides of (1.1) reduce to zero. For $p \geq 1$,

$$AS - SA^* = \sum_{\substack{\mu+\nu=p+1\\ \mu,\nu\geq 1}} \left(A^{-\mu+1}\Phi_2 X\Phi_2^* A^{*-\nu} - A^{-\mu}\Phi_2 X\Phi_2^* A^{*-\nu+1}\right)$$
$$= \Phi_2 X\Phi_2^* A^{*-p} - A^{-p}\Phi_2 X\Phi_2^*. = i\left[\Phi_1\Phi_2^* + \Phi_2\Phi_1^*\right],$$

which verifies (1.1). The other type has the form

$$S = \sum_{\substack{\mu+\nu=p+1\\ \mu,\nu\geq 1}} \Big[A^{\mu-1}(I-\lambda A)^{-\mu}\Phi_2 X\Phi_2^*(I-\lambda A^*)^{-\nu}A^{*\nu-1}$$
$$+ A^{\mu-1}(I-\bar\lambda A)^{-\mu}\Phi_2 X^*\Phi_2^*(I-\bar\lambda A^*)^{-\nu}A^{*\nu-1}\Big],$$
$$\Phi_1 = -i\left[A^p(I-\lambda A)^{-p}\Phi_2 X + A^p(I-\bar\lambda A)^{-p}\Phi_2 X^*\right],$$

where $p \geq 1$ and X is not necessarily selfadjoint. We verify (1.1) in this case in a similar way.

We omit a proof that $\varkappa_{S_v} < \infty$ here because a more general result will be proved later (independently) in Theorem 4.5. □

Proof of Theorem 3.7. The Assumptions 3.1, Case 1, are met. It is sufficient to prove the formula for $\Phi_{1,v}$. For if this is known and

$$\widetilde{S} = \left[\frac{w(z_\mu) - w(z_\nu)^*}{z_\mu - \bar z_\nu}\right]_{\mu,\nu=1}^n,$$

then $A\widetilde{S} - \widetilde{S}A^* = i\left[\Phi_{1,v}\Phi_2^* + \Phi_2\Phi_{1,v}^*\right] = AS_v - S_vA^*$ by Theorem 3.5. Hence $A(\widetilde{S} - S_v) - (\widetilde{S} - S_v)A^* = 0$. Since A and A^* have disjoint spectra, $\widetilde{S} - S_v = 0$, and the formula for S_v follows. We prove the formula for $\Phi_{1,v}$ for corresponding parts of (3.3) and

$$w(z) = \sum_{j=0}^r \int_{\Delta_j}\left[\frac{z}{1-tz} + S_j(t,z^{-1})\right]d\tau(t) - R_0(z^{-1})$$
$$+ \sum_{j=1}^r R_j\left(\frac{z}{1-\alpha_j z}\right) + \sum_{k=1}^s \left[M_k\left(\frac{\bar z}{1-\beta_k\bar z}\right)^* + M_k\left(\frac{z}{1-\beta_k z}\right)\right].$$

Suppose first that

$$i\Phi_{1,v} = \int_{-\infty}^{\infty} \left\{ A(I-At)^{-1} - \sum_{j=0}^{r} \mathfrak{S}_j(t;A) \right\} \Phi_2\,[d\tau(t)],$$

$$w(z) = \int_{-\infty}^{\infty} \left\{ \frac{z}{1-zt} + \sum_{j=0}^{r} S_j(t,z^{-1}) \right\} d\tau(t).$$

For each $\mu = 1,\dots,n$ let P_μ be the projection of $\mathfrak{H} = \mathbf{C}^m \oplus \cdots \oplus \mathbf{C}^m$ onto the μ-th component. Then by (3.12),

$$P_\mu(i\Phi_{1,v}) = \int_{-\infty}^{\infty} \left\{ \frac{z_\mu}{1-z_\mu t} + \sum_{j=0}^{r} S_j(t,z_\mu^{-1}) \right\} d\tau(t) = w(z_\mu),$$

yielding the formula for $\Phi_{1,v}$. Next let

$$i\Phi_{1,v} = \widehat{\mathfrak{R}}_0\,, \qquad w(z) = -R_0(z^{-1})\,.$$

If $R_0(z) = \sum_{p=0}^{2\rho_0+1} C_p z^p$, then by (3.7),

$$\begin{aligned} P_\mu\big(i\Phi_{1,v}\big) &= -\operatorname*{Res}_{\lambda=0}\, P_\mu A(A-\lambda I)^{-1}\Phi_2 R_0(\lambda^{-1})\lambda^{-1} \\ &= -\operatorname*{Res}_{\lambda=0} \sum_{n=0}^{\infty} z_\mu^{-n}\lambda^n \sum_{p=0}^{2\rho_0+1} C_p\lambda^{-p-1} = -\sum_{p=0}^{2\rho_0+1} C_p z_\mu^{-p} = w(z_\mu)\,, \end{aligned}$$

as required. The remaining cases

$$i\Phi_{1,v} = \widehat{\mathfrak{R}}_j\,, \qquad w(z) = R_j\left(\frac{z}{1-\alpha_j z}\right), \qquad \text{and}$$

$$i\Phi_{1,v} = \widehat{\mathfrak{M}}_{1k} + \widehat{\mathfrak{M}}_{2k}\,, \qquad w(z) = M_k\left(\frac{\bar z}{1-\beta_k \bar z}\right)^* + M_k\left(\frac{z}{1-\beta_k z}\right),$$

are handled similarly. □

Proof of Theorem 4.1. We prove (3.11) as in the definite case [12]:

$$\begin{aligned} &\mathfrak{A}(\bar\zeta)J\mathfrak{A}(\bar z)^* - J \\ &\quad= \big[I - i\bar\zeta\Pi^*(I-\bar\zeta A^*)^{-1}S^{-1}\Pi J\big]J\big[I + izJ\Pi^* S^{-1}(I-zA)^{-1}\Pi\big] - J \\ &\quad= izJ\Pi^*S^{-1}(I-zA)^{-1}\Pi - i\bar\zeta\Pi^*(I-\bar\zeta A^*)^{-1}S^{-1}\Pi \\ &\qquad\quad + \bar\zeta z\Pi^*(I-\bar\zeta A^*)^{-1}S^{-1}\frac{AS-SA^*}{i}S^{-1}(I-zA)^{-1}\Pi \\ &\quad= izJ\Pi^*S^{-1}(I-zA)^{-1}\Pi - i\bar\zeta\Pi^*(I-\bar\zeta A^*)^{-1}S^{-1}\Pi \\ &\qquad\quad - i\bar\zeta\Pi^*(I-\bar\zeta A^*)^{-1}S^{-1}(zA-I+I)(I-zA)^{-1}\Pi \\ &\qquad\quad + iz\Pi^*(I-\bar\zeta A^*)^{-1}(\bar\zeta A^*-I+I)S^{-1}(I-zA)^{-1}\Pi \\ &\quad= -i(\bar\zeta - z)\Pi^*(I-\bar\zeta A^*)^{-1}S^{-1}(I-zA)^{-1}\Pi. \end{aligned}$$

To obtain the last statement, apply (3.11) with $\zeta = \bar z$. □

Proof of Theorem 4.2. Setting $\Phi_v(z) = \begin{bmatrix} -iv(z) \\ I \end{bmatrix}$, we obtain

$$\frac{v(z) - v(\zeta)^*}{z - \bar{\zeta}} = i\,\frac{\Phi_v(\zeta)^* J \Phi_v(z)}{z - \bar{\zeta}}, \qquad B_v(z) = (I - zA)^{-1}\Pi J \Phi_v(z),$$

by (4.3). Hence by Theorem 4.1,

$$\begin{aligned} D_v(z,\zeta) &= i\,\frac{\Phi_v(\zeta)^* J \Phi_v(z)}{z - \bar{\zeta}} - \Phi_v(\zeta)^* J \Pi^* (I - \bar{\zeta}A^*)^{-1} S^{-1} (I - zA)^{-1} \Pi J \Phi_v(z) \\ &= \Phi_v(\zeta)^* \,\frac{J\mathfrak{A}(\bar{\zeta}) J \mathfrak{A}(\bar{z})^* J}{i(\bar{\zeta} - z)}\, \Phi_v(z) \\ &= \Phi_v(\zeta)^* \,\frac{\mathfrak{A}(\zeta)^{*-1} J \mathfrak{A}(z)^{-1}}{i(\bar{\zeta} - z)}\, \Phi_v(z) \\ &= i\,\frac{P(\zeta)^* Q(z) + Q(\zeta)^* P(z)}{z - \bar{\zeta}}, \end{aligned}$$

which is (4.12). Again by Theorem 4.1,

$$P(\bar{z})^* Q(z) + Q(\bar{z})^* P(z) = \begin{bmatrix} P(\bar{z})^* & Q(\bar{z})^* \end{bmatrix} J \begin{bmatrix} P(z) \\ Q(z) \end{bmatrix}$$

$$= \begin{bmatrix} iv(\bar{z})^* & I \end{bmatrix} \mathfrak{A}(\bar{z})^{*-1} J \mathfrak{A}(z)^{-1} \begin{bmatrix} -iv(z) \\ I \end{bmatrix} = \begin{bmatrix} iv(\bar{z})^* & I \end{bmatrix} J \begin{bmatrix} -iv(z) \\ I \end{bmatrix} = 0,$$

and the result follows. □

Proof of Theorem 4.4. The function $v(z)$ is defined and analytic on $\mathbf{C}_+ \cup \mathbf{C}_-$ except at isolated points. Set

$$\begin{bmatrix} H(z) \\ K(z) \end{bmatrix} = \mathfrak{A}(z) \begin{bmatrix} P(z) \\ Q(z) \end{bmatrix} = \begin{bmatrix} a(z)P(z) + b(z)Q(z) \\ c(z)P(z) + d(z)Q(z) \end{bmatrix}.$$

Then $v(z) = iH(z)K(z)^{-1}$,

$$\mathfrak{A}(z) \begin{bmatrix} P(z) \\ Q(z) \end{bmatrix} = \begin{bmatrix} -iv(z) \\ I \end{bmatrix} K(z), \tag{6.7}$$

and so

$$\begin{bmatrix} P(z) \\ Q(z) \end{bmatrix} K(z)^{-1} = \mathfrak{A}(z)^{-1} \begin{bmatrix} -iv(z) \\ I \end{bmatrix} = J\mathfrak{A}(\bar{z})^* \begin{bmatrix} I \\ -iv(z) \end{bmatrix}$$

on $\mathbf{C}_+ \cup \mathbf{C}_-$ except at isolated points. We obtain

$$\begin{aligned} &\begin{bmatrix} I & iv(\zeta)^* \end{bmatrix} \mathfrak{A}(\bar{\zeta}) J J J \mathfrak{A}(\bar{z})^* \begin{bmatrix} I \\ -iv(z) \end{bmatrix} \\ &\qquad = K(\zeta)^{*-1} \begin{bmatrix} P(\zeta)^* & Q(\zeta)^* \end{bmatrix} J \begin{bmatrix} P(z) \\ Q(z) \end{bmatrix} K(z)^{-1} \\ &\qquad = K(\zeta)^{*-1} \left[P(\zeta)^* Q(z) + Q(\zeta)^* P(z) \right] K(z)^{-1}. \end{aligned}$$

On the other hand, by (4.9),

$$\begin{aligned}&[I \quad iv(\zeta)^*]J\begin{bmatrix} I \\ -iv(z)\end{bmatrix}\\&\quad = [I \quad iv(\zeta)^*]\,\mathfrak{A}(\bar\zeta)J\mathfrak{A}(\bar z)^*\begin{bmatrix} I \\ -iv(z)\end{bmatrix}\\&\qquad + i(\bar\zeta - z)\,[I \quad iv(\zeta)^*]\,\Pi^*(I-\bar\zeta A^*)^{-1}S^{-1}(I-zA)^{-1}\Pi\begin{bmatrix} I \\ -iv(z)\end{bmatrix}.\end{aligned}$$

It follows that

$$\frac{v(z)-v(\zeta)^*}{z-\bar\zeta} = K(\zeta)^{*-1}\, i\,\frac{P(\zeta)^*Q(z)+Q(\zeta)^*P(z)}{z-\bar\zeta}K(z)^{-1} + \Lambda(\zeta)^*S^{-1}\Lambda(z),$$

where

$$\Lambda(z) = (I-zA)^{-1}\Pi\begin{bmatrix} I \\ -iv(z)\end{bmatrix}.$$

To see that $v(z) = v(\bar z)^*$, multiply the last identity by $z-\bar\zeta$, then take $\zeta = \bar z$ and use the condition (i) in Definition 4.3. By our assumption that $\varkappa_S < \infty$ and condition (iii) in Definition 4.3, we deduce that $v(z) \in \mathbf{N}_{\varkappa_v}$, where $\varkappa_v \le \varkappa_{P,Q}+\varkappa_S$. The equality $\varkappa_{P,Q}+\varkappa_S = \varkappa_{L_v}$ follows from (4.11).

By (4.3), $\Lambda(z) = B_v(z)$ and (4.16) follows. The identity (4.17) is proved by a straightforward algebraic calculation, which we omit. □

Proof of Theorem 4.5. First assume Case 1: $0 \notin \sigma(A)$. In the proof, for brevity we drop the subscript v and write (4.2) more simply as

$$L(z,\zeta) = \begin{bmatrix} S & B(z) \\ B(\zeta)^* & C(z,\zeta)\end{bmatrix}.$$

It is sufficient to show that $\varkappa_L < \infty$ when $L(z,\zeta)$ is calculated from corresponding parts of (2.1). We distinguish five subcases (a)–(e).

Case 1: (a) Fix $j = 1,\dots,r$, and let

$$\begin{aligned} v(z) &= \int_{\Delta_j}\left\{\frac{1}{t-z} - S_j(t,z)\right\}d\tau(t)\,,\\ S &= \int_{\Delta_j}\left\{(I-At)^{-1}\Phi_2[d\tau(t)]\Phi_2^*\,(I-A^*t)^{-1} - d\tau_j(t;A,\Phi_2)\right\},\\ \Phi_1 &= -i\int_{\Delta_j}\left\{A(I-At)^{-1} - \mathfrak{S}_j(t;A)\right\}\Phi_2\,[d\tau(t)]\,.\end{aligned}$$

To calculate $L(z,\zeta)$ in this subcase, first use Theorem 3.2 to obtain

$$\begin{aligned} S = \int_{\Delta_j}\Bigg\{&(I-At)^{-1}\Phi_2[d\tau(t)]\Phi_2^*\,(I-A^*t)^{-1}\\ &- \sum_{\ell=0}^{2\rho_j-1}(t-\alpha_j)^\ell\sum_{p+q=\ell}A_p(\alpha_j)\Phi_2[d\tau(t)]\Phi_2^*A_q(\alpha_j)^*\Bigg\}.\end{aligned}$$

Using Lemmas 6.4 and 6.3, we get

$$\begin{aligned} B(z) &= (I-zA)^{-1}\left[\Phi_{1,v} - i\Phi_2 v(z)\right] \\ &= -i\int_{\Delta_j}(I-zA)^{-1}\Big\{A(I-At)^{-1} + \frac{I}{t-z} \\ &\qquad\qquad - \mathfrak{S}_j(t;A) - S_j(t,z)I\Big\}\,\Phi_2\,[d\tau(t)] \\ &= -i\int_{\Delta_j}\Big\{\frac{(I-At)^{-1}-(I-Az)^{-1}}{t-z} + \frac{(I-Az)^{-1}}{t-z} \\ &\qquad + \operatorname*{Res}_{\lambda=\alpha_j}\left[\frac{(I-A\lambda)^{-1}-(I-Az)^{-1}}{\lambda-z}S(t,\lambda)\right] \\ &\qquad - (I-Az)^{-1}S_j(t,z)\Big\}\,\Phi_2\,[d\tau(t)] \\ &= -i\int_{\Delta_j}\left\{\frac{(I-tA)^{-1}}{t-z} + \operatorname*{Res}_{\lambda=\alpha_j}\left[\frac{(I-\lambda A)^{-1}}{\lambda-z}S_j(t,\lambda)\right]\right\}\Phi_2\,[d\tau(t)] \\ &= -i\int_{\Delta_j}\left\{\frac{(I-tA)^{-1}}{t-z} + \sum_{\ell=0}^{2\rho_j-1}(t-\alpha_j)^\ell\sum_{p+q=\ell}\frac{A_q(\alpha_j)}{(z-\alpha_j)^{p+1}}\right\}\Phi_2\,[d\tau(t)]\,. \end{aligned}$$

We obtain

$$\begin{aligned} C(z,\zeta) = \int_{\Delta_j}\Big\{&\frac{1}{(t-z)(t-\bar\zeta)} \\ &- \sum_{\ell=0}^{2\rho_j-1}(t-\alpha_j)^\ell\sum_{p+q=\ell}\frac{1}{(z-\alpha_j)^{q+1}(\bar\zeta-\alpha_j)^{p+1}}\Big\}\,d\tau(t) \end{aligned}$$

by means of the identity

$$\begin{aligned} \frac{S_j(t,z)-\overline{S_j(t,\zeta)}}{z-\bar\zeta} &= \sum_{\ell=0}^{2\rho_j-1}\frac{(t-\alpha_j)^\ell}{(z-\alpha_j)^{\ell+1}(\bar\zeta-\alpha_j)^{\ell+1}}\,\frac{(z-\alpha_j)^{\ell+1}-(\bar\zeta-\alpha_j)^{\ell+1}}{(z-\alpha_j)-(\bar\zeta-\alpha_j)} \\ &= \sum_{\ell=0}^{2\rho_j-1}(t-\alpha_j)^\ell\sum_{p+q=\ell}\frac{1}{(z-\alpha_j)^{q+1}(\bar\zeta-\alpha_j)^{p+1}}\,. \end{aligned}$$

Thus

$$\begin{aligned} L(z,\zeta) = \int_{\Delta_j}\Bigg\{&\begin{bmatrix}(I-At)^{-1}\Phi_2 \\ \dfrac{-iI}{\bar\zeta-t}\end{bmatrix} d\tau(t)\begin{bmatrix}\Phi_2^*(I-A^*t)^{-1} & \dfrac{iI}{z-t}\end{bmatrix} \\ &- \sum_{\ell=0}^{2\rho_j-1}(t-\alpha_j)^\ell\sum_{p+q=\ell}\begin{bmatrix}A_p(\alpha_j)\Phi_2 \\ \dfrac{-iI}{(\bar\zeta-\alpha_j)^{p+1}}\end{bmatrix} d\tau(t)\begin{bmatrix}\Phi_2 A_q(\alpha_j)^* & \dfrac{iI}{(z-\alpha_j)^{q+1}}\end{bmatrix}\Bigg\}. \end{aligned}$$

To see that $\varkappa_L < \infty$, approximate $\tau(t)$ by functions $\tau_\varepsilon(t)$ that are constant in intervals $(\alpha_j - \varepsilon, \alpha_j)$ and $(\alpha_j, \alpha_j + \varepsilon)$ and define $L_\varepsilon(z, \zeta)$ by the same expression with $\tau(t)$ replaced by $\tau_\varepsilon(t)$. Then

$$\lim_{\varepsilon \downarrow 0} L_\varepsilon(z, \zeta) = L(z, \zeta)$$

pointwise. Since $\tau_\varepsilon(t)$ is constant to the left and right of α_j, the integrations in $L_\varepsilon(z, \zeta)$ can be carried out term by term. Then the first summand in $L_\varepsilon(z, \zeta)$ is nonnegative, and the number of negative squares in what remains has a finite bound independent of ε by the matrix identity

$$\sum_{\ell=0}^{\nu} \sum_{\substack{p+q=\ell+1 \\ p,q \geq 1}} X_p^* H_\ell Y_p = \begin{bmatrix} X_1 \\ X_2 \\ \cdots \\ X_\nu \end{bmatrix}^* \begin{bmatrix} H_1 & H_2 & \cdots & H_{\nu-1} & H_\nu \\ H_2 & H_3 & \cdots & H_\nu & 0 \\ & & \cdots & & \\ H_\nu & 0 & \cdots & 0 & 0 \end{bmatrix} \begin{bmatrix} Y_1 \\ Y_2 \\ \vdots \\ Y_\nu \end{bmatrix}. \tag{6.8}$$

Therefore $\varkappa_L < \infty$.

Case 1: (b) Next assume that

$$\begin{aligned} v(z) &= \int_{\Delta_0} \left\{ \frac{1}{t-z} - S_0(t, z) \right\} d\tau(t), \\ S &= \int_{\Delta_0} \left\{ (I - At)^{-1} \Phi_2 [d\tau(t)] \Phi_2^* (I - A^* t)^{-1} - d\tau_0(t; A, \Phi_2) \right\}, \\ \Phi_1 &= -i \int_{\Delta_0} \left\{ A(I - At)^{-1} - \mathfrak{S}_0(t; A) \right\} \Phi_2 \, [d\tau(t)]. \end{aligned}$$

By Theorem 3.2,

$$\begin{aligned} S = \int_{\Delta_0} \Bigg\{ & (I - At)^{-1} \Phi_2 [d\tau(t)] \Phi_2^* (I - A^* t)^{-1} \\ & - \sum_{\ell=0}^{\rho_0 - 1} \frac{t}{(1+t^2)^{\ell+2}} \sum_{p=1}^{\ell+1} \binom{\ell+1}{p} \sum_{\substack{j+k=2p+1 \\ j,k \geq 1}} A^{-j} \Phi_2 [d\tau(t)] \Phi_2^* A^{*-k} \\ & - \sum_{\ell=0}^{\rho_0 - 1} \frac{1}{(1+t^2)^{\ell+1}} \sum_{p=0}^{\ell} \binom{\ell}{p} \sum_{\substack{j+k=2p+2 \\ j,k \geq 1}} A^{-j} \Phi_2 [d\tau(t)] \Phi_2^* A^{*-k} \Bigg\}. \end{aligned}$$

Calculating as above, we get by (2.6) and Lemma 6.4,

$$\begin{aligned} B(z) &= (I - zA)^{-1} \big[\Phi_{1,v} - i \Phi_2 v(z) \big] \\ &= -i \int_{\Delta_0} \left\{ \frac{(I - tA)^{-1}}{t - z} + \operatorname*{Res}_{\lambda=0} \left[\frac{(A - \lambda I)^{-1}}{1 - \lambda z} \, S_0(t, \lambda^{-1}) \right] \right\} \Phi_2 \, [d\tau(t)] \end{aligned}$$

$$= -i\int_{\Delta_0}\Bigg\{\frac{(I-tA)^{-1}}{t-z} + \sum_{\ell=0}^{\rho_0-1}\frac{t}{(1+t^2)^{\ell+2}}\sum_{p=1}^{\ell+1}\binom{\ell+1}{p}\sum_{\substack{j+k=2p+1\\ j,k\geq 1}} A^{-j}z^{k-1}$$

$$+\sum_{\ell=0}^{\rho_0-1}\frac{1}{(1+t^2)^{\ell+1}}\sum_{p=0}^{\ell}\binom{\ell}{p}\sum_{\substack{j+k=2p+2\\ j,k\geq 1}} A^{-j}z^{k-1}\Bigg\}\Phi_2\,[d\tau(t)]\,.$$

For $C(z,\zeta) = [v(z) - v(\zeta)^*]/(z-\bar{\zeta})$, we get

$$C(z,\zeta) = \int_{\Delta_0}\Bigg\{\frac{1}{(t-z)(t-\bar{\zeta})} - \sum_{\ell=0}^{\rho_0-1}\frac{t}{(1+t^2)^{\ell+2}}\sum_{p=1}^{\ell+1}\binom{\ell+1}{p}\sum_{\substack{j+k=2p+1\\ j,k\geq 1}} z^{k-1}\bar{\zeta}^{j-1}$$

$$-\sum_{\ell=0}^{\rho_0-1}\frac{1}{(1+t^2)^{\ell+1}}\sum_{p=0}^{\ell}\binom{\ell}{p}\sum_{\substack{j+k=2p+2\\ j,k\geq 1}} z^{k-1}\bar{\zeta}^{j-1}\Bigg]\Bigg\}\,d\tau(t)$$

from the identity

$$\frac{S_0(t,z) - \overline{S_0(t,\zeta)}}{z-\bar{\zeta}} = \sum_{\ell=0}^{\rho_0}\frac{t}{(1+t^2)^{\ell+1}}\,\frac{(1+z^2)^\ell - (1+\bar{\zeta}^2)^\ell}{z-\bar{\zeta}}$$

$$+\sum_{\ell=0}^{\rho_0-1}\frac{1}{(1+t^2)^{\ell+1}}\,\frac{z\,(1+z^2)^\ell - \bar{\zeta}\,(1+\bar{\zeta}^2)^\ell}{z-\bar{\zeta}}$$

$$= \sum_{\ell=0}^{\rho_0}\frac{t}{(1+t^2)^{\ell+1}}\sum_{p=0}^{\ell}\binom{\ell}{p}\frac{z^{2p}-\bar{\zeta}^{2p}}{z-\bar{\zeta}}$$

$$+\sum_{\ell=0}^{\rho_0-1}\frac{1}{(1+t^2)^{\ell+1}}\sum_{p=0}^{\ell}\binom{\ell}{p}\frac{z^{2p+1}-\bar{\zeta}^{2p+1}}{z-\bar{\zeta}}$$

$$= \sum_{\ell=0}^{\rho_0-1}\frac{t}{(1+t^2)^{\ell+2}}\sum_{p=1}^{\ell+1}\binom{\ell+1}{p}\sum_{\substack{j+k=2p+1\\ j,k\geq 1}} z^{k-1}\bar{\zeta}^{j-1}$$

$$+\sum_{\ell=0}^{\rho_0-1}\frac{1}{(1+t^2)^{\ell+1}}\sum_{p=0}^{\ell}\binom{\ell}{p}\sum_{\substack{j+k=2p+2\\ j,k\geq 1}} z^{k-1}\bar{\zeta}^{j-1}\,.$$

Thus

$$L(z,\zeta) = \int_{\Delta_0}\Bigg\{\begin{bmatrix}(I-At)^{-1}\Phi_2\\ \dfrac{-iI}{\bar{\zeta}-t}\end{bmatrix} d\tau(t)\begin{bmatrix}\Phi_2^*(I-A^*t)^{-1} & \dfrac{iI}{z-t}\end{bmatrix}$$

$$- \sum_{\ell=0}^{\rho_0-1} \frac{t}{(1+t^2)^{\ell+2}} \sum_{p=1}^{\ell+1} \binom{\ell+1}{p} \sum_{\substack{j+k=2p+1 \\ j,k\geq 1}} \begin{bmatrix} A^{-j}\Phi_2 \\ -i\,\bar{\zeta}^{j-1} \end{bmatrix} d\tau(t) \begin{bmatrix} \Phi_2^* A^{*-k} & i\,z^{k-1} \end{bmatrix}$$

$$- \sum_{\ell=0}^{\rho_0-1} \frac{1}{(1+t^2)^{\ell+1}} \sum_{p=0}^{\ell} \binom{\ell}{p} \sum_{\substack{j+k=2p+2 \\ j,k\geq 1}} \begin{bmatrix} A^{-j}\Phi_2 \\ -i\,\bar{\zeta}^{j-1} \end{bmatrix} d\tau(t) \begin{bmatrix} \Phi_2^* A^{*-k} & i\,z^{k-1} \end{bmatrix} \Bigg\}.$$

As above, we deduce that $\varkappa_L < \infty$.

Case 1: (c) In the case $v(z) = R_0(z)$, $S = \mathfrak{R}_0$, $\Phi_1 = -i\,\widehat{\mathfrak{R}}_0$, the polynomial $R_0(z) = \sum_{\ell=0}^{2\rho_0+1} R_{0\ell} z^\ell$ has selfadjoint matrix coefficients, and

$$S = \sum_{\ell=1}^{2\rho_0+1} \sum_{\substack{j+k=\ell+1 \\ j,k\geq 1}} A^{-j}\Phi_2 R_{0\ell}\Phi_2^* A^{*-k}$$

by Theorem 3.3. By (3.7) and Lemma 6.4,

$$\begin{aligned} B(z) &= (I - zA)^{-1}\big[\Phi_1 - i\Phi_2 v(z)\big] \\ &= i \operatorname*{Res}_{\lambda=0} \Big[(I - zA)^{-1}A(A-\lambda I)^{-1}\Phi_2 R_0(\lambda^{-1})\lambda^{-1}\Big] \\ &\qquad - i\,(I - zA)^{-1}\Phi_2 R_0(z) \\ &= i \operatorname*{Res}_{\lambda=0} \left[\frac{(I - zA)^{-1} - (I - \lambda^{-1}A)^{-1}}{1-\lambda z}\,\Phi_2 R_0(\lambda^{-1})\lambda^{-1}\right] \\ &\qquad - i\,(I - zA)^{-1}\Phi_2 R_0(z) \\ &= -i \operatorname*{Res}_{\lambda=0} \left[\frac{(I - \lambda^{-1}A)^{-1}}{1-\lambda z}\,\Phi_2 R_0(\lambda^{-1})\lambda^{-1}\right] \\ &= i \sum_{\ell=1}^{2\rho_0+1} \sum_{\substack{j+k=\ell+1 \\ j,k\geq 1}} A^{-j}\Phi_2 R_{0\ell} z^{k-1}\,. \end{aligned}$$

A straightforward calculation of $C(z,\zeta)$ yields

$$L(z,\zeta) = \sum_{\ell=1}^{2\rho_0+1} \sum_{\substack{j+k=\ell+1 \\ j,k\geq 1}} \begin{bmatrix} A^{-j}\Phi_2 \\ -i\,\bar{\zeta}^{j-1} \end{bmatrix} R_{0\ell} \begin{bmatrix} \Phi_2^* A^{*-k} & iz^{k-1} \end{bmatrix},$$

and $\varkappa_L < \infty$ by (6.8).

Case 1: (d) Let $j = 1, \dots, r$, and suppose that

$$v(z) = -R_j\Big(\frac{1}{z-\alpha_j}\Big), \qquad S = \mathfrak{R}_j, \qquad \Phi_1 = -i\,\widehat{\mathfrak{R}}_j\,,$$

where $R_j(z) = \sum_{\ell=1}^{2\rho_j+1} R_{j\ell} z^\ell$ has selfadjoint matrix coefficients and constant term zero. By Theorem 3.3,

$$S = \sum_{\ell=1}^{2\rho_j+1} \sum_{\substack{p+q=\ell+1 \\ p,q\geq 1}} A_{p-1}(\alpha_j)\Phi_2 R_{j\ell}\Phi_2^* A_{q-1}(\alpha_j)^* .$$

By (3.8) and Lemmas 6.4 and 6.3,

$$\begin{aligned}
B(z) &= (I - zA)^{-1}\big[\Phi_1 - i\Phi_2 v(z)\big] \\
&= -i \operatorname*{Res}_{\lambda=\alpha_j} \left[(I-zA)^{-1}A(I-\lambda A)^{-1}\Phi_2 R_j\Big(\frac{1}{\lambda-\alpha_j}\Big)\right] \\
&\qquad + i(I-zA)^{-1}\Phi_2 R_j\left(\frac{1}{z-\alpha_j}\right) \\
&= -i \operatorname*{Res}_{\lambda=\alpha_j} \left[\frac{(I-\lambda A)^{-1} - (I-zA)^{-1}}{\lambda - z}\,\Phi_2 R_j\Big(\frac{1}{\lambda-\alpha_j}\Big)\right] \\
&\qquad + i(I-zA)^{-1}\Phi_2 R_j\left(\frac{1}{z-\alpha_j}\right) \\
&= i \operatorname*{Res}_{\lambda=\alpha_j} \left[\frac{(I-\lambda A)^{-1}}{z-\lambda}\,\Phi_2 R_j\Big(\frac{1}{\lambda-\alpha_j}\Big)\right] \\
&= i \sum_{\ell=1}^{2\rho_j+1} \sum_{\substack{p+q=\ell-1 \\ p,q\geq 0}} \frac{A_q(\alpha_j)\Phi_2 R_{j\ell}}{(z-\alpha_j)^{p+1}} \\
&= i \sum_{\ell=1}^{2\rho_j+1} \sum_{\substack{p+q=\ell+1 \\ p,q\geq 1}} \frac{A_{q-1}(\alpha_j)\Phi_2 R_{j\ell}}{(z-\alpha_j)^{p}} .
\end{aligned}$$

A short calculation of $C(z,\zeta)$ yields

$$L(z,\zeta) = \sum_{\ell=1}^{2\rho_j+1} \sum_{\substack{p+q=\ell+1 \\ p,q\geq 1}} \begin{bmatrix} A_{p-1}(\alpha_j)\Phi_2 \\ \dfrac{-iI}{(\bar\zeta - \alpha_j)^p} \end{bmatrix} R_{j\ell} \begin{bmatrix} \Phi_2^* A_{q-1}(\alpha_j)^* & \dfrac{iI}{(z-\alpha_j)^q} \end{bmatrix} ,$$

and we again obtain $\varkappa_L < \infty$ by (6.8).

Case 1: (e) Let $k = 1, \dots, s$, and assume that

$$\begin{aligned}
v(z) &= -M_k\Big(\frac{1}{z-\beta_k}\Big) - M_k\Big(\frac{1}{\bar z - \beta_k}\Big)^*, \\
S &= \mathfrak{M}_{1k} + \mathfrak{M}_{2k}, \\
\Phi_1 &= -i\left[\widehat{\mathfrak{M}}_{1k} + \widehat{\mathfrak{M}}_{2k}\right],
\end{aligned}$$

where $M_k(z) = \sum_{\ell=1}^{\sigma_k} M_{k\ell} z^\ell$ is a polynomial with matrix coefficients and constant term zero. Calculations similar to those above yield

$$S = \sum_{\ell=1}^{\sigma_k} \sum_{\substack{p+q=\ell+1 \\ p,q\geq 1}} \Big[A_{p-1}(\beta_k)\Phi_2 M_{kl}\Phi_2^* A_{q-1}(\bar\beta_k)^* \\ + A_{p-1}(\bar\beta_k)\Phi_2 M_{kl}^*\Phi_2^* A_{q-1}(\beta_k)^* \Big],$$

$$B(z) = i \sum_{\ell=1}^{\sigma_k} \sum_{\substack{p+q=\ell+1 \\ p,q\geq 1}} \left[\frac{A_{p-1}(\beta_k)\Phi_2 M_{kl}}{(z-\beta_k)^q} + \frac{A_{p-1}(\bar\beta_k)\Phi_2 M_{kl}^*}{(z-\bar\beta_k)^q} \right],$$

$$C(z,\zeta) = \sum_{\ell=1}^{\sigma_k} \sum_{\substack{p+q=\ell+1 \\ p,q\geq 1}} \left[\frac{M_{k\ell}}{(z-\beta_k)^q(\bar\zeta-\beta_k)^p} + \frac{M_{k\ell}^*}{(z-\bar\beta_k)^q(\bar\zeta-\bar\beta_k)^p} \right].$$

We again obtain a kernel,

$$L(z,\zeta) = \sum_{\ell=1}^{\sigma_k} \sum_{\substack{p+q=\ell+1 \\ p,q\geq 1}} \begin{bmatrix} A_{p-1}(\beta_k)\Phi_2 & A_{p-1}(\bar\beta_k)\Phi_2 \\ \dfrac{-iI}{(\bar\zeta-\beta_k)^p} & \dfrac{-iI}{(\bar\zeta-\bar\beta_k)^p} \end{bmatrix} \begin{bmatrix} 0 & M_{k\ell} \\ M_{k\ell}^* & 0 \end{bmatrix} \cdot \\ \cdot \begin{bmatrix} \Phi_2^* A_{q-1}(\beta_k)^* & \dfrac{iI}{(z-\bar\beta_k)^q} \\ \Phi_2^* A_{q-1}(\bar\beta_k)^* & \dfrac{iI}{(z-\beta_k)^q} \end{bmatrix},$$

which has a finite number of negative squares. This verifies the conclusion in each of the subcases (a)–(e), and so Theorem 4.5 follows in Case 1.

Assume Case 2: $0 \notin \sigma(A)$. We show that $\varkappa_L < \infty$ in the same subcases (a)–(e). Recall that in Case 2, $\rho_0 = 0$ and $R_0(z) = C_0$ is constant.

Case 2: (a), (d), (e) There is no change here from Case 1.

Case 2: (b) By (3.11), we now have

$$v(z) = \int_{\Delta_0} \left[\frac{1}{t-z} - \frac{t}{1+t^2} \right] d\tau(t),$$

$$S = \int_{\Delta_0} (I-At)^{-1}\Phi_2[d\tau(t)]\Phi_2^*(I-A^*t)^{-1},$$

$$\Phi_1 = -i \int_{\Delta_0} \left[A(I-At)^{-1} + \frac{tI}{1+t^2} \right] \Phi_2\,[d\tau(t)].$$

A short calculation gives

$$B(z) = -i \int_{\Delta_0} \frac{(I-tA)^{-1}}{t-z}\,\Phi_2\,[d\tau(t)],$$

$$C(z,\zeta) = \int_{\Delta_0} \frac{d\tau(t)}{(t-z)(t-\bar\zeta)}.$$

The kernel

$$L(z,\zeta) = \int_{\Delta_0} \begin{bmatrix} (I-zA)^{-1}\Phi_2 \\ \dfrac{iI}{t-\bar{\zeta}} \end{bmatrix} d\tau(t) \begin{bmatrix} \Phi_2^*(I-zA^*)^{-1} & \dfrac{-iI}{t-z} \end{bmatrix}$$

is nonnegative in this subcase.

Case 2: (c) Here $v(z) = R_0(z) = C_0$ is constant, $S = \mathfrak{R}_0 = 0$, and $\Phi_1 = -i\widehat{\mathfrak{R}}_0 = i\Phi_2 C_0$ by (3.11). Thus $L(z,\zeta) = 0$ is a nonegative kernel.

So $\varkappa_L < \infty$ in all subcases (a)–(e) in Case 2, and the result follows. □

Proof of Theorem 4.6. We use the notation in the proof of Theorem 4.5 and verify the conclusion in the same subcases (a)–(e).

Case 1: (a) In this subcase, our previous formula for $B(z)$ can be written

$$B(z) = \int_{\Delta_j} F(t,z)\, d\tau(t), \tag{6.9}$$

where

$$\begin{aligned} F(t,z) &= -i\left[\frac{(I-tA)^{-1}}{t-z} + \sum_{\ell=0}^{2\rho_j-1} (t-\alpha_j)^\ell \sum_{p+q=\ell} \frac{A_q(\alpha_j)}{(z-\alpha_j)^{p+1}}\right]\Phi_2 \\ &= i\sum_{\ell=2\rho_j}^{\infty} (t-\alpha_j)^\ell \sum_{p+q=\ell} \frac{A_q(\alpha_j)}{(z-\alpha_j)^{p+1}}\, \Phi_2\,. \end{aligned} \tag{6.10}$$

The last series converges uniformly for t in a neighborhood of α_j for z in any compact subset of $\mathbf{C}_+ \cup \mathbf{C}_-$. It follows that $B(z)$ is analytic on $\mathbf{C}_+ \cup \mathbf{C}_-$.

Case 1: (b) We now obtain a representation (6.9) with $j = 0$ and

$$\begin{aligned} F(t,z) &= -i\Bigg[\frac{(I-tA)^{-1}}{t-z} \\ &\qquad + \sum_{\ell=0}^{\rho_0-1} \frac{t}{(1+t^2)^{\ell+2}} \sum_{p=1}^{\ell+1} \binom{\ell+1}{p} \sum_{\substack{j+k=2p+1\\ j,k\ge 1}} A^{-j} z^{k-1} \\ &\qquad + \sum_{\ell=0}^{\rho_0-1} \frac{1}{(1+t^2)^{\ell+1}} \sum_{p=0}^{\ell} \binom{\ell}{p} \sum_{\substack{j+k=2p+2\\ j,k\ge 1}} A^{-j} z^{k-1}\Bigg]\Phi_2 \\ &= i\Bigg[\sum_{\ell=\rho_0}^{\infty} \frac{t}{(1+t^2)^{\ell+2}} \sum_{p=1}^{\ell+1} \binom{\ell+1}{p} \sum_{\substack{j+k=2p+1\\ j,k\ge 1}} A^{-j} z^{k-1} \\ &\qquad + \sum_{\ell=\rho_0}^{\infty} \frac{1}{(1+t^2)^{\ell+1}} \sum_{p=0}^{\ell} \binom{\ell}{p} \sum_{\substack{j+k=2p+2\\ j,k\ge 1}} A^{-j} z^{k-1}\Bigg]\Phi_2 \end{aligned} \tag{6.11}$$

The two series in the last expression converge in a neighborhood of infinity for z in any compact subset of $\mathbf{C}_+ \cup \mathbf{C}_-$. Again $B(z)$ is analytic on $\mathbf{C}_+ \cup \mathbf{C}_-$.

Case 1: (c), (d), (e) Our previous expressions for $B(z)$ here are rational functions whose only nonreal poles are at the points $\beta_k, \bar{\beta}_k$, $k = 1, \ldots, s$.

The conclusion holds in all subcases, and the result follows. □

Proof of Theorem 4.7. Notation is as in the proof of Theorem 4.5. We check (4.18) and (4.19) in each of the subcases (a)–(e).

We first prove (4.18).

Case 2: (a), (c), (d), (e) Here, in fact, $\|B(z)\| = \mathcal{O}(1/|z|)$ as $|z| \to \infty$. For (a) this follows from (6.9) and (6.10). In (c), $B(z) \equiv 0$. The assertion is clear for (d) and (e).

Case 2: (b) By the proof of Theorem 4.5, Case 2,

$$B(z) = -i \int_{\Delta_0} \frac{(I - tA)^{-1}}{t - z} \Phi_2 \left[d\tau(t)\right]. \tag{6.12}$$

Define $L^2(d\tau) = L^2(\Delta_0, d\tau)$ as in Appendix 1. If $g \in \mathbf{C}^m$ and $h \in \mathfrak{H}$, then

$$\langle B(z)g, h \rangle = -i \left\langle \frac{\sqrt{t^2+1}}{t - z} \frac{g}{\sqrt{t^2+1}}, \Phi_2^*(I - tA^*)^{-1} h \right\rangle_{L^2(d\tau)}. \tag{6.13}$$

Here $g/\sqrt{t^2+1} \in L^2(d\tau)$ by Theorem 2.1(2°), and $\Phi_2^*(I - tA^*)^{-1}h \in L^2(d\tau)$ by (3.1). For $z = x + iy \in D_\delta$, $x^2 \le cy^2$ for some $c > 0$ and

$$\begin{aligned} \left|\frac{z}{t - z}\right|^2 &= \frac{x^2 + y^2}{(t - x)^2 + y^2} \\ &\le \frac{(c+1)y^2}{y^2} \le c + 1. \end{aligned} \tag{6.14}$$

If also $|y| \ge 1$, then

$$\begin{aligned} \left|\frac{\sqrt{t^2+1}}{t - z}\right| &\le \frac{|t| + 1}{|t - z|} = \frac{|t - z + z| + 1}{|t - z|} \\ &\le 1 + \left|\frac{z}{t - z}\right| + \frac{1}{|t - z|} \\ &\le 1 + \sqrt{c+1} + 1\,. \end{aligned}$$

To deduce that $\|B(z)\| = \mathcal{O}(1)$ as $|z| \to \infty$ in D_δ, we apply the Schwarz inequality in (6.13) for fixed g and h and then appeal to the principle of uniform boundedness. This completes the proof of (4.18).

$$+ \sum_{\substack{\mu+m+n=p-1\\ \mu,m,n\geq 0}} \frac{A_\mu(\bar\beta_k)\Phi_2 M_{kp}^*\Phi_2^* A_m(\beta_k)^*}{(z-\bar\beta_k)^{n+1}}\Bigg].$$

Again clearly this is $\mathcal{O}(1/|z|)$ as $|z|\to\infty$. □

Proof of Theorem 5.1. (1) Suppose $v(z)\in \mathbf{N}(\mathfrak{A})$ and satisfies conditions (i) and (ii). By Theorem 4.4, $v(z)$ is a generalized Nevanlinna function. We verify Assumptions 3.1, Case 1, for any representation (2.1). By assumption, $\sigma(A)$ is a finite set, and $\sigma(A)$ contains no real point because $\sigma(A)\cap\sigma(A^*)=\emptyset$. We claim that $\sigma(A)$ contains no point $1/\beta_k$, $1/\bar\beta_k$, $k=1,\dots,s$. In fact, if $1/\beta_k\in\sigma(A)$ (resp. $1/\bar\beta_k\in\sigma(A)$), then by (i), $v(z)$ has at most a removable singularity at β_k (resp. $\bar\beta_k$). But β_k and $\bar\beta_k$ are poles of $v(z)$ by condition (4°) in Theorem 2.1, which is impossible. The claim follows, and therefore operators S_v and $\Phi_{1,v}$ are defined under Case 1 of Assumptions 3.1.

We show that $S=S_v$ and $\Phi_1=\Phi_{1,v}$. By Theorem 3.5, in addition to the given identity (1.1), we also have

$$AS_v - S_vA^* = i\left[\Phi_{1,v}\Phi_2^* + \Phi_2\Phi_{1,v}^*\right]. \tag{6.18}$$

Define $\widetilde{L}_v(z,\zeta)$ and $\widetilde{B}_v(z)$ by (4.2) and (4.3) but with S and Φ_1 replaced by S_v and $\Phi_{1.v}$. Since

$$B_v(z) = (I-zA)^{-1}\big[\Phi_1 - i\Phi_2 v(z)\big],$$
$$\widetilde{B}_v(z) = (I-zA)^{-1}\big[\Phi_{1,v} - i\Phi_2 v(z)\big],$$

the function

$$F(z) \stackrel{\text{def}}{=} B_v(z) - \widetilde{B}_v(z) = (I-zA)^{-1}\big[\Phi_1 - \Phi_{1,v}\big] \tag{6.19}$$

is analytic in the complex plane except perhaps at a finite number of nonreal points λ such that $1/\lambda\in\sigma(A)$, and it vanishes at infinity because A is invertible. By (ii), $B_v(z)$ has at most a removable singularity at any point λ such that $1/\lambda\in\sigma(A)$, and by Theorem 4.6, $\widetilde{B}_v(z)$ is analytic at any point λ such that $1/\lambda\in\sigma(A)$. It follows that $F(z)$ is entire and therefore $F(z)\equiv 0$. This is only possible if $\Phi_1=\Phi_{1,v}$. Finally, by (1.1) and (6.18),

$$A(S-S_v)-(S-S_v)A^* = i\big(\Phi_1\Phi_2^*+\Phi_2\Phi_1^*\big) - i\big(\Phi_{1,v}\Phi_2^*+\Phi_2\Phi_{1,v}^*\big) = 0.$$

Since $\sigma(A)\cap\sigma(A^*)=\emptyset$, by [6] the operator equation $AX-XA^*=0$ has only the trivial solution, and hence $S-S_v=0$.

(2) Conversely, let $S=S_v$ and $\Phi_1=\Phi_{1,v}$ for some generalized Nevanlinna function $v(z)$ having a representation (2.1) which satisfies Assumptions 3.1, Case 1. We show that $v(z)$ belongs to $\mathbf{N}(\mathfrak{A})$ and satisfies conditions (i) and (ii) in (1). Define $P(z)$ and $Q(z)$ for $v(z)$ as in Theorem 4.2. Then $P(\bar z)^*Q(z)+Q(\bar z)^*P(z)\equiv 0$. By (4.10), $v(z)$ has the representation (4.14) with $c(z)P(z)+d(z)Q(z)\equiv I$. Conditions (i) and (ii) in Definition 4.3 thus hold. To see that condition (iii) in

$$= G_3(z) + \frac{1}{2i}(I - xA)^{-1}\Phi_2\Big[v(z) - v(z)^*\Big]\Phi_2^*(I - xA^*)^{-1}.$$

Using [8, Proposition 3.4], we see that $G_3(z)$ is bounded for $a \leq x \leq b$ and $0 < y \leq 1$. Since also $G_3(x + i0) = 0$ strongly a.e. on $(-\infty, \infty)$, the result follows with $G(z) = G_1(z) + G_3(z)$. □

Proof of Theorem 5.3. (1) Assume that $v(z) \in \mathbf{N}(\mathfrak{A})$ and conditions (i)–(iii) hold. By Theorem 4.4, $v(z)$ is a generalized Nevanlinna function. We show that a representation (2.1) can be chosen such that the conditions of Assumptions 3.1, Case 2, are satisfied.

By [8, Theorem 4.1], (i) allows us to choose a representation (2.1) such that $\rho_0 = 0$ and $R_0(z)$ is constant. We verify (3.1). Define $L_v(z, \zeta)$ and $L_{v,T}(z, \zeta)$ by (4.2) and (4.4). Fix h in $\mathfrak{H}$, and set

$$v_h(z) = \langle B_{v,T}(z)h, h\rangle\ .$$

We show that $v_h(z)$ belongs to $\mathbf{N}_\varkappa$ for some $\varkappa \geq 0$. Since $v(z) \in \mathbf{N}(\mathfrak{A})$, it has a representation (4.14), and hence

$$\begin{bmatrix} P(z) \\ Q(z) \end{bmatrix} K(z)^{-1} = \mathfrak{A}(z)^{-1} \begin{bmatrix} -iv(z) \\ I \end{bmatrix},$$

where $K(z) = c(z)P(z) + d(z)Q(z)$. Therefore by Theorem 4.2,

$$L_v(z, \zeta) = \begin{bmatrix} I & 0 \\ B_v(\zeta)^* S^{-1} & I \end{bmatrix} \begin{bmatrix} S & 0 \\ 0 & D_v(z, \zeta) \end{bmatrix} \begin{bmatrix} I & S^{-1}B_v(z) \\ 0 & I \end{bmatrix},$$

where

$$D_v(z, \zeta) = K(\zeta)^*\ i\ \frac{P(\zeta)^*Q(z) + Q(\zeta)^*P(z)}{z - \bar\zeta}\ K(z)\,.$$

Hence $\varkappa_{L_v} = \varkappa_S + \varkappa_{P,Q}$. By (4.7), if

$$C_{v,T}(z, \zeta) = \frac{B_{v,T}(z) - B_{v,T}(\zeta)^*}{z - \bar\zeta},$$

then

$$\begin{bmatrix} S & -iB_{v,T}(z) \\ iB_{v,T}(\zeta)^* & C_{v,T}(z, \zeta) \end{bmatrix} = L_{v,T}(z, \zeta)$$

$$= \begin{bmatrix} I & 0 \\ L_0(\bar\zeta) & L_2(\bar\zeta) \end{bmatrix} L_v(z, \zeta) \begin{bmatrix} I & L_0(\bar z)^* \\ 0 & L_2(\bar z)^* \end{bmatrix},$$

where $L_0(z)$ and $L_2(z)$ are defined by (4.6). It then follows from (4.15) that

$$\varkappa_{C_{v,T}} \leq \varkappa_{L_{v,T}} \leq \varkappa_{L_v} \leq \varkappa_S + \varkappa_{P,Q} < \infty\,.$$

In particular, $v_h(z)$ belongs to $\mathbf{N}_\varkappa$ for some $\varkappa \geq 0$.

By condition (iii) and [8, Theorem 4.2], the Kreĭn-Langer representation of $v_h(z)$ can be reduced to the form

$$v_h(z) = \left\{ \sum_{j=1}^{r_h} \int_{\Delta_{j,h}} \left[\frac{1}{t - z} - S_j(t, z)\right] d\tau_h(t) - \sum_{j=1}^{r_h} R_{j,h}\left(\frac{1}{z - \alpha_j}\right)\right.$$

$$- \sum_{k=1}^{s} \left[M_{k,h} \left(\frac{1}{z - \beta_k} \right) + M_{k,h} \left(\frac{1}{\bar{z} - \beta_k} \right)^* \right] \Bigg\} + \int_{\Delta_{0,h}} \frac{d\sigma_h(t)}{t - z}$$

$$= \tilde{v}_h(z) + \int_{\Delta_{0,h}} \frac{d\sigma_h(t)}{t - z},$$

where $\tilde{v}_h(z)$ is analytic across the interior of $\Delta_{0,h}$ and real on this set, and $\sigma_h(t)$ is a nondecreasing function satisfying

$$\int_{\Delta_{0,h}} d\sigma_h(t) < \infty. \tag{6.20}$$

Suppose that $[a, b] \subseteq \Delta_{0,h} \cap \Delta_0$ and a and b are points of continuity of $\sigma_h(t)$ and $\tau(t)$. By the Stieltjes inversion formula and Lemma 6.5,

$$\int_a^b d\sigma_h(t) = \lim_{y \downarrow 0} \frac{1}{\pi} \int_a^b \operatorname{Im} v_h(t + iy)\, dt$$

$$= \lim_{y \downarrow 0} \frac{1}{\pi} \int_a^b \left\langle \frac{B_{v,T}(t+iy) - B_{v,T}(t+iy)^*}{2i} h, h \right\rangle dt$$

$$= \lim_{y \downarrow 0} \frac{1}{\pi} \int_a^b \left\langle (I - tA)^{-1} \Phi_2 \frac{v(t+iy) - v(t+iy)^*}{2i} \Phi_2^* (I - tA^*)^{-1} h, h \right\rangle dt.$$

Hence by [8, Theorem 3.1],

$$\int_{\Delta_{0,h} \cap \Delta_0} \langle d\tau(t)\, \Phi_2^*(I - A^*t)^{-1} h, \Phi_2^*(I - A^*t)^{-1} h \rangle = \int_{\Delta_{0,h} \cap \Delta_0} d\sigma_h(t) < \infty,$$

and therefore

$$\int_{\Delta_0} \langle d\tau(t)\, \Phi_2^*(I - A^*t)^{-1} h, \Phi_2^*(I - A^*t)^{-1} h \rangle < \infty.$$

This verifies (3.1), and therefore operators S_v and $\Phi_{1,v}$ are defined.

It remains to show that $\Phi_1 = \Phi_{1,v}$ and $S = S_v$. Recall that $B_v(z)$ and $B_{v,T}(z)$ are defined using the operators A, S, Φ_1, Φ_2 from the given operator identity (1.1). By Theorem 3.5, we have a second operator identity,

$$AS_v - S_v A^* = i \left[\Phi_{1,v} \Phi_2^* + \Phi_2 \Phi_{1,v}^* \right]. \tag{6.21}$$

Define $\widetilde{L}_v(z, \zeta)$ and $\widetilde{B}_v(z)$ by (4.2) and (4.3) but with S and Φ_1 replaced by S_v and $\Phi_{1.v}$. Analogously, define $\widetilde{L}_{v,T}(z, \zeta)$ and $\widetilde{B}_{v,T}(z)$ using the transformed kernel (4.4) with S and Φ_1 replaced by S_v and $\Phi_{1.v}$. Thus

$$\widetilde{B}_v(z) = (I - zA)^{-1} \left[\Phi_{1,v} - i \Phi_2 v(z) \right],$$

$$\widetilde{B}_{v,T}(z) = i \left[S_v L_0(\bar{z})^* + \widetilde{B}_v(z) L_2(\bar{z})^* \right],$$

where $L_0(z)$ and $L_2(z)$ are given by (4.6). In particular,

$$B_v(z) - \widetilde{B}_v(z) = (I - zA)^{-1} \left[\Phi_1 - \Phi_{1,v} \right].$$

Hence for any $g \in \mathbf{C}^m$, by (ii) and Theorem 4.7,

$$(I - iyA)^{-1} \left[\Phi_1 - \Phi_{1,v} \right] g = \mathcal{O}(1), \qquad |y| \to \infty.$$

Since we assume that the only f in $\mathfrak{H}$ such that $\|(I-iyA)^{-1}f\| = \mathcal{O}(1)$ as $|y| \to \infty$ is $f = 0$, we deduce that $[\Phi_1 - \Phi_{1,v}]g = 0$, and so $\Phi_1 = \Phi_{1,v}$.

By what we have shown so far, $B_v(z) = \widetilde{B}_v(z)$. Therefore

$$B_{v,T}(z) - \widetilde{B}_{v,T}(z) = i(S - S_v)L_0(\bar{z})^* = (S - S_v)A^*(I - zA^*)^{-1}.$$

Hence for any $h \in \mathfrak{H}$, by (iii) and Theorem 4.7,

$$A(I - iyA)^{-1}(S - S_v)h = \mathcal{O}(1/|y|), \qquad |y| \to \infty.$$

By the identity $(I - iyA)^{-1} = iyA(I - iyA)^{-1} + I$, we get

$$(I - iyA)^{-1}(S - S_v)h = \mathcal{O}(1), \qquad |y| \to \infty.$$

As above, our assumptions on A imply that $(S - S_v)h = 0$, and therefore $S = S_v$.

(2) Conversely, assume that $S = S_v$ and $\Phi_1 = \Phi_{1,v}$, where $v(z)$ is a generalized Nevanlinna function having a representation (2.1) satisfying Assumptions 3.1, Case 2. To see that $v(z)$ belongs to $\mathbf{N}(\mathfrak{A})$, set

$$\begin{bmatrix} P(z) \\ Q(z) \end{bmatrix} = \mathfrak{A}(z)^{-1} \begin{bmatrix} -iv(z) \\ I \end{bmatrix}.$$

Then $P(z)$ and $Q(z)$ are meromorphic on $\mathbf{C}_+ \cup \mathbf{C}_-$ and by (4.13),

$$\begin{aligned} v(z) &= i\big[a(z)P(z) + b(z)Q(z)\big], \\ I &= c(z)P(z) + d(z)Q(z), \end{aligned}$$

and hence (4.14) holds. By Theorem 4.2, $P(\bar{z})^*Q(z) + Q(\bar{z})^*P(z) = 0$ on $\mathbf{C}_+ \cup \mathbf{C}_-$. Since $S = S_v$ and $\Phi_1 = \Phi_{1,v}$ by assumption, the kernel $L_v(z,\zeta)$ has a finite number of negative squares by Theorem 4.5. Hence by (4.11) and (4.12), the kernel

$$i\,\frac{P(\zeta)^*Q(z) + Q(\zeta)^*P(z)}{z - \bar{\zeta}}$$

has a finite number of negative squares. It follows that $v(z)$ belongs to $\mathbf{N}(\mathfrak{A})$.

We obtain (i) by the converse part of [8, Theorem 4.1]. Conditions (ii) and (iii) follow from Theorem 4.7. □

Proof of Theorem 5.4. The only change in the proof of Theorem 5.3 is in the proofs that $\Phi_1 = \Phi_{1,v}$ and $S = S_v$ in part (1) of the theorem. As before, for any $g \in \mathbf{C}^m$, $(I - iyA)^{-1}\big[\Phi_1 - \Phi_{1,v}\big]g = \mathcal{O}(1)$ as $|y| \to \infty$. By the identity $(I - iyA)^{-1} = iyA(I - iyA)^{-1} + I$,

$$y\,(I - iyA)^{-1}A\big[\Phi_1 - \Phi_{1,v}\big]g = \mathcal{O}(1), \qquad |y| \to \infty.$$

Our assumptions on A imply that $A\big[\Phi_1 - \Phi_{1,v}\big]g = 0$. Since $\ker A = \{0\}$, we deduce that $\Phi_1 = \Phi_{1,v}$. Again as before, for any $h \in \mathfrak{H}$, $A(I - iyA)^{-1}(S - S_v)h = \mathcal{O}(1/|y|)$ as $|y| \to \infty$. By our assumptions on A, $A(S - S_v)h = 0$. Since $\ker A = \{0\}$, $S = S_v$. □

Acknowledgement

The authors thank A.L. Sakhnovich for useful discussions of this paper.

References

[1] D. Alpay, A. Dijksma, J. Rovnyak, and H.S.V. de Snoo, *Schur functions, operator colligations, and reproducing kernel Pontryagin spaces*, Oper. Theory Adv. Appl., vol. 96, Birkhäuser Verlag, Basel, 1997.

[2] K. Daho and H. Langer, *Matrix functions of the class N_κ*, Math. Nachr. **120** (1985), 275–294.

[3] V.E. Katsnelson, *Methods of J-theory in continuous interpolation problems of analysis. Part I*, T. Ando Hokkaido University, Sapporo, 1985, Translated from the Russian and with a foreword by T. Ando.

[4] I.V. Kovalishina and V.P. Potapov, *Integral representation of Hermitian positive functions*, T. Ando Hokkaido University, Sapporo, 1982, Translated from the Russian by T. Ando.

[5] M.G. Kreĭn and H. Langer, *Über einige Fortsetzungsprobleme, die eng mit der Theorie hermitescher Operatoren im Raume Π_κ zusammenhängen. I. Einige Funktionenklassen und ihre Darstellungen*, Math. Nachr. **77** (1977), 187–236.

[6] M. Rosenblum, *On the operator equation $BX - XA = Q$*, Duke Math. J. **23** (1956), 263–269.

[7] J. Rovnyak and L.A. Sakhnovich, *Some indefinite cases of spectral problems for canonical systems of difference equations*, Linear Algebra Appl. **343/344** (2002), 267–289.

[8] ———, *On the Kreĭn-Langer integral representation of generalized Nevanlinna functions*, Electron. J. Linear Algebra **11** (2004), 1–15 (electronic).

[9] ———, *Spectral problems for some indefinite cases of canonical differential equations*, J. Operator Theory **51** (2004), 115–139.

[10] A.L. Sakhnovich, *Modification of V. P. Potapov's scheme in the indefinite case*, Matrix and operator valued functions, Oper. Theory Adv. Appl., vol. 72, Birkhäuser, Basel, 1994, pp. 185–201.

[11] L.A. Sakhnovich, *Integral equations with difference kernels on finite intervals*, Oper. Theory Adv. Appl., vol. 84, Birkhäuser Verlag, Basel, 1996.

[12] ———, *Interpolation theory and its applications*, Kluwer, Dordrecht, 1997.

[13] ———, *Spectral theory of canonical differential systems. Method of operator identities*, Oper. Theory Adv. Appl., vol. 107, Birkhäuser Verlag, Basel, 1999.

James Rovnyak
University of Virginia
Department of Mathematics
P. O. Box 400137
Charlottesville, VA 22904, USA
e-mail: rovnyak@Virginia.EDU

Lev A. Sakhnovich
University of Connecticut at Storrs
735 Crawford Avenue
Brooklyn, NY 11223, USA
e-mail: Lev.Sakhnovich@verizon.net

Operator Theory:
Advances and Applications, Vol. 171, 323–347

Singular Integral Operators in Weighted Spaces of Continuous Functions with Oscillating Continuity Moduli and Oscillating Weights

Natasha Samko

Abstract. We present a survey of some results on the theory of singular integral operators with piece-wise continuous coefficients in the weighted spaces of continuous functions with a prescribed continuity modulus (generalized Hölder spaces $H^{\omega}(\Gamma,\rho)$) together with some new results related to oscillating (non-equilibrated) characteristics and oscillating weights.

Mathematics Subject Classification (2000). Primary 45E05; Secondary 47B35, 47B38.

Keywords. Singular operator, Fredholm operators, index, Hölder space, continuity modulus, Bari-Stechkin class, Boyd indices.

1. Introduction

The characterization of Fredholmness of the singular integral operators

$$Nf := \mathcal{A}(t)(P_+f)(t) + \mathcal{B}(t)(P_-f)(t), \tag{1.1}$$

with piecewise continuous coefficients, where $P_\pm = \frac{1}{2}(I \pm S)$ are the projection operators generated by the singular operator

$$Sf(t) = \frac{1}{\pi i}\int_\Gamma \frac{f(\tau)}{\tau - t}d\tau,$$

is well known in various spaces of integrable functions; for example in weighted Lebesgue or Orlicz spaces, or even in more general Banach function spaces. We refer to [14], [15], ([3], Subsection 9.6), as well as [19], [20], and [21]. In particular, the phenomenon of "massiveness" of the essential spectra of singular integral operators for some choices of weights or curves, is noteworthy.

The corresponding theory of singular integral operators in weighted spaces of continuous functions is much less well developed. In Hölder spaces $H_0^\lambda(\Gamma,\rho)$ with power weights, results on Fredholmness of singular integral operators on Lyapunov curves were obtained in the papers of R.Duduchava [7], [8], [9] (see also [15]). Algebras of singular operators in the spaces $H^\lambda(\Gamma,\rho)$ were considered in [10] and [43]. The papers [4], [5], looked at Fredholmness of singular operators in the weighted Hölder-Zygmund spaces of the type $\mathbb{Z}^\lambda(\Gamma,\rho)$, even allowing for terms with the complex conjugation operator. Results on Fredholmness of singular operator in Hölder spaces may be also found in [44].

In this paper we survey some results obtained for singular integral operators in the generalized Hölder spaces $H_0^\omega(\Gamma,\rho)$, drawing primarily on work in [37], [38], [42]. There characteristics ω and weights ρ more general than simple powers were considered. A number of results from [37], [38], [42] are obtained here in a more general setting and we also simplify some proofs. In [37] and [38], statements on Fredholmness, together with an index formula, are given for singular integral operators on Lyapunov type curves in such spaces when neither the characteristic nor the weight oscillate. In this case the Boyd-type indices m_ω and M_ω of the characteristic $\omega(h)$ coincide, and no massive spectra appear. Consequently, the results on Fredholmness in [37] and [38] are in a certain sense similar to the Gohberg-Krupnik result on Fredholmness of singular integral operators in Lebesgue spaces with power weights.

The appearance of "lunes" (i.e, regions bounded by circular arcs) due to the presence of oscillations of either the characteristic or weight, makes one speculate that massive spectra will appear in Hölder type spaces even in non-weighted cases and on nice curves, as in Orlicz spaces (see [3], Subsection 10.5). The possibility of oscillation of the weights was investigated in [42], and is looked at here as well.

The coefficients $\mathcal{A}(t)$ and $\mathcal{B}(t)$ are taken to be multipliers in $H_0^\omega(\Gamma,\rho)$ with a finite number of discontinuity points, and the weight $\rho(t)$ is assumed to be "fixed" to the discontinuity points of the coefficients, oscillating in general, near those points. We give sufficient conditions for the operator (1.1) to be Fredholm in the space $H_0^\omega(\Gamma,\rho)$. These conditions are in the terms of the upper and lower indices of the characteristic and the weight functions. They are natural, and though it is not proved, they are expected to be necessary as well. It is felt that new techniques will be required to prove necessity.

The following are the main topics of the paper:

1. The Zygmund-Bari-Stechkin class Φ of continuity moduli and generalized Hölder spaces.
2. Boundedness of the singular operator with oscillating characteristic ω and oscillating weight ρ in generalized Hölder spaces.
3. Compactness theorems.
4. Fredholmness of singular integral operators in generalized Hölder spaces.
5. Fredholmness of such operators with a Carleman shift in generalized Hölder spaces.

We use the following notation:

$\Gamma = \{t : t = t(s), 0 \le s \le \ell\}$ is a curve on the complex plane;

$\Pi = \{t_1, t_2, \dots, t_n\}$ is a finite set of points on Γ;

$PC(\Gamma)$ is the space of piecewise continuous functions on Γ with jumps at a finite number of points;

$PH^\omega(\Gamma)$ is the subspace in $PC(\Gamma)$ of functions which belong to the class H^ω on any closed arc between two consecutive points of discontinuity;

M_ω and m_ω are the index numbers of a function ω (see Definition 2.12);

a.i. stands for "almost increasing" and a.d. for "almost decreasing" (see Subsection 2.1.1).

2. The Zygmund-Bari-Stechkin class Φ of continuity moduli and generalized Hölder spaces $H^\omega(\Gamma, \rho)$ with characteristic $\omega \in \Phi$

2.1. Definitions and preliminaries

2.1.1. The Zygmund-Bari-Stechkin class Φ. First we recall that a non-negative function φ on $[0, \ell]$ is said to be *almost increasing* (or *almost decreasing*) – abbreviated *a.i.* and *a.d.* – if there exists a constant $C \ge 1$ such that $\varphi(x) \le C\varphi(y)$ for all $x \le y$ (or $x \ge y$, respectively). Let

$$W = \{\varphi \in C([0,\ell]) : \varphi(0) = 0,\ \varphi(x) > 0 \ \text{ for } \ x > 0,\ \ \varphi(x) \ \text{ is almost increasing}\}.$$

Definition 2.1. A function $\varphi \in W$ is called a *continuity modulus*, if it is nondecreasing and *subadditive*: $\varphi(x_1 + x_2) \le \varphi(x_1) + \varphi(x_2)$.

Definition 2.2. ([2], [16]) We define the *Zygmund-Bari-Stechkin class* Φ_s as the class of functions $\varphi \in W$ satisfying the Zygmund conditions

$$\int_0^h \frac{\varphi(x)}{x}\,dx \le c\varphi(h) \tag{Z}$$

and

$$\int_h^\ell \frac{\varphi(x)}{x^{1+s}}\,dx \le c\frac{\varphi(h)}{h^s}, \qquad s = 1, 2, \dots \tag{Z_s}$$

By $\mathcal{Z}$ we denote the subclass of functions in W, which satisfy condition (Z) and similarly $\mathcal{Z}_s$ consists of functions in W, which satisfy condition (Z_s), so that $\Phi_s = \mathcal{Z} \cap \mathcal{Z}_s$. We write $\Phi_1 = \Phi$ in the case $s = 1$.

The class Φ_s can also be characterized in other terms. Namely, consider the following conditions (B), (B_s), (L), (L_s) and (S), (S_s):

$$\sum_{k=n+1}^{\infty} \frac{1}{k}\varphi\left(\frac{1}{k}\right) \le c\varphi\left(\frac{1}{n}\right), \tag{B}$$

$$\text{there exists a } \xi > 1 \text{ such that } \varliminf_{h\to 0}\frac{\varphi(\xi h)}{\varphi(h)} > 1, \tag{L}$$

$$\text{there exists} \quad \text{a} \ \ \delta_1 > 0 \ \ \text{such that} \ \ \frac{\omega(x)}{x^{\delta_1}} \ \ \text{is almost increasing,} \qquad (S)$$

$$\sum_{k=1}^{n} k^{s-1}\varphi\left(\frac{1}{k}\right) \le cn^s\varphi\left(\frac{1}{n}\right), \qquad (B_s)$$

$$\text{there exists} \ \ \text{a} \ \ \xi > 1 \ \ \text{such that} \ \ \overline{\lim}_{h\to 0}\frac{\varphi(\xi h)}{\varphi(h)} < \xi^s, \qquad (L_s)$$

$$\text{there exists} \ \ \text{a} \ \delta_2 \in (0,s) \ \text{such that} \ \frac{\omega(x)}{x^{\delta_2}} \ \text{is almost decreasing,} \qquad (S_s)$$

known as the Bari, Lozinskii and Stechkin conditions (see [2]).

Lemma 2.3. *Let $\varphi(x) \in W$. Conditions (B), (L), (S) and (Z) are all equivalent. Similarly, conditions (B_s), (L_s), (S_s) and (Z_s) are also equivalent.*

This lemma was proved in [2] in the case when in the definition of the class W functions $\varphi(t)$ are increasing, not almost increasing. However, the lemma remains true in this more general case (see the proof for almost monotonous φ in [34], p. 5, [36] and [17]).

Observe that functions $\varphi \in \Phi_s$ in general oscillate between two power functions whose exponents are defined by the Boyd-type indices of φ. In particular, we give a characterization of the class Φ_2 in such terms in Theorem 2.14.

2.1.2. Hölder type spaces $H_0^\omega(\Gamma,\rho)$. Many of the results from this section appear in [42]. We repeat some of the proofs for completeness.

Let $\Gamma = \{t = t(s), 0 \le s \le \ell\}$, s the arc length, be a rectifiable curve in the complex plane and let

$$H^\omega(\Gamma) = \left\{ f(t) : \max_{\substack{t\in\Gamma,\tau\in\Gamma \\ |t-\tau|\le h}} |f(t) - f(\tau)| \le c\omega(h), \ \ 0 < h < \ell \right\}$$

and

$$H_0^\omega(\Gamma) = \{f(t) : f \in H^\omega(\Gamma), \ \ f(t_0) = 0, \ \ t_0 \in \Pi\}$$

where Π is a given finite set of points on Γ. The function $\omega(h)$, is referred to in the sequel as *the characteristic function of the space*, or simply, *characteristic*, and will be supposed to belong to Φ.

Whenever necessary, we define $\omega(x)$ for $x \ge \ell$ by $\omega(x) \equiv \omega(\ell)$.

We will use some examples of functions in $H^\omega(\Gamma)$ provided by the following in Section 3.

Lemma 2.4. *Let a function $\omega \in W$ satisfy the condition $|\omega(x)-\omega(y)| \le A\omega(|x-y|)$ with $A > 0$ not depending on x and y, and let $z_j \in \mathbb{C}$, $j = 1,2,\dots,n$. Then*

$$\psi(t) := \omega\left(\prod_{j=1}^{n} |t - z_j|\right) \in H^\omega(\Gamma).$$

Proof. We have

$$|\psi(t)-\psi(\tau)| \leq A\,\omega\left(\left|\prod_{j=1}^{n}|t-z_j| - \prod_{j=1}^{n}|\tau-z_j|\right|\right) \leq A_1\omega(c|t-\tau|) \leq A_1\,c_1\omega(|t-\tau|),$$

where the property $\omega(cx) \leq c_1\omega(x)$ follows from the assumption $|\omega(x)-\omega(y)| \leq A\omega(|x-y|)$. Indeed, we have $\omega(2^n x) \leq c_0\omega(x)$ and then $\omega(cx) \leq c_1\omega(x)$, $c_1 = c_1(c)$. □

For a given $\omega \in W$ we define the following Zygmund type function

$$\omega^*(x) = \int_0^x \frac{\omega_1(s)}{s}\,ds + x\int_x^\ell \frac{\omega_1(s)}{s^2}\,ds, \qquad \text{where} \qquad \omega_1(x) = \int_0^x \frac{\omega(t)}{t}\,dt \tag{2.1}$$

The following theorem is known ([16], Theorem 1 on p. 55).

Theorem 2.5. *For $\omega \in \Phi$ the functions ω and ω^* are equivalent: $c_1\omega(x) \leq \omega^*(x) \leq c_2\omega(x)$, $c_1 > 0, c_2 > 0$, and $\omega^*(x)$ is a continuity modulus, so that it is subadditive.*

Corollary 2.6. *Let $\omega \in \Phi$ and ω^* be defined by* (2.1)*. Then $\omega^*\left(\prod_{j=1}^{n}|t-z_j|\right) \in H_0^\omega(\Gamma)$.*

Indeed, the statement of the corollary follows from Lemma 2.4 because ω^* is continuity modulus by Theorem 2.5 and then according to Definition 2.1 it satisfies the assumption of Lemma 2.4.

Recall that a curve Γ is said to satisfy the *chord-arc condition* if there exists a constant $k > 0$ such that $|s_1 - s_2| \leq k|t_1 - t_2|$ for all $t_1 = t(s_1) \in \Gamma$ and $t_2 = t(s_2) \in \Gamma$ with $k > 0$ not depending on t_1 and t_2.

Lemma 2.7. *Let Γ be a closed curve satisfying the chord-arc condition. Let $\Gamma_1 = \{t \in \Gamma : t_1 \preccurlyeq t \preccurlyeq t_2\}$ and $\Gamma_2 = \{t \in \Gamma : t_2 \preccurlyeq t \preccurlyeq t_1\}$ be two parts of Γ between arbitrary points $t_1 \in \Gamma$ and $t_2 \in \Gamma$. If $f(t) \in H^\omega(\Gamma_1)$ and $f(t) \in H^\omega(\Gamma_2)$ where $\omega \in W$ and $\omega(\lambda h) \leq C\omega(h)$ for any $\lambda > 1$, and $f(t_k - 0) = f(t_k + 0), k = 1, 2$, then $f(t) \in H^\omega(\Gamma)$.*

Proof. Estimation of the difference $f(t) - f(\tau)$ is obvious when $t, \tau \in \Gamma_1$ or $t, \tau \in \Gamma_2$. Let $t \in \Gamma_1$ and $\tau \in \Gamma_2$. It suffices to estimate $|f(t) - f(\tau)|$ in a neighborhood of the point t_1 or the point t_2. We have $|f(t)-f(\tau)| \leq |f(t)-f(t_1-0)|+|f(\tau)-f(t_1+0)| \leq c[\omega(|t-t_1|)+\omega(|\tau-t_1|)]$. Since ω is an almost increasing function, we obtain $|f(t) - f(\tau)| \leq c[\omega(s) + \omega(\sigma)] \leq c_1\omega(s + \sigma)$, where s and σ are the arc lengths of the points t and τ counted from t_1 (in the opposite direction). Since Γ satisfies the chord-arc condition and $\omega(\lambda h) \leq C\omega(h)$, we get $|f(t) - f(\tau)| \leq c\omega(|t - \tau|)$. □

Remark 2.8. Lemma 2.7 is obviously valid also for an open curve with two subarcs Γ_1 and Γ_2.

We need also the following lemma on multipliers in the space $H_0^\omega(\Gamma,\rho)$ which was proved in [35]. We say that a point $t_0\in\Gamma$ is not a *whirling point* of Γ, if $\sup_{\substack{t\in\Gamma\\ t\neq t_0}} |\arg(t-t_0)|<\infty$.

Lemma 2.9. *Let Γ be a curve satisfying the arc-cord condition, $\alpha\in\mathbb{R}^1$, $\gamma\in\mathbb{C}$, and $H_0^\omega(\Gamma)$ be the space related to the point $t_0\in\Gamma$. The functions $e^{i\alpha\arg(t-t_0)}$, $|t-t_0|^{i\alpha}$, $\frac{(t-t_0)^\gamma}{|(t-t_0)^\gamma|}$, $\frac{(t-t_0)^\alpha}{s^\alpha}$ where $s=s(t)$ is the arc length of the arc $\{t_0,t\}$, are multipliers in $H_0^\omega(\Gamma)$. If t_0 is not a whirling point of Γ, then the functions $e^{\alpha\arg(t-t_0)}$, $\frac{(t-t_0)^\gamma}{|t-t_0|^{Re\,\gamma}}$, $\frac{(t-t_0)^\gamma}{s^{Re\,\gamma}}$ are multipliers as well.*

Let $\Pi=\{t_1,t_2,\dots,t_n\}$ be any finite set of points on Γ and $\rho(t)$ any non-negative function on Γ vanishing only at the points of the set Π.

Definition 2.10. By $H_0^\omega(\Gamma,\rho),\omega\in W$, we denote the space

$$H_0^\omega(\Gamma,\rho)=\left\{u(t):\rho(t)u(t)\in H^\omega(\Gamma)\ ,\ \lim_{t\to t_k}[\rho(t)u(t)]=0,\ \ t_k\in\Pi,\ \ k=1,\dots,n\right\}$$

where $t_k\in\Gamma$ are the points at which the non-negative weight function $\rho(t)$ vanishes.

Equipped with the norm

$$\|f\|_{H_0^\omega(\Gamma,\rho)}=\|\rho f\|_{H_0^\omega(\Gamma)}=\|\rho f\|_{C(\Gamma)}+\sup_{h>0}\frac{\omega(\rho f,h)}{\omega(h)},$$

this is a Banach space.

As in [34], [37] and [38], we restrict ourselves to the weights $\rho(t)$ of the form

$$\rho(t)=\prod_{k=1}^{n}\varphi_k(|t-t_k|)\ ;\ \ t,t_k\in\Gamma,\ \ k=1,2,\dots,n\ , \tag{2.2}$$

where $\varphi_k(x)\in\Phi_2\cap K$, the class K defined in (2.6).

2.1.3. Weight functions. Let $\alpha>0$ and $\beta>0$. We find it convenient to use the following notation for some classes of weight functions:

$$W^\alpha=\left\{\varphi\in W:\frac{\varphi(x)}{x^{\alpha-\varepsilon}}\ \text{ is a.i. for any }\ \varepsilon>0\right\}, \tag{2.3}$$

$$W_\beta=\left\{\omega\in W:\frac{\varphi(x)}{x^{\beta+\varepsilon}}\ \text{ is a.d. for any }\ \varepsilon>0\right\} \tag{2.4}$$

In Theorem 2.14 and Corollary 2.17 we show that

$$\Phi_s=\bigcup_{0<\alpha\le\beta<s}W^\alpha\cap W_\beta \tag{2.5}$$

We also denote

$$\overline{W}^\alpha=\left\{\varphi\in W:\frac{\varphi(x)}{x^\alpha}\ \text{ is a.i.}\right\}\subset W^\alpha,\qquad \overline{W}_\beta=\left\{\omega\in W:\frac{\varphi(x)}{x^\beta}\ \text{ is a.d.}\right\}\subset W_\beta,$$

$$V_\beta = \left\{\omega \in W : \frac{\varphi(x)}{x^{\beta+\varepsilon}} \text{ is decreasing for any } \varepsilon > 0\right\} \subset W_\beta$$

and

$$K = \left\{\varphi \in W : |\varphi(x) - \varphi(y)| \leq c|x-y|\left(\frac{\varphi(x)}{x} + \frac{\varphi(y)}{y}\right)\right\}. \tag{2.6}$$

Remark 2.11. It can be shown that $V_\beta \subset K$; when $0 < \beta < 1$, the inequality in (2.6) for $\varphi \in V_\beta$ is valid even in the form $|\varphi(x) - \varphi(y)| \leq |x-y| \min\left\{\frac{\varphi(x)}{x}, \frac{\varphi(y)}{y}\right\}$ (see [31], p. 9).

2.1.4. Curves. We shall deal with curves of Lyapunov type. Since our consideration is within the framework of the generalized Hölder setting, it is natural to assume that the curve Γ is also a generalized Lyapunov curve. We say that a curve $\Gamma = \{t = t(s), 0 \leq s \leq \ell\}$ is *a generalized Lyapunov curve* (*GLC*), if $t'(s) \in H^\mu([0,\ell])$ with some $\mu(h) \in W$. Everywhere below, in dealing with GLC's, we assume that

$$\mu(h) \in W \textit{ is subadditive and } \frac{\mu(h)}{h} \text{ is } \textit{almost decreasing.} \tag{2.7}$$

2.2. Characterization of functions $\omega \in \Phi_s$ in terms of their indices m_ω and M_ω

2.2.1. Upper and lower indices of functions $\omega \in W$.

Definition 2.12. Let $\omega \in W$. The numbers

$$m_\omega = \sup_{x>1} \frac{\ln\left[\underline{\lim}_{h\to 0} \frac{\omega(xh)}{\omega(h)}\right]}{\ln x}, \quad M_\omega = \inf_{x>1} \frac{\ln\left[\overline{\lim}_{h\to 0} \frac{\omega(xh)}{\omega(h)}\right]}{\ln x}$$

will be referred to as the *lower and upper indices* of a function $\omega(x) \in W$. We call a characteristic $\omega(x)$ *equilibrated*, if $M_\omega = m_\omega$. These indices were introduced for functions $\omega \in W$ in such a form in [34], [37], where they are used to characterize the validity of the Lozinski conditions [2]. In fact they are the Matuszewska-Orlicz indices introduced earlier in [28] in a slightly different setting (see also [26], p. 20).

Note that the indices m_ω, M_ω may be expressed in terms of the upper limit $\overline{\Omega}(x) = \overline{\lim\limits_{h\to 0}} \frac{\omega(xh)}{\omega(h)}$ as well (this fact was drawn to our attention by A. Karlovich). Submultiplicativity of $\overline{\Omega}(x)$ (see [23], p. 75, or [3], p. 13), yields the following representation of the index numbers m_ω and M_ω:

$$m_\omega = \sup_{0<x<1} \frac{\ln \overline{\Omega}(x)}{\ln x} = \lim_{x\to 0} \frac{\ln \overline{\Omega}(x)}{\ln x}$$

$$M_\omega = \inf_{x>1} \frac{\ln \overline{\Omega}(x)}{\ln x} = \lim_{x\to\infty} \frac{\ln \overline{\Omega}(x)}{\ln x}.$$

Note that $0 \leq m_\omega \leq M_\omega \leq \infty$ for $\omega \in W$ (compare this with (2.9)).

Remark 2.13. It is easily seen that for $\omega_\lambda(x) = \frac{\omega(x)}{x^\lambda}$ one has

$$m_{\omega_\lambda} = m_\omega - \lambda, \qquad M_{\omega_\lambda} = M_\omega - \lambda. \tag{2.8}$$

The following statement characterizes the class Φ_s in terms of the indices m_ω and M_ω. A proof may be found in ([37], p. 125; see also [34]) for $s = 1$ and [17] for $s = 1, 2, \ldots$.

Theorem 2.14. *A function $\omega(x) \in W([0, \ell])$ is in the Bari-Stechkin class Φ_s if and only if*

$$0 < m_\omega \leq M_\omega < s, \tag{2.9}$$

and for $\omega \in \Phi_s$ and any $\varepsilon > 0$ there exist constants $c_1 = c_1(\varepsilon) > 0$ and $c_2 = c_2(\varepsilon) > 0$ such that

$$c_1 x^{M_\omega + \varepsilon} \leq \omega(x) \leq c_2 x^{m_\omega - \varepsilon}, \qquad 0 \leq x \leq \ell. \tag{2.10}$$

Besides this, condition (Z) is equivalent to $m_\omega > 0$ and yields the first inequality in (2.10), while condition (Z_s) is equivalent to $M_\omega < s$ and yields the second inequality in (2.10).

A statement similar to (2.9) was known in another situation – for similarly defined indices of increasing unbounded functions ω defined on $(0, \infty)$ in the context of the Orlicz type spaces ([27], p. 90).

2.2.2. The indices m_ω and M_ω as bounds for (S) and (S_s).

Lemma 2.15. ([37], *p.* 125, [34] *for $s = 1$ and* [17] *for arbitrary $s > 0$). Let $\omega \in W$. If $\omega \in \mathcal{Z}$, then $\frac{\omega(x)}{x^\alpha}$ is almost increasing for any $\alpha < m_\omega$. If $\omega \in \mathcal{Z}_s$, then $\frac{\omega(x)}{x^\beta}$ is almost decreasing for any $\beta > M_\omega$.*

The following lemma provides a converse to Lemma 2.15.

Lemma 2.16. *Let $\omega \in W$. If $\omega \in \mathcal{Z}$ and $\frac{\omega(x)}{x^\alpha}$ is almost increasing for some $\alpha > 0$, then $m_\omega \geq \alpha$. If $\omega \in \mathcal{Z}_s$ and $\frac{\omega(x)}{x^\beta}$ is almost decreasing for some $0 < \beta < s$, then $M_\omega \leq \beta$.*

Proof. Let $\omega \in \mathcal{Z}$. Suppose to the contrary that $m_\omega < \alpha$. Then the function $\omega_1(x) = \frac{\omega(x)}{x^{m_\omega}}$ is also almost increasing and $\omega_1(0) = 0$ since $m_\omega < \alpha$. Therefore, $\omega_1 \in W$. But the function $\frac{\omega_1(x)}{x^{\delta_1}} = \frac{\omega(x)}{x^\alpha}$, $\delta_1 = \alpha - m_\omega$ is almost increasing, that is, the function $\omega_1(x)$ satisfies condition (S) of Subsection 2.1. Then, by Lemma 2.3, the function $\omega_1(x)$ satisfies the (Z)-condition. Therefore, $m_{\omega_1} > 0$ by Theorem 2.14, which is impossible since $m_{\omega_1} = m_\omega - m_\omega = 0$ by (2.8).

The statement $M_\omega \leq \beta$ is similarly obtained. □

From Lemmas 2.15 and 2.16 the following important statement follows.

Corollary 2.17. ([17]) *For any function $\omega \in \mathcal{Z}$ the lower index m_ω may be calculated by the formula*

$$m_\omega = \sup\left\{\alpha > 0 : \ \frac{\omega(x)}{x^\alpha} \ \ \textit{is almost increasing}\right\}, \tag{2.11}$$

while for any $\omega \in \mathcal{Z}_s$ the upper index M_ω is calculated by the formula

$$M_\omega = \inf\left\{\beta \in (0,s) : \frac{\omega(x)}{x^\beta} \text{ is almost decreasing}\right\}.$$

2.3. Examples of oscillating (non-equilibrated) characteristics $\omega \in \Phi$

Examples of equilibrated characteristics ω are easy to find. Besides the trivial examples $\omega(x) = x^\lambda, \omega(x) = x^\lambda \left(\ln \frac{1}{x}\right)^\alpha,\ x^\lambda \left(\ln\ln \frac{1}{x}\right)^\alpha$, $0 < \lambda < 1$, for which $m_\omega = M_\omega = \lambda$, one may also find others in [41]. Examples of monotonic non-equilibrated characteristics belonging to Φ_s are less trivial [41], [42].

2.3.1. Examples of non-equilibrated characteristics $\omega \in W$. It is easy to give sufficient conditions for a function $\omega \in W$ to be non-equilibrated. For instance, if

$$c_1 x^\beta \le \omega(x) \le c_2 x^\alpha, \qquad 0 < \alpha < \beta < 1, \tag{2.12}$$

and there exist sequences $a_n \to 0$, $b_n \to 0$ such that

$$\omega(b_n) = c_1 b_n^\beta, \quad \omega(a_n) = c_2 a_n^\alpha, \tag{2.13}$$

then ω is non-equilibrated. Indeed, under conditions (2.12)–(2.13) we have $m_\omega \le \alpha$ and $M_\omega \ge \beta$. However, conditions (2.12)–(2.13) do not guarantee yet that $\omega \in \Phi_s$, because it may happen that $m_\omega = 0$ and/or $M_\omega \ge s$.

An example of a function $\omega \in \Phi_s$ with different indices m_ω and M_ω was given in [1]; in the context of submultiplicative convex functions another example of functions with different Matuszewska-Orlicz indices was given in [24], the latter example may also be found in [26], p.93.

Below we briefly dwell on an explicit construction of a family of non-equilibrated characteristics $\omega \in \Phi$ which generalize the example from [1]. This family is studied in detail in [41].

2.3.2. On a class of functions oscillating between x^β and x^α. Let

$$\cdots < a_n < a_{n-1} < \cdots < a_1 < a_0 = 1 \quad \text{and} \quad \lim_{n\to\infty} a_n = 0$$

Given $0 < \alpha < \beta < s$, we introduce a function

$$\omega(x) = \begin{cases} c_{2n+1} x^\beta, & \text{if } x \in [a_{2n+2}, a_{2n+1}], \\ c_{2n}\, x^\alpha, & \text{if } x \in [a_{2n+1}, a_{2n}], \quad n = 0,1,2,\dots \end{cases} \tag{2.14}$$

where we take $c_0 = 1$ and subsequently choose $c_1, c_2, \dots$ in such a way that $\omega(x)$ is continuous. Then

$$c_{2n} = \left(\frac{a_0 a_2 a_4 \cdots a_{2n}}{a_1 a_3 \cdots a_{2n-1}}\right)^{\beta-\alpha}, \quad c_{2n+1} = \left(\frac{a_0 a_2 a_4 \cdots a_{2n}}{a_1 a_3 \cdots a_{2n+1}}\right)^{\beta-\alpha}, \quad n = 0,1,2,3,\dots \tag{2.15}$$

and

$$c_{2n} = a_{2n}^{\beta-\alpha} c_{2n-1}, \quad c_{2n} = a_{2n+1}^{\beta-\alpha} c_{2n+1}, \quad c_{2n+2} < c_{2n}, \quad c_{2n+1} > c_{2n-1}. \tag{2.16}$$

We call any function of form (2.14)–(2.16) an *(α,β)-function.*

Observe that, conversely, given any decreasing sequence $c_{2n} > 0$ and an increasing sequence $c_{2n+1} > 0$ with the property $\lim\limits_{n\to\infty} \frac{c_{2n}}{c_{2n-1}} = 0$, there exists the partition $\{\dots, a_n, a_{n-1} \dots, a_1, a_0\}$, $a_n \to 0$, for which the corresponding (α,β)-function has the given coefficients c_{2n} and c_{2n+1}:

$$a_{2n} = \left(\frac{c_{2n}}{c_{2n-1}}\right)^{\frac{1}{\beta-\alpha}}, \qquad a_{2n+1} = \left(\frac{c_{2n}}{c_{2n+1}}\right)^{\frac{1}{\beta-\alpha}}. \tag{2.17}$$

It is easy to check (see [41]) that any (α,β)-function ω with $0 < \alpha \le \beta < s$ belongs to Φ_s and has the properties

$$a_1^{\alpha-\beta} x^\beta \le \omega(x) \le x^\alpha, \quad 0 \le x \le 1, \quad \text{and} \quad m_\omega \ge \alpha, \ M_\omega \le \beta.$$

In [42] the following statement was proved.

Theorem 2.18. *Let $A > 1$ and u_n and v_n be arbitrary positive increasing sequences with $\lim\limits_{n\to\infty} u_n = \lim\limits_{n\to\infty} v_n = \infty$ such that*

$$\lim_{n\to\infty}(v_n - u_n) = \lim_{n\to\infty}(u_n - v_{n-1}) = \infty.$$

Set

$$c_{2n} = e^{-A^{u_n}}, \qquad c_{2n+1} = e^{A^{v_n}},$$

and the corresponding points a_n calculated by formulas (2.17) with $0 < \alpha < \beta < 1$. Then (α,β)-functions $\omega(x)$ functions may be constructed such that

$$m_\omega = \alpha \qquad \text{and} \qquad M_\omega = \beta.$$

3. Boundedness of the singular operator S in the spaces $H_0^\omega(\Gamma,\rho)$

In [31], [32], [34], [37] the following two statements were proved, one given in terms of Zygmund conditions (Z) and (Z_1), another in terms of the indices m_ω and M_ω.

Let

$$\widetilde{W} = \{\varphi \in W : \exists \alpha \in (0,2) \ \text{such that} \ \frac{\varphi(x)}{x^\alpha} \ \text{is a.d.}\} \ (= \mathcal{Z}_2).$$

Theorem 3.1. ([33], *Th. 2′, p. 8;* [37], *Th. 8.5*). *Let*

(i) $\omega(h) \in \Phi$;

(ii) *the curve Γ be a GLC satisfying assumptions* (2.7) *and the condition*

$$\mu(h)(1 + |\ln h|) \le c\omega(h), \quad 0 \le h \le \ell;$$

(iii) *the weight function $\rho(t)$ have the form* (2.2) *where $\varphi_k \in \widetilde{W} \bigcap K$.*

Then the operator S is bounded in $H_0^\omega(\Gamma,\rho)$, if there exist functions $a(x), b(x) \in W$ such that

(1) $\frac{\omega(x)}{a(x)} \in \mathcal{Z}$, $\frac{x\omega(x)}{b(x)} \in \mathcal{Z}_1$;

(2) *the functions $\frac{x}{b(x)}$ and $\frac{\varphi_k(x)}{b(x)}$ are almost increasing;*

(3) *the functions $\frac{\varphi_k(x)}{xa(x)}$ are almost decreasing, $k = 1, 2, \dots, n$.*

As a consequence of this theorem we have

Theorem 3.2. ([32], *Th.* 2′, *p.* 10; [37], *Th. A, p.* 111). *Let conditions* (i) *and* (ii) *of Theorem* 3.1 *be satisfied, and the weight function* $\rho(t)$ *have the form* (2.2) *with* $\varphi_k \in W^{\alpha_k} \bigcap \overline{W}_{\alpha_k} \bigcap K, 0 < \alpha_k < 2, k = 1, 2, \ldots, n$. *Then the operator* S *is bounded in the space* $H_0^\omega(\Gamma, \rho)$, *if*

$$M_\omega < \alpha_k < 1 + m_\omega \ , \quad k = 1, 2, \ldots, n \ .$$

Theorem 3.2 in the case of the usual Hölder spaces with $\omega(h) \equiv h^\lambda$, $0 < \lambda < 1$, and power weights $\varphi_k(x) = x^{\alpha_k}$ is due to R. Duduchava [6] (where boundedness of the singular operator was proved on smooth curves).

Now we wish to show that a statement more general than that given in Theorem 3.2, follows from Theorem 3.1. Namely, the following reinforcement of Theorem 3.2 admitting oscillating weights is valid.

In the sequel for $\varphi_k \in \Phi_2 = \Phi_s\big|_{s=2}$ we denote

$$\alpha_k := m_{\varphi_k} = \sup_{x>1} \frac{\ln \left[\underline{\lim}_{h\to 0} \frac{\varphi_k(xh)}{\varphi_k(h)} \right]}{\ln \ x} = \sup \left\{ \alpha > 0 : \ \frac{\omega(x)}{x^\alpha} \quad \text{is a.i.} \right\}$$

and

$$\beta_k := M_{\varphi_k} = \inf_{x>1} \frac{\ln \left[\overline{\lim}_{h\to 0} \frac{\varphi_k(xh)}{\varphi_k(h)} \right]}{\ln \ x} = \inf \left\{ \beta > 0 : \ \frac{\varphi_k(x)}{x^\beta} \quad \text{is a.d.} \right\}$$

so that

$$0 < \alpha_k \le \beta_k < 2, k = 1, 2, \ldots, n.$$

Theorem 3.3. *Let conditions* (i) *and* (ii) *of Theorem* 3.1 *be satisfied, and the weight function* $\rho(t)$ *have form* (2.2) *with* $\varphi_k \in \Phi_2 \bigcap K$. *Then the operator* S *is bounded in the space* $H_0^\omega(\Gamma, \rho)$, *if*

$$M_\omega < \alpha_k \le \beta_k < 1 + m_\omega \ , \quad k = 1, 2, \ldots, n \ . \tag{3.1}$$

Proof. Obviously, $\varphi_k \in \overline{W}^{\gamma_k}$ for any $\gamma_k > \beta_k$. Therefore, it suffices only to check conditions (1)–(3) of Theorem 3.1.

To this end, we choose $a(x) = x^{m_\omega - \varepsilon}$ and $b(x) = x^{M_\omega + \varepsilon}$ with $\varepsilon > 0$ sufficiently small. Then the function $\frac{\omega(x)}{a(x)x^{\delta_1}} = \frac{\omega(x)}{x^{m_\omega - \varepsilon + \delta_1}}$ is almost increasing under the choice $\delta_1 < \varepsilon$ by (2.11). Also $\frac{\omega(x)}{a(x)} \in \mathcal{Z}$ by Lemma 2.3. Similarly, $\frac{x\omega(x)}{b(x)x^{1-\delta_2}} = \frac{\omega(x)}{x^{M_\omega - \delta_2 + \varepsilon}}$ is almost decreasing under the choice $\delta_2 < \varepsilon$, implying $\frac{x\omega(x)}{b(x)x^{1-\delta_2}} \in \mathcal{Z}_1$ by the same Lemma 2.3.

Checking condition (2) of Theorem 3.1, we see that $\frac{x}{b(x)} = x^{1-m_\omega+\varepsilon}$ is obviously increasing and $\frac{\varphi_k(x)}{b(x)} = \frac{\varphi_k(x)}{x^{M_\omega+\varepsilon}}$ is almost increasing since $\varphi_k \in W^{\alpha_k}$ and $\alpha_k > M_\omega + \varepsilon$ (under the choice $\varepsilon < \alpha_k - M_\omega$). Similarly, for condition (3) we observe that the function $\frac{\varphi_k(x)}{xa(x)} = \frac{\varphi_k x}{x^{m_\omega + 1 - ve}}$ is almost decreasing since $\varphi_k \in W_{\beta_k}$ and $\beta_k < 1 + m_\omega - \varepsilon$ (under the choice $\varepsilon < m_\omega + 1 - \beta_k$). □

Remark 3.4. In connection with the assumption $\varphi_k \in \Phi_2 \bigcap K$ of Theorem 3.3, observe that according to Remark 2.11 one need not check the condition $\varphi_k \in K$ if one knows that $\frac{\varphi_k(x)}{x^b}$ is decreasing for some $b > 0$, not just almost decreasing.

Theorem 3.6 below gives some indication as to whether the bounds for α_k given in (3.1) are also necessary for the operator S to be bounded in the space $H_0^\omega(\Gamma, \rho)$.

First we introduce some definitions. An open set $E \subset [0, \ell]$ with $0 \in \overline{E}$ will be called *thick near the origin* if $\int\limits_E \frac{dx}{x^\gamma} = \infty$ for any $\gamma > 1$. Obviously, a set $E = \cup_{m=1}^\infty (a_m, b_m)$, where $b_{m+1} < a_m < b_m, m = 1, 2, 3, \dots$ and $\lim\limits_{m\to\infty} b_m = 0$, is thick in the above defined sense if and only if $\sum\limits_{m=1}^{\infty} \left(\frac{1}{a_m^p} - \frac{1}{b_m^p}\right) = \infty$ for any $p > 0$.

Let $E_\omega(\varepsilon, c) = \{x \in [0, \ell] : \omega(x) \geq cx^{m_\omega+\varepsilon}\}$ where $c > 0$ may depend on ε.

Definition 3.5. By Φ^* we denote the subset of functions $\omega \in \Phi$ such that for every $\varepsilon > 0$ there exists $c = c(\varepsilon)$ such that the set $E_\omega(\varepsilon, c)$ is thick near the origin, while by Φ_* we denote the subset of functions $\omega \in \Phi$ such that for every $\varepsilon > 0$ there exists a sequence $x_n \to 0$ with the property $\omega(x_n) \leq C x_n^{M_\omega-\varepsilon}, C = C(\varepsilon)$.

Evidently, any equilibrated characteristics $\omega \in \Phi$ is in $\Phi^* \cap \Phi_*$.

Theorem 3.6. ([42]) *Let Γ be a GLC with $\mu(h) \leq c\omega(h)$ and let ρ be the weight (2.2) with $\varphi_k \in \Phi_2$, $k = 1, 2, \dots, n$. For the operator S to be bounded in the space $H_0^\omega(\Gamma, \rho)$ with $\omega \in \Phi$, it is necessary that $\alpha_k \leq M_\omega + 1$ and $\beta_k \geq m_\omega$, $k = 1, 2, \dots, n$.*

In the case $\omega \in \Phi^*$, it is necessary that $\alpha_k \leq m_\omega + 1$ and $\beta_k \geq m_\omega$, and in the case $\omega \in \Phi_*$, it is necessary that $\alpha_k \leq M_\omega + 1$, $\beta_k \geq M_\omega$, $k = 1, 2, \dots, n$. In particular, when $\omega \in \Phi^* \cap \Phi_*$, it is necessary that

$$\alpha_k \leq m_\omega + 1, \ \beta_k \geq M_\omega, \quad k = 1, 2, \dots, n.$$

Proof. **Case 1:** $\alpha_k \leq M_\omega + 1$ or $\alpha_k \leq m_\omega + 1$ when $\omega \in \Phi^*$.

Let

$$f_0(t) = \frac{\psi(t)}{\rho(t)} \quad \text{with} \quad \psi(t) = \omega^*\left(\prod_{k=1}^{n} |t - t_k|\right), \quad t \in \Gamma \tag{3.2}$$

where $\omega^*(h)$ is the function defined in (2.1). The function $f_0(t)$ belongs to $H_0^\omega(\Gamma, \rho)$ according to Corollary 2.6. However, this function is not integrable when $\alpha_k > M_\omega + 1$ (or $\alpha_k > m_\omega + 1$ and $\omega \in \Phi^*$). To show this, we observe that $f_0(t) \sim c\frac{\omega(|t-t_k|)}{\varphi(|t-t_k|)}$ as $t \to t_k$. Since $\frac{\varphi_k(x)}{x^{\alpha_k-\varepsilon}}$ is almost increasing, we have $\varphi_k(x) \leq cx^{\alpha_k-\varepsilon}$ and then for t in a neighborhood of the point t_k we obtain

$$f_0(t) \geq c\frac{\omega(|t - t_k|)}{|t - t_k|^{\alpha_k-\varepsilon}} \geq \frac{c}{|t - t_k|^{\alpha_k-M_\omega-2\varepsilon}} \tag{3.3}$$

according to the left-hand side inequality in (2.10). Therefore $f_0(t)$ is not integrable when $\alpha_k < M_\omega + 1$ (choose $0 < 2\varepsilon < \alpha_k - M_\omega - 1$).

Let now $\alpha_k > m_\omega + 1$ and $\omega \in \Phi^*$. Then we have to modify the estimation from below made in (3.3). Let $E_k(\Gamma, \varepsilon)$ be the portion of the curve Γ defined by $E_k(\Gamma, \varepsilon) = \{t \in \Gamma : |t - t_k| \in E_\omega(\varepsilon, c)\}$ where $c = c(\varepsilon)$ is chosen in accordance with Definition 3.5. Obviously, $|E_k(\Gamma, \varepsilon)| \geq |E_\omega(\Gamma, c)|$, where $|E_k(\Gamma, \varepsilon)|$ stands for the arc length of the set $E_k(\Gamma, \varepsilon)$. Then by the definition of the class Φ^*, instead of (3.3) we have

$$\int_\Gamma |f_0(t)|\,|dt| \geq c \int_{E_k(\Gamma,\varepsilon)} \frac{\omega(|t-t_k|)}{|t-t_k|^{\alpha_k-\varepsilon}}|dt| \geq c \int_{E_k(\Gamma,\varepsilon)} \frac{|dt|}{|t-t_k|^{\alpha_k-m_\omega-2\varepsilon}},$$

the last integral being divergent when $\alpha_k > m_\omega + 1$ (take $2\varepsilon < \alpha_k - m_\omega - 1$), by the definition of the class Φ^*.

Therefore, $H_0^\omega(\Gamma, \rho) \nsubseteq L_1(\Gamma)$ if $\alpha_k > M_\omega + 1$, or $\alpha_k > m_\omega + 1$ and $\omega \in \Phi^*$ and consequently the operator S cannot be bounded in the space $H_0^\omega(\Gamma, \rho)$ in these cases.

Case 2: $\beta_k \geq M_\omega$, or $\beta_k \geq m_o m$ in the case $\omega \in \Phi_*$.

Suppose that $\beta_k < M_\omega$. We choose

$$f_0(t) = \frac{t - t_k}{|t-t_k|\rho(t)t'(s)}\psi(t)$$

where $\psi(t)$ is the function from (3.2). The function $f_0(t)$ is in the space $H_0^\omega(\Gamma, \rho)$. Indeed, $\rho(t)f_0(t) = \frac{t-t_k}{|t-t_k|}\frac{1}{t'(s)}\psi(t)$, where both $\frac{t-t_k}{|t-t_k|}$ and $\frac{1}{t'(s)}$ are multipliers in this space, the former by Lemma 2.9 and the latter by the fact that $t'(s) \neq 0$, the property $t'(s) \in H^\omega(\Gamma)$ following from the assumption that Γ is a GLC with $\mu(h) \leq \omega(h)$.

However, Sf_0 is not in $H_0^\omega(\Gamma, \rho)$ when $\beta_k < M_\omega$. To prove this, we observe that

$$(Sf_0)(t_k) = \frac{1}{\pi i}\int_\Gamma \frac{\psi(t)}{|t-t_k|\rho(t)}\,ds \neq 0$$

since $\psi(t) \geq 0$. Suppose now that $Sf_0 \in H_0^\omega(\Gamma, \rho)$. Then $\rho(t)(Sf_0)(t) = \varphi_k(|t - t_k|)\rho_k(t)(Sf_0)(t) \in H_0^\omega(\Gamma)$, where $\rho_k(t) = \prod_{\substack{j=1\\ j\neq k}}^{n} \varphi_j(|t-t_j|)$. This implies $\varphi_k(|t-t_k|) \in H_0^\omega(\Gamma)$ since $(Sf_0)(t_k) \neq 0$ and $\rho_k(t_k) \neq 0$. Then $\omega(\varphi_k, h) \leq c\omega(h)$. But $\omega(\varphi_k, h) = \sup_{|x-y|\leq h} |\varphi_k(x) - \varphi_k(y)| \geq \sup_{|x|\leq h} |\varphi_k(x)| \geq ch^{\beta_k+\delta}$ where the last inequality is a consequence of the fact that $\frac{\varphi_k(x)}{x^{\beta_k+\delta}}$ is almost decreasing. Hence

$$h^{\beta_k+\delta} \leq c\omega(h) \tag{3.4}$$

which is impossible when $\beta_k < m_\omega$ according to the right-hand side inequality in (2.10). Similarly, it is also impossible when $\beta_k < M_\omega$ and $\omega \in \Phi_*$ by the definition of the class Φ_*. □

4. On compactness of operators in $H_0^\omega(\Gamma, \rho)$

We dwell on compactness results – within the frameworks of the spaces $H^\omega(\Gamma, \rho)$ – for the operators of the form

$$(T\varphi)(t) = \int_\Gamma \frac{\mathcal{K}(t,\tau) - \mathcal{K}(t,t)}{\tau - t} \varphi(\tau)\, d\tau, \tag{4.1}$$

where the function $\mathcal{K}(t,\tau)$ has some "generalized" Hölder behavior, this compactness playing a key role in the investigation of Fredholm nature of singular operators.

Observe that compactness of operators with a weak singularity has been more thoroughly considered. We refer to the book [11] on compactness results for various kinds of operators in weighted Lebesgue or Orlicz spaces. Compactness of the operators of form (4.1) in the usual Hölder space $H^\lambda(\Gamma)$ corresponding to the case $\omega(h) = h^\lambda$, when the function $\mathcal{K}(t,\tau)$ has the Hölder behavior, is known from [29], see also [30]. Compactness of operators of form (4.1) in $H^\omega(\Gamma)$ in the non-weighted case was considered in [46]–[47]. The general case of weighted spaces $H^\omega(\Gamma, \rho)$ was considered in [33], [36], [39]. We give a formulation of the most general result for the spaces $H^\omega(\Gamma, \rho)$ as found in [39].

Let

$$H^{\mu_1,\mu_2}(\Gamma \times \Gamma) = \{f(t,\tau) : |f(t,\tau) - f(u,v)| \leq A[\mu_1(|t-u|) + \mu_2(|\tau - v|)]\} \tag{4.2}$$

where $\mu_1 \in W,\ \mu_2 \in W$ and $A > 0$ does not depend on $t, \tau, u, v \in \Gamma$.

Theorem 4.1. ([39]). *Let $\omega \in \Phi$ and $\rho(t)$ be the weight (2.2) with $\varphi_k \in \Phi_2 \bigcap K,\ k = 1, 2, \ldots, n$, and let $\mathcal{K}(t,\tau) \in H^{\mu_1,\mu_2}(\Gamma \times \Gamma)$, where*

$$\mu_1(h)(1 + |ln\ h|) \leq c\omega(h), \quad \frac{\mu_2(h)}{h} \quad \text{is almost decreasing and} \quad \mu_2(h) \leq c\omega(h). \tag{4.3}$$

Then the operator T is compact in the space $H^\omega(\Gamma, \rho)$.

Corollary 4.2. *The commutator $aS - Sa$ of the singular operator is compact in the space $H_0^\omega(\Gamma, \rho)$, where $a \in H^\omega$, $\omega \in \Phi$ and ρ is the weight (2.2) with $\varphi_k \in \Phi_2 \bigcap K$, $k = 1, 2, \ldots, n$.*

5. Fredholmness of the operator $N = \mathcal{A}P_+ + \mathcal{B}P_-$

In this section we will prove a Fredholmness theorem when both the characteristic $\omega(x)$ and the weight functions $\varphi_k(x)$ are allowed to oscillate (Theorem 5.5). The curve Γ is assumed to be closed.

5.1. On (ω,ρ)-non-singularity (ω,ρ)-index

5.1.1. (ω,ρ)-non-singularity. The result on Fredholmness in $H_0^\omega(\Gamma,\rho)$ of the operator N with piecewise continuous coefficients will use the notion of the (ω,ρ)-index, which generalizes the μ-index introduced in [9] for $\omega(h)\equiv h^\mu$ when $m_\omega = M_\omega = \mu$. The (ω,ρ)-index, introduced below, is determined by the index numbers m_ω and M_ω of the characteristic $\omega\in\Phi=\Phi_1$, $(0<m_\omega<M_\omega<1)$ and the index numbers $\alpha_k,\beta_k, k=1,2,\dots,n$, of the weight functions $\varphi_k\in\Phi_2$, $0<\alpha_k\le\beta_k<2$, and plays the same role for the spaces $H_0^\omega(\Gamma,\rho)$ as the (p,ρ)-index ([15], p. 62) does for the weighted Lebesgue $L^p(\Gamma,\rho)$.

Let $a(t)\in PC(\Gamma)$ with jumps at the points $t_1,t_2,\dots,t_n\in\Gamma$ and $a(\Gamma)$ be its range. Following the ideas ([15]), we add some circular arcs (or "lunes") to the range $a(\Gamma)$ to obtain a closed set. This set in general includes lunes which turn into arcs when $m_\omega = M_\omega$. For any $\theta\in(0,1)$ we define

$$\delta(\theta)=\begin{cases}2\pi\theta, & \text{if } 0<\theta\le\frac{1}{2}\\ 2\pi(1-\theta), & \text{if } \frac{1}{2}\le\theta<1.\end{cases}$$

For every point t_k of discontinuity of $a(t)$ we introduce two circular arcs $a_k(\theta_k^1)$ and $a_k(\theta_k^2)$ connecting the points $a(t_k-0)$ and $a(t_k+0)$ which have the angles $\delta(\theta_k^1)$ and $\delta(\theta_k^2)$, respectively, where $\theta_k^1=\alpha_k-M_w$, $\theta_k^2=\beta_k-m_w$ and where it is supposed that

$$M_\omega<\alpha_k\le\beta_k<m_\omega+1,\qquad k=1,2,\dots,n.$$

We make the usual choice ([15], p. 62) of two such possible arcs: running from $z_-=a(t_k-0)$ to $z_+=a(t_k+0)$ along the arc, the straight line connecting z_- and z_- is located at the left-hand side if $\theta_k^j\le\frac{1}{2}$ and at the right-hand side if $\theta_k^j\ge\frac{1}{2}$, $j=1,2$. As is known ([15], p.63), the parametric representation of the arcs $a_k(\theta_k^j)$ is, for $j=1,2$,

$$z^j=a(t_k-0)[1-f_k(\xi)]+a(t_k+0)f_k(\xi),\quad f_k(\xi)=\frac{\sin(\pi-\theta_k^j)\xi}{\sin(\pi-\theta_k^j)}e^{i(\pi-\theta_k^j)(\xi-1)},\tag{5.1}$$

where $0\le\xi\le1$.

Let now $\mathcal{L}_k$ be the closed lune having the arcs $a_k(\alpha_k-M_w)$ and $a_k(\beta_k-m_w)$ as boundaries, that is, the set of points of the form (5.1) with θ_k^j replaced by θ running over all the values in the interval $[\theta_k^1,\theta_k^2]=[\alpha_k-M_\omega,\beta_k-m_\omega]$.

By $a_{\omega,\rho}(\Gamma)$, where the weight ρ has form (2.2) with $\varphi_k(x)\in W^{\alpha_k}\cap W_{\beta_k}$ we denote by $\mathcal{L}_k$ the closed set obtained from the range $a(\Gamma)$ by adding the lunes.

Definition 5.1. A function $a\in PC(\Gamma)$ is called (ω,ρ)-non-singular, if $0\notin a_{\omega,\rho}(\Gamma)$. An equivalent reformulation of (ω,ρ)-non-singularity of a is $\inf\limits_{t\in\Gamma}|a(t)|>0$ and

$$\mu_k:=\frac{1}{2\pi}\arg\frac{a(t_k-0)}{a(t_k+0)}\in(\beta_k-m_\omega-1,\alpha_k-M_\omega)(\operatorname{mod}1).\tag{5.2}$$

5.1.2. (ω,ρ)-index of a non-singular function $a \in PC(\Gamma)$**.** For every (ω,ρ)-non-singular function a, the winding number of the set $a_{\omega,\rho}(\Gamma)$ is naturally defined. This winding number will be called (ω,ρ)-index of the function a and denoted as $\text{ind}_{\omega,\rho}\ a$.

Remark 5.2. Obviously, for an (ω,ρ)-non-singular function a the winding number of the set $a_{\omega,\rho}(\Gamma)$ coincides with the winding number of the closed curve obtained from $a(\Gamma)$ by adding just one of the arcs $a_k(\alpha_k - M_w)$ and $a_k(\beta_k - m_w)$ instead of the whole lune $\mathcal{L}_k$. Therefore, $ind_{\omega,\rho}\ a$ in the case of the weight $\rho(t) = \prod_{k=1}^{n} \varphi_k(|t - t_k|)$, $\varphi_k \in W^{\alpha_k} \cap W_{\beta_k}$, is the same as the (p,ρ_1)-index of $a(t)$ with the weight $\rho_1(t) = \prod_{k=1}^{n} |t-t_k|^{\gamma_k}$, under the change of $\frac{1+\gamma_k}{p}$ by $\alpha_k - M_\omega$ or by $\beta_k - m_\omega$ or by any value in between (under preservation of the (ω,ρ)-non-singularity of a).

Consequently, making use of the known formulas for the (p,ρ)-index ([15], p. 66, or [45], p. 134, see also [18], p. 17), we arrive at the following formula for the (ω,ρ)-index

$$\text{ind}_{\omega,\rho}\ a = \frac{1}{2\pi}\int_\Gamma d\arg\ a(t)\ - \sum_{k=1}^{m} \mu_k \tag{5.3}$$

where the numbers μ_k defined in (5.2) are chosen in the interval $\beta_k - m_\omega - 1 < \mu_k < \alpha_k - M_\omega,\quad k = 1,2,\dots,m$.

To reformulate the (p,ρ)-index in terms of the Cauchy index, we use the standard procedure [12] of multiplication of a piecewise continuous function a by a power function to obtain a continuous function:

$$a_0(t) := \frac{a(t)}{\Lambda_a(t)} \in C(\Gamma), \quad \Lambda_a(t) = \prod_{k=1}^{n} (t-z_0)^{\gamma_k}, \quad z_0 \in \text{int}\,\Gamma, \tag{5.4}$$

where $(t-z_0)^{\gamma_k}$ is the power function defined by the cut from z_0 to infinity which crosses Γ only at the point t_k, and

$$\gamma_k = \gamma_k(a) = \frac{1}{2\pi i}\ln\frac{a(t_k-0)}{a(t_k+0)}.$$

The function $a_0(t)$ is continuous on Γ independently of the choice of a branch of the logarithm. From the known results on (p,ρ_1)-index [15] and Remark 5.2 there follows a relation between the Cauchy index (winding number) of the function $a_0(t)$ and the (ω,ρ)-index of the function $\frac{1}{a(t)}$:

$$-\,\text{ind}_{\omega,\rho}\frac{1}{a} = \text{ind}\ a_0, \tag{5.5}$$

where it is assumed that $\omega \in \Phi$, $\rho(t) = \prod_{k=1}^{n} \varphi(|t-t_k|)$, $\varphi_k \in \Phi_2$, $M_\omega < \alpha_k \le \beta_k < m_\omega + 1$, the function $a(t)$ is (ω,ρ)-non-singular, and the numbers $\text{Re}\,\gamma_k = \frac{1}{2\pi}\arg\frac{a(t_k-0)}{a(t_k+0)}$ are chosen in the interval $(\beta_k - m_\omega - 1, \alpha_k - M_\omega)$.

5.2. On the imbedding $H_0^\omega(\Gamma,\rho) \subset L^p(\Gamma,\rho_1)$

Weighted spaces $H_0^\omega(\Gamma,\rho)$ may be imbedded into weighted Lebesgue spaces

$$L^p(\Gamma,\rho_1) = \left\{ f : \int_\Gamma \rho_1(t)|f(t)|^p \, |dt| < \infty \right\}.$$

Theorem 5.3. ([42], *Theorem 6.5*) *Let Γ be a Jordan curve, $\omega \in \Phi$ and ρ be the weight function* (2.2) *with $\varphi_k \in W$ and such that $\varphi_k(x) \geq cx^{\lambda_k}, \lambda_k \in \mathbb{R}^1_+, k = 1,2,\ldots,n$. Then*

$$H_0^\omega(\Gamma,\rho) \subset L^p(\Gamma,\rho_1), \quad 1 \leq p < \infty, \tag{5.6}$$

where $\rho_1(t) = \prod_{k=1}^{n} |t-t_k|^{\nu_k}$ and $\nu_k > p(\lambda_k - m_\omega) - 1, \quad k = 1,2,\ldots,n$.

Proof. Let $n=1$ for simplicity. (The case of multiple points is treated by introducing the standard partition of unity.) Let $f \in H_0^\omega(\Gamma,\rho)$ and $\psi(t) = \rho(t)f(t) \in H_0^\omega(\Gamma)$. Then $|\psi(t)| = |\psi(t) - \psi(t_1)| \leq \omega(\psi, |t-t_1|) \leq \omega(|t-t_1|)\|\psi\|_{H_0^\omega(\Gamma)} = \omega(|t-t_1|)\|f\|_{H_0^\omega(\Gamma,\rho)}$, and we obtain

$$|f(t)| \leq \frac{\omega(|t-t_1|)}{\varphi(|t-t_1|)}\|f\|_{H_0^\omega(\Gamma,\rho)} \leq C\frac{\|f\|_{H_0^\omega(\Gamma,\rho)}}{|t-t_1|^{\gamma_k - m_\omega + \varepsilon}}$$

by (2.10) and the assumption that $\varphi_k(x) \geq cx^{\lambda_k}$. Since ε is arbitrary, we arrive at the imbedding in (5.6) under the condition $\nu_k > p(\lambda_k - m_\omega) - 1$. □

5.3. The case of continuous coefficients

The Fredholmness of the operators $\mathcal{A}P_+ + \mathcal{B}P_-$ in the space $H_0^\omega(\Gamma,\rho)$ in the case of continuous coefficients ($\mathcal{A},\mathcal{B} \in H^\omega(\Gamma)$) is obtained in the usual manner, not depending much on the space under consideration, as seen by the following theorem.

Theorem 5.4. *Let $\mathcal{A},\mathcal{B} \in H^\omega(\Gamma)$, $\omega \in \Phi$ and the curve Γ satisfy assumption* (ii) *of Theorem* 3.1. *The operator $N = \mathcal{A}P_+ + \mathcal{B}P_-$ is Fredholm in the spaces $H^\omega(\Gamma)$ and $H_0^\omega(\Gamma,\rho)$ with the weight* (2.2) *where $\varphi_k(x) \in \Phi_2 \cap K$ and $M_\omega < \alpha_k \leq \beta_k < m_\omega + 1$, if and only if $\mathcal{A}(t) \neq 0, \mathcal{B}(t) \neq 0$. In this case, its index in this space is equal to $\varkappa$, the winding number of the function $\frac{\mathcal{B}(t)}{\mathcal{A}(t)}$, and the defect numbers of the operator N are $\varkappa$ and 0, if $\varkappa \geq 0$ and 0 and $|\varkappa|$ if, $\varkappa \leq 0$.*

Such a statement is well known for many other spaces, its proof follows the same line as for the spaces $H^\lambda(\Gamma)$, for instance, since any $a \in H^\omega$ is factorable in the usual way with factors in H^ω. For the spaces $H_0^\omega(\Gamma,\rho)$), Theorem 5.4 may be found in [37] (sufficiency) and in [38], Lemma 3.3 (necessity). Note that, as usual, Fredholmness is a consequence of the existence of the regularizer $\frac{1}{ab}(bP_+ + a_+P_-)$ thanks to the compactness of the commutator $aS - Sa$ in $H_0^\omega(\Gamma,\rho)$, see Theorem 4.2.

Note that the statement of Theorem 5.4 is also valid within the framework of a class of arbitrary function spaces on Γ in which the anti-commutant $QS+SQ$, where $(Qf)(t)=\overline{f(t)}$, is compact. See Lemma 3.3 from [38].

5.4. The case of discontinuous coefficients

In this case we only have a sufficient condition for Fredholmness.

Theorem 5.5. *Assume the following conditions hold:*

(i) $\omega \in \Phi$;
(ii) *a closed curve* Γ *satisfies assumption* (ii) *of Theorem* 3.1;
(iii) *the weight function* $\rho(t)$ *has the form* (2.2) *with* $\varphi_k(x) \in \Phi_2 \cap K$ *and*

$$M_\omega < \alpha_k \le \beta_k < 1 + m_\omega, \quad k = 1, 2, \ldots, n.$$

Then the operator N *with* $\mathcal{A}, \mathcal{B} \in PH^\omega(\Gamma)$ *is Fredholm in* $H_0^\omega(\Gamma, \rho)$ *if*

(1) $\inf_{t\in\Gamma} |\mathcal{A}(t)| \neq 0, \quad \inf_{t\in\Gamma} |\mathcal{B}(t)| \neq 0,$
(2) *The function* $\frac{\mathcal{A}(t)}{\mathcal{B}(t)}$ *is* (ω, ρ)*-non-singular.*

In this case, the function $\frac{\mathcal{A}(t)}{\mathcal{B}(t)}$ *is* (ω,ρ)*-non-singular, the index of the operator* N *in* $H_0^\omega(\Gamma,\rho)$ *is equal to* $\varkappa_{H_0^\omega(\Gamma,\rho)}(N) = -ind_{\omega,\rho}\, \frac{\mathcal{A}(t)}{\mathcal{B}(t)}$, *and the defect numbers of the operator* N *are* κ *and* 0, *if* $\varkappa \ge 0$ *and* 0 *and* $|\varkappa|$ *if* $\varkappa \le 0$.

Remark 5.6. It is natural to expect that conditions (1)–(2) of Theorem 5.5 are also necessary for the operator N to be Fredholm in $H_0^\omega(\Gamma,\rho)$. However the proof requires some new ideas. The difficulty is due to the impossibility of using approximations in the space $H_0^\omega(\Gamma,\rho)$ which is not a separable space (see the proof of necessity in the case of equilibrated characteristics and non oscillating weights in [38]).

5.4.1. Model representations. We follow the standard procedure of reducing the problem of Fredholmness to that of a model operator with power type discontinuous coefficient. Let $\Lambda(t) = \Lambda_a(t) = \prod_{k=1}^{n} (t-z_0)^{\gamma_k}$ be the power function introduced in (5.4), where we take $a(t) = \frac{\mathcal{A}(t)}{\mathcal{B}(t)}$ and $a_0(t) = \frac{a(t)}{\Lambda(t)}$. The function $a_0(t)$ is continuous on Γ independently of the choice of a branch of the logarithm. Then $a_0(t) \in H^\omega(\Gamma)$ by Lemma 2.7. We put

$$(t-z_0)^{\gamma_k} = \frac{\Lambda_k^+(t)}{\Lambda_k^-(t)}, \quad k = 1, 2, \ldots, n,$$

where $\Lambda_k^\pm(t)$, $t \in \Gamma$, are the limiting values of the functions $\Lambda_k^+(z) = (z-t_k)^{\gamma_k}$, $\Lambda_k^-(z) = \left(\frac{z-t_k}{t-z_0}\right)^{\gamma_k}$, analytical in $D^+ = int\,\Gamma$ and $D^- = ext\,\Gamma$, respectively.

Under appropriate conditions we can write

$$Nf = a(t)\left(cI + d\Lambda^+ S \frac{1}{\Lambda^+}\right)(P_+ + \Lambda P_-)f \tag{5.7}$$

where $\Lambda^+(t) = \prod_{k=1}^n \Lambda_k^+(t)$, $c(t) = \frac{1+a_0(t)}{2}$, $d(t) = \frac{1-a_0(t)}{2}$. The following lemma justifies the validity of (5.7) for $\varphi \in H_0^\omega(\Gamma, \rho)$. (See Lemmas 10.1 and 10.2 in [38] and [42], Lemma 6.8.)

Lemma 5.7. *Under the assumptions* (i)–(iii) *of Theorem* 5.5, *the representation* (5.7) *is valid for* $\varphi \in H_0^\omega(\Gamma, \rho)$ *under the choice*

$$M_\omega - \alpha_k < Re\,\gamma_k < 1 + m_\omega - \beta_k, \quad k = 1, 2, \ldots, n. \tag{5.8}$$

Proof. First we observe that all the operators involved in (5.7) are bounded in the space $H_0^\omega(\Gamma, \rho)$. Thus, $d(t)$ and $v(t)$ are obvious multipliers in $H^\omega(\Gamma, \rho)$ and $a(t)$ and $\Lambda(t)$ are multipliers in $H_0^\omega(\Gamma, \rho)$ by Lemma 2.7. The operators $P_\pm$ are bounded by Theorem 3.3. The boundedness of the operator $\Lambda^+ S \frac{1}{\Lambda^+}$ in $H_0^\omega(\Gamma, \rho)$ is equivalent to that of the operator S in the space $H_0^\omega(\Gamma, \widetilde{\rho})$, where

$$\widetilde{\rho}(t) = \Lambda^+(t)\rho(t) = \prod_{k=1}^n (t - t_k)^{\gamma_k} \varphi_k(|t - t_k|).$$

Since Γ has no whirling points, by Lemma 2.9 we conclude that

$$H_0^\omega(\Gamma, \widetilde{\rho}) = H_0^\omega(\Gamma, \rho_0), \quad \text{where} \quad \rho_0(t) = \prod_{k=1}^n |t - t_k|^{\mathrm{Re}\,\gamma_k} \varphi_k(|t - t_k|).$$

Here $x^{\mathrm{Re}\,\gamma_k} \varphi_k(x)$ has the index numbers equal to $\alpha_k + \mathrm{Re}\,\gamma_k$ and $\beta_k + \mathrm{Re}\,\gamma_k$. Therefore, by Theorem 3.3 the operator S is bounded in the space $H_0^\omega(\Gamma, \rho_0)$ if $M_\omega < \alpha_k + \mathrm{Re}\,\gamma_k \leq \beta_k + \mathrm{Re}\,\gamma_k < m_\omega + 1$, which is our condition (5.8).

To show that representation (5.7) is valid on $H_0^\omega(\Gamma, \rho)$, we first note that it is valid, as is known, on "nice" functions $\varphi(t)$. This is not yet sufficient, although all the operators involved in (5.7) are bounded in $H_0^\omega(\Gamma, \rho)$, since no set of functions (different from $H_0^\omega(\Gamma, \rho)$) may be dense in $H_0^\omega(\Gamma, \rho)$. To fill in the gap in the argument, we make use of the imbedding (5.6). Since rational functions are dense in $L_p(\Gamma, \rho_1)$, $1 \leq p < \infty$, $\nu_k > -1$ (see for instance [14]), it suffices to guarantee boundedness of S and $\Lambda^+ S \frac{1}{\Lambda^+}$ in $L_p(\Gamma, \rho_1)$ which requires the condition

$$-1 < \nu_k < p - 1 \tag{5.9}$$

for the operator S, and the condition

$$-1 < \nu_k + p\mathrm{Re}\,\gamma_k < p - 1, \quad k = 1, 2, \ldots, n \tag{5.10}$$

for the operator $\Lambda^+ S \frac{1}{\Lambda^+}$, see [14]. It remains to show that both these conditions on ν_k are compatible with the condition

$$\nu_k > p(\lambda_k - m_\omega) - 1, \quad \lambda_k = \beta_k + \varepsilon, \quad \varepsilon > 0, \tag{5.11}$$

of Theorem 5.3. Conditions (5.9) and (5.11) are compatible since $\beta_k - m_\omega < 1$. Conditions (5.10) and (5.11) are also compatible because $p - 1 - p\mathrm{Re}\,\gamma_k > p(\beta_k - m_\omega) - 1 \iff \mathrm{Re}\,\gamma_k < 1 + m_\omega - \beta_k$, the latter inequality being satisfied by (5.8). Thus, it remains to choose any ν_k in the non-empty interval

$$p(\beta_k - m_\omega) - 1 < \nu_k < \min(p - 1, p - 1 - p\mathrm{Re}\,\gamma_k).$$

For such a choice, representation (5.7) is valid on $L_p(\Gamma,\rho_1)$ and then it is valid on $H_0^\omega(\Gamma,\rho)$ by (5.6). □

Lemma 5.8. ([42], *Lemma* 6.9) *Under assumptions* (i)–(iii) *of Theorem* 5.5 *and condition* (5.8), *the operator* $P_+ + \Lambda P_-$ *is invertible in* $H_0^\omega(\Gamma,\rho)$.

Proof. The equalities

$$\Lambda^+\left(P_+ + \frac{1}{\Lambda}P_-\right)\frac{1}{\Lambda^+}(P_+ + \Lambda P_-)f = (P_+ + \Lambda P_-)\Lambda^+\left(P_+ + \frac{1}{\Lambda}P_-\right)\frac{1}{\Lambda^+}f = f \tag{5.12}$$

are well known (see [15]), being easily verified on nice functions. Extension to $H_0^\omega(\Gamma,\rho)$ can be done in the same way as in the proof of Lemma 5.7, basing on the imbedding (5.6). Hence the invertibility of the operator $P_+ + \Lambda P_-$ in $H_0^\omega(\Gamma,\rho)$ follows by the boundedness in $H_0^\omega(\Gamma,\rho)$ of all the operators involved in (5.12). □

5.4.2. Proof of Theorem 5.5. We use Theorem 5.4, and Lemmas 5.7 and 5.8. Since $a(t) = \frac{\mathcal{A}(t)}{\mathcal{B}(t)}$ is (ω,ρ)-non-singular, by (5.2) we can choose the numbers $\operatorname{Re}\gamma_k = \frac{1}{2\pi}\arg\frac{a(t_k+0)}{a(t_k-0)}$ in the interval $(\alpha_k - m_\omega - 1, \alpha_k - M_\omega)$. Then by Lemma 5.7, the representation (5.7) is valid. To obtain the statements of Theorem 5.5 for the operator $N = \mathcal{A}P_+ + \mathcal{B}P_-$, it remains to refer to Theorem 5.4 and Lemma 5.8. In particular, to arrive at the formula for the index, we observe that by (5.7) and Lemma 5.8

$$\operatorname{Ind} N = \operatorname{ind}\left(cI + d\Lambda^+ S\frac{1}{\Lambda^+}\right) = -\operatorname{ind} a_0(t)$$

and then the formula for the index follows from (5.5).

6. Singular integral operators with a shift in the spaces $H_0^\omega(\Gamma,\rho)$

Finally we mention results obtained on Fredholmness of the singular integral operators

$$N\varphi := a(t)\varphi(t) + b(t)\varphi[\alpha(t)] + c(t)(S\varphi)(t) + d(t)(S\varphi)[\alpha(t)], \quad t\in\Gamma \tag{6.1}$$

Γ is a a closed or open curve and $\alpha(t)$ a Carleman shift on Γ:

$$\alpha[\alpha(t)] \equiv t, \quad t\in\Gamma,$$

and $a(t), b(t), c(t), d(t) \in PH^\omega(\Gamma)$.

The Fredholm nature of operators (6.1) in weighted Lebesgue spaces $L^p(\Gamma,\rho)$ is well known, see for instance, [14]–[15] for operators without shift and [13], [18], [22], [25], for operators with Carleman shifts. In the case of the spaces $H_0^\omega(\Gamma,\rho)$ the question of Fredholmness was open even in the case of the usual Hölder spaces, that is, $\omega(h)\equiv h^\lambda$, $0<\lambda<1$).

The result on Fredholmness given below concerns the case of non-oscillating (equilibrated) characteristics and non-oscillating weights.

The results of this section appear in [40].

6.1. Assumptions

We assume that the following conditions are satisfied:

(i) *conditions on* $\omega(h)$: $\omega \in \Phi$, $m_\omega = M_\omega$;

(ii) *conditions on the curve*: Γ is a GLC with $\mu(x)$ satisfying the assumptions in (2.7) and condition $\sup\limits_{x\in[0,\ell]} \frac{\mu(x)}{\omega(x)}(1+|ln\ x|) < \infty$;

(iii) *conditions on the shift* $\alpha(t)$: $\alpha(t) \in C^1(\Gamma)$ and $\alpha'(t) \in H^\nu(\Gamma)$ where $\nu(x)$ satisfies the same conditions as $\gamma(x)$;

(iv) *conditions on the weight*: the weight $\rho(t)$ has the form (2.2) with $\varphi_k(x) \in \Phi_2 \cap K$ with $\alpha_k = \beta_k$ and $m_\omega < \alpha_k < 1 + m_\omega, k = 1, 2, \ldots, n$;

(v) *conditions on the coefficients*: $a, b, c, d \in PH^\omega(\Gamma)$ with discontinuities at the points $t_1, \ldots, t_m$, $\alpha(t_1), \ldots, \alpha(t_m)$; in the case where the shift changes the orientation on Γ, we assume that its fixed point(s) do not coincide with any of the points $t_1, t_2, \ldots, t_m$.

6.2. Notation

Let

$$\mathfrak{A}(t) = a(t) + c(t), \quad \mathfrak{B}(t) = b(t) + d(t), \quad \mathfrak{C}(t) = a(t) - c(t), \quad \mathfrak{D}(t) = b(t) - d(t).$$

We use the standard notation

$$\Delta(t) = \widetilde{\mathfrak{A}}(t)\mathfrak{C}(t) - \mathfrak{B}(t)\widetilde{\mathfrak{D}}(t), \tag{6.2}$$

$$\Delta_1(t) = \mathfrak{A}(t)\widetilde{\mathfrak{A}}(t) - \mathfrak{B}(t)\widetilde{\mathfrak{B}}(t), \quad \Delta_2(t) = \mathfrak{C}(t)\widetilde{\mathfrak{C}}(t) - \mathfrak{D}(t)\widetilde{\mathfrak{D}}(t), \tag{6.3}$$

where $\widetilde{\mathfrak{A}}(t) = \mathfrak{A}[\alpha(t)], \widetilde{\mathfrak{B}}(t) = \mathfrak{B}[\alpha(t)]$ etc, and also

$$\Delta_+(t) = \widetilde{\mathfrak{B}}(t)\mathfrak{C}(t) - \mathfrak{A}(t)\widetilde{\mathfrak{D}}(t), \quad \Delta_-(t) = \mathfrak{A}(t)\widetilde{\mathfrak{D}}(t) - \widetilde{\mathfrak{B}}(t)\mathfrak{C}(t). \tag{6.4}$$

Let also

$$M_+(t) = \frac{1}{\Delta_2(t)} \begin{pmatrix} \widetilde{\Delta}(t) & \widetilde{\Delta}_+(t) \\ \Delta_+(t) & \Delta_2(t) \end{pmatrix}, \qquad M_-(t) = \frac{1}{\Delta(t)} \begin{pmatrix} \Delta_1(t) & \Delta_-(t) \\ -\Delta_-(t) & \Delta_2(t) \end{pmatrix}$$

under the assumption that $\Delta_2(t) \neq 0, \Delta(t) \neq 0$.

As usual, we have to distinguish the cases where the shift $\alpha(t)$ preserves or changes orientation on Γ. We put

$$M(t) = \begin{cases} M_+(t), & \text{if the shift changes orientation} \\ M_-(t), & \text{if it changes orientation.} \end{cases}$$

6.3. Fredholmness theorem

In Theorem 6.1 below the curve Γ may be closed or open. In the latter case the shift may only be orientation changing on Γ and the end-points of the curve are assumed to be included into the weight (2.2). In this case, in notation (6.2)–(6.4) one should take into account that $A(t_1 - 0) = C(t_1 - 0) = A(t_m + 0) = C(t_m + 0) = 1$, $B(t_1 - 0) = D(t_1 - 0) = B(t_m + 0) = D(t_m + 0) = 0$ and $\Delta_1(t_1 - 0) = \Delta_1(t_m + 0) = \Delta_2(t_1 - 0) = \Delta_2(t_m + 0) = 1$, $\Delta(t_1 - 0) = \Delta(t_m + 0) = 1$, $\Delta_\pm(t_1 - 0) = \Delta_\pm(t_m + 0) = 0$.

The notion of (ω, ρ)-index (see (5.3)) is used in Theorem 6.1 below. The $+$ sign in (6.5) corresponds to the case of preservation of orientation, and $-$ corresponds to the case when $\alpha(t)$ changes orientation.

We have the following statement on Fredholmness of (6.1):

Theorem 6.1. *Suppose assumptions* (i)–(v) *are satisfied.*

Then the operator N defined in (6.1) *is Fredholm in the weighted space $H_0^\omega(\Gamma, \rho)$ if and only if the following matrix singular operator*

$$\mathbb{N} = \begin{pmatrix} aI + cS & bI \pm dS \\ \tilde{b}I + \tilde{d}S & \tilde{a}I \pm \tilde{c}S \end{pmatrix} \tag{6.5}$$

is Fredholm in the space

$$H_0^{\omega,2}(\Gamma, \rho) = H_0^\omega(\Gamma, \rho) \times H_0^\omega(\Gamma, \rho),$$

in which case

$$\operatorname{Ind}_{H_0^\omega(\Gamma,\rho)} N = \tfrac{1}{2} \operatorname{Ind}_{H_0^{\omega,2}(\Gamma,\rho)} \mathbb{N}.$$

The operator (6.5) *is Fredholm in $H_0^{\omega,2}(\Gamma, \rho)$, if*

(1) $\inf_{t\in\Gamma} |\Delta_j(t)| \neq 0, j = 1, 2,$ *in case the shift preserves orientation, and* $\inf_{t\in\Gamma} |\Delta(t)| \neq 0$ *in case the shift changes orientation;*

(2) $\arg \lambda_{1,2}^{(k)} \neq 2\pi(\alpha_k - m_\omega) \pmod{2\pi}$, $k = 1, 2, \ldots, m$, *where $\lambda_{1,2}^{(k)}$ are the eigenvalues of the matrix $M^{-1}(t_k + 0)M(t_k - 0)$.*

Under conditions (1)–(2),

$$\operatorname{Ind}_{H_0^\omega(\Gamma,\rho)} N = -\frac{1}{2} \operatorname{ind}_{\omega,\rho} \frac{\Delta_1(t)}{\Delta_2(t)}$$

in case $\alpha(t)$ preserves orientation and

$$\operatorname{Ind}_{H_0^\omega(\Gamma,\rho)} N = -\frac{1}{2} \operatorname{ind}_{\omega,\rho} \frac{\Delta[\alpha(t)]}{\Delta(t)}$$

in case it changes orientation.

We refer to [40] for the proof of Theorem 6.1.

Note that in the paper [48] which recently appeared, algebras of singular operators with shift were considered in the usual Hölder H^λ-spaces with weight.

Acknowledgements

The author expresses her gratitude to the referee for the comments which helped to improve the presentation in the paper.

References

[1] V.D. Aslanov and Yu.I. Karlovich. One-sided invertibility of functional operators in reflexive Orlicz spaces. *Akad. Nauk Azerbaidzhan. SSR Dokl.*, 45(11-12):3–7, 1989.

[2] N.K. Bari and S.B. Stechkin. Best approximations and differential properties of two conjugate functions (in Russian). *Proceedings of Moscow Math. Soc.*, 5:483–522, 1956.

[3] A. Böttcher and Yu. Karlovich. *Carleson Curves, Muckenhoupt Weights, and Toeplitz Operators.* Basel, Boston, Berlin: Birkhäuser Verlag, 1997. 397 pages.

[4] L.P. Castro, R. Duduchava, and F.-O. Speck. Singular integral equations on piecewise smooth curves in spaces of smooth functions. In *Toeplitz matrices and singular integral equations* (*Pobershau,* 2001), volume 135 of *Oper. Theory Adv. Appl.*, pages 107–144. Birkhäuser, Basel, 2002.

[5] R. Duduchava and F.-O. Speck. Singular integral equations in special weighted spaces. *Georgian Math. J.*, 7(4):633–642, 2000.

[6] R.V. Duduchava. On boundedness of the operator of singular integration in weighted Hölder spaces (in Russian). *Matem. Issledov.*, 5(1):56–76, 1970.

[7] R.V. Duduchava. Singular integral equations in Hölder spaces with weight. I. Hölder coefficients. *Mat. Issled.*, 5(2(16)):104–124, 1970.

[8] R.V. Duduchava. Singular integral equations in Hölder spaces with weight. II. Piecewise Hölder coefficients. *Mat. Issled.*, 5(3(17)):58–82, 1970.

[9] R.V. Duduchava. Singular integral equations in weighted Hölder spaces (in Russian). *Matem. Issledov.*, 5(3):58–82, 1970.

[10] R.V. Duduchava. Algebras of one-dimensional singular integral operators in space of Hölder functions with weight. *Sakharth. SSR Mecn. Akad. Math. Inst. Šrom.*, 43:19–52, 1973. A collection of articles on the theory of functions, 5.

[11] D.E. Edmunds, V. Kokilashvili, and A. Meskhi. *Bounded and compact integral operators*, volume 543 of *Mathematics and its Applications.* Kluwer Academic Publishers, Dordrecht, 2002.

[12] F.D. Gakhov. *Boundary value problems.* (Russian), 3rd ed. Moscow: Nauka, 1977. 640 pages. (Transl. of 2nd edition in Oxford: Pergamon Press, 1966, 561p.).

[13] I. Gohberg and N. Krupnik. On one-dimensional singular integral operators with shift (in Russian). *Izv. Akad. Nauk Armyan. SSR, Matematika*, 8(1):3–12, 1973.

[14] I. Gohberg and N. Krupnik. *One-Dimensional Linear Singular Integral equations, Vol. I. Introduction.* Operator theory: Advances and Applications, **53**. Basel-Boston: Birkhäuser Verlag, 1992. 266 pages.

[15] I. Gohberg and N. Krupnik. *One-Dimensional Linear Singular Integral equations, Vol. II. General Theory and Applications.* Operator theory: Advances and Applications, **54**. Basel-Boston: Birkhäuser Verlag, 1992. 232 pages.

[16] A.I. Guseinov and H.Sh. Mukhtarov. *Introduction to the theory of nonlinear singular integral equations* (in Russian). Moscow, Nauka, 1980. 416 pages.

[17] N.K. Karapetiants and N.G. Samko. Weighted theorems on fractional integrals in the generalized Hölder spaces $H_0^{\omega}(\rho)$ via the indices m_ω and M_ω. *Fract. Calc. Appl. Anal.*, 7(4), 2005.

[18] N.K. Karapetiants and S.G. Samko. *Equations with Involutive Operators*. Birkhäuser, Boston, 2001. 427 pages.

[19] A.Yu. Karlovich. Algebras of singular integral operators with PC coefficients in rearrangement-invariant spaces with Muckenhoupt weights. *J. Operator Theory*, 47(2):303–323, 2002.

[20] A.Yu. Karlovich. Fredholmness of singular integral operators with piecewise continuous coefficients on weighted Banach function spaces. *J. Integr. Eq. and Appl.*, 15(3):263–320, 2003.

[21] V. Kokilashvili and S. Samko. Singular Integrals in Weighted Lebesgue Spaces with Variable Exponent. *Georgian Math. J.*, 10(1):145–156, 2003.

[22] V.G. Kravchenko and G.S. Litvinchuk. *Introduction to the theory of singular integral operators with shift.* London: Kluwer Academic Publishers, 1994. 286 pages.

[23] S.G. Krein, Yu.I. Petunin, and E.M. Semenov. *Interpolation of linear operators.* Moscow: Nauka, 1978. 499 pages.

[24] K. Lindberg. On subspaces of Orlicz sequence spaces. *Studia Math.*, 45:119–146, 1973.

[25] G.S. Litvinchuk. *Boundary Value Problems and Singular Integral Equations with Shift.* (in Russian). Moscow: Nauka, 1977. 448 pages.

[26] L. Maligranda. Indices and interpolation. *Dissertationes Math.* (Rozprawy Mat.), 234:49, 1985.

[27] L. Maligranda. *Orlicz spaces and interpolation.* Departamento de Matemática, Universidade Estadual de Campinas, 1989. Campinas SP Brazil.

[28] W. Matuszewska and W. Orlicz. On some classes of functions with regard to their orders of growth. *Studia Math.*, 26:11–24, 1965.

[29] N.I. Muskhelishvili. *Singular Integral Equations.* (Russian). Moskow: Nauka, 1968. 511 pages.

[30] S. Prössdorf. *Some classes of singular equtions.* (Russian). Moscow: Mir, 1979. 493 pages.

[31] N.G. Samko. Weighted zygmund estimate for the singular operator and a theorem on its boundedness in $H_0^\omega(\rho)$ in the case of general weights (in Russian). Deponierted in VINITI, Moscow, 1989. no. 7559-B89, 49 pp.

[32] N.G. Samko. Fredholmness of singular integral equations with discontinuous coefficients in weighted generalized Hölder spaces $H_0^\omega(\gamma, \rho)$ (in Russian). Deponierted in VINITI, Moscow, 1990. no. 6230-B90, 32p.

[33] N.G. Samko. Singular operator and operators with a weak singularity in weighted generalized Hölder spaces (in Russian). Deponierted in VINITI, Moscow, 1990. no. 2798-B90, 38p.

[34] N.G. Samko. On boundedness of singular operator in weighted generalized Hölder spaces $H_0^\omega(\Gamma, \rho)$ in terms of upper and lower indices of these spaces (in Russian). Deponierted in VINITI, Moscow, 1991. no. 349-B91, 28p.

[35] N.G. Samko. On multipliers in $H_0^\omega(\Gamma)$ and coincidence of the spaces $H_0^\omega(\Gamma, \rho_1)$, $H_0^\omega(\Gamma, \rho_2)$ with different weights. Deponierted in VINITI, Moscow, 1991. no. 350-B91, 21p.

[36] N.G. Samko. *Singular Integral Operators with Discontinuous Coefficients in the Generalized Hölder Spaces (in Russian)*. PhD thesis, Voronezh State University, Voronezh, 150 p., 1991.

[37] N.G. Samko. Singular integral operators in weighted spaces with generalized Hölder condition. *Proc. A. Razmadze Math. Inst*, 120:107–134, 1999.

[38] N.G. Samko. Criterion of Fredholmness of singular operators with piece-wise continuous coefficients in the generalized Hölder spaces with weight. In *Proceedings of IWOTA 2000, Setembro 12-15, Faro, Portugal*, pages 363–376. Birkhäuser, In: "Operator Theory: Advances and Applications", v. 142, 2002.

[39] N.G. Samko. On compactness of Integral Operators with a Generalized Weak Singularity in Weighted Spaces of Continuous Functions with a Given Continuity Modulus. *Proc. A. Razmadze Math. Inst*, 136:91 – 113, 2004.

[40] N.G. Samko. Singular integral operators with Carleman shift and discontinuous coefficients in the spaces $H_0^\omega(\Gamma, \rho)$. *Integr. Equat. Operator Theory*, 51(3):417, 2004.

[41] N.G. Samko. On non-equilibrated almost monotonic functions of the Zygmund-Bary-Stechkin class. *Real Anal. Exch.*, 30(2), 2005.

[42] N.G. Samko. Singular integral operator in weighted spaces of continuous functions with non-equilibrated continuity modulus. *Math. Nachr.*, 2006. (to appear).

[43] H. Schulze. On singular integral operators on weighted Hoelder spaces. *Wiss. Z. Tech. Univ. Chemnitz*, 33(1):37–47, 1991.

[44] A.P. Soldatov. *One dimensional singular operators and boundary value problems of function theory*. Aktual′nye Voprosy Prikladnoĭ i Vychislitel′noĭ Matematiki. [Current Problems in Applied and Computational Mathematics]. (Russian), "Vyssh. Shkola", Moscow, 1991.

[45] I.M. Spitkovsky. Singular integral operators with PC symbols on the spaces with general weights. *J. Funct. Anal.*, 105(1):129–143, 1992.

[46] B.M. Tursunkulov. Completely continuous operators in generalized Hölder spaces. *Dokl. Akad. Nauk UzSSR*, (12):4–6, 1982.

[47] B.M. Tursunkulov. *Noether theory of singular integral equations with a non-Carleman shift*. PhD thesis, Samarkand State University, 1983.

[48] G.Yu. Vinogradova. On algebras of singular integral operators with shift in Hölder weighted spaces. *J. Math. Sciences*, 126:1574–1579, 2005.

Natasha Samko
Departamento de Matemática
Faculdade de Ciências e Tecnologia
Universidade do Algarve
Campus de Gambelas
8005-139 Faro, Portugal
e-mail: `nsamko@ualg.pt`

Operator Theory:
Advances and Applications, Vol. 171, 349–359

Poly-Bergman Spaces and Two-dimensional Singular Integral Operators

Nikolai Vasilevski

Abstract. We describe a direct and transparent connection between the poly-Bergman type spaces on the upper half-plane and certain two-dimensional singular integral operators.

Mathematics Subject Classification (2000). 30G30, 45P05, 47B38.

Keywords. Poly-Bergman spaces, singular integral operators, poly-Bergman projections.

1. Introduction

We show that there is a direct and transparent connection between the poly-Bergman type spaces and certain two-dimensional singular integral operators.

Recall that the poly-Bergman spaces $\mathcal{A}_n^2(\Pi)$ and $\widetilde{\mathcal{A}}_n^2(\Pi)$ on the upper half-plane Π, of analytic and anti-analytic functions respectively, are defined as the subspaces of $L_2(\Pi)$, endowed with the standard Lebesgue plane measure $dv(z) = dxdy$, $z = x + iy$, and consist of functions satisfying the following equations

$$\left(\frac{\partial}{\partial \overline{z}}\right)^n \varphi = \frac{1}{2^n}\left(\frac{\partial}{\partial x} + i\frac{\partial}{\partial y}\right)^n \varphi = 0, \qquad n \in \mathbb{N},$$

and

$$\left(\frac{\partial}{\partial z}\right)^n \varphi = \frac{1}{2^n}\left(\frac{\partial}{\partial x} - i\frac{\partial}{\partial y}\right)^n \varphi = 0, \qquad n \in \mathbb{N},$$

respectively.

We introduce as well the following singular integral operators bounded on $L_2(\Pi)$:

$$(S_\Pi \varphi)(z) = -\frac{1}{\pi}\int_\Pi \frac{\varphi(\zeta)}{(\zeta - z)^2}\, dv(\zeta)$$

This work was partially supported by CONACYT Project U40654-F, México.

and its adjoint

$$(S_\Pi^*\varphi)(z) = -\frac{1}{\pi}\int_\Pi \frac{\varphi(\zeta)}{(\overline{\zeta}-\overline{z})^2}\, dv(\zeta).$$

A. Dzhuraev [4, 5] showed that for a bounded domain D with smooth boundary the orthogonal projections $B_{D,n}$ and $\widetilde{B}_{D,n}$ of $L_2(D)$ onto the spaces $\mathcal{A}_n^2(D)$ and $\widetilde{\mathcal{A}}_n^2(D)$, respectively, can be expressed in the form

$$B_{\Pi,n} = I - (S_D)^n(S_D^*)^n + K_n \qquad \text{and} \qquad \widetilde{B}_{\Pi,n} = I - (S_D^*)^n(S_D)^n + \widetilde{K}_n,$$

where K_n and $\widetilde{K}_n$ are compact operators. Recently J. Ramírez and I. Spitkovsky [8] proved that in the case of the upper half-plane Π the compact summands K_n and $\widetilde{K}_n$ in the above formulas are equal to zero. Using this result Yu. Karlovich and L. Pessoa [7] described the action of the operators S_Π and S_Π^* on the poly-Bergman spaces, obtaining the statements of Theorem 3.5 below.

In this paper we propose another, more direct and transparent, approach to the problem, which follows the ideas of [9, 10] and gives precise information about the structure of S_Π and S_Π^*. In Section 2 we present necessary facts from [9, 10]. The core result of the paper is contained in Theorems 3.1 and 3.2 and gives a simple (functional) model for the operators S_Π and S_Π^*: *each of them is unitary equivalent to the direct sum of two unilateral shifts, forward and backward, both taken with the infinite multiplicity.* This fact permits us an easy access to the majority of the properties of these operators. The most important properties, in the context of the paper, are given by the subsequent Theorems 3.5 and 3.7.

2. Poly-Bergman spaces

Let Π be the upper half-plane in $\mathbb{C}$, consider the space $L_2(\Pi)$ endowed with the usual Lebesgue plane measure $dv(z) = dxdy$, $z = x + iy$. Denote by $\mathcal{A}^2(\Pi)$ its Bergman subspace, i.e., the subspace which consists of all functions analytic in Π. It is well known that the Bergman projection B_Π of $L_2(\Pi)$ onto $\mathcal{A}^2(\Pi)$ has the form

$$(B_\Pi\varphi)(z) = -\frac{1}{\pi}\int_\Pi \frac{\varphi(\zeta)}{(z-\overline{\zeta})^2}\, dv(\zeta).$$

In addition to the Bergman space $\mathcal{A}^2(\Pi)$ introduce the space $\widetilde{\mathcal{A}}^2(\Pi)$ as the subspace of $L_2(\Pi)$ consisting of all functions anti-analytic in Π.

Further, analogously to the Bergman spaces $\mathcal{A}^2(\Pi)$ and $\widetilde{\mathcal{A}}^2(\Pi)$, introduce the spaces of poly-analytic and poly-anti-analytic functions (see, for example, [1, 2, 4, 5]), the poly-Bergman spaces.

We define the space $\mathcal{A}_n^2(\Pi)$ of n-analytic functions as the subspace of $L_2(\Pi)$ of all functions $\varphi = \varphi(z,\overline{z}) = \varphi(x,y)$, which satisfy the equation

$$\left(\frac{\partial}{\partial \overline{z}}\right)^n \varphi = \frac{1}{2^n}\left(\frac{\partial}{\partial x} + i\frac{\partial}{\partial y}\right)^n \varphi = 0.$$

Similarly, we define the space $\widetilde{\mathcal{A}}_n^2(\Pi)$ of n-anti-analytic functions as the subspace of $L_2(\Pi)$ of all functions $\varphi = \varphi(z, \overline{z}) = \varphi(x, y)$, which satisfy the equation

$$\left(\frac{\partial}{\partial z}\right)^n \varphi = \frac{1}{2^n}\left(\frac{\partial}{\partial x} - i\frac{\partial}{\partial y}\right)^n \varphi = 0.$$

Of course, we have $\mathcal{A}_1^2(\Pi) = \mathcal{A}^2(\Pi)$ and $\widetilde{\mathcal{A}}_1^2(\Pi) = \widetilde{\mathcal{A}}^2(\Pi)$, for $n = 1$, as well as $\mathcal{A}_n^2(\Pi) \subset \mathcal{A}_{n+1}^2(\Pi)$ and $\widetilde{\mathcal{A}}_n^2(\Pi) \subset \widetilde{\mathcal{A}}_{n+1}^2(\Pi)$, for each $n \in \mathbb{N}$.

Finally introduce the space $\mathcal{A}_{(n)}^2(\Pi)$ of true-n-analytic functions by

$$\mathcal{A}_{(n)}^2(\Pi) = \mathcal{A}_n^2(\Pi) \ominus \mathcal{A}_{n-1}^2(\Pi),$$

for $n > 1$, and by $\mathcal{A}_{(1)}^2(\Pi) = \mathcal{A}_1^2(\Pi)$; and, symmetrically, introduce the space $\widetilde{\mathcal{A}}_{(n)}^2(\Pi)$ of true-n-anti-analytic functions by

$$\widetilde{\mathcal{A}}_{(n)}^2(\Pi) = \widetilde{\mathcal{A}}_n^2(\Pi) \ominus \widetilde{\mathcal{A}}_{n-1}^2(\Pi),$$

for $n > 1$, and by $\widetilde{\mathcal{A}}_{(1)}^2(\Pi) = \widetilde{\mathcal{A}}_1^2(\Pi)$, for $n = 1$.

We have, of course,

$$\mathcal{A}_n^2(\Pi) = \bigoplus_{k=1}^{n} \mathcal{A}_{(k)}^2(\Pi) \qquad \text{and} \qquad \widetilde{\mathcal{A}}_n^2(\Pi) = \bigoplus_{k=1}^{n} \widetilde{\mathcal{A}}_{(k)}^2(\Pi).$$

To formulate the main result of this section we need more definitions. We start by introducing two unitary operators. Define the unitary operator

$$U_1 = F \otimes I : L_2(\Pi) = L_2(\mathbb{R}) \otimes L_2(\mathbb{R}_+) \longrightarrow L_2(\mathbb{R}) \otimes L_2(\mathbb{R}_+), \tag{1}$$

where the Fourier transform $F : L_2(\mathbb{R}) \to L_2(\mathbb{R})$ is given by

$$(Ff)(x) = \frac{1}{\sqrt{2\pi}} \int_{\mathbb{R}} e^{-ix\xi} f(\xi)\, d\xi. \tag{2}$$

The second unitary operator

$$U_2 : L_2(\Pi) = L_2(\mathbb{R}) \otimes L_2(\mathbb{R}_+) \longrightarrow L_2(\mathbb{R}) \otimes L_2(\mathbb{R}_+)$$

is given by

$$(U_2\varphi)(x, y) = \frac{1}{\sqrt{2|x|}}\, \varphi(x, \frac{y}{2|x|}). \tag{3}$$

Then the inverse operator $U_2^{-1} = U_2^* : L_2(\mathbb{R}) \otimes L_2(\mathbb{R}_+) \longrightarrow L_2(\mathbb{R}) \otimes L_2(\mathbb{R}_+)$ acts as follows,

$$(U_2^{-1}\varphi)(x, y) = \sqrt{2|x|}\, \varphi(x, 2|x| \cdot y).$$

Recall (see, for example, [3]), that the Laguerre polynomial $L_n(y)$ of degree n, $n = 0, 1, 2, \ldots$, and type 0 is defined by

$$\begin{aligned} L_n(y) \quad &= \quad L_n^0(y) = \frac{e^y}{n!}\frac{d^n}{dy^n}(e^{-y}\, y^n) \\ &= \quad \sum_{k=0}^{n} \frac{n!}{k!(n-k)!}\, \frac{(-y)^k}{k!}, \qquad y \in \mathbb{R}_+, \end{aligned} \tag{4}$$

and that the system of functions

$$\ell_n(y) = e^{-y/2} L_n(y), \qquad n = 0, 1, 2, \dots \tag{5}$$

forms an orthonormal basis in the space $L_2(\mathbb{R}_+)$.

Denote by L_n, $n = 0, 1, 2, \dots$, the one-dimensional subspace of $L_2(\mathbb{R}_+)$ generated by the function $\ell_n(y)$.

The main result of the section reads as follows.

Theorem 2.1. *The unitary operator*

$$U = U_2 U_1 : L_2(\mathbb{R}) \otimes L_2(\mathbb{R}_+) \to L_2(\mathbb{R}) \otimes L_2(\mathbb{R}_+)$$

provides the following isometrical isomorphisms of the above spaces:

1. *Isomorphic images of poly-analytic spaces*

$$U \; : \; \mathcal{A}^2_{(n)}(\Pi) \longrightarrow L_2(\mathbb{R}_+) \otimes L_{n-1},$$

$$U \; : \; \mathcal{A}^2_{n}(\Pi) \longrightarrow L_2(\mathbb{R}_+) \otimes \bigoplus_{k=0}^{n-1} L_k,$$

$$U \; : \; \bigoplus_{k=1}^{\infty} \mathcal{A}^2_{(k)}(\Pi) \longrightarrow L_2(\mathbb{R}_+) \otimes L_2(\mathbb{R}_+).$$

2. *Isomorphic images of poly-anti-analytic spaces*

$$U \; : \; \widetilde{\mathcal{A}}^2_{(n)}(\Pi) \longrightarrow L_2(\mathbb{R}_-) \otimes L_{n-1},$$

$$U \; : \; \widetilde{\mathcal{A}}^2_{n}(\Pi) \longrightarrow L_2(\mathbb{R}_-) \otimes \bigoplus_{k=0}^{n-1} L_k,$$

$$U \; : \; \bigoplus_{k=1}^{\infty} \widetilde{\mathcal{A}}^2_{(k)}(\Pi) \longrightarrow L_2(\mathbb{R}_-) \otimes L_2(\mathbb{R}_+).$$

3. *Furthermore we have the following decomposition of the space* $L_2(\Pi)$

$$\begin{aligned} L_2(\Pi) &= \bigoplus_{k=1}^{\infty} (\mathcal{A}^2_{(k)}(\Pi) \oplus \widetilde{\mathcal{A}}^2_{(k)}(\Pi)) \\ &= \bigoplus_{k=1}^{\infty} \mathcal{A}^2_{(k)}(\Pi) \oplus \bigoplus_{k=1}^{\infty} \widetilde{\mathcal{A}}^2_{(k)}(\Pi). \end{aligned}$$

3. Two-dimensional singular integral operators

We introduce the following singular integral operators bounded on $L_2(\Pi)$:

$$(S_\Pi \varphi)(z) = -\frac{1}{\pi} \int_\Pi \frac{\varphi(\zeta)}{(\zeta - z)^2}\, dv(\zeta)$$

and its adjoint

$$(S^*_\Pi\varphi)(z) = -\frac{1}{\pi}\int_\Pi \frac{\varphi(\zeta)}{(\overline{\zeta}-\overline{z})^2}\,dv(\zeta).$$

Note, that the operators S_Π and S^*_Π are the restrictions onto the upper half-plane Π of the following classical two-dimensional singular integral operators over $\mathbb{C} = \mathbb{R}^2$,

$$(S_{\mathbb{R}^2}\varphi)(z) = -\frac{1}{\pi}\int_{\mathbb{R}^2} \frac{\varphi(\zeta)}{(\zeta-z)^2}\,dv(\zeta) \quad \text{and} \quad (S^*_{\mathbb{R}^2}\varphi)(z) = -\frac{1}{\pi}\int_{\mathbb{R}^2} \frac{\varphi(\zeta)}{(\overline{\zeta}-\overline{z})^2}\,dv(\zeta),$$

which are given in terms of the Fourier transform as follows,

$$S_{\mathbb{R}^2} = F^{-1}\frac{\overline{\zeta}}{\zeta}F \qquad \text{and} \qquad S^*_{\mathbb{R}^2} = S^{-1}_{\mathbb{R}^2} = F^{-1}\frac{\zeta}{\overline{\zeta}}F, \tag{6}$$

where $\zeta = \xi + i\eta = (\xi,\eta)$, and the Fourier transform F is given by

$$(F\varphi)(\zeta) = \frac{1}{2\pi}\int_{\mathbb{R}^2} e^{-i\zeta\cdot z}\,\varphi(z)\,dv(z),$$

where $z = x+iy = (x,y)$, and $\zeta\cdot z = \xi x + \eta y$.

By (6) these operators admit the following representations:

$$\begin{aligned} S_\Pi &= (I\otimes\chi_+I)S_{\mathbb{R}^2}(I\otimes\chi_+I) \\ &= (I\otimes\chi_+I)(F^{-1}\otimes F^{-1})\frac{\xi-i\eta}{\xi+i\eta}(F\otimes F)(I\otimes\chi_+I) \end{aligned} \tag{7}$$

and

$$\begin{aligned} S^*_\Pi &= (I\otimes\chi_+I)S^*_{\mathbb{R}^2}(I\otimes\chi_+I) \\ &= (I\otimes\chi_+I)(F^{-1}\otimes F^{-1})\frac{\xi+i\eta}{\xi-i\eta}(F\otimes F)(I\otimes\chi_+I), \end{aligned}$$

where $\xi,\ \eta\in\mathbb{R}$, and the one-dimensional Fourier transform F is given by (2).

Let us introduce the following integral operators

$$\begin{aligned} (S_+f)(y) &= -f(y) + e^{-\frac{y}{2}}\int_0^y e^{\frac{t}{2}}\,f(t)\,dt, \\ (S_-f)(y) &= -f(y) + e^{\frac{y}{2}}\int_y^\infty e^{-\frac{t}{2}}\,f(t)\,dt, \end{aligned}$$

which, as we will see later on, are bounded on $L_2(\mathbb{R}_+)$ and are mutually adjoint.

As in Section 2 we will use the unitary operator

$$U = U_2U_1 : L_2(\mathbb{R})\otimes L_2(\mathbb{R}_+) \longrightarrow L_2(\mathbb{R})\otimes L_2(\mathbb{R}_+),$$

where the operators U_1 and U_2 are given by (1) and (3) respectively.

Theorem 3.1. *The unitary operator $U = U_2U_1$ gives an isometrical isomorphism of the space $L_2(\Pi) = [L_2(\mathbb{R}_+)\otimes L_2(\mathbb{R}_+)]\oplus[L_2(\mathbb{R}_-)\otimes L_2(\mathbb{R}_+)]$ under which the* *two-dimensional singular integral operators S_Π and S_Π^* are unitary equivalent to the following operators*

$$\begin{aligned} U\,S_\Pi\,U^{-1} &= (I\otimes S_+)\oplus(I\otimes S_-),\\ U\,S_\Pi^*\,U^{-1} &= (I\otimes S_-)\oplus(I\otimes S_+).\end{aligned}$$

Proof. By the representation (7) we have

$$\begin{aligned} S_1 &= U_1S_\Pi U_1^{-1} = (F\otimes I)S_\Pi(F^{-1}\otimes I)\\ &= (I\otimes\chi_+I)(I\otimes F^{-1})\frac{\xi-i\eta}{\xi+i\eta}(I\otimes F)(I\otimes\chi_+I).\end{aligned}$$

The operator U_2 is unitary on both $L_2(\mathbb{R}_+)$ and $L_2(\mathbb{R})$, and furthermore it commutes with $\chi_{\mathbb{R}_+}I$. Direct calculation shows that

$$U_2(I\otimes F^{-1})\frac{\xi-i\eta}{\xi+i\eta}(I\otimes F)U_2^{-1} = (I\otimes F^{-1})\frac{\frac{1}{2}\mathrm{sign}\,x-i\eta}{\frac{1}{2}\mathrm{sign}\,x+i\eta}(I\otimes F).$$

Thus

$$\begin{aligned} S_2 &= US_\Pi U^{-1} = U_2S_1U_2^{-1}\\ &= (\chi_+I\otimes\chi_+I)(I\otimes F^{-1})\frac{\frac{1}{2}-i\eta}{\frac{1}{2}+i\eta}(I\otimes F)(\chi_+I\otimes\chi_+I)\\ &+ (\chi_-I\otimes\chi_+I)(I\otimes F^{-1})\frac{\frac{1}{2}+i\eta}{\frac{1}{2}-i\eta}(I\otimes F)(\chi_-I\otimes\chi_+I)\end{aligned}$$

and

$$\begin{aligned} S_2^* &= US_\Pi^*U^{-1}\\ &= (\chi_+I\otimes\chi_+I)(I\otimes F^{-1})\frac{\frac{1}{2}+i\eta}{\frac{1}{2}-i\eta}(I\otimes F)(\chi_+I\otimes\chi_+I)\\ &+ (\chi_-I\otimes\chi_+I)(I\otimes F^{-1})\frac{\frac{1}{2}-i\eta}{\frac{1}{2}+i\eta}(I\otimes F)(\chi_-I\otimes\chi_+I).\end{aligned}$$

The symbols of the two convolution operators

$$\widetilde{S}_+ = F^{-1}\frac{\frac{1}{2}-i\eta}{\frac{1}{2}+i\eta}F \quad\text{and}\quad \widetilde{S}_- = F^{-1}\frac{\frac{1}{2}+i\eta}{\frac{1}{2}-i\eta}F,$$

which are obviously bounded on $L_2(\mathbb{R})$, admit the following representations,

$$\frac{\frac{1}{2}-i\eta}{\frac{1}{2}+i\eta} = -1-\frac{i\eta}{\frac{1}{4}+\eta^2}+\frac{\frac{1}{2}}{\frac{1}{4}+\eta^2} \quad\text{and}\quad \frac{\frac{1}{2}+i\eta}{\frac{1}{2}-i\eta} = -1+\frac{i\eta}{\frac{1}{4}+\eta^2}+\frac{\frac{1}{2}}{\frac{1}{4}+\eta^2},$$

respectively.

Using the formulas 17.23.14 and 17.23.15 of [6] we have

$$F\left(\frac{\frac{1}{2}}{\frac{1}{4}+\eta^2}\right) = \sqrt{\frac{\pi}{2}}\,e^{-\frac{|y|}{2}}, \qquad F\left(\frac{i\eta}{\frac{1}{4}+\eta^2}\right) = \sqrt{\frac{\pi}{2}}\,\mathrm{sign}\,y\,e^{-\frac{|y|}{2}},$$

and thus

$$\begin{aligned}(\widetilde{S}_+ f)(y) &= -f(y) + \frac{1}{\sqrt{2\pi}} \int_{\mathbb{R}} \sqrt{\frac{\pi}{2}} e^{-\frac{|t-y|}{2}} (1 - \operatorname{sign}(t-y)) f(t) dt \\ &= -f(t) + \int_{\mathbb{R}} e^{-\frac{|t-y|}{2}} \chi_-(t-y) f(t) dt\end{aligned}$$

and

$$\begin{aligned}(\widetilde{S}_- f)(y) &= -f(y) + \frac{1}{\sqrt{2\pi}} \int_{\mathbb{R}} \sqrt{\frac{\pi}{2}} e^{-\frac{|t-y|}{2}} (1 + \operatorname{sign}(t-y)) f(t) dt \\ &= -f(t) + \int_{\mathbb{R}} e^{-\frac{|t-y|}{2}} \chi_+(t-y) f(t) dt.\end{aligned}$$

Then the operators $S_+ = \chi_+ \widetilde{S}_+ \chi_+ I_{|L_2(\mathbb{R}_+)}$ and $S_- = \chi_+ \widetilde{S}_- \chi_+ I_{|L_2(\mathbb{R}_+)}$, acting on $L_2(\mathbb{R}_+)$, are as follows:

$$\begin{aligned}(S_+ f)(y) &= -f(t) + \int_{\mathbb{R}_+} e^{-\frac{|t-y|}{2}} \chi_-(t-y) f(t) dt \\ &= -f(y) + e^{-\frac{y}{2}} \int_0^y e^{\frac{t}{2}} f(t) dt\end{aligned}$$

and

$$\begin{aligned}(S_- f)(y) &= -f(t) + \int_{\mathbb{R}_+} e^{-\frac{|t-y|}{2}} \chi_+(t-y) f(t) dt \\ &= -f(y) + e^{\frac{y}{2}} \int_y^\infty e^{-\frac{t}{2}} f(t) dt.\end{aligned}$$

Thus finally

$$\begin{aligned}U S_\Pi U^{-1} &= (\chi_+ I \otimes \chi_+ I)(I \otimes \widetilde{S}_+)(\chi_+ I \otimes \chi_+ I) \\ &+ (\chi_- I \otimes \chi_+ I)(I \otimes \widetilde{S}_-)(\chi_- I \otimes \chi_+ I) \\ &= \chi_+ I \otimes S_+ + \chi_- I \otimes S_- \\ &= (I \otimes S_+) \oplus (I \otimes S_-)\end{aligned}$$

and

$$\begin{aligned}U S_\Pi^* U^{-1} &= (\chi_+ I \otimes \chi_+ I)(I \otimes \widetilde{S}_-)(\chi_+ I \otimes \chi_+ I) \\ &+ (\chi_- I \otimes \chi_+ I)(I \otimes \widetilde{S}_+)(\chi_- I \otimes \chi_+ I) \\ &= \chi_+ I \otimes S_- + \chi_- I \otimes S_+ \\ &= (I \otimes S_-) \oplus (I \otimes S_+),\end{aligned}$$

where the last lines in both representations are written according to the splitting

$$L_2(\Pi) = [L_2(\mathbb{R}_+) \otimes L_2(\mathbb{R}_+)] \oplus [L_2(\mathbb{R}_-) \otimes L_2(\mathbb{R}_+)] \, . \qquad \square$$

We continue to use an orthonormal basis

$$\ell_n(y) = e^{-y/2} L_n(y), \qquad n = 0, 1, 2, \ldots,$$

of the space $L_2(\mathbb{R}_+)$, where the Laguerre polynomials $L_n(y)$ are given by (4).

Theorem 3.2. *For each admissible* n*, the following equalities hold:*

$$(S_+\ell_n)(y) = -\ell_{n+1}(y), \qquad (S_-\ell_n)(y) = -\ell_{n-1}(y), \qquad \text{and} \qquad (S_-\ell_0)(y) = 0.$$

Proof. By [6], formula 8.971.1, we have

$$L'_n(y) - L'_{n+1}(y) = L_n(y). \tag{8}$$

Taking into account that $L_n(0) = 1$, for all n, the integral form of the above formula is as follows:

$$L_n(y) - L_{n+1}(y) = \int_0^y L_n(t)dt.$$

Calculate now

$$\begin{aligned}(S_+\ell_n)(y) &= -e^{-\frac{y}{2}} L_n(y) + e^{-\frac{y}{2}} \int_0^y L_n(t)dt \\ &= e^{-\frac{y}{2}} \left(-L_n(y) + L_n(y) - L_{n+1}(y)\right) = -\ell_{n+1}(y).\end{aligned}$$

Integrating by parts twice and using (8), we have

$$\begin{aligned}\int_y^\infty e^{-t} L_n(t)dt &= e^{-y} L_n(y) + \int_y^\infty e^{-t} L'_{n-1}(t)dt - \int_y^\infty e^{-t} L_{n-1}(t)dt \\ &= e^{-y} L_n(y) - \int_y^\infty e^{-t} L_{n-1}(t)dt \\ &- e^{-y} L_{n-1}(y) + \int_y^\infty e^{-t} L_{n-1}(t)dt \\ &= e^{-y} L_n(y) - e^{-y} L_{n-1}(y).\end{aligned}$$

Thus

$$\begin{aligned}(S_-\ell_n)(y) &= -e^{-\frac{y}{2}} L_n(y) + e^{\frac{y}{2}} \int_y^\infty e^{-t} L_n(t)dt \\ &= -e^{-\frac{y}{2}} L_n(y) + e^{\frac{y}{2}} \left(e^{-y} L_n(y) - e^{-y} L_{n-1}(y)\right) = -\ell_{n-1}(y).\end{aligned}$$

Finally,

$$(S_-\ell_0)(y) = -e^{-\frac{y}{2}} + e^{\frac{y}{2}} \int_y^\infty e^{-t} dt = 0. \qquad \square$$

It is convenient to change the previously used basis $\{\ell_n(y)\}_{n=0}^\infty$ of $L_2(\mathbb{R}_+)$ to the new basis $\{\widetilde{\ell}_n(y)\}_{n=0}^\infty$, where

$$\widetilde{\ell}_n(y) = (-1)^n \ell_n(y), \qquad n = 0, 1, 2, \ldots.$$

We note that the previously defined one-dimensional spaces L_n are generated by the new basis elements $\widetilde{\ell}_n(y)$ as well, and that the statements of Theorem 2.1 remain valid without any change.

Remark 3.3. *As the previous theorem shows, the operator S_+ is an isometric operator on $L_2(\mathbb{R}_+)$ and is nothing but the unilateral forward shift with respect to the basis $\{\widetilde{\ell}_n(y)\}_{n=0}^{\infty}$. Its adjoint operator S_- is the unilateral backward shift with respect to the same basis, and its kernel coincides with the one-dimensional space L_0 generated by $\widetilde{\ell}_0(y) = e^{-\frac{y}{2}}$.*

The above, together with Theorem 3.1, permits us to give a simple functional model for both operators S_Π and S_Π^*. Each of them is unitary equivalent to the direct sum of two unilateral shifts, forward and backward, both taken with the infinite multiplicity.

Let

$$L_n^{\oplus} = \bigoplus_{k=0}^{n} L_k$$

be the direct sum of the first $(n+1)$ L_k-spaces. We denote by P_n and $P_n^{\oplus}$ the orthogonal projections of $L_2(\mathbb{R}_+)$ onto L_n and $L_n^{\oplus}$, respectively.

Corollary 3.4. *For all admissible indices, we have*

$$\begin{aligned}
P_0 &= I - S_+S_-,\\
P_n &= S_+^n P_0 S_-^n,\\
P_n^{\oplus} &= I - S_+^{n+1} S_-^{n+1},\\
S_+^k|_{L_n} &: L_n \longrightarrow L_{n+k},\\
S_-^k|_{L_n} &: L_n \longrightarrow L_{n-k}.
\end{aligned}$$

The next result was obtained in [7] (see Theorem 2.4 and Corollary 2.6 therein) and shows that the action of both operators S_Π and S_Π^* is extremely transparent according to the decomposition

$$L_2(\Pi) = \bigoplus_{k=1}^{\infty} \mathcal{A}_{(k)}^2(\Pi) \oplus \bigoplus_{k=1}^{\infty} \widetilde{\mathcal{A}}_{(k)}^2(\Pi).$$

In our approach it is just a straightforward corollary of Theorems 2.1, 3.1, and Corollary 3.4.

Theorem 3.5. *For all admissible indices, we have*

$$\begin{aligned}
(S_\Pi)^k|_{\mathcal{A}_{(n)}^2(\Pi)} &: \mathcal{A}_{(n)}^2(\Pi) \longrightarrow \mathcal{A}_{(n+k)}^2(\Pi),\\
(S_\Pi)^k|_{\widetilde{\mathcal{A}}_{(n)}^2(\Pi)} &: \widetilde{\mathcal{A}}_{(n)}^2(\Pi) \longrightarrow \widetilde{\mathcal{A}}_{(n-k)}^2(\Pi),\\
(S_\Pi^*)^k|_{\widetilde{\mathcal{A}}_{(n)}^2(\Pi)} &: \widetilde{\mathcal{A}}_{(n)}^2(\Pi) \longrightarrow \widetilde{\mathcal{A}}_{(n+k)}^2(\Pi),\\
(S_\Pi^*)^k|_{\mathcal{A}_{(n)}^2(\Pi)} &: \mathcal{A}_{(n)}^2(\Pi) \longrightarrow \mathcal{A}_{(n-k)}^2(\Pi),
\end{aligned}$$

$$\ker(S_\Pi)^n = \widetilde{\mathcal{A}}_n^2(\Pi), \qquad (\mathrm{Im}\,(S_\Pi)^n)^\perp = \mathcal{A}_n^2(\Pi),$$

$$\ker(S_\Pi^*)^n = \mathcal{A}_n^2(\Pi), \qquad (\mathrm{Im}\,(S_\Pi^*)^n)^\perp = \widetilde{\mathcal{A}}_n^2(\Pi).$$

Corollary 3.6. *Each true-n-analytic function ψ admits the following representation,*

$$\psi = (S_\Pi)^{n-1}\varphi,$$

where $\varphi \in \mathcal{A}^2(\Pi)$.

Each true-n-anti-analytic function g admits the following representation,

$$g = (S_\Pi^*)^{n-1} f,$$

where $f \in \widetilde{\mathcal{A}}^2(\Pi)$.

We denote by $B_{\Pi,(n)}$ and $\widetilde{B}_{\Pi,(n)}$ the orthogonal projections of $L_2(\Pi)$ onto the spaces $\mathcal{A}^2_{(n)}(\Pi)$ and $\widetilde{\mathcal{A}}^2_{(n)}(\Pi)$, consisting of true-$n$-analytic and true-$n$-anti-analytic functions respectively. Let $B_{\Pi,n}$ and $\widetilde{B}_{\Pi,n}$ be the orthogonal projections of $L_2(\Pi)$ onto the spaces $\mathcal{A}^2_n(\Pi)$ and $\widetilde{\mathcal{A}}^2_n(\Pi)$, consisting of n-analytic and n-anti-analytic functions respectively.

We summarize now some important properties of the above projections in terms of singular operators.

Theorem 3.7. *For all admissible indices, we have*

$$\begin{aligned}
B_\Pi &= I - S_\Pi S_\Pi^*,\\
\widetilde{B}_\Pi &= I - S_\Pi^* S_\Pi,\\
B_{\Pi,n} &= I - (S_\Pi)^n (S_\Pi^*)^n,\\
\widetilde{B}_{\Pi,n} &= I - (S_\Pi^*)^n (S_\Pi)^n,\\
B_{\Pi,(n)} = (S_\Pi)^{n-1} B_\Pi (S_\Pi^*)^{n-1} &= (S_\Pi)^{n-1}(S_\Pi^*)^{n-1} - (S_\Pi)^n (S_\Pi^*)^n,\\
\widetilde{B}_{\Pi,(n)} = (S_\Pi^*)^{n-1} \widetilde{B}_\Pi (S_\Pi)^{n-1} &= (S_\Pi^*)^{n-1}(S_\Pi)^{n-1} - (S_\Pi^*)^n (S_\Pi)^n,\\
B_{\Pi,(n+1)} &= S_\Pi B_{\Pi,(n)} S_\Pi^*,\\
\widetilde{B}_{\Pi,(n+1)} &= S_\Pi^* \widetilde{B}_{\Pi,(n)} S_\Pi.
\end{aligned}$$

Proof. Follows directly from Theorems 2.1, 3.1, and Corollary 3.4. □

As direct corollaries of the above results we mention as well the following statements.

Corollary 3.8. *For all $n \in \mathbb{N}$, we have*

$$(S_\Pi)^n (S_\Pi^*)^n (S_\Pi)^n = (S_\Pi)^n, \qquad (S_\Pi^*)^n (S_\Pi)^n (S_\Pi^*)^n = (S_\Pi^*)^n.$$

Introduce the operator

$$(S_\mathbb{R}\varphi)(x) = \frac{1}{\pi i}\int_\mathbb{R} \frac{\varphi(\xi)}{\xi - x}\, d\xi,$$

acting on $L_2(\mathbb{R})$. It is well known that the operators

$$P_\pm = \frac{1}{2}(I \pm S_\mathbb{R})$$

are the Szegő projections of $L_2(\mathbb{R})$ onto the Hardy spaces on the upper and lower half-planes, respectively.

Theorem 3.9. *The following holds:*

$$\begin{aligned} \text{s-}\lim_{n\to\infty}(S_\Pi^*)^n(S_\Pi^*)^n &= P_+\otimes I,\\ \text{s-}\lim_{n\to\infty}(S_\Pi)^n(S_\Pi^*)^n &= P_-\otimes I,\\ \text{s-}\lim_{n\to\infty}[(S_\Pi^*)^n(S_\Pi^*)^n,(S_\Pi)^n(S_\Pi^*)^n] &= S_\mathbb{R}\otimes I,\\ \text{s-}\lim_{n\to\infty}((S_\Pi^*)^n(S_\Pi^*)^n+(S_\Pi)^n(S_\Pi^*)^n) &= I \end{aligned}$$

Corollary 3.10. *We have*

$$\begin{aligned} \text{s-}\lim_{n\to\infty} B_{\Pi,n} &= P_+\otimes I,\\ \text{s-}\lim_{n\to\infty} \widetilde{B}_{\Pi,n} &= P_-\otimes I. \end{aligned}$$

References

[1] M.B. Balk. *Polyanalytic functions and their generalizations, Complex analysis I. Encycl. Math. Sci*, volume 85, pages 195–253. Springer Verlag, 1997.

[2] M.B. Balk and M.F. Zuev. On polyanalytic functions. *Russ. Math. Surveys*, 25(5):201–223, 1970.

[3] Harry Bateman and Arthur Erdélyi. *Higher transcendental functions, Vol. 2.* McGraw-Hill, 1954.

[4] A. Dzhuraev. Multikernel functions of a domain, kernel operators, singular integral operators. *Soviet Math. Dokl.*, 32(1):251–253, 1985.

[5] A. Dzhuraev. *Methods of Singular Integral Equations.* Longman Scientific & Technical, 1992.

[6] I.S. Gradshteyn and I.M. Ryzhik. *Tables of Integrals, Series, and Products.* Academic Press, New York, 1980.

[7] Yu.I. Karlovich and Luís Pessoa. C^*-algebras of Bergman type operators with piecewise continuous coefficients. *Submitted for publication.*

[8] Josue Ramírez and Ilya M. Spitkovsky. *On the algebra generated by a poly-Bergman projection and a composition operator.* Factorization, singular operators and related problems (Funchal, 2002), pages 273–289. Kluwer Acad. Publ., Dordrecht, 2003.

[9] N.L. Vasilevski. On the structure of Bergman and poly–Bergman spaces. *Integr. Equat. Oper. Th.*, 33:471–488, 1999.

[10] N.L. Vasilevski. Toeplitz operators on the Bergman spaces: Inside-the-domain effects. *Contemp. Math.*, 289:79–146, 2001.

Nikolai Vasilevski
Departamento de Matemáticas
CINVESTAV del I.P.N.
Apartado Postal 14-740
07000 México, D.F., México
e-mail: nvasilev@math.cinvestav.mx

Operator Theory:
Advances and Applications, Vol. 171, 361–388

Weak Mixing Properties of Vector Sequences

László Zsidó

Dedicated to the memory of our colleague Gert K. Pedersen

Abstract. Notions of weak and uniformly weak mixing (to zero) are defined for bounded sequences in arbitrary Banach spaces. Uniformly weak mixing for vector sequences is characterized by mean ergodic convergence properties. This characterization turns out to be useful in the study of multiple recurrence, where mixing properties of vector sequences, which are not orbits of linear operators, are investigated. For bounded sequences, which satisfy a certain domination condition, it is shown that weak mixing to zero is equivalent with uniformly weak mixing to zero.

Mathematics Subject Classification (2000). Primary 47A35; Secondary 37A25, 37A55.

Keywords. Weak mixing, uniformly weak mixing, relatively dense set, non-zero upper Banach density, shift-bounded sequence.

1. Introduction

We recall that the *upper density* $D^*(\mathcal{A})$ and the *lower density* $D_*(\mathcal{A})$ of some $\mathcal{A} \subset \mathbb{N} := \{ 0 , 1 , 2 , \ldots \}$ are defined by

$$D^*(\mathcal{A}) := \overline{\lim_{n\to\infty}} \frac{1}{n+1} \operatorname{card}\big(\mathcal{A} \cap [0,n]\big) , \quad D_*(\mathcal{A}) := \underline{\lim_{n\to\infty}} \frac{1}{n+1} \operatorname{card}\big(\mathcal{A} \cap [0,n]\big)$$

(see, e.g., [7], Chapter 3, §5 or [12], §2.3). If upper and lower densities coincide then $\mathcal{A}$ is called having *density* $D(\mathcal{A}) := D^*(\mathcal{A}) = D_*(\mathcal{A})$.

Clearly, for $\mathcal{A} \subset \mathbb{N}^* := \mathbb{N} \setminus \{0\} = \{ 1 , 2 , \ldots \}$ we can use also the formulas

$$D^*(\mathcal{A}) = \overline{\lim_{n\to\infty}} \frac{1}{n} \operatorname{card}\big(\mathcal{A} \cap [1,n]\big) , \quad D_*(\mathcal{A}) = \underline{\lim_{n\to\infty}} \frac{1}{n} \operatorname{card}\big(\mathcal{A} \cap [1,n]\big) .$$

Supported by MIUR, GNAMPA-INDAM and EU.

The upper (resp. lower) density of a sequence $(k_j)_{j\geq 1}$ in $\mathbb{N}^*$ means the upper (resp. lower) density of the subset $\{ k_j ; j \geq 1 \}$ of $\mathbb{N}^*$. It is easy to see that the lower density of a strictly increasing $(k_j)_{j\geq 1}$ is > 0 if and only if $\sup_{j\geq 1} \frac{k_j}{j} < +\infty$.

Let X be a Banach space with dual space X^*. We shall say that a sequence $(x_k)_{k\geq 1}$ in X is *weakly mixing to zero* if

$$\lim_{n\to\infty} \frac{1}{n} \sum_{k=1}^{n} |\langle x^*, x_k \rangle| = 0 \text{ for all } x^* \in X^*, \tag{1.1}$$

and we shall say that it is *uniformly weakly mixing to zero* if

$$\lim_{n\to\infty} \sup \left\{ \frac{1}{n} \sum_{k=1}^{n} |\langle x^*, x_k \rangle| \, ; \, x^* \in X^*, \, \|x^*\| \leq 1 \right\} = 0. \tag{1.2}$$

A linear operator $U : X \to X$ is usually called *weakly mixing to zero at* $x \in X$ if the orbit $\big(U^k(x)\big)_{k\geq 1}$ is weakly mixing to 0.

The following characterization of weak mixing to zero for power bounded linear operators, which is a counterpart of the Blum-Hanson theorem [3] for weak mixing, was proved by L.K. Jones and M. Lin [10] :

Theorem 1.1. *Let U be a power bounded linear operator on a Banach space X, $x \in X$, and $x_k = U^k(x)$, $k \geq 1$. Then the following conditions are equivalent* :

(i) *The sequence $(x_k)_{k\geq 1}$ is weakly mixing to zero.*
(j) *The sequence $(x_k)_{k\geq 1}$ is uniformly weakly mixing to zero.*
(jj) *For every sequence $k_1 < k_2 < \cdots$ in $\mathbb{N}^*$ of lower density > 0,*

$$\lim_{n\to\infty} \left\| \frac{1}{n} \sum_{k=1}^{n} x_{k_j} \right\| = 0.$$

One main goal of this paper is to prove in the next section that conditions (j) and (jj) in Theorem 1.1 are equivalent for any bounded sequence $(x_k)_{k\geq 1}$ in the Banach space X, not only for the points of an orbit of some power bounded linear operator on X (Theorem 2.3). Therefore, for any bounded sequence in a Banach space, uniformly weak mixing to zero is equivalent with the mean ergodic convergence property from (jj) (in particular, for bounded sequences in Hilbert spaces, our notion of "uniformly weak mixing to zero" coincides with the notion of "weak mixing" considered in [2]).

We note that this result was used by C. Niculescu, A. Ströh and L. Zsidó to prove that if Φ is a $*$-endomorphism of a C^*-algebra A, leaving invariant a state φ of A, whose support in A^{**} belongs to the center of A^{**}, and Φ is weakly mixing with respect to φ, then Φ is automatically weakly mixing of order 2 ([13], Theorem 1.3): this is a partial extension to the non-commutative C^*-dynamical systems of a classical result of H. Furstenberg, according to which every weakly mixing measure preserving transformation of a probability measure space is weakly mixing of any order ([7], Theorem 4.11).

For general bounded sequences in Banach spaces (or even in Hilbert spaces), condition (i) in Theorem 1.1 does not imply the equivalent conditions (j) and (jj) (Examples 3.1 and 3.2). Nevertheless, we shall prove in Section 5 that (i) implies (j) and (jj) provided that the sequence satisfies some appropriate domination condition, called "convex shift-boundedness", which of course holds if the sequence is an orbit of some power bounded linear operator. Actually it will be proved that if a convex shift-bounded sequence $(x_k)_{k\geq 1}$ in the Banach space X is weakly mixing to zero, then

$$\lim_{\substack{a,b\in\mathbb{N}^* \\ b-a\to\infty}} \sup\left\{ \frac{1}{b-a+1} \sum_{k=a}^{b} |\langle x^*, x_k\rangle| \,;\, x^* \in X^*,\ \|x^*\| \leq 1 \right\} = 0 \tag{1.3}$$

(Theorem 5.2). Its proof depends upon a structure theorem for sets of natural numbers of non-zero upper Banach density (Theorem 4.2), which is of interest in and of itself. We notice that if $(x_k)_{k\geq 1}$ is an orbit of some power bounded linear operator on X, then (1.3) is an immediate consequence of (1.2).

Finally, in Section 6 it will be shown that in uniformly convex Banach spaces the above implication holds for sequences which satisfy a condition weaker than convex shift-boundedness (Theorem 6.3).

We note that a short investigation of the ergodicity, that is of the Cesaro norm-convergence to zero, of convex shift-bounded sequences is postponed to an appendix.

2. Uniformly weak mixing to zero

A subset $\mathcal{N}$ of $\mathbb{N}^*$ is called *relatively dense* if there exists $L > 0$ such that every interval of natural numbers of length $\geq L$ contains some element of $\mathcal{N}$. In this case $D_*(\mathcal{N}) \geq \frac{1}{L}$ clearly holds, so relatively dense sets are of lower density > 0.

A sequence $(k_j)_{j\geq 1}$ in $\mathbb{N}^*$ is called relatively dense if the subset $\{ k_j \,;\, j \geq 1 \}$ of $\mathbb{N}^*$ is relatively dense. It is easy to see that a strictly increasing sequence $(k_j)_{j\geq 1}$ is relatively dense if and only if $\sup_{j\geq 1} (k_{j+1} - k_j) < +\infty$.

The proof of the following lemma is immediate and we give it only for the sake of completeness:

Lemma 2.1. *For any sequence $(x_k)_{k\geq 1}$ in a Banach space and for any sequence $k_1 < k_2 < \cdots$ in $\mathbb{N}^*$ of lower density > 0 we have*

$$\Big\| \frac{1}{n} \sum_{j=1}^{n} x_{k_j} \Big\| \longrightarrow 0 \iff \Big\| \frac{1}{n} \sum_{\substack{k\in\{k_1,k_2,\ldots\} \\ k\leq n}} x_k \Big\| \longrightarrow 0 .$$

Proof. For $\Longrightarrow$: with $n \geq k_1$, defining $j(n) \in \mathbb{N}^*$ by $k_{j(n)} \leq n < k_{j(n)+1}$, we have

$$\Big\| \frac{1}{n} \sum_{\substack{k\in\{k_1,k_2,\ldots\} \\ k\leq n}} x_k \Big\| = \Big\| \frac{1}{n} \sum_{j=1}^{j(n)} x_{k_j} \Big\| = \underbrace{\frac{j(n)}{n}}_{\leq 1} \cdot \Big\| \frac{1}{j(n)} \sum_{j=1}^{j(n)} x_{k_j} \Big\| \xrightarrow{n\to\infty} 0 .$$

The converse implication $\Longleftarrow$ follows by using

$$\Big\| \frac{1}{n} \sum_{j=1}^{n} x_{k_j} \Big\| = \Big\| \frac{1}{n} \sum_{\substack{k \in \{k_1, k_2, \dots\} \\ k \leq k_n}} x_k \Big\| = \frac{k_n}{n} \cdot \Big\| \frac{1}{k_n} \sum_{\substack{k \in \{k_1, k_2, \dots\} \\ k \leq k_n}} x_k \Big\| . \qquad \square$$

The next lemma is the main ingredient in the proof of the main result of the section:

Lemma 2.2. *Let Ω be a compact Hausdorff topological space, and $f_1, f_2, \dots$ continuous complex functions on Ω of uniform norm $\|f_k\|_\infty \leq 1$. If*

$$\Big\| \frac{1}{n} \sum_{j=1}^{n} f_{k_j} \Big\|_\infty \longrightarrow 0 \ \textit{ for every relatively dense sequence } k_1 < k_2 < \cdots \textit{ in } \mathbb{N}^* ,$$

then

$$\Big\| \frac{1}{n} \sum_{k=1}^{n} |f_k| \Big\|_\infty \longrightarrow 0 .$$

Proof. Without loss of generality we can assume that the functions f_k are real. Furthermore, since $|f_k| = 2\, {f_k}^+ - f_k$, it is enough to prove that

$$\Big\| \frac{1}{n} \sum_{k=1}^{n} {f_k}^+ \Big\|_\infty \longrightarrow 0 .$$

Let us assume the contrary, that is the existence of some $\varepsilon_o > 0$ for which

$$\mathcal{J} := \Big\{ n \geq 1 \, ; \, \Big\| \frac{1}{n} \sum_{k=1}^{n} {f_k}^+ \Big\|_\infty \geq \varepsilon_o \Big\} \text{ is infinite.}$$

For every $n \in \mathcal{J}$ there exists $\omega_n \in \Omega$ such that

$$\text{the cardinality of } \mathcal{N}_n := \{ 1 \leq k \leq n \, ; \, {f_k}^+(\omega_n) \geq \frac{\varepsilon_o}{2} \} \text{ is } \geq \frac{n\, \varepsilon_o}{2} .$$

Indeed, if $\omega_n \in \Omega$ is chosen such that

$$\frac{1}{n} \sum_{k=1}^{n} {f_k}^+(\omega_n) = \Big\| \frac{1}{n} \sum_{k=1}^{n} {f_k}^+ \Big\|_\infty \geq \varepsilon_o$$

then

$$\begin{aligned} \varepsilon_o &\leq \frac{1}{n} \Big(\sum_{k \in \mathcal{N}_n} {f_k}^+(\omega_n) + \sum_{\substack{1 \leq k \leq n \\ k \notin \mathcal{N}_n}} {f_k}^+(\omega_n) \Big) \\ &\leq \frac{1}{n} \Big(\operatorname{card}(\mathcal{N}_n) + \frac{\varepsilon_o}{2} \big(n - \operatorname{card}(\mathcal{N}_n) \big) \Big) \\ &\leq \frac{1}{n} \operatorname{card}(\mathcal{N}_n) + \frac{\varepsilon_o}{2} . \end{aligned}$$

Denoting now the least element of $\mathcal{J}$ by k_1, we can recursively construct a sequence $k_1 < k_2 < \cdots$ in $\mathcal{J}$ such that for

$$\text{the cardinality of } \mathcal{N}'_{k_{j+1}} := \{ k \in \mathcal{N}_{k_{j+1}} \, ; \, k > k_j \} \text{ is } \geq \frac{k_{j+1} \cdot \varepsilon_o}{4} , \quad j \geq 1 .$$

Indeed, it is enough to choose $k_{j+1} \geq \frac{4\,k_j}{\varepsilon_o}$, because then

$$\operatorname{card}(\mathcal{N}'_{k_{j+1}}) \geq \operatorname{card}(\mathcal{N}_{k_{j+1}}) - k_j \geq \frac{k_{j+1} \cdot \varepsilon_o}{2} - k_j \geq \frac{k_{j+1} \cdot \varepsilon_o}{4} .$$

Putting

$$\mathcal{N} := \bigcup_{j \geq 2} \mathcal{N}'_{k_j} ,$$

we have for every $j \geq 2$, $\mathcal{N} \cap (k_{j-1}, k_j] = \mathcal{N}'_{k_j} \subset \mathcal{N}_{k_j}$, so in particular,

$$\begin{aligned} k \in \mathcal{N} , \, k_{j-1} < k \leq k_j \;&\Longrightarrow\; {f_k}^+(\omega_{k_j}) \geq \frac{\varepsilon_o}{2} \\ &\Longrightarrow\; f_k(\omega_{k_j}) = {f_k}^+(\omega_{k_j}) \geq \frac{\varepsilon_o}{2} . \end{aligned}$$

Let us choose some integer $p \geq \frac{16}{{\varepsilon_o}^2}$. Since

$$\mathcal{N}^{(p)} := \mathcal{N} \cup \{ p, \, 2\,p, \, 3\,p \ldots \} \subset \mathbb{N}^*$$

is relatively dense, by the assumption on the functions f_k and by Lemma 2.1 there exists $m_o \geq 1$ such that

$$m \geq m_o \Longrightarrow \Big\| \frac{1}{m} \sum_{\substack{k \in \mathcal{N}^{(p)} \\ k \leq m}} f_k \Big\|_\infty \leq \frac{{\varepsilon_o}^2}{34} .$$

Then we get for any $j \geq 2$ with $k_{j-1} \geq m_o$

$$\begin{aligned} \frac{{\varepsilon_o}^2}{17} = 2\,\frac{{\varepsilon_o}^2}{34} &\geq \Big\| \frac{1}{k_j} \sum_{\substack{k \in \mathcal{N}^{(p)} \\ k \leq k_j}} f_k \Big\|_\infty + \Big\| \frac{1}{k_{j-1}} \sum_{\substack{k \in \mathcal{N}^{(p)} \\ k \leq k_{j-1}}} f_k \Big\|_\infty \\ &\geq \Big\| \frac{1}{k_j} \sum_{\substack{k \in \mathcal{N}^{(p)} \\ k_{j-1} < k \leq k_j}} f_k \Big\|_\infty \geq \Big| \frac{1}{k_j} \sum_{\substack{k \in \mathcal{N}^{(p)} \\ k_{j-1} < k \leq k_j}} f_k(\omega_{k_j}) \Big| \\ &\geq \Big| \frac{1}{k_j} \sum_{\substack{k \in \mathcal{N} \\ k_{j-1} < k \leq k_j}} f_k(\omega_{k_j}) \Big| - \Big| \frac{1}{k_j} \sum_{\substack{k_{j-1} < k \leq k_j \\ k \text{ a multiple of } p}} f_k(\omega_{k_j}) \Big| \\ &\geq \frac{1}{k_j} \cdot \frac{\varepsilon_o}{2} \cdot \operatorname{card}(\mathcal{N}'_{k_j}) - \frac{1}{k_j} \cdot \operatorname{card}(\{ 1 \leq k \leq k_j \, ; \, k \text{ a multiple of } p \}) \\ &\geq \frac{1}{k_j} \cdot \frac{\varepsilon_o}{2} \cdot \frac{k_j \cdot \varepsilon_o}{4} - \frac{1}{k_j} \cdot \frac{k_j}{p} = \frac{{\varepsilon_o}^2}{8} - \frac{1}{p} \geq \frac{{\varepsilon_o}^2}{16} , \end{aligned}$$

which is absurd. □

We can now characterize uniformly weak mixing to zero for bounded sequences in Banach spaces by mean ergodic convergence properties:

Theorem 2.3 (Mean ergodic description of uniformly weak mixing). *For a bounded sequence $(x_k)_{k\geq 1}$ in a Banach space X, the following conditions are equivalent :*

(j) $(x_k)_{k\geq 1}$ *is uniformly weakly mixing to zero, that is*

$$\lim_{n\to\infty} \sup \Big\{ \frac{1}{n} \sum_{k=1}^{n} |\langle x^*, x_k\rangle| \, ; \, x^* \in X^* \, , \, \|x^*\| \leq 1 \Big\} = 0 \, .$$

(jj) *For every sequence $k_1 < k_2 < \cdots$ in $\mathbb{N}^*$ of lower density > 0,*

$$\lim_{n\to\infty} \Big\| \frac{1}{n} \sum_{j=1}^{n} x_{k_j} \Big\| = 0 \, .$$

(jjj) *For every relatively dense sequence $k_1 < k_2 < \cdots$ in $\mathbb{N}^*$,*

$$\lim_{n\to\infty} \Big\| \frac{1}{n} \sum_{j=1}^{n} x_{k_j} \Big\| = 0 \, .$$

Proof. Implication (j) $\Rightarrow$ (jj) follows immediately from Lemma 2.1 and (jj) $\Rightarrow$ (jjj) is trivial.

For (jjj) $\Rightarrow$ (j) we recall that the closed unit ball B_{X^*} of X^* is weak*-compact and the evaluation functions $f_x \, : \, B_{X^*} \ni x^* \mapsto \langle x^*, x\rangle$, $x \in X$ are weak*-continuous. Since

(j) means $\displaystyle\frac{1}{n} \sum_{k=1}^{n} |f_{x_k}| \xrightarrow{\text{uniformly}} 0$ and

(jjj) means that, for every relatively dense sequence $k_1 < k_2 < \cdots$ in $\mathbb{N}^*$,

$$\frac{1}{n} \sum_{k=1}^{n} f_{x_{k_j}} \xrightarrow{\text{uniformly}} 0 \, ,$$

implication (jjj) $\Rightarrow$ (j) follows from Lemma 2.2. □

Theorem 2.3 yields a similar characterization of weak mixing to zero:

Corollary 2.4. *For a bounded sequence $(x_k)_{k\geq 1}$ in a Banach space X and $x^* \in X^*$, the following conditions are equivalent :*

$(\mathrm{i})_{x^*}$
$$\lim_{n\to\infty} \frac{1}{n} \sum_{k=1}^{n} |\langle x^*, x_k\rangle| = 0 \, .$$

$(\mathrm{ii})_{x^*}$ *For every sequence $k_1 < k_2 < \cdots$ in $\mathbb{N}^*$ of lower density > 0,*

$$\lim_{n\to\infty} \Big\langle x^*, \frac{1}{n} \sum_{j=1}^{n} x_{k_j} \Big\rangle = 0 \, .$$

$(\text{iii})_{x^*}$ *For every relatively dense sequence* $k_1 < k_2 < \cdots$ *in* $\mathbb{N}^*$,

$$\lim_{n\to\infty} \left\langle x^*, \frac{1}{n}\sum_{j=1}^{n} x_{k_j} \right\rangle = 0 .$$

Proof. We have just to apply Theorem 2.3 to the bounded scalar sequence

$$\big(\langle x^*, x_k\rangle\big)_{k\geq 1} .$$

□

3. Comparison of weak and uniformly weak mixing to zero

Let us first give an example of a bounded sequence in the Banach space $C\big([0,1]\big)$ of all continuous functions on $[0,1]$, which satisfies (i) but not (j) in Theorem 1.1. $\|\cdot\|_\infty$ will stand for the uniform norm on $C\big([0,1]\big)$ and $\operatorname{supp}(f)$ will denote the support of $f \in C\big([0,1]\big)$.

Example 3.1. *Let* $1 = n_1 < n_2 < \cdots$ *be a sequence in* $\mathbb{N}^*$ *such that*

$$\frac{n_j - 1}{n_{j+1} - 1} \leq \frac{1}{2}, \qquad j \geq 1$$

(for example, $n_1 = 1$, $n_2 = 2$ *and* $n_{j+1} = 2\,n_j - 1$ *for* $j \geq 2$ *),*

$$1 > t_1 > t_2 > \cdots > 0, \quad t_j \longrightarrow 0$$

real numbers, and $g_j : [0,1] \longrightarrow [0,1]$, $j \geq 1$, *continuous functions such that*

$$\operatorname{supp}(g_j) \subset [t_{j+1}, t_j] \ \textit{and}\ \|g_j\|_\infty = 1 \ \textit{for all}\ j \geq 1 .$$

If we set

$$f_k = g_j \ \textit{for}\ n_j \leq k < n_{j+1} ,$$

then $(f_k)_{k\geq 1}$ *is a bounded sequence in* $C\big([0,1]\big)$, *which is weakly convergent to zero, and so is weakly mixing to zero, but which is not uniformly weakly mixing to zero.*

Proof. Since $0 \leq f_k \leq 1$ for every $k \geq 1$, according to the Riesz representation theorem and the Lebesgue dominated convergence theorem, the weak convergence of $(f_k)_{k\geq 1}$ to zero is equivalent to

$$f_k \xrightarrow{\text{pointwise}} 0 , \tag{3.1}$$

while (1.2) for $(f_k)_{k\geq 1}$ is equivalent with

$$\frac{1}{n}\sum_{k=1}^{n} f_k \xrightarrow{\text{uniformly}} 0 . \tag{3.2}$$

For (3.1) let $t \in [0,1]$ be arbitrary. If $t = 0$ then (3.1) holds obviously because $f_k(0) = 0$ for all $k \geq 1$. On the other hand, if $0 < t \leq 1$ then there exists some $j \geq 1$ with $t_j < t$ and so

$$f_k(t) = 0 , \qquad n \geq n_j .$$

Now, by the positivity of the functions g_j and f_k, we have for every $j \geq 1$:

$$\frac{1}{n_{j+1}-1}\sum_{k=1}^{n_{j+1}-1} f_k \geq \frac{1}{n_{j+1}-1}\sum_{k=n_j}^{n_{j+1}-1} f_k = \frac{n_{j+1}-n_j}{n_{j+1}-1}\, g_j = \Big(1-\frac{n_j-1}{n_{j+1}-1}\Big) g_j$$
$$\geq \frac{1}{2}\, g_j\,.$$

Consequently

$$\Big\| \frac{1}{n_{j+1}-1}\sum_{k=1}^{n_{j+1}-1} f_k \Big\|_\infty \geq \frac{1}{2}\, \|g_j\|_\infty = \frac{1}{2} \text{ for all } j \geq 1$$

and so (3.2) does not hold. □

A similar counterexample can also be given in the Hilbert space $L^2\big([0,1]\big)$, with inner product and norm denoted by $(\,\cdot\,|\,\cdot\,)$ and $\|\cdot\|_2$, respectively:

Example 3.2. *Let* $1 = n_1 < n_2 < \cdots$ *be a sequence in* $\mathbb{N}^*$ *such that*

$$\frac{n_j-1}{n_{j+1}-1} \leq \frac{1}{2}\,, \qquad j \geq 1$$

(for example, $n_1 = 1$, $n_2 = 2$ *and* $n_{j+1} = 2\,n_j - 1$ *for* $j \geq 2$*),*

$$1 > t_1 > t_2 > \cdots > 0\,, \quad t_j \longrightarrow 0$$

real numbers, and $g_j : [0,1] \longrightarrow [0,+\infty)$, $j \geq 1$, *continuous functions such that*

$$\mathrm{supp}\,(g_j) \subset [t_{j+1}, t_j] \text{ and } \|g_j\|_2 = 1 \text{ for all } j \geq 1\,.$$

If we set

$$f_k = g_j \text{ for } n_j \leq k < n_{j+1}\,,$$

then $(f_k)_{k\geq 1}$ *is a bounded sequence in* $L^2\big([0,1]\big)$, *which is weakly convergent to zero, and so is weakly mixing to zero, but which is not uniformly weakly mixing to zero.*

Proof. Since the functions g_j are mutually orthogonal, by the Bessel inequality we have for every $f \in L^2\big([0,1]\big)$:

$$\sum_{j=1}^{\infty} |(g_j|f)|^2 \leq \|f\|_2^2 < +\infty\,,$$

Therefore $f_k \xrightarrow{\text{weakly}} 0$.

On the other hand, for every $j \geq 1$,

$$\Big\| \frac{1}{n_{j+1}-1} \sum_{k=1}^{n_{j+1}-1} f_k \Big\|_2^2 = \frac{1}{(n_{j+1}-1)^2} \Big\| \sum_{l=1}^{j} \sum_{k=n_l}^{n_{l+1}-1} f_k \Big\|_2^2$$

$$= \frac{1}{(n_{j+1}-1)^2} \sum_{l=1}^{j} (n_{l+1}-n_l)^2$$

$$\geq \Big(\frac{n_{j+1}-n_j}{n_{j+1}-1} \Big)^2 = \Big(1 - \frac{n_j - 1}{n_{j+1}-1} \Big)^2 \geq \frac{1}{4} .$$

Consequently, $\Big\| \dfrac{1}{n} \sum_{k=1}^{n} f_k \Big\|_2 \not\to 0$, so (1.2) does not hold for $(f_k)_{k\geq 1}$. □

In spite of the above examples, Theorem 1.1 entails that for orbits of power bounded linear operators, weak mixing to zero and uniformly weak mixing to zero are equivalent. We now consider a larger class of vector sequences for which weak mixing to zero and uniformly weak mixing to zero are still equivalent.

Let us call a sequence $(x_k)_{k\geq 1}$ in a Banach space X *convex shift-bounded* if there exists a constant $c > 0$ such that

$$\Big\| \sum_{j=1}^{p} \lambda_j \, x_{j+k} \Big\| \leq c \Big\| \sum_{j=1}^{p} \lambda_j \, x_j \Big\| , \qquad k \geq 1 \tag{3.3}$$

holds for any choice of $p \in \mathbb{N}^*$ and $\lambda_1, \ldots, \lambda_p \geq 0$. Clearly:

- the convex shift-boundedness of a sequence implies its boundedness;
- if $U : X \longrightarrow X$ is a power bounded linear operator and $x \in X$, then the sequence $\big(U^k(x)\big)_{k\geq 1}$ is convex shift-bounded.

We note that not every convex shift-bounded sequence, even in a Hilbert space, is the orbit of a bounded linear operator:

Example 3.3. *Let us define the sequence $(f_k)_{k\geq 1}$ in $L^2\big([0,1]\big)$ by setting for every $k \in \mathbb{N}^*$ with $k \equiv 1 \pmod 4$*

$$f_k(t) := t^k , \quad f_{k+1}(t) := t^{k+\frac{1}{4(k+2)}} , \quad f_{k+2}(t) := t^{k+1} , \quad f_{k+3}(t) := t^{k+1+\frac{1}{2}} .$$

Then $(f_k)_{k\geq 1}$ is convex shift-bounded, but there exists no bounded linear operator $U : L^2\big([0,1]\big) \to L^2\big([0,1]\big)$ such that

$$f_k = U^k(f) , \qquad k \geq k_o$$

for some $f \in L^2\big([0,1]\big)$ and $k_o \in \mathbb{N}^*$.

Proof. First of all, if $0 < \alpha_1 < \alpha_2 < \cdots$ are real numbers and $g_k \in L^2\big([0,1]\big)$ is defined by $g_k(t) := t^{\alpha_k}$, then the sequence $(g_k)_{k\geq 1}$ is convex shift-bounded.

Indeed, for any $p \in \mathbb{N}^*$ and $\lambda_1, \dots, \lambda_p \geq 0$, the function

$$\mathbb{N}^* \ni k \longmapsto \Big\| \sum_{j=1}^{p} \lambda_j \, g_{j+k} \Big\|_2^2 = \sum_{j,j'=1}^{p} \lambda_j \, \lambda_{j'} \, \frac{1}{\alpha_{j+k} + \alpha_{j'+k} + 1}$$

is decreasing. In particular, the sequence $(f_k)_{k\geq 1}$ is convex shift-bounded.

On the other hand, if $\alpha, \varepsilon > 0$ and we define $h \in L^2\big([0,1]\big)$ by $h(t) := t^{\alpha} - t^{\alpha+\varepsilon}$, then

$$\|h\|_2^2 = \int_0^1 \big(t^{2\alpha} + t^{2\alpha+2\varepsilon} - 2\,t^{2\alpha+\varepsilon}\big)\, \mathrm{d}t = \frac{2\,\varepsilon^2}{(2\alpha+1)\,(2\alpha+\varepsilon+1)\,(2\alpha+2\varepsilon+1)}\,.$$

It is easy to verify that

$$\|h\|_2^2 \leq \frac{1}{2\,(2\alpha+1)} \Big(\frac{\varepsilon}{\alpha+\varepsilon}\Big)^2 \leq \frac{1}{2\,(2\alpha+1)} \Big(\frac{\varepsilon}{\alpha}\Big)^2 \qquad \text{if } \varepsilon \leq 1\,, \tag{3.4}$$

$$\|h\|_2^2 \geq \frac{1}{4\,(2\alpha+1)} \Big(\frac{\varepsilon}{\alpha+\varepsilon}\Big)^2 \geq \frac{1}{4\,(2\alpha+1)} \Big(\frac{\varepsilon}{\alpha+1}\Big)^2 \quad \text{if } \varepsilon \leq 1\,, \alpha \geq 2\,. \tag{3.5}$$

Now let $k \in \mathbb{N}^*$ be arbitrary such that $k \equiv 1 \,(\mathrm{mod}\, 4)$. Then we have by (3.4)

$$\|f_k - f_{k+1}\|_2^2 \leq \frac{1}{2\,(2k+1)} \Big(\frac{1}{4\,(k+2)\,k}\Big)^2 = \frac{1}{32\,k^2\,(k+2)^2\,(2k+1)}\,,$$

while (3.5) yields

$$\|f_{k+2} - f_{k+3}\|_2^2 \geq \frac{1}{4\,(2k+3)} \Big(\frac{1}{2\,(k+2)}\Big)^2 = \frac{1}{16\,(k+2)^2\,(2k+3)}\,.$$

Consequently $\|f_{k+2} - f_{k+3}\|_2^2 \geq k^2 \|f_k - f_{k+1}\|_2^2$, and so

$$\|f_{k+2} - f_{k+3}\|_2 \geq k\, \|f_k - f_{k+1}\|_2\,. \tag{3.6}$$

Let us assume that there is a bounded linear operator $U : L^2\big([0,1]\big) \to L^2\big([0,1]\big)$ such that

$$f_k = U^k(f)\,, \qquad k \geq k_o$$

for some $f \in L^2\big([0,1]\big)$ and $k_o \in \mathbb{N}^*$. Then, for every $k \geq k_o$ with $k \equiv 1 \,(\mathrm{mod}\, 4)$, (3.6) yields

$$k\, \|f_k - f_{k+1}\|_2 \leq \|f_{k+2} - f_{k+3}\|_2 = \big\| U^2(f_k - f_{k+1}) \big\|_2 \leq \|U\|^2\, \|f_k - f_{k+1}\|_2\,,$$

hence $\|U\| \geq \sqrt{k}$. But this contradicts the boundedness of U. □

We shall prove (in this section in the realm of reflexive Banach spaces and in Section 5 in full generality) that weak mixing to zero is equivalent with uniformly weak mixing to zero for any convex shift-bounded sequence. First we prove an easy implication of weak mixing to zero:

Lemma 3.4. *Let $(x_k)_{k\geq 1}$ be a bounded sequence in a Banach space X, which is weakly mixing to zero, and $\mathcal{A} \subset \mathbb{N}^*$ with $D^*(\mathcal{A}) > 0$. Then the norm-closure of the convex hull* $\mathrm{conv}\big(\{x_k\,;\, k \in \mathcal{A}\}\big)$ *of* $\{x_k\,;\, k \in \mathcal{A}\}$ *contains* 0.

Proof. Let us assume that 0 is not in the norm-closure of $\mathrm{conv}\big(\{x_k\,;\,k\in\mathcal{A}\}\big)$. Then the Hahn-Banach theorem yields the existence of some $\varepsilon_o>0$ and $x^*\in X^*$ such that

$$\Re\,\langle\, x^*, x_k\rangle \geq \varepsilon_o\,, \qquad k\in\mathcal{A}\,. \tag{3.7}$$

Further, by a classical result of B.O. Koopman and J. von Neumann (see, e.g., [12], Chapter 2, (3.1) or [13], Lemma 9.3), there is a zero density set $\mathcal{E}\subset\mathbb{N}^*$ such that

$$\lim_{\mathcal{E}\not\ni k\to\infty}\langle\, x^*, x_k\rangle = 0\,. \tag{3.8}$$

Then $\mathcal{A}\setminus\mathcal{E}$ is infinite, because otherwise we would get the contradiction

$$0<D^*(\mathcal{A})\leq D^*(\mathcal{A}\setminus\mathcal{E})+D^*(\mathcal{E})=0$$

Let $k_1<k_2<\cdots$ be the elements of $\mathcal{A}\setminus\mathcal{E}$. Then (3.8) implies that $\langle\, x^*, x_{k_j}\rangle\to 0$, in contradiction with (3.7). □

For weakly relatively compact sequences a stronger statement holds, which is essentially [9], Corollary 2:

Lemma 3.5. *A weakly relatively compact sequence $(x_k)_{k\geq 1}$ in a Banach space X is weakly mixing to zero if and only if there exists a zero density set $\mathcal{E}\subset\mathbb{N}^*$ such that*

$$\lim_{\mathcal{E}\not\ni k\to\infty} x_k = 0 \ \textit{with respect to the weak topology of } X\,.$$

Proof. An inspection of the proof of [9], Corollary 2 shows that it works for any weakly relatively compact sequence in a Banach space, not only for those which are orbits of power bounded linear operators. □

We notice that, if $(x_k)_{k\geq 1}$ is a weakly relatively compact sequence in a Banach space X, which is weakly mixing to zero, and $\mathcal{E}\subset\mathbb{N}^*$ is as in Lemma 3.5, then, according to the classical Mazur theorem on the equality of the weak and norm closure of a convex subset of X, the norm-closure of the convex hull of every infinite subset of $\mathbb{N}^*\setminus\mathcal{E}$ contains 0. In particular, for any $\mathcal{A}\subset\mathbb{N}^*$ with $D^*(\mathcal{A})>0$, the norm-closure of the convex hull of the infinite set $\mathcal{A}\setminus\mathcal{E}$ contains 0.

Now we prove a consequence of the negation of uniformly weak mixing to zero (cf. the first part of the proof of [8], Theorem IV):

Lemma 3.6. *Let $(x_k)_{k\geq 1}$ be a sequence in the closed unit ball of a Banach space X, which is not uniformly weakly mixing to zero. Then there exist*

$$0<\varepsilon_o\leq 1\,,$$
$$\mathcal{B}\subset\mathbb{N}^* \ \textit{with}\ D^*(\mathcal{B})\geq\varepsilon_o\,,$$
$$k_1\,,\,k_2\,,\,\ldots\in\mathbb{N}^* \ \textit{with}\ k_j-k_{j-1}>j\,,$$
$$x_1^*\,,\,x_2^*\,,\,\ldots\in X^* \ \textit{with}\ \|x_j^*\|\leq 1\,,$$

such that

$$\mathcal{B} \cap \bigcup_{j\geq 2} (\, k_{j-1}\, ,\, k_{j-1}+j\,] = \emptyset\, ,$$

$$\Re\,\langle\, x_j^*, x_k\rangle > 2\,\varepsilon_o\, , \qquad k\in\mathcal{B}\cap(\, k_{j-1}+j\, ,\, k_j\,]\, ,\, j\geq 2\, .$$

Proof. For any complex number z we shall use the notation

$$\Re^+ z := \begin{cases} \Re\, z & \text{if } \Re\, z\geq 0 \\ 0 & \text{if } \Re\, z\leq 0 \end{cases}, \qquad \Re^- z := \begin{cases} 0 & \text{if } \Re\, z\geq 0 \\ -\Re\, z & \text{if } \Re\, z\leq 0 \end{cases}.$$

Then $\Re\, z = \Re^+ z - \Re^- z = \Re^+ z - \Re^+(-z)$.

Since $(x_k)_{k\geq 1}$ is not uniformly weakly mixing to zero, there is $0<\varepsilon_o\leq 1$ such that

$$\mathcal{J} := \Big\{ n\geq 1\, ;\, \sup\Big\{ \frac{1}{n}\sum_{k=1}^{n} |\langle\, x^*, x_k\rangle|\, ;\, x^*\in X^*\, ,\, \|x^*\|\leq 1\Big\} > 16\,\varepsilon_o\Big\}$$

is infinite. Using (in the complex case) $\langle x^*, x_k\rangle = \Re\,\langle\, x^*, x_k\rangle - i\,\Re\,\langle\, i\, x^*, x_k\rangle$, it follows that also

$$\mathcal{J}_{\Re} := \Big\{ n\geq 1\, ;\, \sup\Big\{ \frac{1}{n}\sum_{k=1}^{n} |\,\Re\,\langle\, x^*, x_k\rangle|\, ;\, x^*\in X^*\, ,\, \|x^*\|\leq 1\Big\} > 8\,\varepsilon_o\Big\}$$

is infinite. Now, since $\Re\,\langle x^*, x_k\rangle = \Re^+\langle\, x^*, x_k\rangle - \Re^+\langle\, -x^*, x_k\rangle$, we obtain that

$$\mathcal{J}_+ := \Big\{ n\geq 1\, ;\, \sup\Big\{ \frac{1}{n}\sum_{k=1}^{n} \Re^+\langle\, x^*, x_k\rangle\, ;\, x^*\in X^*\, ,\, \|x^*\|\leq 1\Big\} > 4\,\varepsilon_o\Big\}$$

is infinite.

Let $n\in\mathcal{J}_+$ be arbitrary. Then there exists $y_n^*\in X^*$ with $\|y_n^*\|\leq 1$ such that

$$\frac{1}{n}\sum_{k=1}^{n} \Re^+\langle\, y_n^*, x_k\rangle > 4\,\varepsilon_o\, .$$

Denoting $\mathcal{B}_n := \{1\leq k\leq n\, ;\, \Re^+\langle\, y_n^*, x_k\rangle > 2\,\varepsilon_o\}$, we have

$$4\,\varepsilon_o < \frac{1}{n}\Big(\sum_{k\in\mathcal{B}_n} \Re^+\langle\, y_n^*, x_k\rangle + \sum_{\substack{1\leq k\leq n\\ k\notin\mathcal{B}_n}} \Re^+\langle\, y_n^*, x_k\rangle\Big) \leq \frac{1}{n}\,\text{card}\,(\mathcal{B}_n) + 2\,\varepsilon_o\, ,$$

hence $\text{card}\,(\mathcal{B}_n)\geq 2\, n\,\varepsilon_o$.

Letting k_1 be the least element of $\mathcal{J}_+$, we can construct recursively a sequence $k_1\, ,\, k_2\, ,\, \ldots\in\mathcal{J}_+$ such that, for every $j\geq 2$,

$$k_j - k_{j-1} > j \text{ and}$$

$$\text{the cardinality of } \mathcal{B}'_{k_j} := \{\, k\in\mathcal{B}_{k_j}\, ;\, k > k_{j-1}+j\,\} \text{ is } \geq k_j\,\varepsilon_o$$

Indeed, if we choose k_j in the infinite set $\mathcal{J}_+$ such that $k_j > \frac{k_{j-1}+j}{\varepsilon_o}$, then

$$\text{card}\,(\mathcal{B}'_{k_j}) \geq \text{card}\,(\mathcal{B}_{k_j}) - (k_{j-1}+j) \geq 2\, k_j\,\varepsilon_o - (k_{j-1}+j) > k_j\,\varepsilon_o\, .$$

Putting

$$\mathcal{B} := \bigcup_{j\geq 2} \mathcal{B}'_{k_j} ,$$

we have for every $j \geq 2$

$$\mathcal{B} \cap (\, k_{j-1} \, , \, k_{j-1} + j \,] = \emptyset \, , \qquad \mathcal{B} \cap (\, k_{j-1} + j \, , \, k_j \,] = \mathcal{B}'_{k_j} \subset \mathcal{B}_{k_j} \, ,$$

and so

$$\Re \langle \, y^*_{k_j} , x_k \rangle = \Re^+ \langle \, y^*_{k_j} , x_k \rangle > 2\, \varepsilon_o \;\text{ for all }\; k \in \mathcal{B} \cap (\, k_{j-1} + j \, , \, k_j \,] .$$

On the other hand,

$$D^*(\mathcal{B}) = \varlimsup_{n\to\infty} \frac{1}{n} \text{ card} \big(\mathcal{B} \cap [1,n] \big) \geq \varlimsup_{j\to\infty} \frac{1}{k_j} \text{ card} \big(\underbrace{\mathcal{B} \cap (k_{j-1}+j, k_j]}_{=\mathcal{B}'_{k_j}} \big) \geq \varepsilon_o .$$

Therefore, setting $x^*_j := y^*_{k_j}$, the proof is complete. □

We recall the following lemma of L.K. Jones on sequences of integers (see [8], Lemma 3 or [9], Lemma):

Lemma 3.7. *Let $\mathcal{A}_o$, $\mathcal{B}$ be subsets of $\mathbb{N}^*$ with $D^*(\mathcal{A}_o) = 1$ and $D^*(\mathcal{B}) > 0$. Then there exists an infinite subset $\mathcal{I} \subset \mathcal{A}_o$ such that*

$$\{k \in \mathcal{B} \, ; \, \mathcal{F} + k \subset \mathcal{B}\}\big) > 0 \; \textit{for any finite } \mathcal{F} \subset \mathcal{I} .$$

Now, using the idea of the proof of [8], Theorem IV, we can prove that weak mixing to zero and uniformly weak mixing to zero are equivalent for any convex shift-bounded sequence in a reflexive Banach space:

Proposition 3.8. *For a convex shift-bounded sequence in a reflexive Banach space, weak mixing to zero is equivalent to uniformly weak mixing to zero.*

Proof. Let $(x_k)_{k\geq 1}$ be a convex shift-bounded sequence in the closed unit ball of a reflexive Banach space X , which is weakly mixing to zero, and let us assume that it is not uniformly weakly mixing to zero. Let $c > 0$ be such that (3.3) holds for any choice of $p \in \mathbb{N}^*$ and $\lambda_1 , \ldots , \lambda_p \geq 0$.

By Lemma 3.6 there exist

$$0 < \varepsilon_o \leq 1 ,$$
$$\mathcal{B} \subset \mathbb{N}^* \text{ with } D^*(\mathcal{B}) \geq \varepsilon_o ,$$
$$k_1 , k_2 , \ldots \in \mathbb{N}^* \text{ with } k_j - k_{j-1} > j ,$$
$$x^*_1 , x^*_2 , \ldots \in X^* \text{ with } \|x^*_j\| \leq 1 ,$$

such that

$$\mathcal{B} \cap \bigcup_{j\geq 2} (\, k_{j-1} \, , \, k_{j-1} + j \,] = \emptyset ,$$

$$\Re \langle \, x^*_j , x_k \rangle > 2\, \varepsilon_o , \qquad k \in \mathcal{B} \cap (\, k_{j-1} + j \, , \, k_j \,] \, , \, j \geq 2 .$$

On the other hand, since any bounded set in a reflexive Banach space is weakly relatively compact, by Lemma 3.5 there exists $\mathcal{A}_o \subset \mathbb{N}^*$ with $D_*(\mathcal{A}_o) = 1$ such that $\lim\limits_{\mathcal{A}_o \ni k \to \infty} x_k = 0$ with respect to the weak topology of X.

Finally, by Lemma 3.7 there exists an infinite subset $\mathcal{I} \subset \mathcal{A}_o$ such that

$$\{k \in \mathcal{B}\,;\ \mathcal{F} + k \subset \mathcal{B}\}\big) > 0 \ \text{ for any finite } \mathcal{F} \subset \mathcal{I}\,.$$

Since $\lim\limits_{\mathcal{I} \ni k \to \infty} x_k = 0$ with respect to the weak topology of X, there are $p \in \mathbb{N}^*$, $n_1 < \cdots < n_p$ in $\mathcal{I}$ and $\lambda_1\,,\,\ldots\,,\,\lambda_p \geq 0\,,\ \lambda_1 + \cdots + \lambda_p = 1$, such that

$$\Big\| \sum_{j=1}^{p} \lambda_j\, x_{n_j} \Big\| \leq \frac{\varepsilon_o}{c}\,.$$

By (3.3) it follows that

$$\Big\| \sum_{j=1}^{p} \lambda_j\, x_{n_j+k} \Big\| \leq c \Big\| \sum_{j=1}^{p} \lambda_j\, x_{n_j} \Big\| \leq \varepsilon_o\,, \qquad k \geq 1\,. \tag{3.9}$$

Now set $j_o := \max\big(n_p - n_1, 2\big)$. Since the set $\big\{k \in \mathcal{B}\,;\ \{n_1\,,\,\ldots\,,\,n_p\} + k \subset \mathcal{B}\big\}$ has strictly positive upper density and so is infinite, it contains some k such that $n_1 + k \geq k_{j_o}$. Then there is a unique $j_1 \in \mathbb{N}^*$ with $k_{j_1 - 1} < n_1 + k \leq k_{j_1}$, for which we have $k_{j_o} \leq n_1 + k \leq k_{j_1}$, hence $j_o \leq j_1$. We claim that

$$k_{j_1-1} + j_1 < n_1 + k \leq n_p + k \leq k_{j_1}\,. \tag{3.10}$$

Indeed, $k_{j_1-1} < n_1 + k\,,\ n_1 + k \in \mathcal{B}$ and $\mathcal{B} \cap (k_{j_1-1}, k_{j_1-1} + j_1] = \emptyset$ imply that $k_{j_1-1} + j_1 < n_1 + k$. Similarly, $n_p + k = n_1 + k + (n_p - n_1) \leq k_{j_1} + j_o < k_{j_1} + j_1 + 1$, $n_p + k \in \mathcal{B}$ and $\mathcal{B} \cap (k_{j_1}, k_{j_1} + j_1 + 1] = \emptyset$ yield $n_p + k \leq k_{j_1}$.

By (3.10) we have $n_j + k \in \mathcal{B} \cap (k_{j_1-1} + j_1, k_{j_1}]\,, 1 \leq j \leq p$, so

$$\Re\,\langle\, x^*_{j_1}, x_{n_j+k}\rangle > 2\,\varepsilon_o\,, \qquad 1 \leq j \leq p\,.$$

Since $\|x^*_{j_1}\| \leq 1$, it follows that

$$\Big\| \sum_{j=1}^{p} \lambda_j\, x_{n_j+k} \Big\| \geq \Re \left\langle x^*_{j_1}, \sum_{j=1}^{p} \lambda_j\, x_{n_j+k} \right\rangle = \sum_{j=1}^{p} \lambda_j\, \Re\,\langle\, x^*_{j_1}, x_{k_j}\rangle > 2\,\varepsilon_o\,,$$

in contradiction with (3.9). □

If in Lemma 3.7 the set $\mathcal{I}$ were not only infinite, but with $D^*(\mathcal{I}) > 0$, then in the proof of Proposition 3.8 we could use Lemma 3.4 instead of Lemma 3.5 and so we would get a proof of Proposition 3.8 without the reflexivity assumption. In the next section we shall prove a result like Lemma 3.7 (Theorem 4.2), which implies that for every $\mathcal{B} \subset \mathbb{N}^*$ with $D^*(\mathcal{B}) > 0$ there is a set $\mathcal{A} \subset \mathbb{N}^*$ with $D^*(\mathcal{A}) > 0$ such that any finite subset of $\mathcal{A}$ has infinitely many translates contained in $\mathcal{B}$. This result will enable us to eliminate the reflexivity condition in Proposition 3.8.

4. Sets of non-zero upper Banach density

We recall that the *upper Banach density* $BD^*(\mathcal{B})$ of some $\mathcal{B} \subset \mathbb{N} = N^* \cup \{0\}$ is defined by

$$BD^*(\mathcal{B}) := \overline{\lim_{\substack{a,b\in\mathbb{N}\\ b-a\to\infty}}} \frac{1}{b-a+1} \operatorname{card}\big(\mathcal{B}\cap[a,b]\big) = \overline{\lim_{\substack{a,b\in\mathbb{N}^*\\ b-a\to\infty}}} \frac{1}{b-a+1} \operatorname{card}\big(\mathcal{B}\cap[a,b]\big)$$

(see, e.g., [7], Chapter 3, §5). For any $\mathcal{B} \subset \mathbb{N}^*$ we have $BD^*(\mathcal{B}) \geq D^*(\mathcal{B})$, but it is easily seen that $BD^*(\mathcal{B}) > D^*(\mathcal{B})$ can happen. In this section we investigate the structure of the sets $\mathcal{B} \subset \mathbb{N}^*$ with $BD^*(\mathcal{B}) > 0$ by proving a precise version of the theorem of R. Ellis [7], Theorem 3.20. The proof is based on the ergodic theoretical methods of H. Furstenberg found in [7], Chapter 3, §5.

Let us consider $\Omega := \{0,1\}^{\mathbb{N}}$ and endow it with the metrizable compact product topology of the discrete topologies on $\{0,1\}$. We shall denote the components of $\omega \in \Omega$ by ω_k, so that $\omega = (\omega_k)_{k\in\mathbb{N}}$. For every $\mathcal{B} \subset \mathbb{N}$ we define $\omega^{(\mathcal{B})} \in \Omega$ by setting

$$\omega_k^{(\mathcal{B})} := \begin{cases} 1 & \text{if } k \in \mathcal{B} \\ 0 & \text{if } k \notin \mathcal{B} \end{cases}.$$

In other words, $\omega^{(\mathcal{B})}$ is the characteristic function of $\mathcal{B}$, considered an element of Ω. Clearly, $\mathcal{B} \longmapsto \omega^{(\mathcal{B})}$ is a bijection of the set of all subsets of $\mathbb{N}$ onto Ω.

Let $s_{\leftarrow}$ denote the backward shift on Ω, defined by

$$s_{\leftarrow}\big((\omega_k)_{k\in\mathbb{N}}\big) = (\omega_{k+1})_{k\in\mathbb{N}}$$

and set, for every $\mathcal{B} \subset \mathbb{N}$,

$$\Omega^{(\mathcal{B})} := \overline{\{s_{\leftarrow}^{\,n}(\omega^{(\mathcal{B})})\,;\, n \geq 0\}}\,.$$

The following result is the one-sided version of [7], Lemma 3.17 and it establishes a link between upper Banach density and the ergodic theory of the dynamical system $(\Omega, s_{\leftarrow})$. Its proof is almost identical to the proof of [7], Lemma 3.17 and we sketch it only for the sake of completeness:

Lemma 4.1. *For every $\mathcal{B} \subset \mathbb{N}$ and every $\varepsilon > 0$ there exists an ergodic $s_{\leftarrow}$-invariant probability Borel measure μ on $\Omega^{(\mathcal{B})}$ such that*

$$\mu\big(\{\omega \in \Omega^{(\mathcal{B})}\,;\, \omega_0 = 1\}\big) > BD^*(\mathcal{B}) - \varepsilon\,.$$

Proof. Choose some $a_1, b_1, a_2, b_2, \ldots \in \mathbb{N}$ with $b_j - a_j \geq j$, $j \geq 0$, such that

$$\frac{1}{b_j - a_j + 1} \operatorname{card}\big(\mathcal{B}\cap[a_j,b_j]\big) \longrightarrow BD^*(\mathcal{B})\,.$$

Passing to a subsequence if necessary, we can assume that, for any continuous function $f \in C\big(\Omega^{(\mathcal{B})}\big)$, the limit

$$I(f) := \lim_{j\to\infty} \frac{1}{b_j - a_j + 1} \sum_{n=a_j}^{b_j} f\big(s_{\leftarrow}^{\,n}(\omega^{(\mathcal{B})})\big) \text{ exists}\,.$$

Then I is a positive linear functional on $C(\Omega^{(\mathcal{B})})$ and $I(1)=1$. Moreover,

$$I(f\circ s_{\leftarrow})=I(f)\,,\qquad f\in C(\Omega^{(\mathcal{B})})\,. \tag{4.1}$$

Indeed, for every $f\in C(\Omega^{(\mathcal{B})})$,

$$\begin{aligned} I(f\circ s_{\leftarrow})-I(f) &= \lim_{j\to\infty}\frac{1}{b_j-a_j+1}\Big(\sum_{n=a_j}^{b_j} f\big(s_{\leftarrow}^{\,n+1}(\omega^{(\mathcal{B})})\big)-\sum_{n=a_j}^{b_j} f\big(s_{\leftarrow}^{\,n}(\omega^{(\mathcal{B})})\big)\Big)\\ &= \lim_{j\to\infty}\frac{1}{b_j-a_j+1}\Big(f\big(s_{\leftarrow}^{\,b_j+1}(\omega^{(\mathcal{B})})\big)-f\big(s_{\leftarrow}^{\,a_j}(\omega^{(\mathcal{B})})\big)\Big)\\ &\le \lim_{j\to\infty}\frac{2\,\|f\|_\infty}{b_j-a_j+1}=0\,. \end{aligned}$$

By the Riesz representation theorem there exists a probability Borel measure ν_I on $\Omega^{(\mathcal{B})}$ such that

$$I(f)=\int_{\Omega^{(\mathcal{B})}} f(\omega)\,\mathrm{d}\nu_I(\omega)\,,\qquad f\in C(\Omega^{(\mathcal{B})})\,.$$

Property (4.1) of I implies that ν_I is $s_{\leftarrow}$-invariant. Moreover, since the characteristic function χ of $\{\omega\in\Omega^{(\mathcal{B})}\,;\,\omega_0=1\}$ is continuous, we have

$$\begin{aligned} \nu_I\big(\{\omega\in\Omega^{(\mathcal{B})}\,;\,\omega_0=1\}\big) &= \int_{\Omega^{(\mathcal{B})}}\chi(\omega)\,\mathrm{d}\nu_I(\omega)=\lim_{j\to\infty}\frac{1}{b_j-a_j+1}\sum_{n=a_j}^{b_j}\chi\big(s_{\leftarrow}^{\,n}(\omega^{(\mathcal{B})})\big)\\ &= \lim_{j\to\infty}\frac{1}{b_j-a_j+1}\,\mathrm{card}\,\big(\mathcal{B}\cap[a_j,b_j]\big)=BD^*(\mathcal{B})\,. \end{aligned}$$

The convex set $\mathcal{P}^{s_{\leftarrow}}(\Omega^{(\mathcal{B})})$ of all $s_{\leftarrow}$-invariant probability Borel measures on $\Omega^{(\mathcal{B})}$, considered imbedded in the dual space of $C(\Omega^{(\mathcal{B})})$, is weak*-compact and its extreme points are the ergodic measures in $\mathcal{P}^{s_{\leftarrow}}(\Omega^{(\mathcal{B})})$ (see, for example, [7], Proposition 3.4). According to the Krein-Milman theorem, it follows that ν_I is a weak*-limit of convex combinations of ergodic measures in $\mathcal{P}^{s_{\leftarrow}}(\Omega^{(\mathcal{B})})$. Therefore, since $\nu_I\big(\{\omega\in\Omega^{(\mathcal{B})}\,;\,\omega_0=1\}\big)=BD^*(\mathcal{B})$, we conclude that there exists an ergodic measure $\mu\in\mathcal{P}^{s_{\leftarrow}}(\Omega^{(\mathcal{B})})$ such that $\mu\big(\{\omega\in\Omega^{(\mathcal{B})}\,;\,\omega_0=1\}\big)>BD^*(\mathcal{B})-\varepsilon$. □

We now prove the announced extension of [7], Theorem 3.20:

Theorem 4.2. *If $\mathcal{B}\subset\mathbb{N}$ and $0<\varepsilon<BD^*(\mathcal{B})$, then there exist*

$$\mathcal{A}\subset\mathbb{N}\ \textit{having density}\ D(\mathcal{A})>BD^*(\mathcal{B})-\varepsilon\,,$$
$$0\le m_1<m_2<\cdots\ \textit{and}\ 0\le n_1<n_2<\cdots\ \textit{in}\ \mathbb{N}\,,$$

for which

$$\mathcal{A}\cap[0,m_j]=\big\{k\in[0,m_j]\,;\,k+n_j\in\mathcal{B}\big\}\,,\qquad j\ge 1\,;$$

that is

$$k\in\mathcal{A}\iff k+n_j\in\mathcal{B}\ \textit{whenever}\ 0\le k\le m_j\,,\,j\ge 1\,.$$

Proof. For every $\omega \in \Omega$ we set $\mathcal{A}_\omega = \{k \in \mathbb{N}\,;\, \omega_k = 1\}$, so that $\omega = \omega^{(\mathcal{A}_\omega)}$. Clearly, $\mathcal{A}_{\omega^{(\mathcal{B})}} = \mathcal{B}$.

By Lemma 4.1 there exists an ergodic $s_{\leftarrow}$-invariant probability Borel measure μ on $\Omega^{(\mathcal{B})}$ such that

$$\mu_{\mathcal{B}} := \mu\big(\{\omega \in \Omega^{(\mathcal{B})}\,;\, \omega_0 = 1\}\big) > BD^*(\mathcal{B}) - \varepsilon\,.$$

Let χ denote the characteristic function of $\{\omega \in \Omega^{(\mathcal{B})}\,;\, \omega_0 = 1\} \subset \Omega^{(\mathcal{B})}$. Then, by the Birkhoff ergodic theorem, for μ-almost every $\omega \in \Omega^{(\mathcal{B})}$ we have

$$\frac{1}{n+1}\,\mathrm{card}\,\big(\mathcal{A}_\omega \cap [0,n]\big) = \frac{1}{n+1}\sum_{k=0}^{n} \chi\big(s_{\leftarrow}^{\,k}(\omega)\big) \longrightarrow \mu_{\mathcal{B}}\,. \tag{4.2}$$

Let $\Omega^{(\mathcal{B})}_{\text{Birkhoff}}$ be the set of all $\omega \in \Omega^{(\mathcal{B})}$, for which (4.2) holds. Then

- $\mathcal{A}_\omega$ has density $D(\mathcal{A}_\omega) = \mu_{\mathcal{B}} > BD^*(\mathcal{B}) - \varepsilon$ for every $\omega \in \Omega^{(\mathcal{B})}_{\text{Birkhoff}}$, and
- $\Omega^{(\mathcal{B})}_{\text{Birkhoff}}$ is μ-measurable and $\mu\big(\Omega^{(\mathcal{B})} \setminus \Omega^{(\mathcal{B})}_{\text{Birkhoff}}\big) = 0$.

Case 1: There exists $\omega \in \Omega^{(\mathcal{B})}_{\text{Birkhoff}} \setminus \{s_{\leftarrow}^{\,n}(\omega^{(\mathcal{B})})\,;\, n \geq 0\}$.

Set $\mathcal{A} := \mathcal{A}_\omega$ and choose some $m_1 \geq 0$. Since

$$\omega \in \Omega^{(\mathcal{B})} = \overline{\{s_{\leftarrow}^{\,n}(\omega^{(\mathcal{B})})\,;\, n \geq 0\}}\,, \tag{4.3}$$

there exists a smallest $n_1 \geq 0$ such that

$$\omega_k = s_{\leftarrow}^{\,n_1}(\omega^{(\mathcal{B})})_k = \omega^{(\mathcal{B})}_{k+n_1} = \begin{cases} 1 & \text{if } k+n_1 \in \mathcal{B} \\ 0 & \text{if } k+n_1 \notin \mathcal{B} \end{cases}, \qquad 0 \leq k \leq m_1\,,$$

that is $\mathcal{A} \cap [0,m_1] = \big\{k \in [0,m_1]\,;\, k+n_1 \in \mathcal{B}\big\}$.

Next $\omega \neq s_{\leftarrow}^{\,n_1}(\omega^{(\mathcal{B})})$ implies that $\omega_{m_2} \neq s_{\leftarrow}^{\,n_1}(\omega^{(\mathcal{B})})_{m_2}$ for some $m_2 \in \mathbb{N}$. Since $\omega_k = s_{\leftarrow}^{\,n_1}(\omega^{(\mathcal{B})})_k$ for all $0 \leq k \leq m_1$, we have $m_1 < m_2$. Now, again by (4.3), there exists a smallest $n_2 \geq 0$ such that

$$\omega_k = s_{\leftarrow}^{\,n_2}(\omega^{(\mathcal{B})})_k = \omega^{(\mathcal{B})}_{k+n_2} = \begin{cases} 1 & \text{if } k+n_2 \in \mathcal{B} \\ 0 & \text{if } k+n_2 \notin \mathcal{B} \end{cases}, \qquad 0 \leq k \leq m_2\,,$$

that is $\mathcal{A} \cap [0,m_2] = \big\{k \in [0,m_2]\,;\, k+n_2 \in \mathcal{B}\big\}$. By the minimality property of n_1 we have $n_1 \leq n_2$, while $\omega_{m_2} \neq s_{\leftarrow}^{\,n_1}(\omega^{(\mathcal{B})})_{m_2}$ yields $n_1 \neq n_2$. Therefore $n_1 < n_2$.

By induction we obtain $m_1 < m_2 < \cdots$ and $n_1 < n_2 < \cdots$ in $\mathbb{N}$ such that

$$\mathcal{A} \cap [0,m_j] = \big\{k \in [0,m_j]\,;\, k+n_j \in \mathcal{B}\big\} \text{ for all } j \geq 1\,.$$

Case 2: $\Omega^{(\mathcal{B})}_{\text{Birkhoff}} \subset \{s_{\leftarrow}^{\,n}(\omega^{(\mathcal{B})})\,;\, n \geq 0\}$.

We claim that there exists a smallest $n_o \in \mathbb{N}^*$ such that $s_{\leftarrow}^{\,n_o}(\omega^{(\mathcal{B})}) = \omega^{(\mathcal{B})}$. For let us assume that all $s_{\leftarrow}^{\,n}(\omega^{(\mathcal{B})})$ are different. Then, for every $n \geq 0$, since

$$\{s_{\leftarrow}^{\,n}(\omega^{(\mathcal{B})})\} \subset s_{\leftarrow}^{-1}\big(s_{\leftarrow}^{\,n+1}(\omega^{(\mathcal{B})})\big) \subset \{s_{\leftarrow}^{\,n}(\omega^{(\mathcal{B})})\} \cup \big(\Omega^{(\mathcal{B})} \setminus \Omega^{(\mathcal{B})}_{\text{Birkhoff}}\big)\,,$$

by the $s_{\leftarrow}$-invariance of μ we obtain $\mu\big(\{s_{\leftarrow}^{\,n}(\omega^{(\mathcal{B})})\}\big) = \mu\big(\{s_{\leftarrow}^{\,n+1}(\omega^{(\mathcal{B})})\}\big)$. Thus

$$\mu(\Omega^{(\mathcal{B})}) = \sum_{n=0}^{\infty} \mu\big(\{s_{\leftarrow}^{\,n}(\omega^{(\mathcal{B})})\}\big) = \begin{cases} 0 & \text{if } \mu\big(\{\omega^{(\mathcal{B})}\}\big) = 0 \\ +\infty & \text{if } \mu\big(\{\omega^{(\mathcal{B})}\}\big) > 0 \end{cases},$$

in contradiction with $\mu(\Omega^{(\mathcal{B})}) = 1$.

Now $s_{\leftarrow}^{\,n_o}(\omega^{(\mathcal{B})}) = \omega^{(\mathcal{B})}$ means that $k \in \mathbb{N}$ belongs to $\mathcal{B}$ if and only if $k+n_o \in \mathcal{B}$. Therefore, with $\mathcal{A} := \mathcal{B}$, any $0 \leq m_1 < m_2 < \cdots$ and $n_j := j\, n_o$, we have

$$D(\mathcal{A}) = \frac{1}{n_o}\,\text{card}\,\big(\mathcal{B} \cap [0, n_o - 1]\big) = BD^*(\mathcal{B})$$

$$\mathcal{A} \cap [0, m_j] = \big\{k \in [0, m_j]\,;\, k + n_j \in \mathcal{B}\big\}\,, \qquad j \geq 1\,. \qquad \square$$

We recall that a celebrated theorem of E. Szemerédi (answering a conjecture of P. Erdős) states that if $\mathcal{B} \subset \mathbb{N}^*$ has non-zero upper Banach density, then it contains arbitrarily long arithmetic progressions [14]. H. Furstenberg gave a new ergodic theoretical proof of Szemerédi's theorem by deducing it from a far-reaching multiple recurrence theorem [6] (see also [7], Chapter 3, §7). It is interesting to notice, even if it appears not to be relevant, that the proof of Szemerédi's theorem can be reduced via Theorem 4.2 to the case when $\mathcal{B}$ has non-zero density.

The last theorem implies the following counterpart of Lemma 3.7:

Corollary 4.3. *Let $\mathcal{A}_o$, $\mathcal{B}$ be subsets of $\mathbb{N}^*$ with $D^*(\mathcal{A}_o) = 1$ and $0 < \varepsilon < BD^*(\mathcal{B})$. Then there exists $\mathcal{I} \subset \mathcal{A}_o$ with $D^*(\mathcal{I}) > BD^*(\mathcal{B}) - \varepsilon$, such that*

$$\{k \in \mathbb{N}\,;\, \mathcal{F} + k \subset \mathcal{B}\} \text{ is infinite for any finite } \mathcal{F} \subset \mathcal{I}\,.$$

Proof. By Theorem 4.2 there exist $\mathcal{A} \subset \mathbb{N}$ having density $D(\mathcal{A}) > BD^*(\mathcal{B}) - \varepsilon$, as well as $0 \leq m_1 < m_2 < \cdots$ and $0 \leq n_1 < n_2 < \cdots$ in $\mathbb{N}$, such that

$$\mathcal{A} \cap [0, m_j] = \big\{k \in [0, m_j]\,;\, k + n_j \in \mathcal{B}\big\}\,, \qquad j \geq 1\,.$$

Set $\mathcal{I} := \mathcal{A} \cap \mathcal{A}_o$. Since

$$1 = D^*(\mathcal{A}_o) \leq D^*(\mathcal{A} \cap \mathcal{A}_o) + D^*(\mathbb{N} \setminus \mathcal{A}) = D^*(\mathcal{I}) + 1 - D(\mathcal{A})\,,$$

we have $D^*(\mathcal{I}) \geq D(\mathcal{A}) > BD^*(\mathcal{B}) - \varepsilon$. On the other hand, for any $j \geq 1$, the set

$$\big\{k \in \mathbb{N}\,;\, \big(\mathcal{I} \cap [0, m_j]\big) + k \subset \mathcal{B}\big\}$$

contains $\{n_j\,,\, n_{j+1}\,,\, \ldots\}$, hence is infinite. $\square$

5. Weak mixing to zero for convex shift-bounded sequences

Using Theorem 4.2, in this section we show that Proposition 3.8 holds without the reflexivity assumption. Actually we shall prove a slightly more general result, stating that any convex shift-bounded sequence in a Banach space, which is weakly mixing to zero, satisfies (1.3). For the proof we shall use the following counterpart of Lemma 3.6 for the sequences not satisfying (1.3):

Lemma 5.1. *Let $(x_k)_{k\geq 1}$ be a sequence in the closed unit ball of a Banach space X, such that, for some $a_1, b_1, a_2, b_2, \ldots \in \mathbb{N}^*$ with $b_j - a_j \geq j$, $j \geq 1$, we have*

$$\overline{\lim_{j\to\infty}} \sup\left\{ \frac{1}{b_j - a_j + 1} \sum_{k=a_j}^{b_j} |\langle x^*, x_k\rangle| \, ; \, x^* \in X^*, \, \|x^*\| \leq 1 \right\} > 0 .$$

Then there exist

$$0 < \varepsilon_o \leq 1,$$
$$\mathcal{B} \subset \mathbb{N}^* \text{ with } BD^*(\mathcal{B}) \geq \varepsilon_o,$$
$$j_1 < j_2 < \cdots \text{ in } \mathbb{N}^* \text{ with } b_{j_n} - b_{j_{n-1}} > n,$$
$$x_1^*, x_2^*, \ldots \in X^* \text{ with } \|x_n^*\| \leq 1,$$

such that

$$\mathcal{B} \cap \bigcup_{n\geq 2} (b_{j_{n-1}}, b_{j_{n-1}} + n] = \emptyset,$$
$$\Re \langle x_n^*, x_k \rangle > 2\varepsilon_o, \qquad k \in \mathcal{B} \cap (b_{j_{n-1}} + n, b_{j_n}], \, n \geq 2 .$$

Proof. We shall proceed as in the proof of Lemma 3.6.

Let $0 < \varepsilon_o \leq 1$ be such that

$$0 < 16\,\varepsilon_o < \overline{\lim_{j\to\infty}} \sup\left\{ \frac{1}{b_j - a_j + 1} \sum_{k=a_j}^{b_j} |\langle x^*, x_k\rangle| \, ; \, x^* \in X^*, \, \|x^*\| \leq 1 \right\}.$$

Then

$$\mathcal{J} := \left\{ j \geq 1 \, ; \, \sup\left\{ \frac{1}{b_j - a_j + 1} \sum_{k=a_j}^{b_j} |\langle x^*, x_k\rangle| \, ; \, x^* \in X^*, \, \|x^*\| \leq 1 \right\} > 16\,\varepsilon_o \right\}$$

is infinite. Using (in the complex case) $\langle x^*, x_k\rangle = \Re\langle x^*, x_k\rangle - i\,\Re\langle i\,x^*, x_k\rangle$, it follows that also

$$\mathcal{J}_{\Re} := \left\{ j \geq 1 \, ; \, \sup\left\{ \frac{1}{b_j - a_j + 1} \sum_{k=a_j}^{b_j} |\Re\langle x^*, x_k\rangle| \, ; \, x^* \in X^*, \, \|x^*\| \leq 1 \right\} > 8\,\varepsilon_o \right\}$$

is infinite. Now, since $\Re\langle x^*, x_k\rangle = \Re^+\langle x^*, x_k\rangle - \Re^+\langle -x^*, x_k\rangle$, we obtain that

$$\mathcal{J}_{+} := \left\{ j \geq 1 \, ; \, \sup\left\{ \frac{1}{b_j - a_j + 1} \sum_{k=a_j}^{b_j} \Re^+\langle x^*, x_k\rangle \, ; \, x^* \in X^*, \, \|x^*\| \leq 1 \right\} > 4\,\varepsilon_o \right\}$$

is infinite.

Let $j \in \mathcal{J}_+$ be arbitrary. Then there exists $y_j^* \in X^*$ with $\|y_j^*\| \leq 1$ such that

$$\frac{1}{b_j - a_j + 1} \sum_{k=a_j}^{b_j} \Re^+\langle\, y_j^*, x_k \rangle > 4\,\varepsilon_o\,.$$

Denoting $\mathcal{B}_j := \{a_j \leq k \leq b_j\,;\, \Re^+\langle\, y_j^*, x_k\rangle > 2\,\varepsilon_o\}$, we have

$$\begin{aligned} 4\,\varepsilon_o &< \frac{1}{b_j - a_j + 1} \Big(\sum_{k\in\mathcal{B}_j} \Re^+\langle\, y_j^*, x_k\rangle + \sum_{\substack{a_j\leq k\leq b_j \\ k\notin\mathcal{B}_j}} \Re^+\langle\, y_j^*, x_k\rangle\Big) \\ &\leq \frac{1}{b_j - a_j + 1} \operatorname{card}(\mathcal{B}_j) + 2\,\varepsilon_o\,, \end{aligned}$$

hence $\operatorname{card}(\mathcal{B}_j) \geq 2\,(b_j - a_j + 1)\,\varepsilon_o$.

Letting j_1 be the least element of $\mathcal{J}_+$, we can construct recursively a sequence $j_1 < j_2 < \cdots$ in $\mathcal{J}_+$ such that, for every $n \geq 2$,

$$b_{j_n} - b_{j_{n-1}} > n \text{ and}$$
$$\text{the cardinality of } \mathcal{B}'_{j_n} := \{\, k \in \mathcal{B}_{j_n}\,;\, k > b_{j_{n-1}} + n\,\} \text{ is } > (b_{j_n} - a_{j_n})\,\varepsilon_o\,.$$

Indeed, if we choose j_n in the infinite set $\mathcal{J}_+$ such that $j_n > j_{n-1}$ and

$$b_{j_n} - a_{j_n} + 1 \geq j_n + 1 > \frac{b_{j_{n-1}} + n}{\varepsilon_o} \geq b_{j_{n-1}} + n$$

then $b_{j_n} - b_{j_{n-1}} > n$ and

$$\begin{aligned} \operatorname{card}(\mathcal{B}'_{j_n}) &\geq \operatorname{card}(\mathcal{B}_{j_n}) - (b_{j_{n-1}} + n) \geq 2\,(b_{j_n} - a_{j_n} + 1)\,\varepsilon_o - (b_{j_{n-1}} + n) \\ &> (b_{j_n} - a_{j_n} + 1)\,\varepsilon_o\,. \end{aligned}$$

Putting

$$\mathcal{B} := \bigcup_{n\geq 2} \mathcal{B}'_{j_n}\,,$$

we have for every $n \geq 2$

$$\mathcal{B} \cap (\, b_{j_{n-1}}\,,\, b_{j_{n-1}} + n\,] = \emptyset\,, \qquad \mathcal{B} \cap (\, b_{j_{n-1}} + n\,,\, b_{j_n}\,] = \mathcal{B}'_{k_j} \subset \mathcal{B}_{k_j}\,,$$

and so

$$\Re\,\langle\, y_{j_n}^*, x_k\rangle = \Re^+\langle\, y_{j_n}^*, x_k\rangle > 2\,\varepsilon_o \text{ for all } k \in \mathcal{B} \cap (\, b_{j_{n-1}} + n\,,\, b_{j_n}\,]\,.$$

On the other hand,

$$\begin{aligned} BD^*(\mathcal{B}) &\geq \varlimsup_{n\to\infty} \frac{1}{b_{j_n} - a_{j_n} + 1} \operatorname{card}\big(\mathcal{B} \cap [a_{j_n}, b_{j_n}]\big) \\ &\geq \varlimsup_{n\to\infty} \frac{1}{b_{j_n} - a_{j_n} + 1} \operatorname{card}\big(\underbrace{\mathcal{B}'_{j_n} \cap [a_{j_n}, b_{j_n}]}_{=\mathcal{B}'_{j_n}}\big) \geq \varepsilon_o\,. \end{aligned}$$

Therefore, setting $x_n^* := y_{j_n}^*$, the proof is complete. □

For the proof of the next theorem we adapt the proof of Proposition 3.8, in which instead of Lemmas 3.6, 3.5 and 3.7 we use Lemma 5.1, Theorem 4.2 and Lemma 3.4:

Theorem 5.2 (Weak mixing for convex shift-bounded sequences). *For a convex shift-bounded sequence $(x_k)_{k\geq 1}$ in a Banach space X, the following conditions are equivalent :*

(i) $(x_k)_{k\geq 1}$ *is weakly mixing to zero, that is*

$$\lim_{n\to\infty} \frac{1}{n} \sum_{k=1}^{n} |\langle x^*, x_k\rangle| = 0 \textit{ for all } x^* \in X^* .$$

(j) $(x_k)_{k\geq 1}$ *is uniformly weakly mixing to zero, that is*

$$\lim_{n\to\infty} \sup \Big\{ \frac{1}{n} \sum_{k=1}^{n} |\langle x^*, x_k\rangle| \, ; \, x^* \in X^* , \, \|x^*\| \leq 1 \Big\} = 0 .$$

(jw) (1.3) *holds, that is*

$$\overline{\lim_{\substack{a,b\in\mathbb{N}^* \\ b-a\to\infty}}} \sup \Big\{ \frac{1}{b-a+1} \sum_{k=a}^{b} |\langle x^*, x_k\rangle| \, ; \, x^* \in X^* , \, \|x^*\| \leq 1 \Big\} = 0 .$$

Proof. The implications (jw) $\Rightarrow$ (j) $\Rightarrow$ (i) are trivial. For (i) $\Rightarrow$ (jw) we shall show that (i) and the negation of (jw) lead to a contradiction.

Let $c > 0$ be a constant such that (3.3) holds for any choice of $p \in \mathbb{N}^*$ and $\lambda_1 , \ldots , \lambda_p \geq 0$. Since (1.3) does not hold, there exist $a_1 , b_1 , a_2 , b_2 , \ldots \in \mathbb{N}^*$ with $b_j - a_j \geq j$, $j \geq 1$, such that

$$\overline{\lim_{j\to\infty}} \sup \Big\{ \frac{1}{b_j - a_j + 1} \sum_{k=a_j}^{b_j} |\langle x^*, x_k\rangle| \, ; \, x^* \in X^* , \, \|x^*\| \leq 1 \Big\} > 0 .$$

By Lemma 5.1 there exist

$$0 < \varepsilon_o \leq 1 ,$$

$$\mathcal{B} \subset \mathbb{N}^* \text{ with } BD^*(\mathcal{B}) \geq \varepsilon_o$$

$$j_1 < j_2 < \cdots \text{ in } \mathbb{N}^* \text{ with } b_{j_n} - b_{j_{n-1}} > n$$

$$x_1^* , x_2^* , \ldots \in X^* \text{ with } \|x_n^*\| \leq 1 ,$$

for which

$$\mathcal{B} \cap \bigcup_{n\geq 2} (b_{j_{n-1}} , b_{j_{n-1}} + n] = \emptyset ,$$

$$\Re \langle x_n^*, x_k \rangle > 2 \varepsilon_o , \qquad k \in \mathcal{B} \cap (b_{j_{n-1}} + n , b_{j_n}] , \, n \geq 2 .$$

Further, by Theorem 4.2, there exist

$$\mathcal{A} \subset \mathbb{N}^* \text{ having density } D(\mathcal{A}) > 0 ,$$

$$1 \leq m_1 < m_2 < \cdots \text{ and } 1 \leq n_1 < n_2 < \cdots \text{ in } \mathbb{N}^*,$$

such that

$$\mathcal{A} \cap [1, m_j] = \{k \in [1, m_j]\,;\, k + n_j \in \mathcal{B}\}\,, \qquad j \geq 1$$

Finally, (i) and Lemma 3.4 entail that there are $p \in \mathbb{N}^*$, $k_1 < \cdots < k_p$ in $\mathcal{A}$ and $\lambda_1, \ldots, \lambda_p \geq 0$, $\lambda_1 + \cdots + \lambda_p = 1$, such that

$$\Big\| \sum_{j=1}^{p} \lambda_j\, x_{k_j} \Big\| \leq \frac{\varepsilon_o}{c}\,.$$

By (3.3) it follows that

$$\Big\| \sum_{j=1}^{p} \lambda_j\, x_{k_j+n} \Big\| \leq c \Big\| \sum_{j=1}^{p} \lambda_j\, x_{k_j} \Big\| \leq \varepsilon_o\,, \qquad n \geq 1\,. \tag{5.1}$$

Now let $q \in \mathbb{N}^*$ be such that $k_1, \ldots, k_p \leq m_q$. Then

$$k_1 + n_j, \ldots, k_p + n_j \in \mathcal{B}, \qquad j \geq q\,.$$

Choose $j_* \geq q$ with $k_1 + n_{j_*} \geq b_{j_{m_q}}$ and define $n \in \mathbb{N}^*$ by $b_{j_{n-1}} < k_1 + n_{j_*} \leq b_{j_n}$. Since $b_{j_{m_q}} \leq k_1 + n_{j_*} \leq b_{j_n}$ and the sequence $(b_{j_{n'}})_{n' \geq 1}$ is increasing, we have $m_q \leq n$. We claim that

$$b_{j_{n-1}} + n < k_1 + n_{j_*} \leq k_p + n_{j_*} \leq b_{j_n}\,. \tag{5.2}$$

Indeed, $b_{j_{n-1}} < k_1 + n_{j_*}$, $k_1 + n_{j_*} \in \mathcal{B}$ and $\mathcal{B} \cap (b_{j_{n-1}}, b_{j_{n-1}} + n] = \emptyset$ imply $b_{j_{n-1}} + n < k_1 + n_{j_*}$. Further, $k_p + n_{j_*} = k_1 + n_{j_*} + (k_p - k_1) \leq b_{j_n} + m_q < b_{j_n} + n + 1$, $k_p + n_{j_*} \in \mathcal{B}$ and $\mathcal{B} \cap (b_{j_n}, b_{j_n} + n + 1] = \emptyset$ yield $k_p + n_{j_*} \leq b_{j_n}$.

By (5.2) we have $k_1 + n_{j_*}, \ldots, k_p + n_{j_*} \in \mathcal{B} \cap (b_{j_{n-1}} + n, b_{j_n}]$, so

$$\Re\, \langle x_n^*, x_{k_j + n_{j_*}} \rangle > 2\,\varepsilon_o\,, \qquad 1 \leq j \leq p\,.$$

Since $\|x_n^*\| \leq 1$, it follows that

$$\Big\| \sum_{j=1}^{p} \lambda_j\, x_{k_j+n} \Big\| \geq \Re \left\langle x_n^*, \sum_{j=1}^{p} \lambda_j\, x_{k_j+n} \right\rangle = \sum_{j=1}^{p} \lambda_j\, \Re\, \langle x_n^*, x_{k_j+n_{j_*}} \rangle > 2\,\varepsilon_o\,,$$

in contradiction with (5.1). □

6. Weak mixing to zero for Cesaro shift-bounded sequences

If X is a uniformly convex Banach space, then Theorem 5.2 holds under a milder assumption on $(x_k)_{k \geq 1}$ than convex shift-boundedness, since in such a space the classical Mazur theorem about the equality of the weak and norm closure of a convex subset holds in the following sharper form:

Theorem 6.1 (Mazur type theorem in uniformly convex Banach spaces). *Let S be a bounded subset of a uniformly convex Banach space X, and x an element of the weak closure of S. Then there exists a sequence $(x_k)_{k \geq 1} \subset S$ such that*

$$\lim_{n \to \infty} \Big\| x - \frac{1}{n} \sum_{k=1}^{n} x_k \Big\| = 0\,.$$

Proof. Uniformly convex Banach spaces are reflexive (see, e.g., [4], page 131), so S is weakly relatively compact. Consequently, since normed linear spaces are angelic in their weak topology (see, e.g., [5], 3.10.(1)), there exists a sequence $(y_j)_{j\geq 1}$ in S, which is weakly convergent to x. Now, by the "Banach-Saks Theorem" [1] for uniformly convex Banach spaces, (due to S. Kakutani [11]; see also [4], Chapter VIII, Theorem 1), there exists a subsequence $\big(y_{j_k}\big)_{k\geq 1}$ such that

$$\lim_{n\to\infty} \Big\| x - \frac{1}{n}\sum_{k=1}^{n} y_{j_k} \Big\| = 0 . \qquad \square$$

Let us call a sequence $(x_k)_{k\geq 1}$ in a Banach space X *Cesaro shift-bounded* if there exists a constant $c > 0$ such that (3.3) holds for any choice of $p \in \mathbb{N}^*$ and $\lambda_1 , \dots , \lambda_p \in \{0,1\}$, that is

$$\Big\| \sum_{j=1}^{p} x_{n_j+n} \Big\| \leq c \Big\| \sum_{j=1}^{p} x_{n_j} \Big\| , \qquad n \geq 1$$

for any $p \in \mathbb{N}^*$ and $n_1 , \dots , n_p \in \mathbb{N}^*$ with $n_1 < \cdots < n_p$.

Clearly, every convex shift-bounded sequence in X is Cesaro shift-bounded, but the converse does not hold, even in Hilbert spaces:

Example 6.2. *Let H be an infinite-dimensional Hilbert space and choose, for every $k \in \mathbb{N}$, three vectors $u_k , v_k , w_k \in H$ such that*

$$\|u_k\| = \|v_k\| = \|w_k\| = 1 , \, 0 < \|u_k - v_k\| < \frac{1}{k+3} , \, w_k \perp \{u_k , v_k\} \text{ for all } k \in \mathbb{N},$$
$$\{u_k , v_k , w_k\} \perp \{u_l , v_l , w_l\} \text{ whenever } k \neq l .$$

Let us define the sequence $(x_n)_{n\geq 1}$ by

$$x_{3k+1} := 2\,u_k , \, x_{3k+2} := -v_k , \, x_{3k+3} := w_k \text{ for even } k \in \mathbb{N} ,$$
$$x_{3k+1} := 2\,u_k , \, x_{3k+2} := w_k , \, x_{3k+3} := -v_k \text{ for odd } k \in \mathbb{N} .$$

Then $(x_n)_{n\geq 1}$ is Cesaro shift-bounded, but not convex shift-bounded.

Proof. For every $k \in \mathbb{N}$ we denote by $\mathcal{V}_k$ the set of the vectors

$$x_{3k+1} , x_{3k+2} , x_{3k+3}$$
$$x_{3k+1} + x_{3k+2} , x_{3k+1} + x_{3k+3} , x_{3k+2} + x_{3k+3}$$
$$x_{3k+1} + x_{3k+2} + x_{3k+3} .$$

It is easy to verify that $\dfrac{2}{3} \leq \|x\| \leq \sqrt{5}$ for all $x \in \mathcal{V}_k$.

Now let $p \in \mathbb{N}^*$, $n_1 , \dots , n_p \in \mathbb{N}^*$ with $n_1 < \cdots < n_p$, and $n \in \mathbb{N}^*$ be arbitrary. Let q denote the number of all $\mathcal{V}_k$ which contain some x_{n_j}. Then

$$\Big\| \sum_{j=1}^{p} x_{n_j} \Big\| \geq \sqrt{q\left(\frac{2}{3}\right)^2} = \sqrt{\frac{4}{9}\,q} . \tag{6.1}$$

On the other hand, since the number of all $\mathcal{V}_k$ which contain some x_{n_j+n} is $\leq 2\,q$, we have

$$\Big\| \sum_{j=1}^{p} x_{n_j+n} \Big\| \leq \sqrt{2\,q\,(\sqrt{5})^2} = \sqrt{10\,q}\,. \tag{6.2}$$

Now (6.2) and (6.1) entail that

$$\Big\| \sum_{j=1}^{p} x_{n_j+n} \Big\| \leq \sqrt{\frac{45}{2}} \Big\| \sum_{j=1}^{p} x_{n_j} \Big\|$$

and we conclude that the sequence $(x_n)_{n\geq 1}$ is Cesaro shift-bounded.

To show that $(x_n)_{n\geq 1}$ is not convex shift-bounded, let us assume the contrary, that is the existence of some constant $c > 0$ such that

$$\Big\| \sum_{n=1}^{p} \lambda_n\, x_{n+m} \Big\| \leq c \Big\| \sum_{n=1}^{p} \lambda_n\, x_n \Big\|, \qquad m \geq 1$$

for any $p \in \mathbb{N}^*$ and $\lambda_1 , \ldots , \lambda_p \geq 0$. Let k be an arbitrary even number in $\mathbb{N}^*$ and set $p := 3k + 2$, $\lambda_n := 0$ for $1 \leq n \leq 3k$, $\lambda_{3k+1} := 1$ and $\lambda_{3k+2} := 2$. Then

$$\Big\| \sum_{n=1}^{p} \lambda_n\, x_n \Big\| = \big\| 2\, u_k - 2\, v_k \big\| < \frac{2}{k+3}\,,$$

while

$$\Big\| \sum_{n=1}^{p} \lambda_n\, x_{n+3} \Big\| = \big\| 2\, u_{k+1} + 2\, w_{k+1} \big\| = 2\sqrt{2}\,.$$

It follows that $2\sqrt{2} \leq \dfrac{2\,c}{k+3}$, which is not possible for any even $k \in \mathbb{N}^*$. □

Using Theorem 6.1, we can adapt the proof of Theorem 5.2 to the case of Cesaro shift-bounded sequences in uniformly convex Banach spaces:

Theorem 6.3 (Weak mixing for Cesaro shift-bounded sequences). *For a Cesaro shift-bounded sequence $(x_k)_{k\geq 1}$ in a uniformly convex Banach space X, the following conditions are equivalent :*

(i) $(x_k)_{k\geq 1}$ *is weakly mixing to zero,*
(j) $(x_k)_{k\geq 1}$ *is uniformly weakly mixing to zero,*
(jw) (1.3) *holds.*

Proof. Since the implications (jw) ⇒ (j) ⇒ (i) are trivial, to complete the proof we need only to prove that (i) ⇒ (jw).

Let $c > 0$ be a constant such that (3.3) holds for any choice of $p \in \mathbb{N}^*$ and $\lambda_1 , \ldots , \lambda_p \in \{0, 1\}$. If (1.3) does not hold, there exist $a_1 , b_1 , a_2 , b_2 , \ldots \in \mathbb{N}^*$ with $b_j - a_j \geq j$, $j \geq 1$, such that

$$\varlimsup_{j\to\infty} \sup \Big\{ \frac{1}{b_j - a_j + 1} \sum_{k=a_j}^{b_j} |\langle x^*, x_k\rangle| \,;\, x^* \in X^*,\, \|x^*\| \leq 1 \Big\} > 0\,.$$

By Lemma 5.1 there exist

$$0 < \varepsilon_o \le 1\,, \qquad \mathcal{B} \subset \mathbb{N}^* \text{ with } BD^*(\mathcal{B}) \ge \varepsilon_o\,,$$
$$j_1 < j_2 < \cdots \text{ in } \mathbb{N}^* \text{ with } b_{j_n} - b_{j_{n-1}} > n\,,$$
$$x_1^*\,,\, x_2^*\,, \ldots \in X^* \text{ with } \|x_n^*\| \le 1\,,$$

for which

$$\mathcal{B} \cap \bigcup_{n\ge 2} (\, b_{j_{n-1}}\,,\, b_{j_{n-1}} + n\,] = \emptyset\,,$$
$$\Re\,\langle\, x_n^*, x_k \rangle > 2\,\varepsilon_o\,, \qquad k \in \mathcal{B} \cap (\, b_{j_{n-1}} + n\,,\, b_{j_n}\,]\,,\, n \ge 2\,.$$

On the other hand, since X is reflexive and any bounded set in a reflexive Banach space is weakly relatively compact, by Lemma 3.5 there exists $\mathcal{A}_o \subset \mathbb{N}^*$ with $D_*(\mathcal{A}_o) = 1$ such that $\lim\limits_{\mathcal{A}_o \ni k \to \infty} x_k = 0$ in the weak topology of X.

Finally, by Corollary 4.3 there exists $\mathcal{I} \subset \mathcal{A}_o$ with $D^*(\mathcal{I}) > 0$, such that

$$\{n \in \mathbb{N}\,;\, \mathcal{F} + n \subset \mathcal{B}\} \text{ is infinite for any finite } \mathcal{F} \subset \mathcal{I}\,.$$

Then $\lim\limits_{\mathcal{I} \ni k \to \infty} x_k = 0$ with respect to the weak topology of X and by Theorem 6.1 there are $p \in \mathbb{N}^*$ and $k_1 < \cdots < k_p$ in $\mathcal{I}$ such that

$$\Big\| \frac{1}{p} \sum_{j=1}^{p} x_{k_j} \Big\| \le \frac{\varepsilon_o}{c}\,.$$

By (3.3) it follows that

$$\Big\| \frac{1}{p} \sum_{j=1}^{p} x_{k_j + n} \Big\| \le c \Big\| \frac{1}{p} \sum_{j=1}^{p} x_{k_j} \Big\| \le \varepsilon_o\,, \qquad n \ge 1\,. \tag{6.3}$$

Now set $m := \max\big(k_p - k_1, 2\big)$. Since the set $\big\{k \in \mathbb{N}^*\,;\, \{k_1\,, \ldots, k_p\} + n \subset \mathcal{B}\big\}$ is infinite, it contains some n_o such that $k_1 + n_o \ge b_{j_m}$. We define $n_1 \in \mathbb{N}^*$ by $b_{j_{n_1 - 1}} < k_1 + n_o \le b_{j_{n_1}}$. Since $b_{j_m} \le k_1 + n_o \le b_{j_{n_1}}$ and the sequence $(b_{j_n})_{n\ge 1}$ is increasing, we have $m \le n_1$. We claim that

$$b_{j_{n_1 - 1}} + n_1 \le k_1 + n_o < k_p + n_o \le b_{j_{n_1}}\,. \tag{6.4}$$

Indeed, $b_{j_{n_1 - 1}} < k_1 + n_o\,,\, k_1 + n_o \in \mathcal{B}$ and $\mathcal{B} \cap (b_{j_{n_1 - 1}}, b_{j_{n_1 - 1}} + n_1] = \emptyset$ imply $b_{j_{n_1 - 1}} + n_1 < k_1 + n_o$. Further, $k_p + n_o = k_1 + n_o + (k_p - k_1) \le b_{j_{n_1}} + m < b_{j_{n_1}} + n_1 + 1$, $k_p + n_o \in \mathcal{B}$ and $\mathcal{B} \cap (b_{j_{n_1}}, b_{j_{n_1}} + n_1 + 1] = \emptyset$ yield $k_p + n_o \le b_{j_{n_1}}$.

By (6.4) we have $k_1 + n_o\,, \ldots, k_p + n_o \in \mathcal{B} \cap (b_{j_{n_1 - 1}} + n_1\,, b_{j_{n_1}}]$, so

$$\Re\,\langle\, x_{n_1}^*, x_{k_j + n_o} \rangle > 2\,\varepsilon_o\,, \qquad 1 \le j \le p\,.$$

Since $\|x_{n_1}^*\| \le 1$, it follows that

$$\Big\| \frac{1}{p} \sum_{j=1}^{p} x_{k_j + n_o} \Big\| \ge \Re \left\langle x_{n_1}^*, \frac{1}{p} \sum_{j=1}^{p} x_{k_j + n_o} \right\rangle = \frac{1}{p} \sum_{j=1}^{p} \Re\,\langle\, x_{n_1}^*, x_{k_j + n_o} \rangle > 2\,\varepsilon_o\,,$$

in contradiction with (6.3). □

7. Appendix: Ergodic theorem for convex shift-bounded sequences

We can also define *ergodicity* of a bounded sequence $(x_k)_{k\geq 1}$ in a Banach space X by requiring that

$$\lim_{n\to\infty} \Big\| \frac{1}{n} \sum_{k=1}^{n} x_k \Big\| = 0$$

(cf. [2], Section 3). Clearly, if $(x_k)_{k\geq 1}$ is uniformly weak mixing, then it is ergodic. In this section we complete our knowledge about convex shift-bounded sequences by proving a mean ergodic theorem for them (Corollary 7.2).

Let $l^\infty(X)$ denote the vector space of all bounded sequences $(x_j)_{j\geq 1}$ in a Banach space X, endowed with the uniform norm $\|(x_j)_j\|_\infty = \sup_j \|x_j\|$, and let $\sigma_\leftarrow$ be the backward shift on $l^\infty(X)$, defined by

$$\sigma_\leftarrow\big((x_j)_{j\geq 1}\big) = (x_{j+1})_{j\geq 1} .$$

Theorem 7.1 (Mean Ergodic Theorem for sequences). *For a bounded sequence $(x_k)_{k\geq 1}$ in a Banach space X, the following conditions are equivalent:*

(e)
$$\lim_{\substack{0\leq m<n \\ n-m\to\infty}} \Big\| \frac{1}{n-m} \sum_{k=m+1}^{n} x_k \Big\| = 0 .$$

(ee) *The norm-closure of the convex hull*

$$\mathrm{conv}\Big(\big\{\sigma_\leftarrow^k\big((x_j)_{j\geq 1}\big)\,;\, k\geq 0\big\}\Big) \subset l^\infty(X)$$

contains the zero sequence.

Proof. Without loss of generality we can assume that $\|x_k\|\leq 1$ for all $k\geq 1$.

The proof of (e) $\Rightarrow$ (ee) is immediate. Indeed, if $\varepsilon>0$ and $n_\varepsilon\in\mathbb{N}^*$ are such that

$$\Big\| \sum_{k=m+1}^{n} x_k \Big\| \leq (n-m)\,\varepsilon , \qquad 0\leq m<n ,\; n-m\geq n_\varepsilon ,$$

then we have for every $n\geq n_\varepsilon$:

$$\Big\| \frac{1}{n} \sum_{k=1}^{n} \sigma_\leftarrow^k\big((x_j)_{j\geq 1}\big) \Big\| = \Big\| \frac{1}{n} \sum_{k=1}^{n} \big(x_{j+k}\big)_{j\geq 1} \Big\| = \sup_{j\geq 1} \Big\| \frac{1}{n} \sum_{k=1}^{n} x_{k+j} \Big\| \leq \varepsilon .$$

Conversely, let us assume that (ee) is satisfied and let $\varepsilon>0$ be arbitrary. Then there exist $p\in\mathbb{N}^*$ and $\lambda_1,\ldots,\lambda_p\geq 0$, $\lambda_1+\cdots+\lambda_p=1$, such that

$$\sup_{k\geq 1} \Big\| \sum_{j=1}^{p} \lambda_j\, x_{j+k} \Big\| = \Big\| \sum_{j=1}^{p} \lambda_j\, \sigma_\leftarrow^j\big((x_k)_{k\geq 1}\big) \Big\| \leq \frac{\varepsilon}{2} . \tag{7.1}$$

On the other hand, we have for every $0 \leq m < n$ with $n - m \geq p$:

$$\frac{1}{n-m} \sum_{k=m+1}^{n} x_k - \frac{1}{n-m} \sum_{k=m+1}^{n} \Big(\sum_{j=1}^{p} \lambda_j \, x_{j+k} \Big)$$

$$= \frac{1}{n-m} \sum_{k=m+1}^{n} \sum_{j=1}^{p} \lambda_j \, (x_k - x_{j+k}) = \frac{1}{n-m} \sum_{j=1}^{p} \lambda_j \sum_{k=m+1}^{n} (x_k - x_{j+k})$$

$$= \frac{1}{n-m} \sum_{j=1}^{p} \lambda_j \Big(\sum_{k=m+1}^{m+j} x_k - \sum_{k=n+1}^{n+j} x_k \Big) ,$$

hence

$$\Big\| \frac{1}{n-m} \sum_{k=m+1}^{n} x_k - \frac{1}{n-m} \sum_{k=m+1}^{n} \Big(\sum_{j=1}^{p} \lambda_j \, x_{j+k} \Big) \Big\| \leq \frac{2p}{n-m} . \tag{7.2}$$

Now (7.1) and (7.2) yield

$$0 \leq m < n \, , \, n - m \geq \frac{4p}{\varepsilon} \quad \Longrightarrow \quad \Big\| \frac{1}{n-m} \sum_{k=m+1}^{n} x_k \Big\| \leq \varepsilon \, . \qquad \square$$

For convex shift-bounded vector sequences the statement of Theorem 7.1 can be strengthened:

Corollary 7.2 (Mean Ergodic Theorem for convex shift-bounded sequences). *For a convex shift-bounded sequence $(x_k)_{k \geq 1}$ in a Banach space X, the following conditions are equivalent:*

(e)
$$\lim_{\substack{0 \leq m < n \\ n-m \to \infty}} \Big\| \frac{1}{n-m} \sum_{k=m+1}^{n} x_k \Big\| = 0 \, .$$

(f)
$$\lim_{n \to \infty} \Big\| \frac{1}{n} \sum_{k=1}^{n} x_k \Big\| = 0 \, .$$

(ff) *The weak closure of the convex hull* $\operatorname{conv}\big(\{(x_k \, ; \, k \geq 1\}\big) \subset X$ *contains* 0.

Proof. The implications (e) $\Rightarrow$ (f) $\Rightarrow$ (ff) are trivial.

Since the weak closure of $\operatorname{conv}\big(\{(x_k \, ; \, k \geq 1\}\big)$ is equal to its norm closure, (ff) implies that, for any $\varepsilon > 0$, there are $p \in \mathbb{N}^*$ and $\lambda_1 , \ldots , \lambda_p \geq 0$, $\lambda_1 + \cdots + \lambda_p = 1$, such that

$$\Big\| \sum_{j=1}^{p} \lambda_j \, x_j \Big\| \leq \varepsilon \, .$$

Using (3.3), it follows that

$$\Big\| \sum_{j=1}^{p} \lambda_j \, \sigma_{\leftarrow}^{j} \big((x_k)_{k \geq 1} \big) \Big\| = \sup_{k \geq 1} \Big\| \sum_{j=1}^{p} \lambda_j \, x_{j+k} \Big\| \leq c \, \varepsilon \, .$$

By the above (ff) implies condition (ee) in Theorem 7.1, hence (e). $\square$

Acknowledgment

Part of this paper was written while the author was a guest at l'Université de Lille in March 2002. He is grateful to Professors Mustafa Mbekhta and Florian-Horia Vasilescu for their warm hospitality.

References

[1] S. Banach and S. Saks, *Sur la convergence forte dans les champs L^p*, Studia Math. **2** (1930), 51–57.

[2] D. Berend and V. Bergelson, *Mixing sequences in Hilbert spaces*, Proc. Amer. Math. Soc. **98** (1986), 239–246.

[3] J. Blum and D. Hanson, *On the mean ergodic theorem for subsequences*, Bull. Amer. Math. Soc. **66** (1960), 308–311.

[4] J. Diestel, *Sequences and Series in Banach Spaces*, Graduate texts in mathematics 92, Springer-Verlag, 1984.

[5] K. Floret, *Weakly Compact Sets*, Lecture Notes in Mathematics 801, Springer-Verlag, 1980.

[6] H. Furstenberg, *Ergodic behaviour of diagonal measures and a theorem of Szemerédi on arithmetic progressions*, J. d'Analyse Math. **31** (1977), 204–256.

[7] H. Furstenberg, *Recurrence in Ergodic Theory and Combinatorial Number Theory*, Princeton Univ. Press, New Jersey, 1981.

[8] L.K. Jones, *A mean ergodic theorem for weakly mixing operators*, Adv. Math. **7** (1971), 211–216.

[9] L.K. Jones, *A generalization of the Mean Ergodic Theorem in Banach spaces*, Z. Wahrscheinlichkeitstheorie verw. Geb. **27** (1973), 105–107.

[10] L.K. Jones and M. Lin, *Ergodic theorems of weak mixing type*, Proc. Amer. Math. Soc. **57** (1976), 50–52.

[11] S. Kakutani, *Weak convergence in uniformly convex spaces*, Tôhoku Math. J. **45** (1938), 188–193.

[12] U. Krengel, *Ergodic Theorems*, Walter de Gruyter, Berlin-New York, 1985.

[13] C. Niculescu, A. Ströh and L. Zsidó, *Noncommutative extensions of classical and multiple recurrence theorems*, J. Operator Theory **50** (2003), 3–52.

[14] E. Szemerédi, *On sets of integers containing no k elements in arithmetic progression*, Acta Arith. **27** (1975), 199–245.

László Zsidó
Department of Mathematics
University of Rome "Tor Vergata"
Via della Ricerca Scientifica
I-00133 Rome, Italy
e-mail: `zsido@axp.mat.uniroma2.it`